Degrees of Belief

Subjective Probability and Engineering Judgment

Steven G. Vick

I owe much of this to you.

Steve

American Society of Civil Engineers
1801 Alexander Bell Drive
Reston, Virginia 20191-4400

ABSTRACT: *Degrees of Belief: Subjective Probability and Engineering Judgment* explores how the dualities of thinking in engineering and science are manifested in probability and risk analysis, professional judgment, and the nature of expertise. Written from a geotechnical perspective, these topics are woven into professional practice through examples and case studies, and through the lives, careers, and lasting achievements of some legendary civil engineering experts. By examining the cognitive processes that underlie the whole of engineering, this work brings new substance to engineering judgment, showing in the process how probability, judgment, and expertise are all inextricably linked as indispensable elements of professional practice, and indeed of thinking itself.

LIBRARY OF CONGRESS CATALOGING-IN-PUBLICATION DATA

Vick, Steven G., 1950-
Degrees of belief : subjective probability and engineering judgment / Steven G. Vick
p. cm.
Includes bibliographical reference and index.
ISBN 0-7844-0598-0
1. Engineering—Statistical methods. 2. Risk assessment. 3. Probabilities. I. Title.
TA340 .V53 2002
624—dc21 2002074410

Library of Congress Catalog Card No: 2002074410
ISBN 0-7844-0598-0
Manufactured in the United States of America.

Contents

PREFACE vii

1 THE MEANINGS OF PROBABILITY 1

2 THE FREQUENCY/BELIEF DUALITY 19

3 THE THEORY/JUDGMENT DUALITY 55

4 RELIABILITY, RISK, AND PROBABILISTIC METHODS 105

5 SUBJECTIVE PROBABILITY AND COGNITIVE PROCESSES 181

6 ASSESSMENT OF SUBJECTIVE PROBABILITIES 253

7 EXPERTS AND EXPERTISE 321

8 JUDGMENT, PROBABILITY, AND THINKING 395

APPENDIX: PROBABILITY AXIOMS, THEOREMS, AND BOUNDS 407

REFERENCES 431

INDEX 447

Geology has revealed the fact that the crust of the earth is composed of five layers or strata. We exist on the surface of the fifth. Geology teaches, with scientific accuracy, that each of these layers was from ten thousand to two million years forming or cooling. [A disagreement as to a few hundred thousand years is a matter of little consequence to science.] The layer immediately under our layer, is the fourth or "quaternary"; under that is the third, or tertiary, etc. Each of these layers has its peculiar animal and vegetable life, and when each layer's mission was done, it and its animals and vegetables ceased from their labors and were forever buried under the new layer, with its new-shaped and new fangled animals and vegetables. So far, so good. Now the geologists Thompson, Johnson, Jones and Ferguson state that our own layer has been ten thousand years forming. The geologists Herkimer, Hildebrand, Boggs and Walker all claim that our layer has been four hundred thousand years forming. Other geologists just as reliable, maintain that our layer has been from one to two million years forming. Thus we have a concise and satisfactory idea of how long our layer has been growing and accumulating. That is sufficient geology for our present purpose.

Mark Twain
A Brace of Brief Lectures on Science, 1871

Preface

Most books are about things that we know. This one is about all the rest. But every book tells a story, and in this it is no different from any other. It tells a story of probability, a story of judgment, and of how the two are so closely intertwined. Yet every book has its own story too. So by way of introducing the story *in* the book, here is the story *of* the book, of how it was conceived and why it came to be. It relates how a straightforward topic revealed something much more complex as its conceptual layers were peeled away.

The story began at a meeting on risk analysis where civil engineers from many countries came to exchange their views. The discussions started out well enough; that is, until turning to the heart of the matter—probabilistic methods. Some thought it best to find analytical models that everyone could agree on, then add statistical overlays to the input data. Others were concerned with the kinds of uncertainties that can't be analyzed—just characterized, using judgment. Gradually it became clear there was a language problem far beyond those being spoken: A shared language of probability was simply not there. This had nothing to do with its mathematics and everything to do with its meanings. What people were talking about were different notions of probability, one as objective frequency and the other subjective belief, but without a common dialect with these concepts in its lexicon they couldn't be expressed. The frustration was of that uniquely painful variety sometimes afflicting small children who know what they want to say, but just can't manage to say it.

In retrospect, this should not have been so surprising. The conceptual language of probability lies outside science and engineering, and even those fluent in its mathematics are never taught to speak it. Its places of origin are varied and distant, in the history and philosophy of science. Degree-of-belief probability is heard in far-flung regions of cognitive and experimental psychology, in business and management. These are places we seldom visit and their literatures are not easily accessible, least of all in any physical sense. Each has its own research, protocols, and vernaculars that engineers find exceptionally hard to penetrate. Moreover, these separate worlds are sometimes more concerned with articulating their own internal perspectives on probability than adapting them to pragmatic use, a custom especially foreign to those of a problem solving bent. Yet exploring these places and translating their languages could be well worth the trouble. It would certainly be useful, interesting at the least, and perhaps even entertaining as well.

But this would not be enough. The delegates hadn't just been speaking different languages, they were thinking differently too. More than anything this concerned engineering judgment, what it was, and what if any role it had to play. Geotechnical engineers more than any others are so steeped in the traditions of judgment that they use the term and apply the concept in an almost offhanded way. Perhaps then this too is where the difficulties in communication might lie. If judgment were to provide the underpinnings of subjective probability as it must, then it would have to be picked up, dusted off, and presented respectably.

In engineering, judgment has always been elusive, a thing most prized but least understood. Lacking any firm intellectual foundations, judgment has come to be seen as a talisman more myth than reality, a kind of metaphysics for the elderly or the analytically inept. And even if so far escaping outright disrepute, it has surely suffered a certain neglect. Judgment had long been the house of geotechnical worship with cornerstones placed by Terzaghi, but its paint was beginning to peel. While some might still genuflect at the door, few anymore prayed at the altar. Computational methods, the faith of first principles, were the new idols, and the church of judgment was being passed by. Nor was this the occasional wayward straying from the righteous path. An entire basis of reasoning—a way of seeing the engineering world—was being lost, and it badly needed to be recovered for probability to be understood. That this would take more than recitation of the usual liturgy was made apparent by an incident involving a dam.

The dam was an important one, and it had done certain things well-behaved dams aren't supposed to do. An investigation campaign was immediately mounted, with borings drilled, testing performed, and all manner of exploratory techniques brought to bear. But after many months, many more millions spent, and reams of data collected, nothing was forthcoming to explain these things in any comprehensive or unified way. Granted, as often happens, some of this information was vexingly ambiguous and even conflicting. But the investigation failed nonetheless. It was no failure on the part of the engineers, all dedicated and highly competent, nor of the techniques carried out to exquisite perfection. It was an intellectual failure. And more than just a failure of judgment, it was a rejection of the principles that underlie it.

Deduction was the method of reasoning relied upon to provide the answer, but deduction was not up to the task. First of all, direct demonstration didn't succeed: No indisputably incriminating feature, a gaping crack or yawning cavity, was ever found. Nor could first principles get to the heart of the matter—the problem was one of diagnosis, not analysis. Evaluating the evidence became a search for all possible uncertainties it might contain, a game of intellectual "gotcha," and uncertainty became grounds for dismissing it. A true conclusion could never be deductively reached because the premises could not be established as true. With this, the investigation hit a brick wall and a kind of paralysis ensued. Inductive reasoning, weighing and synthesizing the evidence, was not an accepted process, and even as the heap of discarded evidence grew higher it was dismissed as merely circumstantial. The judgmental induction of probable cause was not enough, and only deductive proof would do. No conclusion at all, it seemed, was preferable to one not deductively derived.

So it was that the inductive foundation of the problem became fully revealed. Underlying subjective probability was judgment, and inductive reasoning beneath that. Straightening out the language problem at the risk analysis gathering would be a long trip. It would require visiting distant fields beyond engineering, re-establishing the role of judgment, inquiring into the reasoning processes we as engineers use, and examining the workings of what we call expertise. The chronicle of this journey became the story of subjective probability and engineering judgment, but the two could just as easily have been reversed.

It will be apparent by now that geotechnical engineering is home to this book, not for any lack of relevance to other engineering specialties—and science too, for that matter—but for the deep inductive roots of geology found there. As for probability, this book picks up where most texts leave off, though in many ways it should really precede them. Some background will be useful but not required, and it is especially directed to the growing ranks of those who must apply or interpret risk-based methods, background notwithstanding. How to perform probabilistic manipulations can be found in any number of places. This book is about how to use probability and use it sensibly with appreciation of what it can and cannot do. Because engineers and scientists, whether instructed or not, are called upon more and more to participate in probabilistic and risk-based assessments, to evaluate them, and foremost to know what they mean. But even proficiency in method does not itself produce any deep grasp of the meaning beneath the numbers, and here novice and expert alike are left to their own devices. Geotechnical engineers most of all are adrift between the legitimacy of science and the legacy of judgment, between theory and practice, between the probability of frequency and that of belief. But this is not new. All of these things are related to each other, and their dualities have conflicted the profession from its beginnings. The story in the book is the story of all of them, and regardless of where one's interests may lie, this story will have been worthwhile in the telling if it causes readers to reflect on how it is that they think about what they do.

For ultimately this book is about thinking, something engineers are not taught. This may come as no small surprise. Engineers generally speaking are smart people, problem solvers second to none with the deductive logic of analytical method our stock in trade. But thinking is more than analysis. Defining the problem precedes its solution, and interpretation is what follows. This is where diagnosis, judgment, and all of the other things that distinguish thinking from problem solving, that separate engineers from technicians, come into play. There is no cookbook to instruct in thinking, and those looking for one will not find it here. Even so, most of us manage to learn to think through our own expedients, and there could be no higher ambition than to endorse and encourage this enterprise. Because thinking is central to what it means to be an expert. And experts think differently, as a number of renowned civil engineers bear witness. So whether we require expert skills for subjective probability or anything else, it is expert thinking we seek. And in the end, the nature of expertise and of judgment are so closely connected that subjective probability opens doors to both.

But the story of this book would not be complete without recounting how the manuscript escaped the forest fires that ravaged the West in that desperate summer of 2000, though by only the slimmest of margins. That it survived at all to be printed here stands as a testament to the expertise of the firefighters who saved it—and no less to their judgment, of course. So here is the story of the book. And even if the story in it might lack quite the same drama, readers should find it equally compelling.

STEVEN G. VICK
Bailey, Colorado

Degrees of Belief

I

The Meanings of Probability

There are two kinds of probability.

Such a simple statement belies its profound implications that few engineering users of probabilistic methods ever stop to consider. For most, probability has come to be identified with the characterization of data according to statistical rules, a perspective derived from longstanding tradition in such specialties as structural reliability. But the mathematics (or the *calculus*) of probability does not prescribe this view, or for that matter any other. The probability calculus is merely a mathematical operator and is altogether silent on how probability should be determined or just exactly what it should be taken to mean.

Equally rich traditions in the geotechnical field hold that judgment is an integral part of engineering. It is axiomatic that judgment is a mandatory component of any engineering undertaking and therefore essential for evaluating the various uncertainties that unavoidably affect engineering practice. Sometimes judgments regarding uncertainty are quantified as numerical probability statements and manipulated using the same probability laws applied to data statistics. But even if these laws do not prohibit such a practice, some can find it troubling that their mathematical rigor could be so easily corrupted by the inherently subjective nature of judgment.

Beneath this paradox are deeper divisions in engineering thought that are played out on the probability stage. They have to do with the roles of deduction and induction, of theory and practice, of analysis and judgment. The pull and tug of competing elements of objective and subjective thinking is evident everywhere throughout engineering, but successful engineering practice demands that both, if not reconciled, at least be accommodated. So too then must probability incorporate both objective and subjective aspects of uncertainty if engineering practice is to realize the full benefits it offers. Recognizing that there are two equally useful and legitimate kinds of probability is fundamental to bringing this about. This chapter introduces the dichotomy along with both the prospects and problems it poses. This will set the stage for exploring its origins and manifestations in everyday engineering, and almost everything that follows can be understood in this context.

Probability Interpretations

Even as probabilistic methods have advanced to their current levels of sophistication, exactly what probability is has never been conceptually simple. The interpretations of probability are many, and they have been debated and discussed ever since its mathematics first appeared. Of the various schemes for classifying these ideas, that put forward by Salmon (1966) is typical:

1. *The classical interpretation.* One of the oldest concepts, the classical interpretation holds that probability is the ratio of favorable to equally possible outcomes of an event, as illustrated by a probability of 1/6 for rolling a three in a single throw of a six-sided die. The meaning of "equally possible" is contained in the principle of *probabilistic indifference* deriving from one's state of knowledge about possible outcomes.
2. *The subjective interpretation.* Under this interpretation, probability is a numerically scaled expression of a person's degree of conviction or belief in the outcome of an event or the validity of a proposition, given their state of knowledge. If one is convinced with complete certainty that the sun will come up tomorrow, then a probability of 1.0 expresses the corresponding degree of belief.
3. *The logical interpretation.* This interpretation also defines probability as a measure of confidence or belief, but here the basis for belief must conform to logic rules, hence it is also called *objective belief.* Formally, the logical interpretation holds that probability expresses the logical relationship between statements of evidence and the truth of a hypothesis.
4. *The personalistic interpretation.* This is an outgrowth of the subjective interpretation that also views probability as one's personal degree of belief, with the added stipulation that the relationships among expressed or derived probabilities must conform to the probability calculus. Here it is entirely possible that equally rational persons may hold different beliefs, where beliefs are revealed by preferences inferred from actions or choices under uncertainty.
5. *The frequency interpretation.* According to a common version of this interpretation, probability is the limit of the proportional frequency of an outcome in an infinite series of trials. For example, if one were to toss a coin a sufficient number of times, the proportional frequency of heads would converge to 0.5.

Philosophers of science as a whole have not found any of these interpretations to be singularly satisfying, and the debates promise to occupy generations of their progeny to come. Meanwhile, those who apply probability to pragmatic ends are left to their own devices, and certain of these interpretations have become customary in various fields of the sciences and engineering, notwithstanding the philosophical implications. Among the interpretations presented above, the last two have emerged in various forms as the most useful and relevant in engineering applications. Together they incorporate the distinction that underlies all interpretations: the notion of probability as a direct outgrowth of observed outcomes, and the view of probability as an expression of belief about outcomes, observed or otherwise. More specifically, the following operational definitions embody these two concepts:

Relative frequency approach: The probability of an uncertain event is its relative frequency of occurrence in repeated trials or experimental sampling of the outcome.

Subjective, degree-of-belief approach: The probability of an uncertain event is the quantified measure of one's belief or confidence in the outcome, according to their state of knowledge at the time it is assessed.

In everyday use, probability embraces both of these interpretations, sometimes alone and sometimes combined but almost always without recognizing the difference. In many cases this may be of little practical consequence, but in others it can produce an extraordinary degree of conceptual ensnarlment. These two interpretations of probability mean different things and they do different things, and to understand the conceptual underpinnings of probability we start by first exploring their separate implications.

Relative Frequency

The relative frequency interpretation of probability is concerned with characterizing variability in measured data or observed occurrences. As such, it is fundamentally tied to statistics, which provides prescriptive rules for doing so, and is sometimes referred to as the *statistical approach* as a result. Underlying these procedures are several important concepts.

A fundamental precept of the relative frequency interpretation is that long-run variations in future outcomes are predictable according to past variations; in effect, that the past will faithfully predict the future. With respect to variability in conditions, a similar principle applies: that the variability exhibited at one location reflects the variation at some other. In both cases, variability is produced by some underlying physical process, whether or not we know what it is.

A *stochastic* process is one whose outcomes are too variable to be reliably predicted by ordinary *deterministic* rules of cause and effect, which will always predict the same outcome under any given set of circumstances. If a stochastic process does not change in time or location, it is said to be *stationary*: It still exhibits variability, but the statistical properties which describe this variability (such as mean and variance) remain constant. So the occurrence of earthquakes on a fault could be considered a *temporally stationary* process, provided that the tectonic stress regime is assumed to be invariant over the period of interest. The occurrence of karst features could correspondingly be taken as *spatially stationary* if solutioning is not influenced preferentially by geologic structure. Only for stochastic processes that are stationary can past occurrences predict the future or observations at one location serve as a guide to another.

Under the relative frequency interpretation, drawing inferences about future occurrences or conditions elsewhere requires repeated sampling of data or repeated trials of observed outcomes. A collection of data is referred to as a data *population*. Where the population consists of some subset taken to represent the data as a whole, this is the *sample population* or *statistical sample*. What makes a sample representative is subject to a variety of criteria, but among the most important is that the population it is drawn from be *statistically homogeneous*. Simply put, this means is that the variability reflected in the sample must conform to that in the larger population of interest. This is not revealed by the data, but must be inferred from interpretations outside what the data provide.

For example, laboratory tests on a collection of specimens from a clay layer provide repeated trials of strength measurement and therefore constitute a statistical sample. This sample will reflect strength variation in the actual clay layer to the extent that its strength is statistically homogeneous. This might be a reasonable assumption if the layer had the same depositional origin and stress history throughout, as inferred from geologic information. This would not be the case, however, if the clay were affected by pre-shearing that had reduced its strength to residual values along some induced plane of weakness. Here the population sampled is no longer statistically homogeneous because the clay stratum is not physically homogeneous, and the data obtained will not adequately reflect the weak seam produced by these different physical conditions. This would require that a separate and homogeneous *subpopulation* be sampled, consisting of strength of the weak seam alone, to reflect its own variability.

The relative frequency interpretation always depends on how data populations and subpopulations are defined. This is of critical importance, for it determines whether some data are "outliers" in a given population or representative members of a completely different one. Every automobile insurer compiles data on repeated trials of auto theft. But the insurer is concerned with its total payout and therefore with long-run averages rather than individual thefts. So the relative frequency interpretation and data statistics serve well, to a point. The insurer knows from its claims history that the frequencies of theft for my new four-wheel-drive pickup and my old Volkswagen bus are quite different. So it separates its data into subpopulations according to vehicle type and age, and this is why I pay different premiums. But even long-run averages for these subpopulations cannot reflect my unique circumstances as an individual. If I leave my new pickup in a high school parking lot with the keys in the ignition and the engine running, this is a *single-event occurrence,* and the likelihood of my truck being stolen is something that long-run frequency data cannot reflect. Only if I am foolish enough to engage in repeated trials of this activity and my truck is repeatedly stolen can past occurrences be any guide to the future. Then I become a subpopulation of one—and the insurer cancels my policy.

The relative frequency interpretation of probability is ubiquitous in engineering applications. It is applied to such manufactured components as structural members or electromechanical devices—pumps, switches, or valves. Such components are made to be identical by their manufacturing process, and their probability of failure is determined from their failure frequency in repeated trials consisting of in-service performance or destructive testing. While it is easy to infer after the fact that those components that failed must have been unique in containing some hidden defect or flaw, there is nothing to identify these characteristics beforehand and therefore nothing that would assign them to their own subpopulation. Likewise, the probability of extreme events like floods is customarily treated according to the frequency interpretation. The observed frequency of past floods is taken as the probability of future floods, provided that the process is temporally stationary with no change in such things as runoff characteristics or long-term climate.

But difficulties arise if for some reason repeated trials are not possible or if these trials are not identical to the conditions at hand. This is because under the relative frequency interpretation, probability cannot be applied to such circumstances, and attempts to do so often give rise to much conceptual confusion. Take dam failure. Various compilations provide the annual failure frequency of dams of a given class, say earthfill types, according to the performance of thousands of such dams each with many years of operating history. For some particular earthfill dam, the relative frequency approach would derive its probability of failure in any given year directly from the experimental trials that the historic record represents, in much the same way as a draw from the proverbial urn filled with red and white balls in established proportion.

Additionally, however, we may know something more about the dam than a random draw from the probabilistic urn allows us to express. It might have poor foundation characteristics or poor construction or some other feature that might lead one to suspect it of being a "bad actor" with greater propensity to fail. But the number of red balls in the frequency urn cannot somehow be increased to reflect this additional knowledge. To introduce such specific factors into a relative frequency characterization would require identifying some homogeneous subpopulation of dams, each with the exact characteristics as the particular structure of interest. But because each dam is unique, this is impossible to find. Nor can we fail the same dam more than once, unlike a truck that could in principle be repeatedly stolen. What we are looking for here is the probability of a single-event occurrence that incorporates specific knowledge. What the relative frequency interpretation can provide is the probability of occurrence within some class for which repeated trials are available, and perhaps a category can be found that provides some approximation. But under the relative frequency interpretation, the notion of probability pertains only to long-run averages over repeated trials and does not apply at all to unique, single-event occurrences that cannot be duplicated.

Other situations produce more bizarre conundrums. A precept of geotechnical reliability analysis is that measured data statistics provide the relative frequency with which some specified volume of soil would take on some value of a property, say strength. Each possible strength value is then applied in some analysis that yields a factor of safety. The frequency with which this factor of safety falls below 1.0 is equated to the frequency of the strength values that produce this condition, and the result is said to be the structure's probability of failure.

The literal interpretation of this failure probability according to its relative frequency origin requires that the soil be redeposited, with different strength characteristics each time, and the same structure reconstructed with each such trial to observe the frequency with which it fails. Obviously, neither geology nor construction practices make this very convenient, a conceptual difficulty seldom addressed by reliability researchers with the exception of Baecher (1999) and Harr (1987). Both conclude that the probability of failure produced by such relative frequency techniques can only have meaning if taken in the context of belief. So here a quite remarkable conceptual transformation takes place behind the Wizard's curtain, whereby repeated trials are converted into a single-event occurrence, and the frequency interpretation becomes that of belief. Fanelli (1997) also remarks on this conceptual leap, which produces a subjectively interpreted probability from one objectively derived.

Useful as it may be, restriction of the relative frequency interpretation to repeated sampling or trials limits its application to many situations of engineering importance. This is particularly so in geotechnical engineering, where geologic conditions are often unique. While statistical rules can yield precise and rigorous characterization of uncertainty associated with measured or observed data, the probability statements these rules allow are narrowly confined to these cases. Neither is the application of judgment to interpretation of the data admissible within a strict frequency interpretation. Though the end use of such a probability may be tempered by professional opinion or qualified by judgment, these matters must remain external to its formulation.

Subjective, Degree-of-Belief

Alternatively, the subjective, degree-of-belief approach holds that the probability of an uncertain event is a scaled numerical measure of one's belief about its occurrence. It is all but identical to the personal probability interpretation described earlier, and it is sometimes known by this name, or as "judgmental" probability, or for reasons to be explained later as the "Bayesian" probability approach. Here the nature of the uncertain event is not restricted to those amenable to repeated sampling or trials but is more broadly defined to include in addition:

- A unique or one-time occurrence (i.e., whether failure occurs for some particular structure)
- The validity of an uncertain proposition or hypothesis (i.e., whether behavior is linear)
- The uncertain presence of a condition (i.e., whether there are hidden cracks)
- An uncertain state of nature (i.e., whether a fault exists)

The last two involving a state of nature or condition have particularly important implications. Suppose it is determined from geologic evidence that a cavity may exist at some specified point in the ground. It is not possible to drill many borings at this point and find a cavity in say 40% of them and no cavity in 60% as the relative frequency interpretation would require. Either a cavity will be found or it will not, equally in all borings no matter how many are drilled. This might lead some to say that the probability of such a state of nature must be either zero or unity and nothing in between. But what this means is simply that after the state of nature becomes known, it is no longer uncertain. Only then do the two limiting values apply, albeit trivially. A subjective probability for cavity presence expresses the degree of confidence that one exists prior to knowing for sure.

Subjective probability can be used to describe uncertainty in such cases of unique single-event or one-time occurrences associated with a state of nature or other conditions not subject to repeated sampling. In this way it extends the use of probability to a variety of circumstances that the relative frequency interpretation is unable to address. Central to this concept is that uncertainty derives from one's personal state of knowledge, where this knowledge is derived from sources of all kinds and none—including measured data—is sacrosanct. Certainly "hard" data are admissible, and incorporating observations from previous occurrences can be essential. But in addition, other knowledge from personal experience, professional opinion, intuition, and above all judgment is not just germane but actively solicited in subjective probability formulation. Its relation to these factors allows subjective, degree-of-belief probability to be thought of as simply the quantified expression of judgment or opinion about the likelihood of an uncertain event.

Because such a probability is a function of one's state of knowledge, it is not an invariant quantity nor is it unique. The nature and amount of underlying information, observations, and data may change from time to time. Technology changes, and with it the understanding of processes and mechanisms. Field per-

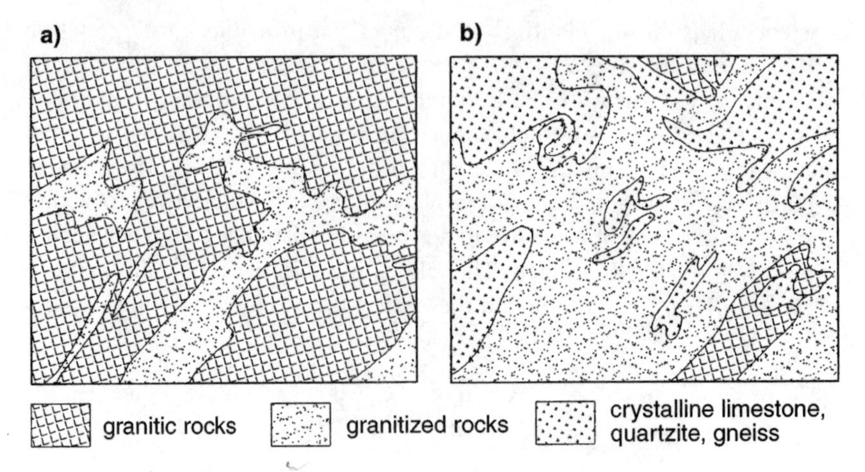

FIGURE 1-1. *Geologic maps of the same area: a, mapped in 1928; b, mapped in 1958.*

formance experience increases as more case histories become available. All of these effects on the external knowledge base will be reflected to one degree or another in the internal knowledge of the probability assessor. At the same time, because the accumulation of experiences and how they are interpreted is so highly personal, the judgment that guides how different sources of knowledge are assimilated and interpreted will vary from one person to another, even (or perhaps especially) among equally qualified professionals. A subjective probability is therefore specific to the person assessing it according to their information and knowledge at the time. As these elements change, so too will the value assessed.

Baecher (1983) illustrates this well with two geologic maps showing the same area of the Canadian Shield drawn 30 years apart (Fig. 1-1). Although both were compiled from the same geologic data, the difference is striking. Clearly geology did not change over this period, but geologic thinking did. In particular, the theory of granitization, by which rocks may take on a granitic character without passing through a magmatic stage, became popular, and the field data were reinterpreted in this light. So had a subjective probability been assigned in 1928 to some state of nature—say, the presence of an intrusive batholith—this value might well have been different if assessed in 1958 according to the different state of knowledge. It was not the data but their interpretation, in other words geologic judgment, that changed.

These characteristics differ markedly from the relative frequency approach, where the probability of an uncertain event will eventually converge on some fixed and constant value if only a sufficient number of trials are performed or enough data obtained. Here, the singular value that is the "true" probability exists somewhere out there in the intellectual ethers waiting to be discovered by the statistician diligent enough to find it. Such a relative frequency probability is an intrinsic property of the event. By contrast, a degree-of-belief probability is not a property of the event but rather of the observer. There exists no uniquely "true" probability, only one's true beliefs.

This is no small difficulty to those who might overlook the distinction, where the quest for the holy grail of a singular, unique, and invariant subjective probability is inevitably futile. If judgment provides the basis for subjective probability, and if judgment varies from person to person as it must, then so too will subjective probabilities vary accordingly. The validity of subjective probability values is determined only by the criteria that they reflect the actual belief of the assessor according to the information and knowledge at hand and that they conform to the basic probability axioms. The latter requirement for *external validity* or *coherence* may dictate that the assessor modify some values in the interest of algebraic consistency, but it does not specify which ones. The mathematical laws of probability are insensible to such matters, and to the frequency/belief distinction itself. They simply operate on the probabilities specified and neither know nor care how these values were obtained.

Knowledge, Belief, and Judgment

It is one thing to speak of the state of knowledge on which subjective probabilities are predicated but quite another to consider just what this knowledge constitutes. To begin with, engineering knowledge is fundamentally different from information or data. Knowledge is the context that makes information meaningful, for without context information cannot be interpreted. Beyond this, an engineer said to possess knowledge ordinarily has some depth of experience in theory, practice, or both. Yet experience per se need not necessarily consist of more than a mass of accumulated facts or theoretical constructs little different in principle from information or data. From this it is apparent that knowledge requires something more than experience alone.

Judgment provides the missing ingredient. If knowledge is information in context, then this demands the synthesis and integration of information as a unified whole, with judgment the vehicle that brings it about. Judgment is the means by which evidence is recognized, supporting evidence compiled, conflicting evidence reconciled, and evidence of all kinds weighed according to its perceived significance. Evidence includes personal experience, and learning from experience is among the most important ways of gaining knowledge. But it often provides little direct information. Using experience requires drawing inferences from it—learning its lessons as we say—and judgment is what makes this possible. Judgment, of course, is specific to the individual and therefore inherently subjective, which is why subjective probability is so inseparable from judgment.

Moreover, if judgment interprets information to produce knowledge, then this requires a different kind of reasoning from the deductive process we often use. Deduction goes from the general to the specific. It starts with true premises

TABLE 1-1. *Characteristics of Probability Interpretations*

Attribute	*Relative frequency*	*Subjective, degree-of-belief*
Applies to	Repeatable occurrences	Single-event or repeatable occurrences
Based on	Data statistics	State of knowledge
Measure of	Stable long-run frequency as trials $\rightarrow \infty$	Belief or confidence
Property of	The event	The observer
Reasoning used	Deductive	Inductive
Information incorporated	Measured data	Data and/or other knowledge
Subjective factors	Implicit or external	Explicitly incorporated
Criteria for validity	Statistical rules	Actual beliefs and coherence with probability axioms
Uniqueness	Singular value exists in principle	No singular value exists

and works through to conclusions that are proven true by some law. But if the premises have only some probability of being true, then so will the conclusions that follow. This is how relative frequency probability works, using data from repeated trials as the premises together with the laws of statistics, and in this it is the probabilistic extension of deduction.

By contrast, judgment uses inductive reasoning that goes from the specific to the general by means of associative inference or analogy. Induction cannot prove a conclusion true, but it can establish it as likely. So inductive reasoning is ready-made for expressing this likelihood numerically. Deduction and induction, therefore, have different jobs and they do different things. They operate on information differently, one to produce proof and the other interpretive knowledge, complementing each other in the same way as their respective probabilistic manifestations of frequency and belief.

So from this, the distinguishing characteristics of the two probability interpretations become apparent. Table 1-1 compares the various attributes of the frequency and belief interpretations, and in this light it is surprising to find that they are used so commonly together despite their differences. This complementary nature also has much to do with why the meanings of probability are so often confused (and confusing). But it also is the reason why probability is such a versatile descriptor of uncertainty, for without this dual nature it would be narrowly restricted to describing only limited sources of uncertainty, and even then only under certain conditions.

Given a choice between the attributes described in Table 1-1, most engineers would feel it wrongheaded to give up something deductive and rule-based, founded on hard data, for something so nonunique and personal—in short, something so subjective. But this is not an either–or proposition; they are simply suited to different tasks. Probability does not require a choice between data and judgment, one to the exclusion of the other, in this being among the few slices of cake served up to engineers they can enjoy eating as well. Relative frequency approaches apply best to uncertainties in data, degree-of-belief to uncertainties in knowledge, and the geotechnical field is uncommonly rich in both varieties. We are fortunate then that the probability calculus is so accommodating of both, as we can now go on to see.

Frequency and Belief in Common Usage

Both frequency and belief meanings are found in routine notions of probability in everyday use. These concepts often underlie probabilistic expressions of "odds" or "chance," but their respective roles are not always obvious. A few examples can help to clarify these contributions.

Coin Tossing

Take the simple toss of a coin. The convergence of the probability of heads or tails to a value of 0.5 in an infinite number of tosses is so common an illustration of the relative frequency interpretation as to have become all but synonymous with its definition. At the same time, the probability of heads as 0.5 is intuitively evident to most people whether they have ever encountered the relative frequency definition or not. But why?

To begin with, infinite trials are a useful thought experiment but impossible to perform. As a practical matter, some large number of coin toss trials might suffice, and perhaps such an experiment has actually been performed by some hapless graduate student. Perhaps too it is even reported in the literature, but few people cite any such study in support of their probability claim. Despite its association with the relative frequency interpretation, the popularly held value for coin tossing clearly resides elsewhere.

The notion of probabilistic indifference has considerable appeal, where this simply means the absence of preference among possible outcomes. We might reason that all of the factors affecting a coin's dynamics could never be established, and even if they were they would be impossible for physics to predict. And since no coin has ever been known to land on its edge, or so we presume, we have no reason to believe that landing heads-up is any more or less likely than landing heads-down—basically a symmetry argument. But ultimately it is our state of knowledge that governs our indifference, and indifference that leads us to equal degrees of belief about the two outcomes. Still, as much as probabilistic indifference provides a convenient rationalization for what we already know, it falls short. Careful observers have noted the proclivity for toast to land butter-side down; coins too are embossed differently on both faces, and no mint certifies that each is equally balanced. One might just as easily suppose that some nanogram's difference in weight could cause the appearance of one side to be marginally more likely than the other. So an Argentine peso could reasonably be expected to behave differently from a French franc, and the Coriolis force might influence them in opposite ways in their hemispheres of origin.[1] Similarly plausible prospects are limited only by imagination, but the point is that probabilistic indifference does not survive this kind of scrutiny.

In the end, the origin of 0.5 as the probability of heads in a coin toss is anecdotal, and people adopt this value because we believe it is so. Countless sources have testified to its validity. It is taught to children as the ultimate arbiter of fairness long before they have any concept of probability, and its fundamental truth is confirmed by no less authority than the National Football League before the kickoff of every game. We hold this accepted belief because it would appear foolish not to, but its popularity makes it no less subjective. Thus, despite its standing as the icon of relative frequency, the established probability for a coin toss derives fundamentally from belief with deep roots in behavioral factors.

Weather Forecasting

Most people are familiar with probability in weather forecasting, say to express the chance of rain tomorrow. Among the sources of information such a forecast incorporates will be climatological data, in this case whether or not there was measurable precipitation on this date over the period of record. Experienced forecasters use the long-run frequency of rain from these data to benchmark their long-run forecasting performance. As they are quick to acknowledge, it is "hard to beat climatology"; and even if one knew nothing at all about meteorology, climatological statistics alone have been shown to provide impressive forecast accuracy.

But meteorologists do know more than this about the likelihood of rain tomorrow, and their probability incorporates other information along with recorded data in evaluating such a one-time occurrence. Reports from nearby weather stations and atmospheric pressure conditions provide indicators of movement in weather patterns. Local conditions such as orographics or lake effects can cause moisture movement and precipitation to vary from these patterns. Local experience helps identify data that may be unreliable or for use in empirical correlations. And no self-respecting forecaster's prediction is ever complete without a look out the window.

Professional judgment has long been used to integrate and synthesize all of these elements of meteorological information and knowledge along with climatological frequency data in precipitation probability forecasting. Every day

[1]A group of Polish mathematicians is said to have recently discovered that the newly introduced euro coin is unbalanced. Which way was not reported.

throughout the United States, regional forecasts are issued by the National Weather Service using satellite imagery and a nationwide system called the Limited Fine Mesh, in which hundreds of weather balloons released twice daily provide temperature, wind speed, dew point, and barometric pressure data from altitudes up to 60,000 feet. This information is used in complex atmospheric models run on Cray supercomputers at the National Meteorological Center in Camp Springs, Maryland. Yet for all their sophistication, data, and analytical rigor, these forecasts are just another component of the information used by local meteorologists. As one put it, "there is still an intuitive element to forecasting that even the most powerful computers cannot duplicate" (Junger, 1997). That intuitive element is judgment, and what is conveyed to us then as the probability of rain remains an expression of the forecaster's subjective degree of belief that it will.

One way of combining statistical data and judgmentally derived information is by using Bayes' theorem, as we will see later. This can be considered a special case because it requires specific kinds of information expressed in particular ways. Another less formal combinatorial technique has been called the *normalized frequency* approach, whereby data statistics are adjusted, or normalized, to account for various departures in the case at hand from those included in the statistical database. So a weather forecaster might start with the frequency of rain on a particular day from the climatological statistics (the *base-rate frequency* as it is sometimes called), then adjust this frequency subjectively for local factors. Though an amalgam of frequency and belief approaches, normalized frequency still must be considered subjective because it introduces nonstatistical information that the frequency interpretation would not otherwise allow. Yet another way of integrating frequency and judgmentally derived information is to use each separately for the uncertainties they best describe, then computationally aggregate the results, as in a familiar application of probability to earthquakes.

Seismic Shaking

Probabilistic seismic hazard analysis, or PSHA, aims to determine the probability of occurrence (or more precisely, exceedance) of any given earthquake ground motion level at some location during a specified period of time.[2] As such, it provides input to seismic safety assessment for many kinds of engineered structures. PSHA has well-developed methodologies and vernaculars of its own, but in essence it consists of the following procedural steps (Idriss, 1985; Panel on Seismic Hazard Analysis, 1988):

---

[2]As a technical term, *hazard* is used in a variety of incompatible and ambiguous ways. A "natural hazard" is an occurrence, such as a landslide or earthquake. But the "hazard" associated with a natural hazard connotes likelihood, as in "seismic hazard." For dam failure, "downstream hazard" refers neither to the occurrence nor its likelihood but instead to the consequences. For liquefaction, "hazard" can mean any of these. For obvious reasons, the term is avoided here except where its meaning is clear from context.

1. Seismic source zones are delineated around the point of interest that might each contain faults, other seismogenic features, or regions of seismotectonic stability. Each zone is defined such that earthquake occurrence is spatially homogeneous within it. This allows for probabilistic specification of the distance between an earthquake and the desired location.
2. For each source zone, a time-rate of earthquake occurrence is specified, along with the distribution of earthquake magnitudes up to some limiting maximum magnitude (m_x), judged possible from geologic evidence.
3. An attenuation function is specified that describes how earthquake ground motions decay with distance as a function of magnitude.
4. A probability analysis integrates overall earthquake sizes and distances, and sums over all source zones to obtain the expected number of exceedances of every possible ground motion level per unit time.

Uncertainties of various kinds are present in the first three steps, some represented using the relative frequency approach. The rate of earthquake occurrence and magnitude in time (step 2) is derived from the frequency of past occurrences observed in the historic record supplemented by geologic evidence. Also, attenuation relationships (step 3) determined from measurements in previous earthquakes exhibit data variability, which can also be characterized statistically.

Other uncertainties are related to geologic judgment and expressed as subjective, degree-of-belief probabilities typically described as "weighting" factors. Source zonation itself and the associated m_x specification are matters of geologic interpretation that lend themselves to this approach (Thenhaus, 1983). Many different attenuation relationships have been derived, and weighting the alternative possibilities for any particular region is also based on professional judgment. Subjective probabilities may also be assigned to uncertainties regarding fault activity, rupture characteristics, length, and related factors, with Youngs, et. al. (1988) providing one such example.

Probabilistic seismic hazard methodology therefore uses both the frequency-derived and subjective probability values for determining ground motion exceedance probability (step 4) in ways best suited to each. Statistically based information from observations of past earthquake occurrences is combined with information from geology, recognizing that both elements are necessary to comprehensively address all of the elements contributing to uncertainty in the occurrence of seismic ground motions.

Believing in Belief

The frequency and belief interpretations illustrated in the three previous examples are summarized in Table 1-2. A relative frequency approach alone might be adopted in each of these cases but with varying degrees of success. For the coin

TABLE 1-2. *Probability Interpretations in Everyday Use*

Example	*Relative frequency*	*Degree of belief*
Coin toss	Repeated trials	Probabilistic indifference on outcomes Anecdotal (accepted belief)
Weather forecast	Climatological data	Climatological data Reports from nearby stations Isobaric patterns Moisture and orographics Calibration to local experience Look out the window
Seismic hazard	Magnitude/recurrence Statistical attenuation variability	Source zonation Maximum magnitudes (m_x) Form of attenuation relationship

toss, repeated trials in sufficient number would be pragmatically dubious, but in any case this is not what people actually do. For the weather forecast, climatology alone would fail to account for more specific meteorological knowledge; and in the case of seismic hazard, only a limited number of the contributing uncertainties could be addressed solely from statistical data. In each instance, however, supplementing relative frequency with the degree-of-belief approach captures more fully all of the relevant uncertainties according to how people conceive them. The two approaches are not mutually exclusive and are often used together in complementary fashion to enhance the information about uncertainty that each expresses. So clearly from these examples, the subjective, degree-of-belief approach—whether it expresses judgment directly, whether it incorporates frequency information, or whether it is used together with the frequency approach—is pervasive in how probability is actually used by people in general, including engineers. Its great virtue is that it allows people to incorporate the full range of knowledge they possess in ways that correspond to how they actually think. And the uncertainties that people think about and deal with every day go far beyond the limitations that frequency imposes.

Odd then that subjective, degree-of-belief concepts receive so little attention from engineering probabilists, with Baecher (1972) and Roberds (1990) all but alone in giving it much serious treatment in the geotechnical field. Even those who adopt subjective probability often do so with thinly disguised reticence, relegating it to the darker shadows of probabilistic legitimacy as frequency's poor relation. Consider these excerpts:

> ...It is practical to assign a probability ... In many cases it will be subjective ... but it is better than not trying.

> Clearly, it would be possible to attempt a more thorough reliability analysis ... in order not to rely so heavily on expert subjective opinion.

> In the absence of a more rigorous approach (i.e., with less subjective input) ... this approach is not any more subjective than current practice....
>
> It may be possible to resort to judgmental probabilities, or if all else fails, assign probabilities based on expert opinion.

At best such apologies see subjective probability as better than nothing, and at least no worse than the alternative. Hardly a ringing endorsement, especially in a field like geotechnical engineering where the necessity of judgment is seldom called into question. This manifestly broad gap between the universal endorsement of judgment and its expression as subjective probability reveals a surprisingly superficial understanding of both. And in diminishing the role of the subjective altogether, others are perhaps more to the point:

> The [subjective] uncertainty statements do not have an objective meaning.
>
> The factual value of [subjective] uncertainty statements remains largely unchecked.
>
> When expert opinion is used ... subjectivity creates additional levels of uncertainty.
>
> These guesses contaminate the risk analysis, degrading quality.

Here it seems that subjective probability is so objectionable precisely because it is ... well, so subjective. Notwithstanding a certain circularity, this holds subjective probability to the objective standard of relative frequency. But the subjective is quite obviously not intended to be objective, nor is judgment a fact to be checked. And to see either as guesswork unmasks an ignorance all the more spectacular for its appearance in print. Subjective probability does not claim to represent a statement of objective fact. But then again, neither does any but the most trivial statement provided by engineering, and this is where much of the difficulty seems to lie.

In the same way as relative frequency probability, science deals in objective demonstration by observable fact. In the former, repeated trials constitute objective fact, and verification by replicable experiment in the latter. The scientist complies with these deductive tenets of science, being concerned with proving why things work. But the engineer does what has been inductively shown to work, and things often work or they don't with considerable disregard for what science has or hasn't proven. It is not a statement of fact to say that a structure will stand, and no engineer's design can objectively prove it so. Rather, it is a statement of belief in the adequacy of its design, given certain facts and the interpretation of these facts in light of customary design assumptions and criteria, past precedent, and personal experience. These subjective elements of induction, interpretation, and judgment will always remain the defining aspects of engineering that distinguish it from science. They are part and parcel of engineering, and they cannot be made to go away. From a probability point of view, the only choice is whether to simply ignore them.

At the root of all this is that science and engineering have different jobs. The scientist's job is to explain natural phenomena. The engineer's is to deal with them whether they are adequately explained or not. And so it seems that to achieve a certain facility with subjective probability, one must first decide whether to be a scientist or an engineer and establish their outlook accordingly.

The Purposes of Probability

By now it should be clear there are two kinds of probability, and why. Sometimes an expressed probability is exclusively one or the other, sometimes it is a degree-of-belief value that incorporates frequency information, and sometimes both approaches are employed together in complementary ways. When we use probability, we seldom make a conscious choice in selecting or even distinguishing them—this usually follows automatically from the kind and quantity of information at hand. If there are repeated trials or observations of a process that faithfully reflect a future occurrence of interest, a relative frequency interpretation of these data serves to advantage. If we know something beyond what the data reflect, we may adopt a subjective, degree-of-belief approach that incorporates this knowledge as well. And if no relevant data exist, if those that do exist are insufficient, or if the process is by nature not repeatable, then judgment itself drives the probability expressed. But if these are essentially pragmatic matters, and if the probability calculus does not concern itself with them, then why should we?

The answer lies in the purposes of probability. The most evident reason for its use in the first place is as a vehicle to externally communicate uncertainty to others. The fertile linguistic lexicon of uncertainty is mostly sufficient to this end, but the mathematics of probability provide a convenient way to derive useful statements about the uncertainty associated with particular events that have not been or cannot be evaluated directly. And to use this mathematics requires that the expression of uncertainty be quantified in numerical form.

Whether derived mathematically or not, if numerical probability is to communicate the degree of uncertainty, it must also convey honestly and fairly the nature of that uncertainty. Otherwise, it is only half complete and worse, subject to gross misinterpretation. If we represent a probability as a statement of fact, then it cannot contain subjective elements of judgment impervious to proof. Such probability statements must be limited to those having some verifiable statistical validity. But if we put forward a probability statement as a more comprehensive description of global uncertainty, then it cannot be limited to only those aspects amenable to statistical validation while ignoring others that reflect relevant information of different kinds, its interpretation and synthesis, or the subjective judgment that this necessarily entails. Nor can such a probability statement be presented as objectively factual or as anything other than belief about uncertainty. Accordingly, the nature of frequency and belief interpretations must be acknowledged along with any corresponding probability statement if it is to communicate uncertainty openly and without misrepresentation. In this sense, the frequency/belief distinction is a kind of truth in probabilistic advertising.

Beyond this, probability in engineering applications also has an underlying internal purpose more important in many ways than its external expression. That is to better understand the sources contributing to uncertainty in all their forms, and the frequency/belief distinction provides an essential framework. Frequency approaches allow the factors contributing to data variability to be identified and distinguished. This opens the door to reducing those sources that can be and better appreciating the implications of those that cannot. Likewise, subjective, degree-of-belief applications submit judgment to internal reflection and scrutiny through its numerical quantification. Carefully weighing and synthesizing information and knowledge produces new insights into uncertainty that might not otherwise be achieved. Through this internal purpose of probability, we can come to better understand not just what the degree of uncertainty is, but also why, and our abilities as engineers cannot help but be enhanced in the process.

So what we have established here is that each of the two interpretations of probability has its own sphere of meaning. But things didn't start out this way, and the separation did not spring up overnight. In what might be called the age of probabilistic innocence, both frequency and belief interpretations were accepted as a matter of course, linked together and integrated so naturally that the distinction was hardly worth mentioning. But a great deal happened along the way, and on another level the meanings of probability cannot be truly appreciated without recounting the story of how they came to be.

2

The Frequency/Belief Duality

The distinction between probability as a measure of stable frequency on one hand and an expression of belief on the other is not new. Its origins predate the probability calculus and are found even in precursors to modern probabilistic thought. In what is widely regarded as the outstanding work on the history of probability, Hacking (1975) calls this duality the Janus face of probability, arguing that it has been present from the very beginning and that it arose almost of historic necessity. Others (Gigerenzer, et al., 1989; Gigerenzer, 1994) show that probability both shaped and was shaped by the various fields in which it came to be applied and that early probabilists accommodated both faces of the duality, moving easily back and forth between them. Only when the social forces of democratization shifted around the time of the French Revolution from faith in the "reasonable man" to the rise of the "average man" did a recognizable dichotomy occur.

Either way, the concepts of belief, opinion, and inductive reasoning provide the underpinnings of subjective probability as we have come to know and use it, and to fully understand it the origins of these ideas in probabilistic thought must be explored. They make for an intriguing story, and if ever there were reason for an engineer to become a student of history a more compelling case could not be found. The story reveals that in one form or another, the roots of subjective probability have been around for a very long time.

The Origin of the Calculus

It is said that probability is one of the few fields of contemporary thinking with an identifiable date of birth, which legend ascribes to 1654 and the Chevalier de

Méré. It was clear from the start that probability could be useful for describing a variety of practical things. The problem was, and in ways still remains, that its mathematics were developed long before anybody figured out exactly what.

Like other French noblemen of his time, gambling provided an important if irregular source of income for Méré. He took pride in placing repeated bets on dice-rolling outcomes with only the narrowest of margins in his favor, boasting "I have discovered in mathematics things so rare that the most learned of ancient times have never thought of them and by which the best mathematicians in Europe have been surprised" (Bernstein, 1996). However, this was a long-run strategy of low profit on high volume, and it could quickly go awry if the game were for some reason interrupted. Then the two players would have to fairly divide the stakes on the table, recognizing that the one ahead at the time would have had a better chance of winning. But just how to do this had never been discovered.

Blaise Pascal was an enigmatic and troubled prodigy whose work in algebra and geometry by age 16 had impressed no less than Descartes. And it speaks to the originality of his experiments on fluid pressure that its unit of measure bears his name. For a time in his mid-twenties, Pascal took a sabbatical from mathematics and physics for religious contemplation with the Jansenists, a rival Catholic order to the Jesuits. According to Gottfried von Leibniz—then a young German law student and correspondent of Pascal's who chronicled the story—it was through a mutual friend and gentleman of Paris named Roannez that Pascal became acquainted with Méré and returned to the learned world, stresses and all (Hacking, 1975).

The exact moment that Méré posed his legendary problem to Pascal remains shrouded, but in 1654 Pascal turned to Pierre de Fermat, a lawyer by trade and mathematician by hobby whose pastime was differential calculus and number theory. Together in famous correspondence they solved the problem by the end of the year, Fermat algebraically and Pascal by an ingenious tabulation. Less than a month later, Pascal by his own account experienced a religious epiphany and retreated once more to the Jansenist monastery at Port Royal in Paris, renouncing for good this time mathematics and physics (Bernstein, 1996). Even so, his influence on probability concepts continued through his later philosophical work.

What Pascal and Fermat developed was the combinatorial mathematics of probability, using what we now know as "Pascal's triangle" for binomial expansion using all the possible combinations and permutations of how an outcome can occur. Although it would take nearly three centuries for the axioms of probability to be formally posed from set theory by the Russian mathematician Kolmogorov, the year 1654 marks not just the birth of the probability calculus but a watershed in probabilistic thinking that occurred at almost the same time.

The Medieval Precursors

Prior to the mid-1600s, probability was not the concept it later became, and the word itself was used in an altogether different way. The medieval view of knowledge held it to a strict standard with a clear demarcation from opinion. Deduction from first principles was the sole source of knowledge, while opinion was an expression of views approved by authorities. *Probability* in the old sense was an attribute of this opinion, and it had no place in knowledge. But as the Renaissance emerged, a new adjunct to knowledge was introduced by actors we will soon meet. Observations or "signs" of the natural world could now be a basis for knowing something about how it worked, even if the underlying causes were obscure. For all intents and purposes, this was the beginning of knowledge by induction from evidence, and it set the stage for probability as we know it.

Knowledge, Opinion, and the Approbation of Authorities

To get to the bottom of the dual nature of probability, we have to go back. Way back. Back to the year 1200 or thereabouts, to a time when the very concept of knowledge was just beginning to be rediscovered. Although the world was filled with unfathomable mysteries, there was no uncertainty about how it worked, at least not where it really counted. So neither was there much need for probability. The world was *deterministic*, and things did not happen by chance. Every effect had its cause and everything happened for a reason, even if known only to God. And one thing for sure, as the ultimate source of truth God did not deal in uncertainty. There was an underlying order to all things ordained by the Almighty, and God's truth was revealed through theological doctrines.[1]

But some remarkable changes were then coming about that nearly overwhelmed the Church. Arab translations of the ancient Greek texts began to flood Europe from Spain. They contained, among other things, Aristotle's concept of logic which opened up nature to be explored as an entity unto itself. The Church, of course, held that God had created the world and everything in it. But Aristotle claimed to have established by the new methods of logical proof that the world had always existed. Should this tide roll in uncontained, it could lead to some, let's say, awkward questions.

It was 13th-century theologian, philosopher, and (not incidentally) saint, Thomas Aquinas (1225–1274), who stepped in to avert the collision, co-opting this duality of thinking by imposing a kind of double standard (Burke, 1985). What he did was give the Aristotelian view the blessing of holy writ, making it possible for science to flourish alongside religion by endorsing the concept of reason.

[1]Albert Einstein's scientific determinism evoked its theological counterpart when he wrote to Max Born, "You believe in a God who plays with dice, and I in complete law and order in a world which objectively exists" (Bernstein, 1996).

As was customary in his line of work, Aquinas was concerned with truth and how to establish it. He now proposed that it could be arrived at in two different ways. First there were truths that were divinely inspired by faith and belief as revealed by authorities and doctrines. These were derived from *opinio* and would remain the province of theology. Then too, there could be truths of reason and logic, which were true by direct demonstration from facts of nature. These produced knowledge, and they would be the purview of science. To put it simply, some things could be proven and others could not, but to Aquinas this made neither any less true. Between 1256 and 1259, he held over 250 two-day short courses on such heady topics as "Is God's knowledge the cause of things?" and "Do angels know the future?" A list of things God could not do included "change Himself," "forget anything," and "commit sins." Among them as well was "God cannot make the sum of the internal angles of a triangle add up to more than two right angles."

Aquinas interests us here for two main reasons. First, he was forced to confront the duality of belief and opinion on one hand, and reason and logic on the other. They operated in different domains of the provable and the unprovable, though they both could arrive at truth in their own ways. Second, it was Aquinas' *opinio* that was the basis for the early notion of probability. But God did not operate by chance, and this still had nothing to do with uncertainty.

Evidence of one kind or another is central to our understanding of probability—evidence which may not be sufficient to prove something true, but which can make it probably true nonetheless. Hacking (1975) argues that this modern notion of probability was not possible in Aquinas' time because the concept of evidence did not yet exist.[2] For Aquinas, science was knowledge, and knowledge was arrived at either by direct demonstration or logically derived from universal principles so fundamental as to be irrefutable. By contrast, Aquinas' *opinio* was from doctrines, derived from argument or reflection of learned authorities. Just as faith was different from reason, opinion did not constitute knowledge. And neither left room for evidence.

Thus, a geotechnical engineer under Aquinas' tutelage might wish to arrive at the truth concerning stability of a 2:1 slope. First off, this might be taken as an article of faith, to the extent that God looks out for the engineer. One could hold an opinion according to widely held doctrines of soil mechanics texts, learned consultants, or the professional engineering community. Or those less inspired might wish to acquire relevant knowledge, in this case from universal principles

---

[2]Franklin (2001) advances a different historical interpretation, showing that precursors to contemporary conceptions of probability, evidence, and nondeductive inference can be found in ancient Greek thinking, medieval legal theories, and Talmudic law. Be this as it may, it must have taken something more for these ideas to gain the currency they did, and Hacking's explanation is adopted here.

of effective stress and equilibrium mechanics. But observational evidence like cracks at the head of the slope or seeps at the toe could not pertain to either knowledge or opinion because the very idea of evidence was not recognized.

The word "probability" would have been understood by Aquinas, but in a very different sense than we would today. Both Aquinas and his successors Galileo Galilei and Francis Bacon in the early development of physics four centuries later would have had no use for probability as we know it.[3] They reasoned from cause to effect, where first principles were the gold standard of knowledge and knowledge stated only universal truths. Opinion did not concern them, though it was opinion that was the basis for *probabilis* as they all would have recognized the term. In its medieval sense, the Latin *probabilis* meant worthy of approval or approbation, something testified to or attested by authority. Thus, our medieval engineer might be said to be a "probable engineer," meaning one who had the approval of others and whose opinion might be held probable in the same sense. Such a "probable opinion" was widely held and approved by those in authority, but this had nothing to do with its likelihood of correctness. Stability of a slope was "probable" if authorities and doctrines said so, notwithstanding how likely it might be to fail or any observational evidence to this effect.

Galileo himself had more than passing acquaintance with authority. Sobel (1999) recounts how he braved trial by the Inquisition, papal ostracism, and ultimately confinement until his death in 1642 because his telescope provided proof by direct demonstration of the Copernican theory that the earth revolves about the sun. To Galileo this knowledge was indisputable, even though it failed to conform to doctrines of the prevailing Ptolemaic and Aristotelian authorities. It therefore illustrates well the medieval origins of the term that Galileo called the Copernican astronomy "improbable," in reference to their fierce opposition (Hacking, 1975). For his part, Bacon in 1620 expressed disgust for the Aristotelians and alchemists, wishing "to banish all authorities and all sciences" that then held sway. Using the term in the same way, this school, he said:

> ...produces dogmas of a more deformed and monstrous nature ... but in the confined obscurity of a few experiments. Hence this species of philosophy appears probable and almost certain to those who are daily practiced in such experiments, and have thus corrupted their imagination, but incredible and futile to others.

We will hear the alchemists' side of the story momentarily. For now, it is sufficient to point out that Bacon thought their dogmas "probable"—albeit manifestly wrong.

---

[3]Bernstein (1996) notes that Galileo did write a short essay on the mathematics of gambling for his patron the Grand Duke of Tuscany, but its publication in pedestrian Italian, not Latin, and a disclaimer that he was ordered to write it showed evident disinterest in a topic he felt would go no further.

In all of these respects, the connection of probability in its old sense to opinion, the opinion of authorities, was well established long before its modern context. Yet even in contemporary use of the term as a measure of likelihood, vestiges of the old meaning can sometimes be found not far beneath the surface, as the influence of widely held and approved opinion in the coin toss example of Chapter 1 showed. It would remain the opinion of authority that would pave the way for the new probability. Only now it would be nature's authority that mattered.

Signs, Evidence, and the Authority of Nature

For the likes of philosophers and theologians, the medieval canons of knowledge, restrictive though they were, served well enough despite the occasional brush with authorities. Less fortunate were those whose everyday problems went far beyond what either this knowledge or doctrines had to offer. Between divine revelation and scientific demonstration was a huge knowledge vacuum crying to be filled. Of pragmatic necessity, people turned to signs.

In the deepest trouble were physicians of the day who had little hope of ever deducing the effects of illnesses from their causes, though this did not stop their patients from demanding cures. But signs were everywhere if one would only look for them: in the wind, the stars, locust swarms, or the piles of dead mice in the streets before an outbreak of plague. Almost at the same time several physicians hit upon their importance, and as it happened, mining and syphilis were the keys.

Girolamo Fracastoro (1478–1553) came from an old family of Verona and received his M.D. in 1502 from the university at Padua in Italy. His medical works culminated with the publication of *De Contagione et Contagiosis Morbis et Curatione* in 1546. In it he suggested that illnesses could be transmitted by invisible living creatures, earning him a lasting place in what would become epidemiology. Among the most deadly such diseases was the inevitably fatal one that Fracastoro named "syphilis" in his book by the same title written in 1530—and in poetry, no less. Hacking (1975) provides another of Fracastoro's contributions, his explanation of signs:

> Contagions have their own particular signs of which some announce beforehand contagions to come, while others indicate that they are already present. The signs that are called premonitory come from the sky or the air or from the vicinity of the soil or water, and among these some are almost always, others are often, to be trusted. Therefore one ought not to consider them all as prognostications, but only as signs of probability.

Here, Fracastoro arrives at once to several key ideas. Signs can be premonitory, that is, diagnostic and possessing predictive value. Signs come from observations of natural processes. And they have varying degrees of reliability, some to be trusted often, others almost always. So signs, at least some of the time, can lead to a new sort of knowledge that bypasses first principles altogether—in short, knowledge derived from evidence. Signs have become the basis for evidence.

No less significant is that they are signs of "probability." Fracastoro still uses the word in its medieval context, but here for the first time it is signs that are probable. They may not be certain and they're not universal truths, but they are still worthy of approbation. What is new here is that the approbation comes from nature, not the authorities. Nature has become the ultimate authority.

It was a contemporary of Fracastoro's known as Paracelsus (1494–1541) who became the chief proponent of signs. Paracelsus was a revolutionary in every sense and was called the "Luther of medicine" in his time for his dramatic and vocal break from its old doctrines. Today, physicians acknowledge him as the grandfather of modern medical science for bringing chemistry to the treatment of disease. But even in an age noted for its heretics and eccentrics, Paracelsus was by all accounts in a class by himself. Hacking (1975) calls him a "bizarre hermetic" prone to drunken ramblings, and Boorstin (1983) a "vagrant and a visionary," the mad prophet who nonetheless led the way. In keeping with custom among the educated, Paracelsus was a pseudonym from the Latin. His real, and more lyrical, name was Theophrastus Phillipus Aureolus Bombastus von Hohenheim. While Bombastus denotes family lineage, the eponymous adjective would not have been far off the mark. So outrageous was he that he remains even today an irresistible character for fictionalized but entirely believable accounts of one of the founding practitioners of alchemy (Connell, 1991).

Paracelsus was born to an old and noble Swiss family. As a boy his father, himself a physician, moved them to a mining town in Austria where he attended the Fugger mining school, for a time working as an apprentice in the mines and smelters and later returning to manage the metallurgical works. Paracelsus also studied medicine in Italy, and although it is unclear whether he ever received a degree, he became a surgeon and traveled with the armies of Venice throughout Europe, as far as Scandinavia and perhaps the Near East (Boorstin, 1983).

The prevailing medical theory of disease at the time went back to the ancient Greek physician Galen. It held that disease was a result of imbalance in the four "humors" of the body: blood, phlegm, choler, and melancholy or black choler. To restore health was to restore the balance among them, but because each person's balance was unique there could be as many diseases as there were patients who had them.[4] This, at least, was how the medical authorities on the faculty of the university at Basel saw things when Paracelsus received an appointment there in 1527. It would not last for long.

[4]Inasmuch as the mapping of the human genome has revealed a genetic basis for some diseases unique to each individual, Galen might have been amused to know that thinking on this topic would eventually come full circle.

Paracelsus had come to believe that diseases had causes outside the body that were not unique to the individual and were revealed by God through signs. To us, some of these signs seem quite absurd. The shape or color of an herb might indicate the organ it could cure—a yellow herb for a yellow liver. Others could come from the stars in a system of names Paracelsus himself devised. But whatever their kind, signs could be found if nature were observed closely enough:

> [Nature] indicates the age of the stag by the ends of his antlers and it indicates the influence of the stars by their names. Thus she made liverwort and kidneywort with the leaves in the shape of the parts she can cure.... Do not the leaves of the thistle prick like needles? Thanks to this sign the art of magic discovered that there is no better herb against internal prickling. (Hacking, 1975)

Not your average run-of-the-mill alchemist, Paracelsus wasn't simply trying out cures at random; and magic or not, his signs led him to some quite sensible observations. He knew the life of the miner and his diseases, and mercury poisoning would surely have been one of them. Mercury, or quicksilver, was greatly prized because it was the sole known agent for liberating gold from its ores by forming an amalgam with it, as with silver in dental fillings today. But extraction of mercury from its own cinnabar ores drives off highly toxic vapors. The woodcut in Figure 2-1 depicts the prevailing practice for direct reduction of elemental mercury, and there can be no mistaking the wind blowing the fumes in the opposite direction.[5] Paracelsus not only recognized mercury poisoning but was also able to distinguish its acute and chronic symptoms. From these observations, he had also discovered by 1529 that quicksilver, administered in doses that wouldn't kill the patient, could cure syphilis (Boorstin, 1983). Indeed, this was known to Fracastoro as well, as he described in his 1530 *Syphilis*.[6]

But none of this sat well with the faculty at Basel. In the face of their censure, Paracelsus mounted a ranting diatribe of assorted insults and fulminations against the medical authorities which for good measure he delivered in vernacu-

---

[5]Agricola (1556) elaborates on the rendering of quicksilver depicted in Figure 2-1: "If the fumes give out a very sweet odour it indicates that the quicksilver is being lost, and since this loosens the teeth, the smelters and others standing by, warned of the evil, turn their backs to the wind, which drives the fumes in the opposite direction...." Dental problems aside, mercury has serious neurological effects from which Newton among many others is believed to have suffered.

[6]Mercury's value at the time has served a curious purpose today. In their study of the seismotectonics of Peru, Dorbath, et al. (1990) rely on accounts of Spanish priests following Pizarro's conquest in 1535, who described the great loss when tsunamis from offshore subduction earthquakes swept the coast and carried away the flasks of mercury (like those in Fig. 2-1) left there by returning ships. Had this mercury not been equally vital to Spain as the gold it produced, contemporary understanding of the tectonics of the Nazca Plate would be greatly diminished.

FIGURE 2-1. *Direct reduction of quicksilver: a, hearth; b, poles; c, hearth without fire in which the pots are placed; d, rocks; e, rows of pots; f, upper pots; and g, lower pots.*

Source: Agricola, 1556.

lar *Schweizer-deutsch* instead of scholarly Latin. This culminated with a bonfire of Galen's texts and a pronouncement that from then on, his courses in medicine would be taught from his own experience instead. Clearly for Paracelsus the old authorities of opinion would not do. As he put it, there were no incurable diseases, only ignorant physicians. Like Fracastoro, Paracelsus had turned to the authority of nature and the testimony its evidence provided. Needless to say, this would not do for the faculty, and Paracelsus was gone from the university just two years after he came. For the rest of his life he became what Boorstin (1983) calls a "medical Don Quixote," wandering through Austria, Bavaria, and Bohemia as an itinerant mystic, physician, and student of miner's diseases, never to have a permanent home again. Still, in so vociferously condemning the theories of Galen, Paracelsus had been the first to reject the notion that disease could be cured from first principles. From then on, it would be cured empirically by observation of nature through the evidence its symptoms provided. For as Connell (1991) would have him say:

> The carpenter thinks out a cottage—how it should be built—and then he goes to work. Not so the physician who does not think out how a disease should be, since he did not make it. Nature invents disease and therefore knows its constitution, so if a doctor would know what to do he must acquaint himself with what she has to teach ... And I say further that men must see by what operation nature conducts her work in order to revise the meaning of what they think.

Much the same could be said of geology. And indeed it was.

In the figure of Georgius Agricola (1494–1555), born Georg Bauer, we find a spokesman for signs and observational evidence manifestly more sane and somewhat closer to home. A contemporary of both Fracastoro and Paracelsus, he is known for his incomparable *De Re Metallica* (Agricola, 1556), with its comprehensive documentation of medieval mining, metallurgy, and geology practices. Although a physician as well, it is not too much to call Agricola a mining engineer, and his accounts and woodcut depictions (Fig. 2-1 among them) ring true to any mining or metallurgical engineer today.[7]

Agricola was born in Germany the same year as Paracelsus. He taught Greek and Latin for a time in Leipzig, then went on to Italy to study science and medicine at several universities, which likely would have included Fracastoro's in Padua. Returning to Joachimsthal in Bohemia, he was appointed resident physician in 1527. When he got there, the place was a booming mining camp in the heart of the most prolific mining district of central Europe and not far from the renowned mining school at Freiberg. By Agricola's own account, he spent all of his off-duty time visiting the mines and smelters, reading the old Greek and Latin mining texts, learning about mining and smelting practices, and eventually becoming a mining consultant of sorts. He was surely acquainted with the same miners' diseases that Paracelsus encountered, but what interests us here are Agricola's signs of geology and how he used them.

No less than medicine at the time, medieval mining had its own authorities, dogmas, and first principles, including how the presence of ores could be deduced. Although Agricola sets up his argument by pretending otherwise, the divining rod had remained the prescribed method since Greek and Roman times:

> Therefore it seems that the divining rod passed to the mines from its impure origin with the magicians. Then when good men shrank with horror from the

[7]We owe much of this to mining engineer Herbert Hoover and his wife Lou, who prior to their better known activities found time in 1912 to translate *De Re Metallica* from Latin and related materials with a wealth of technical and historical background. The accounts here are drawn largely from this remarkable work.

> incantations and rejected them, the twig was retained by the unsophisticated common miners, and in searching for new veins some traces of these ancient usages remain. (Agricola, 1556)

He wasted no words in saying that these miners had been duped:

> ...manipulation is the cause of the twig's twisting motion. It is a conspicuous fact that these cunning manipulators do not use a straight twig, but a forked one cut from a hazel bush, or from some other wood equally flexible, so that if it be held in the hands, as they are accustomed to hold it, it turns in a circle for any man wherever he stands ... Nevertheless, these things give rise to faith among common miners that veins are discovered by the use of twigs, because whilst using these they do accidentally discover some; but more often it happens that they lose their labor, and although they might discover a vein, they become none the less exhausted in digging useless trenches than do miners who prospect in an unfortunate locality.[8]

But if twigs could not do the job, what could? Agricola responds with this list:

> But by skill we can investigate hidden and concealed veins, by observing in the first place the bubbling waters of springs, which cannot be very far distant from the veins because the source of water is from them.
>
> ...secondly, by examining the fragments of the veins which the torrents break off from the earth, for after a long time some of these fragments are again buried in the ground. Fragments of this kind lying about on the ground, if they are rubbed smooth, are a long distance from the veins, because the torrent, which broke them from the vein, polished them while it rolled them a long distance; but if they are fixed in the ground, or if they are rough, they are nearer to the veins...
>
> Further, we search for the veins by observing the hoar-frosts, which whiten all herbage except that growing over the veins, because the veins emit a warm and dry exhalation which hinders the freezing of moisture, for which reason such plants appear rather wet than whitened by the frost.
>
> Lastly, there are trees whose foliage in the spring-time has a bluish or leaden tint ... These phenomena are caused by the intensely hot and dry exhalations which do not spare even the roots. Verily the veins do emit this exhalation ... Likewise along a course where a vein extends, there grows a cer-

[8]Never one to pass up a good controversy, Paracelsus had weighed in earlier on the same matter: "These [divinations] are vain and misleading, and among the first of them are divining rods, which have deceived many miners. If they once point rightly, they deceive ten or twenty times." Both this and Agricola's remarks ("more often") come tantalizingly close to a relative frequency argument.

> tain herb or fungus which is absent from the adjacent space, or sometimes even from the neighborhood of the veins.

And finally, he concludes with his grand summation:

> By these signs of Nature a vein can be found.

Agricola had his own geologic theory about the genesis of orebodies that the Hoover translation describes in some detail. In it, ore channels were created by erosion from subterranean waters then filled with minerals by circulating groundwater solutions, which accounts for Agricola's great attention to these "juices" as springs and vapors. But he did not contend that the presence of ores was deduced from this theory, which nevertheless is remarkably similar to today's explanation of hydrothermal deposition. Instead, he was relying on their signs, the "signs of Nature," and any modern exploration geologist could do worse. Springs can accompany the kind of structural discontinuities often associated with mineralization. Agricola's observations about the geomorphological processes of erosional transport and fragment polishing foreshadow James Hutton's about all of geology two centuries hence ("The present is key to the past"). The warm and dry exhalations unmistakably describe the exothermal reactions that occur when sulfide minerals such as pyrite are exposed to oxygen, which can even produce steam. And his description of vegetation anomalies corresponds to one of the fundamentals of modern geochemical and satellite imagery prospecting.

Agricola's elegant and lucid accounts describe how his signs were used in the everyday engineering of his time. We too have ours and even the word is the same, which we take to mean predictive evidence for some condition or state of nature. Tension cracks are a "sign" of slope instability. Sand boils are a "sign" of liquefaction. Muddy seepage is a "sign" of internal erosion. Agricola took his signs from observation of nature, and Karl Terzaghi his from observation of geology. The correspondence over four centuries' time could hardly be more stunning.

We cannot know whether Fracastoro, Paracelsus, and Agricola had their ideas independently, but some of the convergences are hard to ignore. All three were practicing physicians and probably received medical training at Padua around the same time. As Fracastoro and Paracelsus found almost simultaneously, mercury cured syphilis. Mercury came from mining, and both Paracelsus and Agricola knew mining and treated its diseases, including those caused by mercury. Paracelsus's itinerant wanderings took him to the mining camps of Bohemia where Agricola practiced, and they both scorned the divining rod for remarkably similar reasons. Regardless of whether any of them ever crossed paths or knew of each other's work, their ideas boiled down to the same thing: Knowledge was no longer restricted to deduction from first principles or direct demonstration, and neither was opinion of learned authorities what mattered. From this, two revolutionary concepts emerged.

Medieval physicians and miners were pragmatists, and they needed to do what worked. What worked could be discovered by observing the signs of nature, an offshoot of the old *opinio* that replaced authority's testimony with nature's. The idea that knowledge could be obtained by observing nature had, of course, been around since Aquinas and Aristotle before him. To them, nature offered up facts that gave proof of cause and effect. But this was different. In this new kind of knowledge, signs gave evidence of some cause, and something about how it worked could be understood from its external effects even if its internal operation could not. Signs were evidence of an association between cause and effect, and for many practical problems this was enough when proof was not to be had. Through signs, the concept of evidence had been discovered, unlocking a new dimension of knowledge. This was nothing less than the foundation for inductive reasoning.

At the same time, signs opened the door for probability as we know it and were what made it possible. Signs were premonitory and thus predictive. But evidence from signs was not always correct, and one could trust it to varying degrees. Recall Fracastoro: "...some are almost always, others are often to be trusted." This embodied two things of immense importance. First, trust is a property of the observer indistinguishable from confidence or belief. And as the measure of this belief, "always" and "often" cannot have meaning but through repeated observations—the signs' frequency of correctness. In embracing both frequency and belief concepts, Fracastoro's statement was a defining one, and surely the Janus face of modern probability could not be far behind.

The Classical Probabilists

It was not long after Pascal and Fermat's 1654 calculus of probability became known that people began exploring its uses, and two were immediately apparent. One extended its application to gambling, with the associated kinds of uncertainties becoming known as *aleatory*, from the Latin *aleatorius* or literally "about games of chance." The other was *epistemic*, from the Greek *epistemikos*, or "having to do with knowledge or understanding." One was external and derived from things, the other internal and from the mind. In their own ways, the Classical probabilists dealt equally with both as necessary and complementary aspects of modern probability.

Aleatory and Epistemic Development

Although Holland experienced a brief shining as a center for mathematics, Paris remained the intellectual capitol of the time and Dutchman Christian Huygens went there to study. By then Pascal was ensconced in the Port Royal monastery, and Fermat was elsewhere. Through Roannez, the same Paris gentleman who had introduced Pascal to the Chevalier de Méré, Huygens too met him and learned of the Pascal–Fermat correspondence although he never saw the written solution (Hacking, 1975). Three years after their work, Huygens published his own monograph, *De Ratiociniis in Alae Ludo* (Calculating in Games of Chance), in 1657. His contribution was to determine the worth of any particular gamble, and this he did using the probability of winning p and the stakes to be won V, where:

$$p(V) = E \tag{2-1}$$

The product E of probability and winnings was called the *expectation*, or today *expected value* of the bet. The expectation of one of a thousand lottery tickets with a prize of \$10,000 would be \$10, the long-run average payoff. The aleatory probability Huygens used came from dice. It was therefore a property of things.

But these times of the Enlightenment had a strong moral undercurrent. The breathtaking developments in physics and mathematics had a clear moral purpose in improving the intellect of reasonable men, and probability was no exception. As Laplace would later say in *A Philosophical Essay on Probabilities*, mathematical probability was only "good sense reduced to a calculus," although for many probability's aleatory roots in gambling made it, if not downright unrespectable, then certainly somewhat frivolous as it was for Galileo. But now it could, and did, have a higher moral purpose in helping reasonable men express, and therefore overcome, the gaps in their knowledge (Gigerenzer, et al., 1989). These uncertainties were epistemic in nature. They sprang not from things but from the mind. Just as Paracelsus had said a century earlier, there were no incurable diseases—just ignorant physicians.

The early epistemic concept shares with modern subjective probability the view that probability expresses belief according to one's state of knowledge, and the two are closely related. But deep down, the Classical probabilists were determinists at heart. Insufficient knowledge was just a temporary state of affairs that would inevitably yield to scientific discoveries like Galileo's. All events were predictable in principle and none would be seen to occur by chance when the governing laws were found. So in this respect all uncertainty could ultimately be shown to be epistemic in nature. Some noted that backward peoples still gambled on eclipses, and that someday gambling on dice would seem equally foolish when the science of mechanics was perfected. In this, their probability was subjective—it was a function of belief—but the beliefs of reasonable and suitably well-educated men would eventually and necessarily converge.

Pascal's Wager

Even if Pascal was no longer engaged in mathematics at Port Royal he had not given up probability, now in relation to theology. The very act of bringing probability concepts to the question of the Deity shows how far things had come since Aquinas. Pascal posed his classic question in a wager which dealt with belief in a way that anticipated the concept of utility in modern decision theory. First published in 1662 in a collection of Port Royal work entitled *Logic*, he inquired whether a reasonable man should believe in God:

> God is or is he not? Which way should we incline? Reason cannot answer.

Like Aquinas, Pascal asserted that deductive logic, or reason, could not prove God's existence one way or the other, which was clearly a matter of one's belief. But how should this be established? Pascal framed the problem as a choice between two bets—one that God exists and the other that God does not. He argued that belief in God could be determined only by how one chose to behave. A person who lives a pious life is betting that God exists and stands to receive an eternal reward either way, but the sinner stands a possibility, no matter how small, of eternal damnation. To Pascal, the way we should incline—which wager to choose—was obvious (Bernstein, 1996).

Pascal's wager showed that belief could be revealed from preferences in a decision and that the decision depended in part on the consequences of the outcome. In this way, he extended Huygens's concept of expectation to payoffs having nonmonetary value (to us, *utility*). Equally significant is how Pascal, whose seminal work with Fermat dealt with gambling, had no trouble at all using the classically aleatory activity of betting and the thoroughly epistemic concept of belief both in the same breath. Gigerenzer (1994) maintains that moving so easily between these dual senses of probability without conflict was the hallmark of the Classical probabilists.

In its purest form, the epistemic notion of probability concerned belief in evidence. As we have seen, Fracastoro and his associates took the testimony of nature through its signs as evidence. They recognized that it could have varying degrees of validity (some almost always, others often, to be trusted), but they never got around to determining how evidence should be weighed. It was Leibniz who attacked this problem, and the courtroom was his venue.

Leibniz and the Law

We return to Leibniz where we first met him, the young German Boswell who recorded Pascal's acquaintance with the Chevalier de Méré. Leibniz later became known as the co-founder of differential calculus and equally for his battle with the Royal Society of London's iron-fisted dictator Newton for the honor, which Boorstin (1983) calls the "scientific spectacle" of the 18th century. Leibniz was never much engaged in probability mathematics as such, but his early interest in the evidence of courtroom testimony doubtless came from his father's position as a judge. Hacking (1975) calls Leibniz the first philosopher of probability, and in his Baccalaureate essay *De Conditionibus* in 1665 he originated the use of the numerical probability scale to reflect one's degree of certainty. Just 19 at the time, he knew nothing of the developments in Paris of the previous decade.

Evidence and testimony are the stock in trade of judges, who had long distinguished different kinds that carried different weights. An overheard whisper was quite a different thing from eyewitness testimony of a suspect fleeing the scene with a bloody dagger. The worth of the evidence depended on the judge's degree of belief in it. It had no aleatory component. It made no difference whatsoever how many times the same kind of evidence had convicted suspects in the past; the case at hand was the only one that mattered. This made probability applicable not just for repeatable events but for single-event occurrences as well.

In this, Leibniz found the epistemic view of probability perfectly suited to law. He called it "natural jurisprudence" and said that the whole of judicial procedure is "nothing but a kind of logic applied to questions of law." In the system he developed for this purpose, Leibniz used a numerical scaling from 0 to 1 to reflect what he called "degrees of proof" and "degrees of probability" from the evidence. He took it for granted that probability is "in proportion to what we know" (Hacking, 1975), thus formalizing the concept of probability as a manifestation of one's state of knowledge. And it is of more than passing significance that such a progenitor of modern mathematics would so warmly embrace the epistemic view. Leibniz then must have been keenly interested when Jacob Bernoulli approached the same topic, but from a quite different angle.

Bernoulli and Degrees of Certainty

As a family, it is safe to say that the Bernoullis were a dynasty at the same University of Basel in Switzerland where we left Paracelsus—and he left it—more than a century earlier. Together they were such prolific mathematicians, physicists, and philosophers that it can be hard to keep them straight. Figure 2-2 helps. Jacob (Jacques) Bernoulli (1654–1705) was the family's mathematical patriarch and the first to hold a position at the university.[9] Jacob's nephew Daniel (1700–1782) we know best in engineering as the originator of hydrodynamics. Both Daniel and his cousin Nicolaus (1687–1759) were themselves accomplished in probability and helped formalize the concept of utility that Pascal's wager introduced, extending it to monetary risk aversion (Bernstein, 1996).

But it is Jacob we are most concerned with here because he is taken by many as the father of subjective probability and the first to actually use the term, a fitting honor for one born in the year of the Pascal–Fermat correspondence. How-

[9]This was a continuing source of rivalry between Jacob and his youngest brother Johann that occasionally erupted in public criticism of each other's mathematical work. They both knew Leibniz well, and Johann lobbied for him throughout Europe in his struggle with Newton.

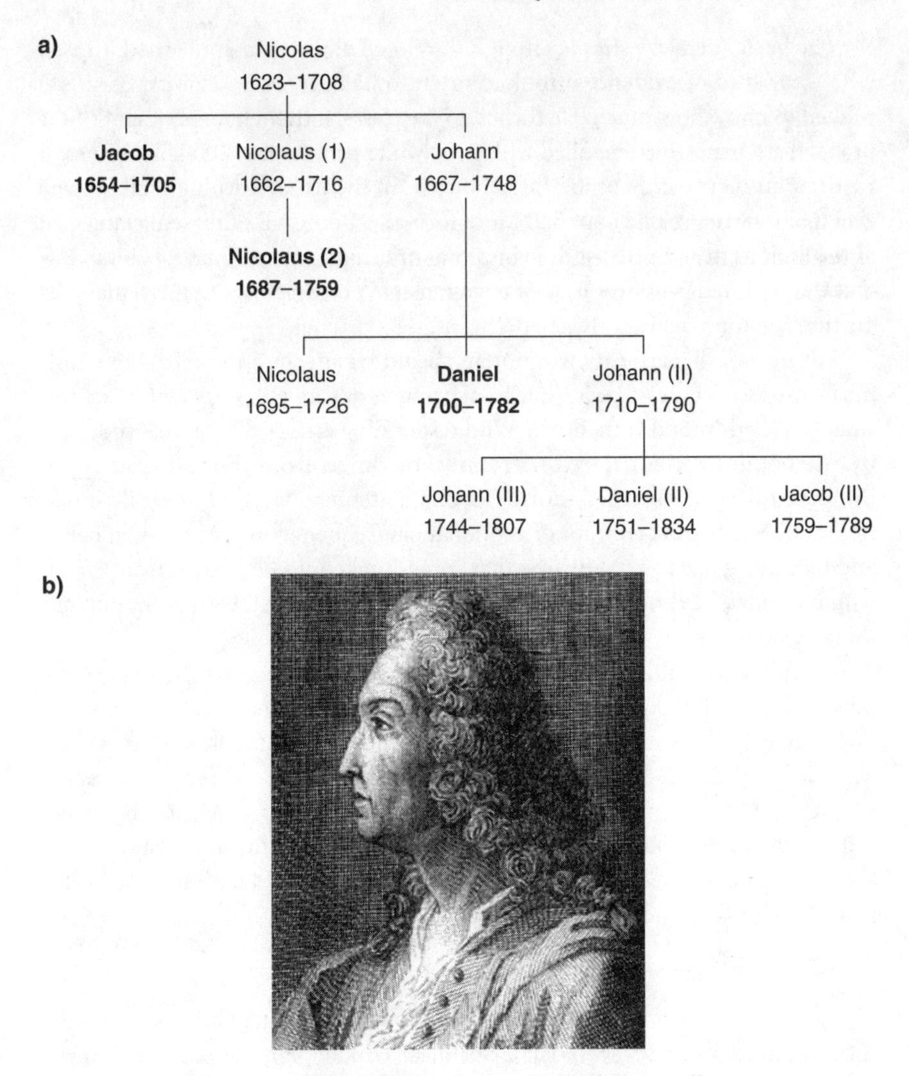

FIGURE 2-2. *The Bernoullis: a) the family tree; b) Jacob Bernoulli.*

Source: David Eugene Smith Collection, Rare Book and Manuscript Library, Columbia University.

ever, as Hacking (1975) is quick to point out, other schools of probabilistic thought justifiably lay claim to him too, no doubt because like the other Classical probabilists he used aleatory and epistemic concepts with such equal facility. Jacob Bernoulli had worked on his ideas over the last 20 years of his life at Basel, and they were published in 1713 eight years after his death with nephew Nicolaus as editor. The book was *Ars Conjectandi* (The Art of Conjecture) with what Hacking (1975) calls its "squirming duality."

The basic idea was simple enough.[10] While Leibniz had concerned himself with what kind of evidence it took to reach some degree of certainty, Bernoulli wanted to know how much. He turned to the classical thought experiment of the probabilists' urn, which he filled with 3000 white pebbles and 2000 black ones. If we draw an increasing number of pebbles from the urn, returning each, we will eventually arrive at a ratio of 3:2. More formally, Bernoulli's theorem states that in the limit as the number of drawings N approaches infinity, the probability P—that the observed proportion of observations m/N corresponds to the actual proportion p—approaches certainty.

For Bernoulli, certainty was not an absolute concept but a moral one intimately tied to degree of belief. For him, being 98% sure was sufficient to achieve what he called "moral certainty."[11] With resounding echoes of Leibniz, "probability," Bernoulli said, "is degree of certainty and differs from absolute certainty as the part differs from the whole." In this statement we find several things. Bernoulli extends the concept of proportional frequency to proportional belief, and he came close to establishing that something could be known from effects without understanding their cause. But as Gigerenzer, et al. (1989) point out, one thing was missing.

Given the probability of the actual proportion p, Bernoulli's theorem shows how likely it is that the observed frequency P will approximate it to some specified degree of certainty. But in order to determine something from effects without knowing their cause, what he needed was the opposite: Given the observed frequency, how likely is it to approximate the unknown probability? Bernoulli called p the *prior* probability, the one that pertains before the observations from the urn are made. Dealing with it more conclusively would have to wait for the arrival of the Reverend Thomas Bayes.

Bayes and His Theorem

Not much is known about Bayes' life or thought except that he was a Nonconformist minister of a Presbyterian chapel near London. His famous theorem was (as most all such work seems to have been) published posthumously in *Essay Towards Solving a Problem in the Doctrine of Chances.* He left the manuscript to Richard Price, who presented it to the Royal Society in 1764:

---

[10]At least so Bernoulli thought. As he told Leibniz, "even the stupidest man knows by some instinct of nature per se and by no previous instruction" that the greater the number of confirming observations the surer the conjecture (Gigerenzer, et al., 1989). In this alone Bernoulli reveals his inclinations toward the internal basis for epistemic belief as well as his intuitive linkage of belief and frequency.

[11]Statisticians today associate an all but identical morality to the 95% confidence level almost invariably, but arbitrarily, adopted as the measure of statistical goodness.

> I now send you an essay which I have found among the papers of our deceased friend Mr. Bayes, and which, in my opinion has great merit ... In an introduction which he has writ to this Essay, he says, that his design at first thinking on the subject of it was, to find out a method by which we might judge concerning the probability that an event has to happen, in given circumstances, upon supposition that we know nothing concerning it but that, under the same circumstances, it had happened a certain number of times, and failed a certain other number of times.

Bayes started with the precept that through frequency observations we could determine the probability that some effect would occur and that we could likewise find the *conditional* probability of the effect given its cause by observing how many times the cause had exhibited that same effect before. We can now determine the "reverse" (or more precisely, *posterior*) probability that Bernoulli had been after, the probability of the cause from the effect. Let:

$$\begin{aligned} \text{p[cause]} &= \text{probability that the cause occurs} \\ \text{p[effect]} &= \text{probability that the effect occurs} \\ \text{p[effect | cause]} &= \text{conditional probability of the effect given the cause} \\ \text{p[cause | effect]} &= \text{conditional probability of the cause given the effect} \end{aligned}$$

In its simplest form, Bayes' theorem states:

$$\text{p[cause | effect]} = \frac{\text{p[effect | cause]} \times \text{p[cause]}}{\text{p[effect]}} \tag{2-2}$$

The power of Bayes' theorem becomes apparent in a more concrete formulation. In engineering, the cause we are often concerned with is the mechanism that may produce failure of a structure. The observed effect of that cause is manifested as some evidence or sign, which in general we can call an "indicator" of failure. We can determine something about how reliable the indicator is by looking to postfailure investigations, for example, to see how often we find it when failure has occurred. We can also, in principle, determine its false-positive rate by looking to how often the indicator appears anyway when failure does not occur. What we want to know is the probability of failure given that we observe the indicator. Hence:

$$\text{p[failure | indicator]} = \frac{\text{p[indicator | failure]} \times \text{p[failure]}}{\{\text{p[indicator | failure]} \times \text{p[failure]}\} + \{\text{p[indicator | no failure]} \times \text{p[no failure]}\}} \tag{2-3}$$

The careful reader will have noticed the missing element not discussed so far in either formulation, the *prior* probability of the cause or of the failure: p[cause] or p[failure]. Bayes' theorem requires that we know something to begin with about the posterior probability we seek, even before invoking the observed evidence. Where this prior probability comes from is something that Bayes seems never to have worked out, and perhaps this is why he never published the theorem himself. The answer, of course, is that the Bayesian prior is a subjective probability. It comes from what we believe, as Bernoulli would have put it, "by some instinct of nature per se, and by no previous instruction." Or as we would say, by judgment.[12]

Bayes' theorem can be viewed as the crowning achievement of the Classical probabilists. It began with the medieval signs, some always and others often to be trusted, and in its complete version could incorporate as many signs as nature might reveal. It showed how Leibniz's evidence in different forms could be reconciled, and it fulfilled Bernoulli's quest for knowing about causes from only their effects. Inductive methods had achieved full partnership with deductive knowledge, and their probabilistic formulation was now complete.

Although Bayes' theorem is simply a way of combining different sources of evidence, for the Classical probabilists it formalized and at the same time reconciled the frequency/belief duality with an elegance that belied its simplicity. It was nothing less than the mathematical statement of the philosophical cord that had always united frequency and belief. By melding evidence from frequency-based observations with *a priori* belief based on other kinds of knowledge and judgment, it rationalizes for us how they could be so apparently adaptable to both. Gigerenzer, et al. (1989) describe a further explanation that in a very real sense anticipated today's cognitive view of subjective probability. By the mid-1700s, philosophers John Locke, David Hartley, and David Hume created and refined a theory of association in which the mind was a kind of counting device that automatically cumulated past events and scaled degrees of belief according to their frequency. Hume felt that this neural judgment machine was exquisitely tuned, for as he wrote in 1739 in *A Treatise on Human Nature*:

> When the chances or experiments on one side amount to ten thousand, and on the other to ten thousand and one, the judgment gives the preference to the latter, upon account of that superiority.

Hartley went so far as to postulate that each repeated sensation set up cerebral vibrations which etched an ever deeper groove in the brain, strengthening belief in proportion. Accordingly, it was perfectly sound for reasonable men to embrace both the aleatory and epistemic concepts of probability. This would change soon enough.

---

[12]Because the subjective prior is such a fundamental element of Bayes' theorem, subjective probability is often called "Bayesian probability," and subjective probabilists "Bayesians." These designations should not be confused with actual application of the theorem itself, which "Bayesian probability" does not necessarily imply.

The Rise of the Statisticians

While the Classical probabilists were interested in different kinds of evidence, both aleatory and epistemic, and how it might be combined, the statisticians were concerned with only one thing: data. The rise of statistics, although proceeding in parallel with the work of the Classical probabilists, had little to do with their thinking. Statistics would come to dominate the frequency/belief duality, eventually triumphing and destroying it for years to come. In this, statistics changed the focus of probability from how the "reasonable man" behaved in formulating inductive knowledge, to the behavior of the new denizen of the masses it created, the "average man." As such, probability was no longer concerned with the outcome of single-event occurrences but only with the trend of repeatable events from their long-run frequencies. Past events would become the only predictor of future events, regardless of evidence, knowledge, beliefs, or anything else.

Mortality

When John Graunt did his pioneering demographic work on births and deaths in London, he as others remained unaware of the Pascal–Fermat correspondence taking place in Paris at almost the same time. In an age of rigid class distinctions, Graunt was isolated from the close-knit mathematical community of the Classical probabilists, instead a simple, hard-working merchant and tradesman once apprenticed to a London haberdasher. Yet even among those of his circumstance and time he seems to have been uncommonly industrious and studious, perhaps almost obsessively so, known for taking detailed notes of sermons in shorthand during church services (Boorstin, 1983). Graunt was simply fascinated with data and what they could reveal, though less for their mathematical constructs than bookkeeping. Graunt collected data like some people collect stamps. Admitting to having found "much pleasure" in his analysis of London's Bills of Mortality, his work, he said, was:

> [T]o know how many people there be of each Sex, State, Age, Religious, Trade, Rank, or Degree, &c. by the knowledge whereof Trade and Government may be made more certain, and Regular; for, if men know the People as foresaid, they might know the consumption they would make, so as Trade might not be hoped for where it is impossible. (Newman, 1988)

Bernstein (1996) notes that Graunt may very well have originated the concept of market research—to say nothing of Internet cookies.

Known as the founder of economics, Graunt's cohort William Petty was of a more pecuniary ilk, with the two as made for each other as Cratchit and Scrooge. A member in good standing of the upper-class elite, Petty had become wealthy by exploiting the subjugated Irish as the Crown's Surveyor there, besides being a shameless war profiteer, and, most discongruously, Professor of Anatomy and

TABLE 2-1. *Graunt's Mortality Table*

Age (years)	*Survivors per 100 births*
6	64
16	40
26	25
36	16
46	10
56	6
66	3
76	1

Arts (Hacking, 1975). Petty was well aware of the value of demographic statistics in government taxation and conscription, and he lobbied for a central statistical office to collect data on the whole Kingdom. We would doubtless recognize Graunt today as that arch-nemesis of the engineer, the bean counter, and Petty as that nefarious schemer, the chief financial officer.

All but unimaginable to us, the devastation of the Plague was so commonplace in Graunt's time as to make it almost inevitable that it would be the grist for anyone so infatuated with figures. Perhaps only the true English spirit could carry on with statistical work amidst such horrors, which might explain Graunt's pleasure in doing so. His database was London's Bills of Mortality, which listed the dead weekly. By 1603 it was being distributed by parish clerks every Thursday morning for a penny per sheet, or four shillings by annual subscription for the faithful readership (Boorstin, 1983).[13] Morbid though they were, these tabulations provided Graunt with a way for the first time to calculate life expectancy. He found that 64 of 100 infants survived to age 6, and 1 in 100 lasted to age 76. His "life table" (Table 2-1), published in 1662, showed the number of survivors through the intervening decades.

The Bills of Mortality also provided causes of death, specious as the data might be, and Graunt observed that their frequency could be quite regular over time. About 22% were from acute diseases, 30% from chronic afflictions, only 2% from "outward griefs" (sores, leprosy, broken bones, and the like), and even fewer from "murthers" in London at the time. Graunt is also credited with being the first to use statistical sampling. To estimate London's population, he started with the number of births and the fertility rate, which gave him a rough estimate of

---

[13]Boorstin (1983) also describes the macabre and bizarre "searchers" or "ancient matrons," the enforcers of quarantines who obtained Graunt's data by inspecting the corpses. Conspicuous by their red wands, they were notorious drunkards and would, for a price, cleanse the record of ignominious causes of death like the ever-present syphilis. Graunt was well aware of this corruption and its like effect on his data, and he tried to correct for the latter.

the number of women, which from his estimate of family size yielded a total population of 384,000. He confirmed this number in two ways, first by sampling the population of three of London's parishes and again from inhabited area and population density. From this he could show that Plague losses had been more than replenished by immigration in London's astonishing growth (Hacking, 1975).

Graunt and Petty's population statistics provided a new way of looking at disease altogether different from Fracastoro's "signs of contagion." Signs and evidence no longer mattered. One no longer needed to reason either deductively from first causes or inductively from effects; knowledge was data and data were knowledge, at least in the aggregate. Of course this would not help the physician diagnose any particular patient's case, but Graunt's statistics were not about the individual. As statistics of populations, not probabilities of one-time occurrences, they considered disease purely from a public health perspective. While the Classical probabilities may have been about individual cases, statistics were about the masses.

Annuities and Insurance

As rival mercantile nations, England and The Netherlands regularly found themselves at war, so it was equally natural that they turn to the same source for funding them. Annuities were for both nations the 17th-century equivalent of deficit financing, but their fiscal footing was ruinous and they tended to lose as much as they made. From English law going back to 1540, one could buy an annuity for a certain amount and receive in return an annual payment for life that, with interest, would equal the purchase amount in seven years (Hacking, 1975). But this was independent of the purchaser's age. Obviously a better deal for the youthful than the elderly, citizens responded in kind much to the detriment of government treasuries.

Johann de Witt was the chief statesman of The Netherlands, his colleague Johannes Hudde the mayor of Amsterdam, and both civic-minded mathematicians who sought to aid their nation's cause in the matter of annuities. In 1671 de Witt proposed an annuity pricing scheme that used Amsterdam mortality statistics compiled by Hudde, together with countryman Huygens's new probability calculus of expected value. Mathematicians all over Europe were quick to recognize that mortality statistics could be used to burnish probability's tawdry gambling image, and they immediately seized on statistical data as aleatory frequency (Gigerenzer, et al., 1989). In 1693, England's Royal Astronomer Edmund Halley working along similar lines, also proposed a method for fair annuity pricing from life expectancy statistics (Bernstein, 1996). Strangely enough, nobody in either government paid much attention.

The same proved to be true in the insurance industry, even though there could hardly have been a more aleatory gamble. Basically an annuity in reverse, any mathematician could easily use the same mortality statistics for pricing insurance premiums by age, and several did. One of the more entrepreneurial, James Dodson, established the Equitable Society for the Assurance of Lives in 1762 in London with just this in mind. But the idea did not catch on, with most insurers still charging flat premiums regardless of age, mortality statistics, or mathematicians.

Gigerenzer, et al. (1989) explain why. Long before the advent of mathematical probability or statistics, parties to aleatory contracts like annuities, life, or maritime insurance had always priced future contingencies in ways directly counter to the new statistical ideas. Their world was one of individual cases, each to be evaluated on its merits by an old hand in the business. Maritime insurance manuals described as the basis for premiums things like the season, the route, the cargo, and rumors of wars or pirates. Only good judgment and long experience could serve to price the risk, a doctrine equally enshrined in publications on annuities of the time.

Statistics would eventually come to change all that. They replaced judgment with arithmetic, and this required a profound change in thinking that went to the insurers' very reason for being. They could no longer take into account their judgments about the spry 60-year-old who might well outlive the ailing 30-year-old or the ship sailing into waters with notoriously treacherous shoals. What might happen in the individual case did not matter; what statistics determined would happen in the long run was all that counted. Judgment was not relevant, and Laplace's probability as "good sense reduced to a calculus" was itself reduced to the calculus and little more. Good sense was something statistics no longer required, thank you, just data. This has much to do with the characterization that acclaimed statistical pundit Samuel Clemens would later immortalize by way of lies and damned lies, attributing it to Prime Minister Disraeli in *Autobiography of Mark Twain*. Such practices prevail in the insurance industry today, with premiums determined all but universally according to past claims history regardless of judged risk. Recently, a well-known insurer duly reported as "statistical fact" that one's credit history predicts their propensity for automobile crashes—while professing to have not the slightest idea why.

The Divorce of Frequency and Belief

Shortly after the end of the 18th century the triumph of statistics over Classical probability was essentially complete, with an intellectual dominance produced in no small part by its moral supremacy in society. Frequency and belief had always been complementary concepts, but now they became incompatible. The former reached celebrity status, while the latter went underground.

Adolphe Quetelet was a Belgian artist, poet, mathematician, and astronomer who made no apologies for hitching his statistical engine to whole human race. In Brussels he began compiling all the data he could find on human beings—physical dimensions of the body, facts on crimes, criminals, and generally as he put it "that which relates to the human species, considered *en masse*" (Boorstin, 1983). His observation that crime rates remained remarkably stable within age groups led him to postulate a kind of implied budget for these acts established by the laws of some "social physics" just as commanding as nature's counterpart. Three types of data were key: crimes, suicides, and marriages—the "moral statistics." While each was produced by individual choice, their impressive statistical regularity served to show how the laws of statistics superseded even the free will of individuals, and their beliefs. In 1835, Quetelet published these ideas as his *Treatise on Man and the Development of His Faculties, An Essay on Social Physics*, which quickly became celebrated throughout Europe. In it he introduced the concept of *l'homme moyen*, the "average man," whose statistical norms governed humans and their behavior.

Quetelet showed that the heights of men were distributed systematically about the mean height—the height of the average man—and there was every reason to expect that statistical principles should apply in the same way to moral and intellectual qualities. To demonstrate this, he compared his distribution of height in the French population to that in the French army, with the mean height of the soldiers proving that about 2000 others had misrepresented their stature to avoid conscription. Moving on to more shady practices in his 1835 *On Man and the Development of His Faculties*, Quetelet asserted that statistics on crime and punishment in the courts of France all but preordained the actions of its citizens:

> We can tell beforehand how many will stain their hands with the blood of their fellow-creatures, how many will be forgers, how many poisoners, almost as one can foretell the numbers of births and deaths.

Just as Graunt had obviated the need for Fracastoro's signs of nature in disease, so Quetelet had done for Leibniz's evidence in law. Even though the moral statistics pertained *en masse* to the "average man," it was not a huge leap to suppose that the statistical laws of society must also rule individual men and their separate cases. Through the social forces of democratization that reached their highest frenzy with the French Revolution only shortly before, the "average man" had guillotined the elite "reasonable man" of the Classical probabilists, and along with him went the co-existing duality in frequency and belief he had maintained for so long (Gigerenzer, et al., 1989). The "moral certainty" of Jacob Bernoulli and its intimate tie to belief had been eclipsed by the "moral statistics" of Quetelet. The statisticians had gained the moral high ground.

As a result, statistical concepts centered on mass phenomena gained currency in areas far beyond demography, epidemiology, and government. If human behavior was erratic and unpredictable at the individual level but regular at the societal level as dictated by statistical laws, then so could James Clerk Maxwell and Ludwig Boltzman use a similar model for the behavior of gas molecules. Social physics pointed the way for the physics of science. From kinetic theory of gases to quantum mechanics, probability in physics came to be about statistics, and statistics were about frequency. The urn of Classical probability was now concerned with these mass phenomena, to the exclusion of subjective degrees of belief pertaining to single-event occurrences (Gigerenzer, 1994).

Locke, Hartley, and Hume's reconciliation of frequency and belief as the mind's mapping of one onto the other had been steadily eroded, and others stepped forward to deliver the *coup de grace*. In 1837, Poisson became the first to sharply differentiate the "objective" probability of data from "subjective" probability. Soon probabilists everywhere hammered out the singular interpretation of probability as frequency, encouraged by the masses of statistical data now accumulated throughout Europe. Subjective probability became denigrated as mathematically deviant, and by 1843 John Stuart Mill was calling the probability of judgments "the real opprobrium of mathematics." Antoine Cournot in the same year was the one to take the final step of eliminating subjective probability from the realm of mathematics altogether (Gigerenzer, et al., 1989; Gigerenzer, 1994), and it would remain banished from mathematical treatment for the next hundred years until Kolmogorov's probability axioms from set theory would allow it to be reinstated. The new view of probability understood it to be about the stable frequency of events in the long run and not about the outcome of any particular event. The exclusionary view of frequency had thus been born. The divorce had become complete.

The Return of Subjective Belief

By systematically characterizing data, statistics became an indispensable tool of science, and scientific thought in turn was shaped in no small part by statistical ideas. Although belief and opinion were to play a major role in the emergence of many controversial scientific concepts, statistics sought to actively eliminate them and thereby preserve the ostensibly objective character of scientific practice. In this it amply succeeded. Statistical methods became science's household seal of approval. But beyond science lay another world concerning how one should act in the face of uncertainty, especially in making decisions. Statistics could be used to draw inferences from data, and this could be useful as far as it went. Yet the most difficult uncertainties in actions and decisions were produced by the very absence of data where statistics could be of little help. Instead, the probability of subjective belief rose from the ashes, this time in the context of human behavior and judgment, allowing fertile areas of decision theory and risk analysis to emerge. Still, the Classical probabilists had been largely forgotten. Their duality of frequency and belief was re-established, but this time the coexistence was more uneasy.

The Reign of Statistics

Statistical and frequency-based probability concepts continued to preside over the reign of science throughout the 19th and well into the 20th centuries.

Firmly grounded in objective facts, statistics became a recognized branch of mathematics, but not before being handed down from one scientific discipline to another. According to the London Statistical Society, statistics did not concern itself with causes and effects, its "first and most essential rule" being to "exclude all opinion" (Gigerenzer, et al., 1989). The statistics of collective regularities were used to model mass phenomena irrespective of their causes, and Quetelet's statistical model of social science proliferated by analogy to a host of other areas.

From their birth in demographics, statistical methods were quickly adopted by population biology, genetics, and evolution. Both the biological and social versions of Darwinism were highly controversial but equally influential. With variation as its keystone idea, Darwinian evolution helped establish that the statistical laws of chance could prevail over the deterministic laws of science that heretofore had governed the natural order.

Although Gauss had applied statistics to experimental errors in astronomical observations in the early 1800s, modern physics was more resistant to the role of chance in its deterministic world. Classical physics—particle mechanics, continuum mechanics, and electrodynamics—had long been the stronghold of scientific determinism. On a discrete level, the principles worked well enough provided that things like the position and momentum of basic constituents could be known, but the difficulties in applying them to, say, 10^{23} molecules of gas proved pragmatically insurmountable. Processes such as radioactive decay, particle interactions, and quantum mechanics could only be understood as mass phenomena and described statistically. So were the deductive tenets of physics transformed from laws that held of necessity to those that held with some statistical probability.

Statistics came into its own as a separate discipline of mathematics around 1900, applying its tools of measurement, sampling, and data description. Its great breakthrough came around 1940 with the ability to draw statistical inferences from data to hypotheses, and two competing methods emerged. R.A. Fisher was the chief proponent of *significance testing* for determining whether the *null hypothesis*—the opposite of that of interest—can be rejected by the data at some *significance level* specifying the discrepancy of the data to the hypothesis. In what became known as the Neyman-Pearson approach, Fisher's single null hypothesis was replaced by alternative hypotheses with a statistical test for choosing between them. These two camps provoked a heated controversy that continues still, but one that Gigerenzer, et al. (1989) argue has been suppressed in almost all modern statistical texts. Along with the hybridization of these methods came ignoring the subjective judgment necessary for selecting the hypotheses, the significance levels, and the statistical tests themselves, despite all of the various developers' warnings against drawing inferences without it. Through this institutionalization of hybrid methods, statistical inference came to be seen as an objective and monolithic body of singular truth devoid of subjective influence.

The heyday of statistics came with the Second World War, as statistical predictions became the basis for gunnery and bombing, quality control, search optimization, and personnel selection. As the defense industry, and later aerospace, widely adopted these procedures, so did many engineering disciplines, with some of the first uses of statistical methods to evaluate structural safety appearing around that time (Freudenthal, 1947). Statistical inference, together with the newly developed game theory of John von Neumann and Oskar Morgenstern, had grown beyond the handmaiden of science to a formalized method for making decisions and taking actions based on measured data.

The Probability Axioms

For their part, mathematicians had not entirely abandoned probability to their statistical offspring. The combinatorial mathematics of probability in games of chance had, of course, been around ever since Pascal, Fermat, and Huygens; the first limit theorem since Jacob Bernoulli; and the theorem for combining probabilities since Bayes. But the various mathematical constructs had never been unified. In 1933, Soviet mathematician A. N. Kolmogorov accomplished this by defining the mathematical properties that probability should have. These were its axioms developed around set theory from the analogy between properties of sets and probabilities of events.

In brief, Kolmogorov's axioms assigned a real number between 0 and 1 to each set in a group of sets. The probability of the set containing all elementary events is equal to 1, and all other sets are subsets of it. If two sets have no common elements, the probability of their union is the sum of their probabilities. Immensely simple, the effect of these axioms was equally huge in establishing a unifying framework for probability theory and in reinvigorating it as a serious mathematical discipline. But they also had the effect of welcoming subjective probability back into the fold of mathematical legitimacy by showing that the mathematical constructs of probability placed no restriction on its interpretation or the means used to determine its values. It is these axioms and the related theorems that we look to today in establishing the mathematical consistency of probability statements, either frequency or degree-of-belief. These are further developed in the Appendix.

Personalistic Probability

The concept of probability as degree of belief had remained moribund for almost a century since being dispatched as mathematically degenerate by Mill and Cournot in the 1840s. Statistics had largely succeeded in "eliminating all opinion," but Kolmogorov's axioms had shown that this was no mathematical directive. A few voices were heard from the wilderness during the 1920s to 1950s, notably Frank Ramsey (1931), Bruno de Finetti (1937), and Leonard Savage (1954) with their *personalistic* interpretation of probability. As Savage (1954) defined it:

> These views hold that probability measures the confidence that a particular individual has in the truth of a particular proposition, for example, the proposition that it will rain tomorrow. These views postulate that the individual concerned is in some ways 'reasonable' but they do not deny the possibility that two reasonable individuals faced with the same evidence may have different degrees of confidence in the truth of the same proposition.

The probability axioms impose conditions on the relationships among these various degrees of conviction, and additionally the expressed measure of confidence must reflect the person's true state of mind. These two things are all that the personalistic interpretation required. Coherence with probability axioms might dictate that some values be modified, but how is not prescribed. Just like Bayes' prior probabilities, personal probabilities were in the mind, and they could be unknown "only insofar as one fails to know one's own mind" (Savage, 1954). In this, the Classical probabilists would have experienced *deja vu*: The rational man had returned from his exile. Probability was no longer the exclusive domain of frequency in mass phenomena and repeatable events. The personalistic approach was nothing less than the revival of probability for single-event occurrences.

The personalists were not unaware of the developments in statistics, the controversies surrounding statistical inference, or the various meanings of probability itself:

> As in other sciences, controversies over the foundations of statistics reflect themselves in everyday practice...here, as elsewhere, catastrophe is avoided primarily because in practical situations common sense saves all but the most pedantic of us from flagrant error. (Savage, 1954)

So their probability addressed the same "everyday practice" and "practical situations" as did statistical inference, but to them "common sense" or judgment—not data—controlled how people actually behaved in these situations. As the quantified measure of belief, personal probability was inseparable from one's behavior under uncertainty as manifested by actions or decisions:

> ... the quantitative aspects of beliefs as the basis of action are evidently more important than the intensities of belief-feelings. (Ramsey, 1931)

It was not enough then to simply express a probability value because this would not evoke a sufficiently introspective or considered response. Instead, this value had to be inferred from actions such as the choice between two bets, one involving the uncertain event in question, the other a hypothetical gamble, and both with specified rewards. Here, the odds of the gamble were varied until the subject had no expressed preference between them, and the desired probability could then be back-calculated from the expected values.

The personal probabilists were not the first to relate probability to belief of the mind, with roots in behavior at least as old as Pascal's wager. But by emphasizing its behavioral aspects as they did, they opened the door to the cognitive view of subjective probability that prevails today. While subjective probabilities can be inferred from revealed preferences in gambles, this is not how they are actually formed. Preferences derive from belief judgments, not the other way around (Tversky and Kahneman, 1974), and these belief judgments derive from the cognitive processes that people use in formulating and expressing them. Just how it is that "knowing one's own mind" comes about has been the subject of a great deal of behavioral research conducted over the past 40 years and is treated more deeply in chapters to follow. Although direct interrogation was considered inadmissible as little as 25 years ago, asking people for their probability judgments directly has now become routine from this improved understanding of cognitive reasoning strategies (Baron and Frisch, 1994). So it is that the statistical norms governing human behavior in the aggregate have been reunited with the cognitive processes that produce it in individuals.

Contemporary Applications of Subjective Probability

The personal probabilists were pioneers in yet other respects. Their rediscovery of degree-of-belief concepts revived the application of Bayes' theorem, particularly to decision problems. This opened up a new era of formal probabilistic decisionmaking under uncertainty as *Bayesian decision analysis* or simply *decision analysis* in areas such as business decisions (Schlaiffer, 1959; Raiffa, 1968) and drilling for oil and gas (Grayson, 1960). These applications were among the first to extensively adopt event tree techniques for decomposing uncertainties into their constituent components. Today these decision analysis methods and the subjective probabilities they adopt have become routine tools for all kinds of corporate and organizational decisionmaking affected by uncertainty, such as the introduction of new products or construction of new facilities; research and development allocations; bidding and pricing strategies; investment options; and even litigation strategies. Many typical decision analysis procedures have become more-or-less standardized in commercially available software. Graunt and Petty would have been surprised but no doubt pleased to know how far the economic uses of probability had come.

Another area of subjective probability application is known variously as *decision support* or *expert systems* within the parent discipline of *artificial intelligence* (Clark, 1992; Gammack, 1992). This involves computer software that seeks to codify decisions under uncertainty by combining computational speed, memory, and search capability with human subjective judgment, common sense, and sensitivity to context. Used in financial forecasting, medical diagnosis, and defense, many of these techniques are rule-based, adopting "if–then" representations of causal or empirical dependencies. Since these relationships may not universally pertain or antecedents may not always be diagnostic of outcomes, subjective probability is one of the ways used to qualify them (Krause and Clark, 1994). This has provided a major impetus to research in cognitive processes, human judgment formation, and their probabilistic expression.

Still, no contemporary account of subjective probability would be complete without exploring the role of probabilistic risk assessment in nuclear safety and the lessons its history provides. Risk analysis in the nuclear industry over the past 25 years has been an underlying but pervasive influence on almost every aspect of subjective probability since then.

Probabilistic risk assessment first gained public attention through opposition to various technologies, including large chemical and liquefied natural gas facilities. Taken at face value, this opposition was due to concerns about safety, encouraged to some extent by the traditional deterministic engineering factor-of-safety approach in design for unexpected or extreme events. The public was told of the catastrophic consequences of some worst-case "maximum credible" or "probable maximum" event but at the same time assured that such an event could never happen, or at least that the possibility was negligibly remote (Otway and Thomas, 1982). In terms of public persuasion, this left much to be desired.

The hunt was on for an objective way to prove that controversial facilities were safe based on indisputable scientific facts and rigorous engineering analysis, especially in the intensely adversarial and highly regulated nuclear safety arena. Here, three main goals emerged: (1) to clarify the nature and magnitude of the risks and narrow the areas of debate, (2) to provide for siting decisions and approval, and (3) to satisfy regulatory requirements for existing or proposed facilities (Linnerooth-Bayer and Wahlström, 1991). All of these required or at least strongly encouraged precise and unambiguous estimates of risk and the underlying probabilities of reactor core radiation release.

Various probabilistic risk studies of reactor safety were carried out in the late 1960s, but the most serious attempt came in 1975 with a generic study known variously as the *Reactor Safety Study*, the *Rasmussen Report*, or *WASH-1400* (Fullwood and Hall, 1988). It relied extensively on relative-frequency probability approaches using failure statistics for various plant components, but many probabilities were based on "expert judgment"—in other words, subjective probability. In one review of this work, the Lewis Committee found its use "necessary and appropriate," providing "reasonable input" to the calculated probability values while at the same time expressing doubts about the absolute values of the numbers (McCormick, 1981; Cooke, 1991). One of the concerns was that subjective probability judgments were often embedded in assumptions, incorporated into other estimates, and generally not explicitly derived or adequately documented.

Partly in response, subsequent efforts became increasingly comprehensive and complex, requiring thousands of probability estimates and like numbers of pages to document. Soon a full probabilistic risk assessment for a nuclear power plant required up to 386 person-months to complete, and even "simplified" methods for seismic aspects alone could take up to 70 (Nuclear Regulatory Commission, 1983; Shieh, et al., 1985), with fault trees that could easily wrap around the walls of a room many times over. Even so, variation in probability estimates from one study to another remained, leading some to suggest that these inconsistencies represented a fundamental problem with risk assessment itself (Government Accounting Office, 1985). Meanwhile, the 1979 events at Three Mile Island had cast things in a somewhat different light. Subsequent study showed that the accident sequence had indeed been accounted for in the Reactor Safety Study, although in hindsight some of the probability values for human error appeared less than realistic (Cooke, 1991). With several influential investigations recommending greater use of probabilistic approaches, they came back into good graces.

In light of the Three Mile Island accident and other events, a major update of the original Reactor Safety Study called *NUREG 1150* was undertaken in 1987 to determine whether existing plants were meeting new probabilistic safety criteria and whether retrofits were necessary (Okrent, 1989). As the study progressed, new methods for *formal elicitation* of subjective probabilities were introduced to better structure and document the process while making experts better aware of cognitive factors that could affect their judgments (Keeney and von Winterfeldt, 1991; Otway and von Winterfeldt, 1992). This process, conducted over a year with some 40 experts and 1000 subjectively assigned probability distributions, addressed many of the previous criticisms. However, it brought about an increase rather than a reduction in the variability of subjective probability estimates as experts became better trained to recognize and acknowledge the limits of their abilities. The quest for unique and stable subjective probabilities would have to continue.

As these efforts went on, it was increasingly apparent that they were not having the desired effect on the public. Nevertheless, the objective and factual status of risk analysis continued unquestioned even as the public remained unmoved. If the risks were low—as they demonstrably were from analyses of such stupefying complexity—and if the public was failing to respond appropriately, then the problem must be that the public was misperceiving the risks. Perhaps then risk could be communicated in a different way, or better yet the public's perception changed to a more rational and objective view. This spawned an entirely new behavioral discipline on the quantified study of uncertainty and risk perception using controlled experiments and measurements, *psychometrics* as it came to be known (Fischhoff, et al., 1981; Slovik, et al., 1982). Still, the more introspective could not help but wonder if risk perception, like judgment itself, wasn't highly personal, subjective, and therefore intrinsically variable, and if so whether the whole matter was more a political question of public policy than a scientific one of objective fact and stable frequency (Linnerooth-Bayer and Wahlström, 1991).

Otway and Thomas (1982) posed this as a fundamental ideological division within the risk analysis field, but it can equally be seen as simply another manifestation of the frequency/belief duality that has existed all along.

Even as the twilight of nuclear power construction in the United States began to glow on the horizon, attempts continued to reduce the "inconsistencies" in subjective probability judgments that so stubbornly persisted, this time for seismic aspects with calls for still greater standardization in the interest of regulatory consistency (Veneziano, 1995). In response, yet another approach to subjective probability elicitation for probabilistic seismic hazard assessment was advanced, using "aleatory" and "epistemic" to refer to frequency and degree-of-belief interpretations, respectively.[14] It had three new features. First, elicitation was conducted in a group setting with the aid of a facilitator in a consensus-building process. Second, the aleatory and epistemic probabilities were propagated separately through the analysis, with their contributions carefully segregated in the results. And third, the probabilities elicited were not directly those of the expert assessors but rather their probability judgments made on behalf of what was called the "informed technical community" (Senior Seismic Hazard Analysis Committee, 1997).

Greater consistency among the experts' subjective probability judgments did come about, no doubt in some measure from the group interactions and structuring of the process. The benefit of so meticulously separating the aleatory and epistemic components was less clear, and some found the distinction itself convoluted, ambiguous, and of questionable value overall (Panel on Seismic Hazard Evaluation, 1997). Nevertheless, it did illustrate the uneasiness in the marriage of frequency and belief concepts that had long pertained in the nuclear safety field.

Yet as a device for improving consistency among experts, the curious sleight-of-hand of interposing the undefinable persona of the "informed technical community" as a probabilistic intermediary showed how far the quest for uniformity had gone. Strictly speaking, this was no longer Savage's personal probability of the individual but the opinion of some phantom cadre unable to speak for itself. Indeed it was the same *opinio* of Thomas Aquinas' medieval *probabilis*, the approval of the community of technical authorities, that pre-dated the modern term. With this, the very notion of probability had come full circle in around about 500 years, and along with it a faith that the doctrines of current authority would remain constant and unchanging in the future (Hanks, 1997).

[14]The terms *aleatory* and *epistemic* have become fashionable in some branches of science, but their ambiguity leads to much unnecessary confusion and they are avoided here except in their archaic usage. An *aleatory* uncertainty is said to derive from a process whose outcome is inherently unknowable, like the toss of a coin or the occurrence of an earthquake. But to say something is unknowable intrinsically conveys a state of knowledge about it, so an aleatory uncertainty must always have an epistemic component. One never rolls aleatory dice without epistemically believing they aren't loaded.

The story of probabilistic risk assessment in the nuclear industry holds many lessons, but it is enough to distill only a few. Through these applications, subjective probability as "expert judgment" was highlighted to a much greater degree than it ever had been before. From the start, it was recognized that comprehensively addressing all of the contributing uncertainties would require not only data-based statistical frequencies but subjectively assessed values as well. Yet degree-of-belief probability never achieved real parity with its frequency cousin. The increasingly complex, mathematically rigorous, and scientifically objective analyses that regulatory criteria and public acceptance seemed to demand were never able to fully come to grips with the inherently variable and nonunique character of individual judgment, its expression as subjective probability, or its effect on risk perception. It was an article of faith for Savage and the personal probabilists that reasonable individuals faced with the same evidence could (and very likely would) hold different degrees of belief. But the driving need for an ostensibly objective result fueled the continuing search for an oxymoronic objective subjective probability that could reduce judgment to a uniform and standardized property. In so doing, these applications looked to subjective probability for something it was never intended to be, and asked of it things it was never intended to achieve.

But in the end, none of this mattered very much. Probability had turned from a vehicle for communication into an instrument of public persuasion, and in such an adversarial environment as nuclear safety this it clearly failed to do. The rational man of the Classical probabilists had indeed made a comeback to join the average man of the statisticians, but he had returned to a decidedly nonrational world.

Summary

The overriding theme in this account remains that subjective, degree-of-belief probability is and always has been a legitimate probability interpretation whose long and rich heritage rivals that of its frequency cousin in every important respect. The development of probabilistic thought traces the ebb and flow of these ideas, and only if one were transported to the present without this historical context could subjective probability be seen as some mutant offspring of "real" probability that somehow appeared overnight. It is impossible for probability to be understood or used sensibly in its present-day context without knowing how it came to be and why it became what it did.

From the very start, opinion was the precursor to modern probability. As the opinion of authorities gave way to Fracastoro's, Paracelsus's, and Agricola's signs of nature, inductive reasoning from evidence became possible for the first time alongside its deductive counterpart. But the evidence of signs was reliable only some of the time, so it could be believed only to some degree. This was the birth of the frequency/belief duality and their coexistence for so long. From Pascal to Leibniz and Bernoulli to Bayes, the concept of probability as a function of one's knowledge and belief had a prominent, albeit not exclusive, place. Whereas the early development of modern science, and along with it statistics, looked to probability as a measure of incontrovertible and objective fact, only the most naive would view either enterprise that way today. If nothing else, the experience of the nuclear industry has shown that technology, engineering, and probability spring from and exist within a world of inherently variable, personal, and ultimately subjective judgment.

Notwithstanding all of this, it has become customary that mainly one version of probability or the other is recognized and accepted in any given field, institutionalizing within each its own proprietary doctrine of probabilistic truth. Frequentists dominate the experimental sciences, subjectivists artificial intelligence, business management, and to a lesser extent economics, with virtually no common language or communication between them. Shafer (1989) called this the "balkanization" of probability and Savage (1954) its "tower of Babel." Even within the civil engineering community, structural engineering and water resources adopt mainly a frequency approach with little direct employment of degree-of-belief. Here too the richness of the language of probability remains garbled and often unheard through the static. But geotechnical engineering can ill afford such indifference. Deductive processes from first principles require frequency approaches for data analysis, engineering judgment invokes the inductive processes of belief, and geotechnical practice could not exist without either one. What is really required then is a return to the frequency/belief duality in its original and mutually complementary sense. The Classical probabilists' equal facility with both provides a most accessible and exemplary parallel.

3

The Theory/Judgment Duality

The duality of frequency and belief probability concepts is not exclusively the artifact of probabilistic thought. It has other origins as well in a much deeper dichotomy in engineering thinking: that between theory and analysis on one hand, and engineering judgment on the other. Perhaps nowhere is this more evident than in geotechnical engineering where, like frequency and belief in probability, the underlying duality between theory and judgment has been present from the very beginning.

Prior to the appearance of Karl Terzaghi's *Erdbaumechanik* in 1925 few mechanics-based principles for understanding the behavior of soils existed, and those that did—notably Coulomb and Rankine's earth pressure theory and Darcy's seepage law—lacked a unifying framework. As Terzaghi wrote in the introduction to his book, earthwork engineering had until then been based on "engineering instinct" or judgment alone. With Terzaghi's principle of effective stress, these assorted theories were linked together in a new field called soil mechanics and what became geotechnical engineering was born.

Engineering judgment, of course, did not die with *Erdbaumechanik*, and Terzaghi remained among its chief proponents throughout his life. But judgment, like degree-of-belief probability, has experienced its own period of dormancy, at least insofar as its stature with theory and analysis is concerned. This is in no small part due to present-day computational efficiency and the profusion of numerical methods it has inspired. Reasoning from first principles has become the coin of the geotechnical realm, and some have come to view judgment with the same suspicion as the faculty at Basel saw Paracelsus. Judgment is taken to be a geotechnical metaphysics of sorts—ill-defined, unreliable, and possessed mainly by those of a certain age. In the past 30 years of the *Journal of Geotechnical and Geoenvironmental Engineering,* the U.S. profession's flagship journal of record, the word "judgment" appears in the title of a paper but once. As an inferior replacement for analytical rigor, judgment has become the last refuge of the analytically inept, or so it might appear.

If judgment provides the basis for subjective probability, then this cannot bode well for either one. It calls for a deeper examination of both theory and judgment in geotechnical practice and how they influence the probabilistic methods it uses today. Subjective probability cannot take its rightful place unless engineering judgment assumes its proper role. Like frequency and belief in probability, theory and judgment must be seen as complementary, not competing, sources of understanding.

To understand the role of theory we must first look to what it is and what it does. This would seem simple enough: Every engineer learns the mechanics of theory and analysis and how to execute them. But what concerns us here is what theory and analysis mean. It will be shown that they are not what they are often taken to be as sources of objective truth and that they could not be formulated in the first place without their own subjective element. One thing they do is to promote consistency. Even so, this consistency fades over time as theories change, sometimes radically. We will see too that many have long called for the integration of theory and judgment, but this is easier said than done. Like the frequency and belief concepts of probability, theory and judgment constitute distinctly different views of the geotechnical world. If we look carefully, however, we will find that geotechnical practice provides a framework for bringing them together.

This leaves us to consider the nature of judgment, the most difficult task of all. Judgment is not something one thinks about, just uses. But several identifiable processes operate unseen within judgment, and bringing them to light is the ultimate goal of this chapter. For only by first understanding judgment can we understand the engineering foundations of subjective probability.

Theory and Analysis

In deductive reasoning, theory is the master and analysis the servant who carries it out. Once established and accepted, theory and analysis become the "first principles" from which conclusions are drawn and predictions made. These conclusions are taken as deductive because if the premises are true, then the conclusion follows necessarily. Here there are actually two premises: first the validity of the analysis, including both the algorithm and the parameters it adopts, and second the parameter values selected as input. So that if the analysis is true and the input is true, then the predictive conclusion could be nothing other than true. In this way a conclusion derived from theory and analysis becomes imbued with a certain purity of rational thought that deduction from first principles confers:

The predictive conclusion is objectively derived and therefore objectively truthful.

Predictions lie at the heart of geotechnical and indeed all of civil engineering practice (Lambe, 1973). They may apply to a constructed facility such as a foundation or embankment or to a natural feature such as a slope or seafloor, but either way predictions involve how the feature or facility will perform under the loads it may experience. Performance is judged by whether some desired function is satisfied, and in one way or another this usually comes down to a question of movement: either whether movement will occur (stability) or how much (settlement or deformation).

Performance predictions involve manipulating data using some method of analysis. These are typically geomechanical models that incorporate first principles of engineering mechanics and effective stress, and thus accorded rational status. In practice, first principles in geotechnical engineering are extended to include obtaining the input for these models, and the analytical exercise in its entirety is seen as a three-step process:

1. Characterize the subsurface conditions by drilling and sampling,
2. Measure the applicable engineering properties by laboratory testing, and
3. Perform relevant analyses using geomechanical models.

Reduced to its basic elements in this way, the drill–test–analyze strategy represents deduction from the general to the specific, from first principles to the prediction. The conclusion follows necessarily from the premises: The predictive conclusion could not be false if the analytical and soil-property premises are true. All engineering models embody this fundamentally deductive character (Baecher, et al., 1984).

Geotechnical theory, analyses, and models are established from the *hypothetico-deductive method* familiar to science. It works in the following way. Given some set of initial conditions, a hypothesis is posited and some predictive statement is deduced from it. Experimentation and observation reveal whether the hypothesis is confirmed, and if so it is accepted, at least tentatively. The confirmed hypothesis becomes a theory from which other predictions can then be more routinely deduced (Salmon, 1966). Science, of course, does not specify where these hypotheses come from or how they are generated in the first place. Their discovery lies outside the realm of science in tacit knowledge or intuition. Some philosophers of science (and some scientists as well) have struggled with the notion that the genesis of scientific theories should be so primordially intuitive and that the rationality of scientific deduction should have such nonrational roots. But once theories become established as valid, their origins no longer matter and deductive inference from them proceeds apace.

For geotechnical theories to be accepted for general use, they must be confirmed by field observations of their predictions. As Terzaghi (1936) put it:

> No honest business man and no self-respecting scientist can be expected to put forth a new scheme or a new theory as a 'working proposition' unless it is sustained by at least fairly adequate evidence.

And:

> ... No new theory in soil mechanics can be accepted for practical use without ample demonstration by field observations that it is reasonably accurate under a variety of conditions. (Terzaghi and Peck, 1948)

No sensible geotechnical practitioner would adopt the drill–test–analyze strategy without acknowledging the associated limitations, usually understood to be its various assumptions and approximations. At face value, this is often taken to mean that the resulting predictions, while influenced by these assumptions and approximations, remain valid within some limits of precision. With this added proviso, analytical predictions remain grounded in the rational tenets of first principles that deduction implies. But the matter goes deeper than this. There are other factors that seriously challenge the objective truth of analysis, and we go on to consider some of them here.

Affirming the Consequent

Models that have come to be accepted from the correspondence of their predictions to observed data are often said to be "verified," or quite literally proven to be true. However, this is a fallacy of logic called *affirming the consequent* because models cannot be proven true, only confirmed by supporting data. Models may be corroborated by data to varying degrees, but this is always a matter of interpretation, not proof. If a model fails to reproduce observed data, then it cannot be confirmed and we know it is somehow faulty. But the reverse is never the case. If a model does provide a correct prediction in some case, there is always the possibility it might not in the next. To presume otherwise is to affirm the consequent.

Correspondence between model results and observed data may come about for any number of reasons, and there may always be some other explanation for the observed outcome quite apart from the model's prediction (Oreskes, et al., 1994). It has been noted, for example, that slopes of earthfill dams may remain stable for a variety of reasons, but a calculated factor of safety is seldom chief among them (Peck, 1982). Moreover, it is a requirement of protocol to confirm a model by comparing its results to some experimental outcome or field case history, but what is taken as sufficient for confirmation may actually be quite meager. For years, virtually all dynamic response models were confirmed using the time–history of the 1940 El Centro earthquake because this was one of the few good records available. Likewise, the Lower San Fernando Dam case history still predominates for liquefaction effects. Like the army that so thoroughly prepares for fighting the previous war, our models may be adequately confirmed for the last earthquake but not necessarily the next. So in the end we can never prove that the model is true or accurate under the full range of conditions it is used to address, at best only that its results are likely to correspond to field behavior.

Were we to attempt to verify in the strict sense some geomechanical model involving stability, this would require demonstrating its ability to predict both failure and nonfailure conditions. Nonfailure conditions seldom pose much of a problem. Because most geotechnical models are developed in the context of design, their essential function is to accurately predict nonfailure conditions, a feat they usually accomplish with the added assurance of some factor of safety. Yet the actual factor of safety mobilized in such cases can never be known, only that it is greater than unity, and a model's ability to successfully predict stable conditions does not necessarily imply any corresponding capability for conditions of failure. For this we must look to failure case histories, but these have some problems of their own. Leonards (1982) considered postfailure investigations of embankments and cut slopes in soft clay and showed that they can yield ambiguous results unable to prove or disprove any particular theory or analytical method. Due to incomplete pore pressure or strength information and disruption of original conditions by the failure itself, a "correct" failure prediction can be merely the result of canceling errors, with the actual cause remaining unrecognized. Focht (1994) noted similar effects of compensating errors in the everyday prediction of pile capacity. So again, these matters preclude verification of a model in any objective sense, and all we can do is confirm it as being likely to provide truthful results. To say that a model's predictions are consistent with observations does not prove the model correct.

Affirming the consequent can also pertain to how model results are applied. A typical interpretation of some regulatory or design criterion might be stated as follows:

(premise)	→	only structures with FS > 1.5 are stable
(premise)	→	this structure has FS > 1.5
(conclusion)	∴	this structure is stable

Most engineers would say on first impression that the above conclusion follows from the premises, but the argument is logically invalid. The first premise does not say that all structures with FS > 1.5 are stable. The unfortunate reality is that for a variety of reasons, some structures with calculated FS > 1.5 have failed just the same. Recasting the argument in equivalent terms:

(premise)	→	for any structure x, if x is stable, then x has FS > 1.5
(premise)	→	this structure has FS > 1.5
(conclusion)	∴	this structure is stable

Here, the problem of affirming the consequent is more readily recognized because the order in which the clauses appear in the first premise makes the mistaken direction of inference more apparent: "if" stable "then" FS > 1.5 does not imply the reverse if–then relationship. But in the original formulation the sense of this relationship is disguised by the order of the clauses in the first premise. Simply put, "if B then A" does not make "A therefore B" a valid conclusion. Cooke (1991) illustrates other such results, which can have important implications for how probabilities are interpreted, as we will see later on. Here it is sufficient to note that affirming the consequent not only precludes verification of models in any strict sense but can also cause their results to be misapplied.

Ordinarily, of course, we do not seek proof that a theory is correct or that an analysis is accurate. It is enough that we confirm them from satisfactory agreement with field observations in a sufficient number of cases, just as Terzaghi said. The point here is that confirmation is an inductive rather than deductive exercise. Inductive confirmation reasons from the specific to the general, and if a sufficient number of field observations confirm the theory we accept it more universally and move on to apply it routinely. But just what constitutes "satisfactory agreement" is always a matter of subjective judgment. So to attribute objective truth to deductively derived predictions the method provides is to ignore the subjective roots of its confirmation.

Uniqueness

Suppose that some model, theory, or analysis procedure has been confirmed by observations to some acceptable degree. Since no such model can be proven to be singularly true, there may well be others that hold similar or indistinguishable degrees of confirmation. Here we cannot know which model to use, and there is no prescribed way to choose among them. Any one of them might be just as good as any other—or just as bad.

The problem of uniqueness has long afflicted geotechnical earthquake engineering and soil dynamics, where debates over methods and procedures are legion. To take one example, Seed, et al. (1988) reported on a re-evaluation of the 1971 liquefaction and resulting slide in the Lower San Fernando Dam performed with the benefit of exceptionally good information. Two competing methods for deriving the post-liquefaction strength of the hydraulic fill were evaluated: residual strength from the Standard Penetration Test (SPT), and steady-state strength from laboratory tests. Suitably interpreted (with emphasis on "suitably"), both methods produced values consistent with the failure. Moreover, the SPT approach is itself subject to differing interpretations, one as undrained shear strength (Seed and Harder, 1990) and another that normalizes this strength by overburden stress (Stark and Mesri, 1992). Here again, the case records themselves are not sufficient to support one method over the other.

Yet another example of procedural proliferation is provided by Cai and

Bathurst (1996) who identified three separate categories of analytical methods for determining seismic displacements of earth structures: (1) force-based pseudostatic approaches, (2) displacement-based sliding block methods, and (3) finite element procedures. Within the second category alone there were fully 15 different methods cited, none of which was universally applicable.

The interpretations used in confirming analytical methods can also be the source of vagaries. Duncan (1996) summarizes published results of some 80 finite-element stress and deformation analyses of slopes and embankments using four kinds of stress–strain relationships. Most of the studies reported reasonable agreement between measured deformations and those predicted by the particular relationship adopted, leaving little for choosing among the several stress–strain representations. Then too, the majority of these comparisons were also performed after the measurements had already been made. Such after-the-fact "type C" predictions can be of dubious value in confirming predictive methods (Lambe, 1973).

All of these ambiguities point to the non-uniqueness of competing theories or analysis methods, with field observations unable to prove or disprove any one in particular. So there is never any single objectively correct theory, analysis, or geomechanical model when competing alternatives exist.

Completeness

Geotechnical theory and analysis are only partial predictors of important quantities. For some movement-related processes, applicable theory and analytical techniques do not exist at all in the geotechnical domain. This pertains especially to particle transport by internal erosion. Although some analytical models for internal erosion have been advanced (Indraratna and Vafai, 1997), the governing processes are poorly understood, with difficulties in formulating theoretical approaches that parallel those confronting classical physics on an individual particle level.

Beyond the occurrence of movement is its magnitude, and here theory and analysis can be even more restricted in scope. Duncan (1996) concluded that finite-element analyses were capable of predicting possible ranges of deformations, but for various reasons they tended to overpredict the actual magnitude. Such techniques apply principally to movements that are small in proportion to the dimensions of the structure, but very large deformations also have important practical consequences for such phenomena as liquefaction flowslides, mudflows and debris flows, and rock avalanches. Theoretical models for these processes are few (Jeyapalan et al., 1983; Ledbetter and Finn, 1993) and are often hindered by difficulties in establishing the governing boundary conditions and constitutive properties; determining the necessary input parameters; or again, insufficient understanding of mechanisms (Hungr, 1988, 1990; Finn, 1998). Most methods of geotechnical analysis and models have no real need to predict large deformations when used in design. For this purpose they need only be able to predict initial movements or incipient conditions that herald the onset of failure, and limiting-equilibrium methods fall in this class. With few exceptions, geotechnical analysis does not speak to large-strain processes or movements because its design purpose is to avoid them to begin with. But should we wish to know something about the progression of failure processes after they have been initiated, analytical methods leave us mostly in the dark.

Two final aspects of movement are time and velocity, considerations largely absent from applicable geotechnical theory and analysis. Except for consolidation, deformation resides outside the fourth dimension imprisoned in a static world. But soil properties can change with time (Mitchell, 1986; Schmertmann, 1991) and, depending on movement rates, the effects of large-deformation phenomena can range from catastrophic to inconsequential (Morgenstern, 1985). For important classes of stability problems, even the occurrence of movement must be predicated on its presumed velocity and corresponding drained or undrained strength behavior (Brinch-Hansen, 1962; Ladd, 1991). Here it is the rate of soil deformation, not necessarily the rate of load application, that governs this fundamental distinction (Eckersley, 1990), and this cannot be analytically predicted.

Virtually every geotechnical method of analysis requires constant boundary conditions, which apply only within some specified domain. But actual processes, especially those that are progressive, rapid, or very large in scale, do not recognize this distinction, with boundary conditions that can change dramatically with time throughout their course. Together these considerations illustrate that as predictors of movement, its magnitude, and its rate, geotechnical theory and analysis are far from complete and may provide only fragmentary representations.

Indeterminacy

Soils are astoundingly complex materials, something easy for geotechnical engineers to forget as we view them through our customary windows of simplification. It is for our own convenience that we separate their engineering properties into the three discrete categories of strength, compressibility, and permeability, but the soils remain ignorant of these distinctions. In reality, none of these properties is independent of any other. Strength is a function of stress state and pore pressure. Pore pressure and stress state are a function of strain and strain rate, with permeability thrown in for good measure; and to come full circle, strain has a marked influence on strength. This recursive nonlinearity, where one aspect folds back into another, can make the properties of soils metamorphose like an M.C. Escher print.

More broadly, the engineering properties of soil are neither constant nor uniquely defined, and this is what truly differentiates geotechnical analysis from that in other fields. Taken as premises, there are no fundamental soil properties that exist in nature independent from how they are represented in the analyses and the tests used to measure them. Stress, deformation, and strength properties are a function of the boundary conditions, state of stress, stress path, strain magnitude, and rate of straining the soil experiences at different locations and times in the ground, and these are seldom faithfully represented by laboratory or field tests customarily performed (Lambe, 1967; Ladd and Foott, 1974; Casagrande, 1980). No geomechanical model yet devised can directly incorporate measured soil properties without certain assumptions about soil behavior. Because the soil properties determined in the laboratory are not those operative during the process the analytical model represents, some measured "reference property" must be transformed to reflect in-situ stress and strain conditions and how they change during the process being modeled. This transformation becomes an integral part of the analysis itself (Lambe, 1973; Kulhawy, 1992).

Though we accept this as a matter of course, it poses yet another hurdle for the objective truthfulness of deductive analysis. Because as rules of deductive inference, theory and analysis are supposed to work like mathematical operators. They must operate on the soil-property premises to produce a predictive conclusion that follows necessarily. If, however, the analysis used to produce the conclusion already contains the premises, the conclusion becomes a tautological restatement of the premises. It is said to be *nonampliative*: The conclusion contains little more than the premises already provide.[1]

To the extent that geomechanical models are forced to incorporate soil behavior premises, their predictive conclusions are necessarily nonampliative. Take the case of a simple slope stability analysis. It will be predicated on either drained or undrained behavior of the soil as this relates to the pore pressures generated by shearing on the failure plane at failure. The calculated factor of safety will vary by roughly a factor of two, depending on which kind of behavior the analysis incorporates (Johnson, 1975; Ladd, 1991). To say then that the factor of safety is 1.5 or any other such number becomes an almost flagrantly circular restatement of the soil behavior premise embedded in the analysis rather than a conclusion that follows necessarily from the soil properties themselves. While such nonampliative conclusions might be true, they are not necessarily true. The predictive conclusion becomes a foregone conclusion by virtue of containing the premise. To put it another way, the analysis is predicated on the very thing we would have it predict.

Invariance

All theories, geotechnical and otherwise, are subject to change. This has long been recognized in geotechnical thinking, as Terzaghi wrote to Tschebotarioff in 1951 (Goodman, 1999):

---

[1]Lewis Carroll has been credited with having shown most decisively the distinction between premises and deductive rules and why the rule cannot contain the premise.

> Try to realize that our findings, no matter how important they may be, are no more than links in a long chain and that the truth of today is almost inevitably the error of tomorrow.

Here Terzaghi addresses truth, and if our predictions from theory and analysis are objectively true today then they should also be true tomorrow. But what Terzaghi recognized was that theories change, analyses change, models change, and to deny this would be to freeze the profession's development in time. Today's publication of some new and improved computational method will inevitably become yesterday's news. Consider a situation that occurs with disturbing regularity. A structure that has been subjected to rigorous and thorough seismic analysis. It demonstrates all required factors of safety in every kind of analysis known and is therefore pronounced safe, these deductive methods having proven it so. Then a major seismic event occurs somewhere. And with each new earthquake, it seems, come new attenuation functions for ground motions, new phenomena of dynamic response, and new seismotectonic interpretations that weren't there before. This changes everything: Predicted seismic effects become more severe, and the structure that was safe yesterday becomes unsafe overnight, though this through no fault of its own. The structure hasn't changed, nor has the geology. The theory has. And if the theory were once objectively truthful, this was only so for the moment because theory and analysis are inevitably a reflection of the state of knowledge at the time. They can remain invariant only if we manage never to learn anything more.

Such changes in theories are a fact of scientific and engineering life, and indeed we often look forward to the advancements we presume they will bring about. A widely held view in science is that the development of theory, while it may initially contain certain simplifications and approximations, gradually but inevitably converges with accreting layers of refinement to some objective and rational truth. This is no less prevalent in geotechnical research, where it is all but obligatory to cite a continuing need for incremental refinement in the opening and closing paragraphs of any doctoral dissertation. But this monotonic progression does not appear to be how the most important advances come about, and just how it is that science makes progress and what directions this takes were the central topics of Thomas Kuhn's groundbreaking work that has long since made "paradigm" a household word. Kuhn showed that cumulative, linear, and progressive advancement of existing theory is not how science actually works. Instead, old theories give way to new ones in a sharp break with old traditions much more revolutionary than evolutionary that supplants previous theories as opposed to refining them. Kuhn's theory of scientific revolutions has itself revolutionized how the workings of science are viewed, and it has such far-reaching implications that we go on to consider it further.

Kuhn's Paradigms

Thomas Kuhn's ideas have had such profound effects on scientific thinking that it is well to know something about him. As he relates it, it was in 1947 as he was completing his dissertation in theoretical physics that he was asked to prepare a series of lectures on the origins of 17th-century mechanics. This led him naturally to its historical figures and from there to how their ideas came about. Until then it had never occurred to him that history should hold the slightest interest for physics. As everyone knew, historians merely collected facts and arranged them in chronological order. So he was astonished to find that this history was more than a chronicle. It was explanatory, and the connections it revealed gave rise to new forms of understanding. It was not until 1962 that *The Structure of Scientific Revolutions* came out, with Kuhn by then both a credentialed physicist and historian of science. In it, he shows that a shared conceptual framework or *paradigm* develops within a professional community centered around the problem solving achievements of its theories:

> Close historical inspection of a given specialty at a given time discloses a set of recurrent and quasi-standard illustrations of various theories in their conceptual, observational, and instrumental applications. These are the community's paradigms, revealed in its textbooks, lectures, and laboratory exercises. (Kuhn, 1962)

He argues that so long as the prevailing paradigm retains its problem solving capabilities, research in the field serves primarily to articulate and amplify it, and again much contemporary geotechnical research can be seen in this light. During this phase, the paradigm determines what questions are asked, what quantities are measured, and what phenomena are observed, all of which are directed toward confirming, not challenging, accepted theory. This provides every appearance of sustained advance, but as Kuhn (1962) observed dryly:

> One of the reasons why normal science seems to progress so rapidly is that its practitioners concentrate on problems that only their own lack of ingenuity should keep them from solving.

Eventually, however, anomalies arise that the old paradigm cannot address. Attempts are made to modify prevailing theories to accommodate these disparities, but at some point there comes a pronounced failure in the paradigm's problem solving ability and it breaks down altogether within a short period of time. A new conceptual framework with different principles and theories emerges to change the entire view of the field and its fundamentals. This is not merely an extension of the old theories but an abrupt discontinuity. A revelation comes about by collapse under the accumulating weight of small discrepancies, and the old paradigm is replaced by the new.

Kuhn cites many examples of these paradigm shifts throughout the history of science. Perhaps the one that comes to mind most readily is Einstein's development of special relativity which overthrew Newton's laws as the generally applicable case, and in so doing fundamentally changed how physics viewed the natural world—as well as itself. Geology too is a field concerned with the natural world, and one having seen even more, and more abrupt, changes in how it is viewed. We have only to look to Agricola to see how his signs of nature so thoroughly changed geologic ideas and along with them the whole of medieval thinking. But a more recent and perhaps more familiar case is the revolution in geologic thinking that came about with the theories of plate tectonics and continental drift. Although many had noted the congruence of juxtaposed outlines of the continents, the prevailing geologic paradigm held that the crustal surface is essentially rigid, with cooling having produced compressive forces and consequent mountain-building processes. It was not until 1966 that research discovered multiple and regular reversals in magnetic polarity of seafloor rocks that this theory could not accommodate. In a virtual contagion of scientific epiphany, seafloor spreading, plate mobility, and tectonically induced mountain-building processes emerged as a new paradigm that completely replaced its predecessor within the remarkable time span of less than five years (Cone, 1991).[2] Seen retrospectively, certain milestone events in geotechnical engineering have produced similar revolutions in thinking and theory as well.

It had long been recognized that loose, saturated sands are particularly vulnerable to liquefaction during earthquakes, leading to flowslides or unstable foundation conditions (Casagrande, 1936). Customary practice represented inertial forces as a sustained lateral force coefficient in various *pseudostatic* methods for seismic analysis of retaining walls, embankments, and slopes. The selection of these coefficients was fairly arbitrary and only peripherally related to actual seismic forces, in part from the prevailing view that structures such as earth dams were inherently stable under earthquake shaking anyway. Few dams had ever failed during earthquakes, and those that did were often old and constructed with outdated practices; so the prevailing paradigm attributed these failures more to poor construction than to any inadequacy of procedures for predicting them (Seed, 1973; 1979).

But several strong earthquakes within a brief period shook these views. Apartment buildings severely tilted in the 1964 Niigata (Japan) earthquake, slope failures occurred in the 1964 Alaska Good Friday earthquake, and a number of tailings dams failed during the La Ligua (Chile) earthquake a year later. All of

[2]Herbert Einstein (1991) notes that in his lectures at Robert College in Istanbul in 1924, Terzaghi came remarkably close to putting together these ideas himself. He further developed them in a 1928 lecture on analysis and opinion in geology, a topic also remarkably close to that treated here.

these incidents involved loose, saturated cohesionless soils, and they would not have been predicted by the prevailing pseudostatic procedures. Over an equally brief period it became recognized that the strength of the soils was not independent of the inertial forces applied by seismic shaking, which produced cyclic-induced pore pressures with consequent loss of strength and large deformations. This spawned new procedures for liquefaction analysis that could explain many of the previously unexpected occurrences (Seed and Lee, 1966; Seed and Idriss, 1967; Seed, 1968). Not long afterward it was demonstrated that these procedures could account for many of the salient features of the slide in the Lower San Fernando Dam that nearly caused it to fail in the earthquake of 1971, notwithstanding that pseudostatic procedures had shown it to be safe only a few years earlier (Seed, et al., 1975). With this, the old procedures were abandoned for liquefiable soils and the new liquefaction paradigm became firmly cemented in geotechnical theory, giving rise to the entirely new subspecialty of soil dynamics.

This is no isolated occurrence. It was in the 1850s that Darcy performed the famous experiments that gave rise to the empirical principles governing one-dimensional seepage flow in soils. This was extended to two dimensions by Forcheimer's analogy to heat flow and Terzaghi's adaptations, and it remained the cornerstone of geotechnical thinking on the topic for the next hundred years, manifested first as flow nets and later refined with finite element and finite difference solutions. It was restricted to saturated flow, but this was how the geotechnical paradigm defined its problem solving domain and therefore that which was relevant. Partially saturated flow problems resided in the shadowy realms of agriculture and irrigation where they were distinguished as "infiltration," not nearly so grand as the "flow through porous media" by which the textbook geotechnical seepage paradigm had come to be called.

The way Kuhn describes it, a field's paradigms are how it defines itself and are developed from within. Mostly the rewards are few for those who venture outside it, and its inhabitants do so but rarely. But such crossover applications, when they manage to happen, provide a cross-pollenization that yields impressive fruit. It was just such an occasion that arose in the early 1980s when it became apparent that geotechnical theories were inadequate for addressing seepage-borne contamination from waste facilities and the liners and covers used to control it. These were matters that had always been encompassed by the paradigm of geotechnical theory, but for what was to shortly become a major area of geotechnical practice the theories of saturated flow no longer served.

Meanwhile, in the mid-1960s and the decade beyond, soil science and agronomy had been busy quantifying how both air and water flow through soils, and some of the first such applications to be adopted by the geotechnical field appeared not long thereafter (McWhorter and Nelson, 1979). Permeability was no longer a constant to be measured in the laboratory and then transferred to the field. In the new framework of unsaturated flow, permeability became a function of water content and matric suction—which themselves depended on flow and therefore on permeability—so that fine-grained soils could enhance seepage and coarse layers impede it, contorting the old seepage principles like reflections in a funhouse mirror. This did more than extend the fundamentals of saturated flow to unsaturated conditions. In the remarkably short time of no more than a decade, it overturned these principles by invalidating their application to many problems. Just like the shift from Newtonian physics to special relativity, what once was the general case of saturated flow now became the special case to be invoked only in specific circumstances.

Like the development of soil dynamics, the adoption of unsaturated flow principles has all the earmarks of a Kuhnian paradigm change. These shifts occur quickly within the larger scheme of things to produce fairly dramatic upheavals, and more than refining previous theories come to largely supplant them altogether. They occur more rapidly and more often than might be apparent to us who experience them, and Kuhn's version of time-lapse photography provides a historical lens that reveals them best. Seen within a limited time frame, it can be easy to view theory as immutable and its development as an inexorable march of uninterrupted progress toward ultimate truth. If theories have certain limitations, then it is only a matter of time until they are stripped away. But theory is not invariant and its development takes steps both forward and back in unpredictable and dramatic lurches. Theories come and theories go, as do the analysis methods and geomechanical models that adopt them. Who is to say where or when the next geotechnical paradigm shift will occur?

Theory, Objective Truth, and Consistency

So where does all of this leave us? What we have seen is that, as vehicles for the deductive process, theory and analysis have some rather serious shortcomings. First, they can never be proven to be singularly true. They may be confirmed to some greater or lesser degree by empirical evidence, but they can never be verified. Moreover, different methods give rise to different results, with no objective way to choose among them or to establish which if any is true. In a geotechnical setting, theory and analysis cannot be separated from the soil behavior interpretations they necessarily incorporate, where it can be impossible to prove which will pertain under field conditions. And to top it all off, theory and analytical methods can change drastically over time, sometimes superseded almost as quickly as they become fashionable.

None of this, obviously, is to advocate abandoning theory and analysis. They provide a means for quantified prediction without which the geotechnical field could hardly be said to be engineering at all. Failing to use them—and use them properly—would take us right back to where Terzaghi said we were at in 1925. Neither is this to condemn geotechnical theory or analysis as singularly deficient, with similar issues in related earth science fields like hydrogeology and geochemistry (Oreskes, et al., 1994). What it does mean is that neither theory and analysis nor their predictive results can be taken as uniquely or objectively true. This is not to belabor the obvious. It is well recognized that theory and analysis always incorporate various assumptions that impose certain limitations on the validity of their results. But these assumptions and approximations are seen to influence mostly the numerical solution procedures the methods use, not their basic truthfulness as faithful representations of field processes. And in principle, these approximations can always be eliminated by some sufficient degree of computational sophistication. So the fallacy of objective truthfulness lives on. Predictive results follow from first principles embodied in the drill–test–analyze strategy. The strategy is deductive, so its answer is proven to be true. The method is objective, so its result could not be otherwise.

Besides, the argument goes, theory and analysis produce results that are consistent, and here lies their biggest attraction. Objective and therefore untouched by subjective human judgment, the same analytical procedures using the same input parameters always give the same answer, more descriptively known as "turning the crank." So if procedures are applied consistently, one need not be concerned with who is doing the turning or their judgment and expertise in these matters. It is sufficient that they simply follow the procedural rules the analysis methodology prescribes.

But consistency is indifferent to truth. One can be entirely consistent and still be entirely wrong, with Einhorn and Hogarth (1981) reminding about the paranoid who finds perfect consistency in the world's actions. The implied promise of consistency in method is correctness of result, and this promise might even be fulfilled were it not for the awkward circumstance that the problems are not consistently the same. Because geology is never constant, each geotechnical problem differs in diagnosis and interpretation, so that applying uniformly consistent solution techniques will guarantee that some answers will be consistently right. And others consistently wrong.

So the virtue of consistency must lie elsewhere, and where it is found is convenience. There is no doubt that life is made easier when everyone's answers agree, so we collectively accept and apply certain theories and analytical procedures by convention as things that we mutually agree to. Still, convention does not produce truth any more than expedience produces correctness. In the end, the convenience that consistency purchases comes at a price. Consistency hides different interpretations that remain undiscovered and therefore ignored. In standardizing interpretation, consistency eliminates judgment.

What comes through in all of this is an irreducibly subjective element in theory and analysis that belies their outwardly objective nature. The subjective resides ultimately in a belief in the adequacy of analytical results which cannot be objectively proven true. If there is a problem with theory and analysis, it lies not in their application per se but in the significance that is attached to them and the roles they are accorded. So it is not our models that are the chief culprit here, but rather those of us who use them. This problem is not new, however, nor is it confined to engineering. At the turn of the past century Jules Henri Poincaré was among the first to recognize that deduction from the principles of modern science was not the objective truth it had been taken to be. And long before Kuhn, Poincaré would shake the objective foundations of science.

The Objective and Subjective in Science

A shambling figure direct from central casting, Poincaré (1854–1912) appeared much like you'd expect for a professor of mathematics at the Sorbonne. Bow tie askew, he was:

> ...a French savant who looked alarmingly like a French savant. He was short and plump, carried an enormous head set off by a thick spade beard and a splendid mustache, was myopic, stooped, distraught in speech, absent-minded and wore pince-nez glasses attached to a black silk ribbon. (Newman, 1988)

In other respects, however, he occupied a class entirely by himself. Pirsig (1974) provides a fascinating account of what Poincaré's thought revealed. This mathematical phenomenon of his time was an international celebrity by age 35, a living legend at 58, and, according to Bertrand Russell, "by general agreement, the most eminent scientific man of his generation." Until then, scientific truth had been taken as infallible and beyond doubt, but an alarmingly deep rift had already begun. It was becoming apparent even then that experimental testing of scientific hypotheses could never really prove them. Confirmation by experiment was not fundamentally deductive but inductive. True proof would require an infinite number of experiments, and there was always some chance, however small, that the next in any finite sequence of experiments would be the disconfirming one. So one could not speak of the truth of science or a scientific theory in any absolute sense but more properly its likelihood of truth given the particular body of experimental evidence available at the moment. But Poincaré took this one step further. He distinguished between the experimental sciences and the "exact" science of mathematics. Placing no reliance on experiment, mathematics was the bedrock of scientific truth. So here Poincaré turned to the dilemma posed by geometry.

Only recently had mathematics recognized the separate alternatives to Euclidian geometry of Lobachevski and Riemann. Starting with their own axioms, each of the three geometries was internally consistent, but they gave wildly different answers. So which one was true? Poincaré (1905) argued that this question had no more meaning than asking if the metric system was "true." Axioms, like scientific hypotheses, were only "definitions in disguise." There could be entire universes of axioms, hypotheses, and facts out there waiting to be discovered, all in infinite number. But science has no time to consider them in infinite variety, and observations made at random could no more produce science than a monkey at the keyboard could compose the Fugue in D minor. Those chosen were chosen for a reason. The axioms, hypotheses, and even the facts that science elects to consider were simply conventions, selected not arbitrarily or by chance but in the interest of mathematic and scientific convenience. Then it must be the scientist, not science—the subjective, not objective—that preordained what to observe. So science was much more than simply applying deductive rules.

Needless to say, this swept in a storm of protest. If experimental facts were preselected, this would destroy the objective truthfulness of the scientific method altogether. If science was not true but merely convenient, then truth could be whatever you like it to be, nothing more than one's subjective belief. But Poincaré thought the matter went deeper than this. Subjective did not mean arbitrary. Something much more was at work here.

Reflecting on his own experiences, Poincaré realized that his hypotheses had come to him in moments of insight, a wave of crystallization since described by so many others.[3] To him it was not arbitrarily or by chance that certain facts came together. To the contrary, certain elements that at first appeared unconnected managed to "arrange themselves in an unexpected order" in the process forming a "harmonious whole" (Poincaré, 1905). It was this order and harmony of facts selected by the subliminal self that governed what to observe, and with it a "true esthetic feeling which all mathematicians know, but of which the profane are so ignorant as often to be tempted to smile." It was not the facts themselves ("*les faits ne parlent pas*"—facts do not speak), but how they fit together, how they complemented each other, their sense of order, their esthetics—in short, their patterns—that had everything to do with how science actually worked. After Poincaré, notions of the factual basis of science and its objective truth would never be quite the same. For what Poincaré had discovered was that it was impossible to do science without melding logic and intuition. In a 1904 essay he wrote:

> It is by logic we prove, it is by intuition we invent.

And in a later article:

> Logic, therefore, remains barren unless fertilised by intuition.

But all too easily this synthesis could become antithesis, and in a battle between the two it would be analytical logic that won out:

> Instead of endeavoring to reconcile intuition and analysis, we are content to sacrifice one of them, and as analysis must be flawless, intuition must go to the wall. (Poincaré, 1905)

---

[3]Including Terzaghi, who described how his knowledge had "crystallized" into judgment.

The harmonious ordering of facts achieved through this intuition is one of the attributes of what we would call judgment, and science now recognizes but isolates it. Judgment is that which comes before. Cooke (1991) credits philosopher of science Hans Reichenbach as the first to distinguish scientific discovery from justification, whereby the admittedly subjective attributes of the former are scrupulously restricted from the latter. Having the idea, in other words, is not the same as going on to use it in accordance with scientific method. Perhaps, but one could not occur absent the other, and without having had the idea in the first place the ruminations of science could never go far. Some have called this act of discovery "problem finding," where often in the great discoveries of science the most important thing is that a certain question is found (Arlin, 1990). It is the question then that is at least as important as the answer, and finding the right problem, asking the right question, is what defines originality in science. This has been said to be the signature of a true scientist, with complex problem solving the work of the highly trained technician (Mackworth, 1965). If so, then the pinnacle of scientific achievement cannot be attained without the subjective, the subliminal self, the judgment that guided Poincaré's esthetic feeling and harmonious whole.

One property that can make things decidedly un-harmonious is anomaly, and Kuhn (1977a) credits anomalies, and the awareness of them, with a major role in scientific discovery. When Roentgen discovered X-rays, it was because of a mysterious glow in his cathode ray tube that should not have been there. When Herschel discovered the planet Uranus it was because the would-be star was so unexpectedly large. And when Priestly established the existence of oxygen, it was because he had isolated a gas that behaved almost, but not quite, like ordinary air. According to Kuhn (1977a), it took a kind of genius in each of these situations to have seen things differently, or to have seen different things, with awareness that certain expectations had been violated. It also took instruments for observing how nature ought to behave—its patterns—to form these expectations in the first place. There will be more to say shortly about the role of patterns and anomalies in judgment and much more later when we examine the attributes of expertise. But all of these things, of course, are subjective and depend on personal abilities and qualities. So even science, it seems, has allowed a crack in the door of objective deduction.

Kuhn (1977b) pries it open a bit wider by extending the role of the subjective not just to scientific discovery but to justification as well. In particular, he examined the problem of uniqueness where more than one possible theory applies, and he went on to consider how the scientist goes about judging a theory against its competitors in selecting one to use. Kuhn established five criteria. *Accuracy* is the extent to which a particular theory agrees with experiments and observations. *Consistency* is its internal agreement and agreement with other theories. Its *scope* is its broadness and generality, while *simplicity* is self-explanatory. The final criterion Kuhn calls *fruitfulness*, or the ability of the theory to disclose previously unnoticed relationships.

While all of these are desirable attributes, they can be ambiguous and conflicting, and there is no objective algorithm for deciding which should have precedence. Take accuracy, the criterion some would most favor. Yet a theory accurate in one respect may not be in another. One method of stability analysis, for instance, might yield an accurate factor of safety but not the critical failure surface location, so a choice must be made concerning which kind of accuracy is more important. Then there are those who might value a theory most for its scope in generalizing to the broadest range of possible conditions, whereas others might say that the most powerful theory is the one most fruitful in pointing the way to new advances. If scientific justification were entirely objective, there would be some formula to determine how these criteria should be valued. Kuhn (1977b) argues that because there are no such rules, there is no objective way for science to prescribe which of its theories should be used, in the same way that Poincaré argued of the geometries. Instead, these matters are subjective—they go to the scientist, not science, to matters of experience, familiarity, value preference, and even personality. And as such, they are matters of the individual scientist's judgment.

Kuhn's ideas have since spawned a whole cottage industry that dissects and interprets his work (Horwich, 1993). Some have raised various objections, or at least qualifications, that Kuhn considers in turn. In acknowledging this subjective element, the charitable might say that it is merely a crude and temporary expedient and that when science reveals more objective facts the cream of theories will inevitably rise to the top. But so long as competing theories satisfy different criteria to different degrees, equally rational scientists may still differ in their choices. It is also said that the subjective obviates the objective, allowing the factual basis of science to be ignored and returning it to superstition. But Kuhn notes that the subjective does not function in the face of, or at variance with, demonstrated facts, and the standards of objective factuality are not compromised. Rather, and again with shades of Poincaré, the subjective determines by communal consensus which facts are given weight and significance, which of them are to be interpreted, and how. If all of this is so, then how is it that science has produced such impressive advances? Here Kuhn rests his case, saying only that science incorporates a formidable inductive element that is necessarily subjective because no rules or procedures govern the inductive process. Each must flesh out their own rules, and each will do so in different ways, even if they ultimately come to the same conclusion.

All of this demonstrates that even science is no monolithic bastion of the objective, with the objective and the subjective working hand-in-hand. Pirsig (1974) goes so far as to say that in any creative endeavor, not just the scientific, there is something else that stands above in the hierarchy of the objective and subjective. This is a preintellectual awareness, a precognitive condition he calls *Quality*, and it is what unites the two so that they work together to produce in the end the harmonious whole and the order that Poincaré was talking about. This is the relationship between the objective and the subjective, and it cannot exist if either is adopted to the exclusion of the other. Engineering works in much the same way. Analyses and calculations are, without doubt, data-driven and objective. But these calculations must be integrated with subjective judgments in deciding what is important enough to calculate, what variables are chosen to represent it, and in the end what the results mean. These are the parts of analysis we call assumptions, and we make them so casually we often forget that they're there. But as Poincaré established for axioms of science, engineers make these assumptions not for their objective truth but for our own convenience. Assumptions are our definitions in disguise.

The Geotechnical Paradigms of Theory and Practice

Poincaré notwithstanding, geotechnical theory and analysis retain the halo of scientific rationality despite the manifold challenges to their deductive truthfulness. Calculated factors of safety are taken as real in the most literal way because first principles have produced them, and calculated deformations are quantities of palpable form and substance cut from the same cloth. The geomechanical model is no longer an idealization of some process; it is the process, with its findings correspondingly construed as statements of fact. The assumptions and approximations appearing as footnotes may qualify the precision of results but do little to shake an unwavering faith in their basic truth. To admit otherwise would be to succumb to the forces of irrationality and subjectivism.

If complaints like these seem familiar to geotechnical ears, it is only because they are so old. As developer of some of the most fundamental geotechnical theories, Terzaghi could speak with some authority on the matter. As a young man in 1918 he undertook a literature review extending as far back as 1850 to learn what he could about the performance of soils.[4] His disappointing conclusion was that even by 1880 theoretical findings had come to be taken as fact to the neglect of all else and indeed that engineers had been overtly encouraged to do so. In his presidential address to the First International Conference of Soil Mechanics and Foundation Engineering in 1936, Terzaghi amplified on the outcome of his literature review, saying:

[4]This and other accounts of Terzaghi's life, work, and thought rely extensively on Goodman's (1999) biography containing materials from correspondence and personal diaries that would otherwise have remained inaccessible. For this, the profession owes perhaps as much to Goodman as to Terzaghi himself.

> ...once a theory appears on the question sheet of a college examination, it turns into something to be feared and believed, and many of the engineers who were benefitted by a college education applied the theories without even suspecting the narrow limits of their validity.

The acceptance of theoretical results without reservation is probably even more entrenched today, over a century since Terzaghi marked the rise of the practice. Once more, Kuhn's notion of paradigms provides a useful framework for understanding why this has persisted for so long.

The Origins

As Kuhn defined it, a profession's paradigm is constructed around its theories as revealed in its textbooks, lectures, and laboratory exercises, and the geotechnical profession's drill–test–analyze strategy clearly belongs in this category. The paradigm assumes such importance because it legitimizes the profession, distinguishing it from other fields and defining its identity in a very real sense. Seen in this light, the elevated standing of geotechnical theory comes about from its deductive link to the legitimacy of science. So the theoretical paradigm is taken for more than it really is—simply a shared context that expresses the geotechnical profession's understanding at the moment of the phenomena in its domain—and instead becomes a source of objective truth arrived at by deduction from first principles. If a calculated quantity or factor of safety is taken unquestioned as real and truthful, this is because the theoretical paradigm encourages and endorses it.

Yet at the same time, another geotechnical paradigm coexists with the theoretical one—the paradigm of practice. It adopts its own protocols that also use theory and analysis but use them in a very different way. Even its methods of reasoning are different, emphasizing inductive processes over deduction in deriving predictions. Thus, inductive empiricism, using correlations that derive the general from specific cases observed, may take precedence over first principles that work in exactly the opposite fashion. And judgment is the embodiment of induction that governs how these specific cases are integrated and synthesized to formulate both general rules and individual predictions.

The paradigm of geotechnical practice often finds itself at a disadvantage to the theoretical. By definition, practice is not self-propagating and cannot be formally taught. It passes down through generations of practitioners through apprenticeship and without written instruction. Practice does have its own literature of sorts, chiefly case-history publications sometimes in entire conferences, albeit subject to less rigorous peer-review standards and accorded lesser overall stature. But without the close deductive ties to science, the paradigm of geotechnical practice can never achieve the scientific legitimacy that identifies and validates the theoretical, especially in the eyes of sister disciplines or regulatory bodies that look to the ostensibly objective certitude and precision of deductive prediction. Perhaps its most serious handicap is that the principles of judgment that practice relies upon so heavily cannot be codified and are seldom elucidated, making it difficult to articulate the most basic element of its paradigm.

Theory and practice are taken here as paradigms in the Kuhnian sense because they constitute distinctly different ways of looking at the geotechnical world, each with their own ways of defining their achievements, their own distinct reward structures, and their own separate notions of what constitutes truth. Indeed this duality has been with the geotechnical profession from its founding, as we will go on to see. For there to be two operative paradigms coexisting in a single profession is perhaps rare but not unknown. The fields of physics and medicine are similarly divided. Physics has its theoretical and experimental branches, which operate almost independently. Biomedical science provides the scientifically deduced understanding of disease and its organic causes, with clinical medicine adopting inductive processes for diagnosis and treatment of individual cases. Other parallels are apparent as well. Clinical medicine is learned firsthand during an extended period of residency under close supervision. Likewise, a geotechnical engineer is not considered competent to practice without a less structured but otherwise similar apprenticeship. While a recent graduate in mathematics might be a neophyte but still basically credentialed, both physicians and geotechnical engineers just emerged from training in theory and science must then acquire an altogether different class of skills for practice. For in both clinical medicine and geotechnical practice, theory may underlie the answer but does not provide the answer itself. By circumstance or proclivity, certain members of the profession may adhere more to one paradigm than the other, but it would be wrong to assume that even the developers of theory reject the paradigm of practice or vice versa. Yet the coexistence of the geotechnical paradigms has not always been an easy one.

Peck (1979) clearly distinguished the two paradigms, calling them engineering science and engineering practice. Much like Kuhn and like Terzaghi in his letter to Tschebotarioff, Peck warned that developments in engineering science do not always translate into progress for engineering practice because scientific advancements do not occur monotonically:

> In short, engineering science and engineering practice are not identical. Advances in science may temporarily appear to run counter to good practice ... but it should by no means be assumed that the latest scientific advancement is always in the right direction. Science has its own ways of making progress, and as evidence accumulates it corrects its errors and improves its predictions.... But science may temporarily mislead the unwary ...

Although this may have been among the first formal statements of the theory/practice duality to make its way into the geotechnical literature, this was by no means the first time it had been recognized. Peck had been considerably more direct in remarks he made fully 35 years earlier, which revealed that the struggle between the paradigms of theory and practice had gone on even then for some time, and with no small degree of intensity. Goodman (1999) provides this and the following remarks:

> Blind application of theory can directly lead to disaster … this is the idea which nearly ruined soil mechanics and against which the best efforts of Terzaghi and others have only recently been able to make headway.

But again it was Terzaghi himself who spoke most eloquently and forcefully to what he considered to be the misuse of theory, much of it his own, and in defense of the paradigm of practice. In 1936 he wrote:

> Unfortunately soils are made by nature and not by man…. As soon as we pass from steel and concrete to earth, the omnipotence of theory ceases to exist…. In soil mechanics the accuracy of computed results never exceeds that of a crude estimate, and the principal function of theory consists in teaching us what and how to observe in the field…

Several years later, he summed up succinctly:

> Nature interferes with extensive applications of theory.

While certainly not dismissing theory, Terzaghi did not subscribe to its paradigm, vigorously insisting that the highest and proper purpose of theory was to guide judgment. Although fully intending his theoretical contributions to be used to this end, he was forced to watch with consternation that grew into regret as those who he called the "victims" of his theories used them to fuel the paradigm of theory, not practice. To a friend he wrote in 1943:

> I do hope that the practical consequences of my excursions into Engineering Geology will be less catastrophic than those of my doings in Soil Mechanics. I blush if I think I am partly responsible for what is being printed in this field.

By 1961 near the end of his life Terzaghi was if anything even stronger in these convictions, going so far as to call it fatal to the profession should theory prevail over judgment:

> I produced my theories and made my experiments for the purpose of establishing an aid in forming a correct opinion and I realized with dismay that they are still considered by the majority as a *substitute* for common sense and experience. As long as I can still crawl, think, and write I shall fight this fatal tendency [emphasis in original].

While far from alone in its adoption, Peck and Terzaghi were unsurpassed in giving voice to the paradigm of practice. Through them, we find that the duality of theory and judgment is as old as the geotechnical profession and the two paradigms it has long contained.

Ferguson (1992) provides an educational perspective that extends this duality across broad areas of engineering. It became entrenched, he says, more than 50 years ago as an outgrowth of the academic research for military applications so vital to the outcome of World War II. The geotechnical field was no exception, with many of the profession's leaders-to-be having first been exposed to soil mechanics for the purpose of airfield construction in a course then taught at Harvard by Arthur Casagrande (Marcuson, 2000). This program and others like it were directed in Washington by Vannevar Bush, an electrical engineer and former dean of engineering at MIT who headed the Office of Scientific Research and Development under Franklin Roosevelt. According to Bush, military officers managed to tolerate scientists, but the same could not be said of engineers.[5] So engineering researchers would simply become scientists, if that's what it took. Bush reflected later that "the business of elevating the scientists to a pedestal probably started with this move, and it has certainly persisted and misled many a youth." It only got worse from there.

The formal education of engineers had always concerned itself with design, this being the engineer's principal role in society. Design is an intuitive process without any single uniquely defined answer that can be deductively ascertained. So most engineering curricula addressed the design elements of engineering practice; although without any "right" answers for such subjective aspects as creativity and imagination, this was more something that students could be exposed to rather than instructed in as such. Even so, the duality of engineering science and engineering practice had been alive and well, existing more or less comfortably together in academia for some time.

A turning point came in 1953 when an accreditation report by the Grinter Committee recommended that courses in practice be dropped in favor of a curriculum in engineering science (Ferguson, 1992). Education would follow funding as surely as water flows downhill, and funding would follow the avenues of research that science directed. Thenceforward engineering education would speak for the theoretical paradigm, and the paradigm of practice would be left to fend for itself. By the 1960s the fruits of this trend had become clear. One no longer needed engineering judgment to become educated as an engineer. Every problem had a single, objectively defined answer, where "neither the data, the applicability of the method, nor the result are open to question" as one follow-up report put it with an unmistakable note of discouragement.

---

[5]This despite the rich tradition of the academy at West Point, which lays claim to the first program of engineering education in the country.

Theory in Geotechnical Practice

From this it can be seen how the theory/judgment duality is largely suppressed in the paradigm of geotechnical theory. Theory and judgment are seen as diametrically opposed, with judgment reduced to a bare minimum of subjective influence. The inexorable and cumulative advance of science will ultimately prevail, and if judgment is tolerated it is only as a temporary expedient until still more rigorous procedures can be devised that succeed in eliminating it altogether. Subjective is pejorative, and much like the London Statistical Society the ultimate goal of the theoretical paradigm is to advance engineering science so as to eventually "exclude all opinion." We look again to Kuhn (1977c) to explain why this should be.

According to Kuhn, in the early stages of a field's development, pressing needs of society guide and direct a profession, determining the problems it addresses. We will see later through the biographies of several civil engineering experts of the late 19th and early 20th centuries how their major works were commissioned at society's behest to produce extraordinary changes in the lives of ordinary people. At this stage, the field's paradigm is heavily influenced by pragmatism and common sense, finding and doing what works. Theory follows along to assist and to explain what has already been learned.

This changes as the field becomes mature, however, as geotechnical engineering is now said to be. With the pressing problems already solved, a field's members become progressively more steeped in a sophisticated body of established theory and technique. This is largely the status of ordinary science, which Kuhn (1962) calls mainly a "mopping up" operation, whose members engage mostly in articulating and confirming the prevailing paradigm. As the field matures they turn inward, a "special subculture" talking mostly to themselves and isolated from the pragmatic import of the problems they once took on. And as they do so, they become the exclusive audience for, and judges of, each other's work. The challenges are no longer external but internally imposed to increase theory's precision and scope (Kuhn, 1977c).

While this arguably could describe the contemporary status of the geotechnical profession, it does not hold for all its members. Just as we have seen with Terzaghi, some of the foremost developers of geotechnical theory and analysis have at the same time been, and remain today, among the most cognizant of the paradigm of practice and the judgment it entails. They reflect the field's historic vitality in their melding of theory and judgment, according proper roles to each:

> The fact that equilibrium analyses of slope stability involve assumptions and limitations does not mean they are valueless. It means that they cannot be used without understanding and judgment. (Duncan, 1996)

As well, they remind that theory departs from reality when uncertainty is thrown into the mix:

> It is essential to expect the unexpected, and to deal with soils as they are, not as we might wish them to be. (Mitchell, 1986)

Unlike its theoretical counterpart, the paradigm of geotechnical practice recognizes and accommodates the theory/judgment duality, viewing the two as mutually complementary. Most practitioners would argue that the application of judgment is necessary at every step of the drill–test–analyze process, and in this respect judgment is supportive of theory and analysis. At the same time but in the opposing sense, theory and analysis are seen as buttressing judgment, as a heuristic aid to understanding:

> ...the true value of the analysis often lies in the insights and understandings that come from careful formulation of the problem.... Engineers use analyses to sharpen their judgment... (Whitman, 1984)

And:

> Judgment is required to set up the right lines for scientific investigation, to select the appropriate parameters for calculations, and to verify the reasonableness of results. What we can calculate enhances our judgment, permits us to arrive at better engineering solutions. (Peck, 1980)

In this context, understanding the problem is distinguished from determining literal truth. Calculated values are not taken as incontrovertible reality but as aids to achieving this understanding. Thus a calculated factor of safety becomes a "useful index" of stability (Duncan, 1996), an "empirical tool" (Morgenstern, 1995) pointing to the need for inductively integrating analysis results with information of other kinds and sources in deriving the predictive conclusion.

Chief among these other factors is precedent, an essential foundation of the paradigm of geotechnical practice. Long before *Erdbaumechanik*, earthwork engineers extrapolated the experience of past precedent to the case at hand. But configurations, soil properties, and loadings are never an exact match. Theory and analysis provide the bridge for extending precedent to differing conditions. As always, Terzaghi put it best, this time in a 1939 consulting report (Goodman, 1999):

> When utilizing past experience in the design of a new structure we proceed by analogy and no conclusion by analogy can be considered valid unless all the vital factors involved in the cases subject to comparison are practically identical. Experience does not tell us anything about the nature of these factors and many engineers who are proud of their experience do not even suspect the conditions required for the validity of their mental operations. Hence our practical experience can be very misleading unless it combines with it a fairly accurate conception of the mechanics of the phenomena under consideration.

Peck amplified on this same topic on a number of occasions. He described theory and analysis as the platform for extrapolation of precedent, in a supporting role to judgment:

> Theory and calculation are not substitutes for judgment, but are the bases for sounder judgment. A theoretical framework into which the known empirical observations and facts can be accommodated permits us to extrapolate to new conditions with far greater confidence than we could justify by empiricism alone. (Peck, 1969a)

Recounting his collaboration with Terzaghi on their landmark *Soil Mechanics in Engineering Practice,* Peck described how its treatment of retaining walls and spread foundations helped solidify their thinking on the application of soil mechanics theory with respect to precedent:

> When we reached the realization that ordinary retaining walls, say those twenty feet or less in height, are not designed on the basis of soil mechanics but on the basis of precedent, and when we realized that this approach is perfectly satisfactory for most retaining walls, the place of soil mechanics suddenly became apparent. Soil mechanics would serve to point out the circumstances under which retaining walls of moderate height might be subject to conditions outside the scope of precedent and would help us recognize when the use of precedent would be dangerous.... The chapters on footings, for example, recognized first that most footings are designed by precedent, and then pointed out the circumstances under which soil mechanics could be of use. (Peck, 1973a)

Later, Peck made similar comments about the role of analysis in the design of earth dams, noting that precedent was more important to certain aspects of their safety:

> ... we might inquire whether the safety of dams is actually related in any significant degree to the results of stability calculations for surfaces of sliding in the embankment materials. I think not ... embankment slopes are established on the basis of precedent ... and a host of other factors. The stability check is just that: a final review to see that the design does not violate precedent with respect to the factor of safety. (Peck, 1982)

The richness and breadth of these excerpts show that rather than seeking to eliminate judgment, the paradigm of practice places judgment at least on a par with, and in many ways preeminent to, theory and analysis. For their part, theory and analysis are incorporated in well-defined roles but not as sacrosanct sources of objective truth or ends in themselves. Both as aids to understanding and as a means for extending precedent, theory and analysis are used in ways that reconcile the theory/judgment duality as opposed to suppressing it. And just as Terzaghi introduced this topic, so should he have the last word with the remarks he made to the Institution of Civil Engineers. Terzaghi spoke of soil mechanics then as an emerging field of engineering science, but still he cautioned about the place of science in engineering:

> To accomplish its mission in engineering, science must be assigned the role of a partner and not that of a master. (Terzaghi, 1939)

The Meaning of Judgment

If judgment assumes the importance that it does, then it is only fair to inquire as to its nature and meaning. As we have already seen, judgment can mean many things to many people, from observation and experience to simple common sense. But to fully grasp the meaning of judgment requires some concept of what it is and how it works. This has never been easy for geotechnical practitioners to articulate, nor have they often found occasion to do so.

Focht (1994) is among the few to have put forward an explicit framework for incorporating judgment as an integral component of the predictive process. In the paradigm of practice, he noted that practitioners are called upon to produce predictive conclusions before the fact—Lambe's (1973) "type A" predictions—rather than after the outcome is already known as for confirmation of analytical models. Focht further noted that predictions using the drill–test–analyze strategy sometimes depart from actual performance, which he attributed more to a procedural deficiency, a problem of "mental approach," than to faults in analytical method. He described a judgmentally based predictive strategy as containing the following stepwise elements:

1. Utilizing sound judgment, select an approach to a problem and define critical components of the process;
2. Develop an adequate model of the stratigraphy, including water level, and an understanding of site and regional geology;
3. Develop an adequate model of applicable soil parameters;
4. Select a method of analysis and a desired factor of safety compatible with the parameters, method of analysis, and problem;
5. Gather and evaluate historical data that are applicable to the problem; and
6. Apply judgment and intuition in review of conclusions, predictions, and recommendations.

These six stages can be quickly recognized as an extension of the drill–test–analyze predictive strategy, but closer examination reveals considerably more. Taken as a whole, this process synthesizes deductive analysis and the inductive elements of diagnosis and interpretation.

Judgment first enters the process at the beginning, where it is used initially to select an approach to the problem and define the critical elements of the mechanisms and conditions at work. Judgment therefore has a diagnostic component—it is used to formulate the problem the predictive process must then address. Judgment is that which comes before the problem solution can begin.

Although adopted explicitly only in the first and last stages, Focht did not intend that judgment be restricted to these elements, noting that they are all interrelated and that "actually, judgment is necessary throughout the process." It can be seen that the intermediate stages each involve different sources and kinds of information, from stratigraphy and geology, to analysis and results, to historical data and their evaluation. All of these factors must be weighed and combined in some way, which points to an inductive aspect of judgment as the means for doing so.

Focht's final stage again invokes judgment in "review" of the predictive conclusions. This means critical review. Implying much more than final confirmation, it goes to retrospective questioning of the applicability of each element of the predictive process. In short, do the prediction and the process make sense? Here, the role of judgment is interpretive—to provide context for what the prediction should be taken to indicate within the larger framework of the problem. This interpretive effort gives meaning to the predictive conclusion, and it involves establishing conformance within some overall structure even if, as Focht said, this a "general automatic, almost subconscious, step in the mental process of the predictor."

We can therefore distill from this examination of the predictive process that at its most basic level judgment possesses three fundamental attributes. It has a diagnostic character in problem definition, an inductive character in combination of evidence, and an interpretive character in providing meaning and context to predictive conclusions. But beyond just their characterization and perhaps more to the point, Focht contended that the absence or under-representation of these aspects of judgment was the single most important cause of failure of the predictive process in practice. With this in mind, we now proceed to consider these attributes individually.

Diagnosis

Any engineering textbook introduces example problems by way of something along the following lines: "Suppose we wish to find..." or "It is desired to find...." The problem is then presented, and the student works through the solution. The student seldom questions why one would want to find the specified quantity in the first place, the most obvious reason being a desire to pass the course. A less obvious reason has to do with diagnosis, what we consider here. The predictive process outside the classroom cannot presuppose that the problem has already been diagnosed.

Diagnosis is the first step in Focht's representation of the predictive process. He calls this the act of selecting an approach to the problem, saying that this requires sound judgment. It is evident that an "approach" requires problem definition, much like that presented to the student, but it also implies something more. Problem diagnosis includes formulating hypotheses, applying intuition, identifying relationships, and visualizing processes. These then become the components of diagnostic judgment.

HYPOTHESIS FORMULATION. The predictive process for every site and every structure begins with a hypothesis. The hypothesis posits the conditions that might be present that could result in adverse performance if they were. These could be geologic features, properties of the soil and its behavior, or the characteristics of the structure itself and the loadings it might experience.

A hypothesis is formulated even before the first boring is drilled or the first element designed. It guides and directs what things to observe, what properties to measure, and what conditions to design against. All subsequent efforts in the predictive process are devoted in one way or another to disconfirming the initial hypothesis ("we looked for but didn't find..."); to affirming and accommodating it ("as a result of finding..."); or to modifying or replacing it ("instead of finding..."). The hypothesis is entirely the product of judgment. It does not spring from theoretical considerations, nor is it arrived at by any deductive process. A closer look at how geotechnical hypotheses come to be has much to say about judgment itself.

Consider some structure to be built on stiff clay. If it is sensitive to settlement, then compressibility might constitute the property to be measured. If excavations are involved, then fissuring with its effects on strength might be the characteristic of interest. And if the site were near a valley slope subject to stress-relief effects, then residual-strength planes of weakness might be the problem to be addressed. Of course, all three situations might be involved, but none can be predictively addressed without first postulating their possible presence and effects.

Problem diagnosis is not contained at all in the theoretical paradigm, which deals exclusively with problem solution. Because it is the paradigm of theory that governs formal instruction, diagnostic techniques are not taught. This is a huge omission. A correct solution cannot be obtained unless the correct problem is solved. So to the extent that problem diagnosis and problem solution are equally important in the predictive process, then fully half of it lies beyond formal training. Indeed, many failures of both structures and the predictive process attributed to inadequate investigation are in fact failures to formulate the related hypothesis in the first place. In this regard, de Mello (1977) said that safety rests:

> ...not upon the accuracy of our calculations, but upon the adequacy of our hypotheses.

Especially in the environment of everyday geotechnical practice, competitive pressures and client expectations ("I want an engineer who will solve my problems, not look for more") can mitigate against formulating hypotheses too prolifically, handicapping judgment in problem formulation from the outset. But Peck (1973b) too maintained that insufficient diagnosis of problems was more pervasive than their inadequate solution:

> I am persuaded that many more failures of foundations or earth structures occur because a potential problem has been overlooked than because the problem has been recognized but incorrectly or imprecisely solved.

How hypothesis formulation works can be found in another foundation of geotechnical practice, the *observational approach.* In contrast to Focht's predictive strategy, the observational approach defers *a priori* predictive conclusions. Reserved for conditions too complex or difficult to fully characterize before the fact, it relies on its own stepwise process (Peck,1969b):

1. Exploration to establish the general nature and properties of the deposit, but not necessarily in detail;
2. Assessment of the most probable conditions and the most unfavorable conceivable deviations from these conditions;
3. Establishment of design premised on the most probable conditions, with performance predictions for both this and the most unfavorable case;
4. Selection, in advance, of actions or modifications to be implemented for every foreseeable significant deviation from the design premise; and
5. Measurement of predicted quantities as construction proceeds, with preplanned modifications as necessary to suit the values obtained and conditions revealed.

To paraphrase, one hopes for the best but plans for the worst, using measurements and observations to distinguish the two as events unfold. Here, it is not just one hypothesis to be formulated but many—the most probable, the most unfavorable, and all the others that together encompass every foreseeable deviation in between. This truly places a heavy burden on diagnostic judgment. Peck (1969b) suggested that these hypotheses would follow from a "detailed inventory" of all possible differences between reality and the design assumptions. But any such inventory would have to be detailed indeed to account for possible differences of infinite number and nature, much like the infinite number of scientific hypotheses that sparked Poincaré's inquiry. Surely there must be other guidance, and Peck (1962) turned by analogy to the iterative way in which geologic hypotheses are formulated. The first element in geology is to gather and organize evidence around a structured format—like a geologic map:

> In the field, the geologist's procedure consists first of all to collect facts and evidence. He makes observations.... He then organizes his information. Often the organization consists of constructing a geologic map on which he simply records all the known information. Only then is he ready to study the map

> and the other organized information in order to form hypotheses regarding the development of the region.

The initial hypotheses (noting here the plural) serve to guide subsequent collection of information that in turn suggests modifications. This process eventually arrives at conclusions about the preferred hypothesis by iterative hypothesis testing, but this always remain tentative:

> After he has formed hypotheses, he tests them against the evidence he has collected. He also sees other tests that should be applied and returns to the field to obtain further evidence which he again organizes and compares with his hypothesis. In this fashion he arrives at a conclusion, which he rarely regards as final because he knows that subsequent information will be obtained that will require alteration of his concept.

So hypothesis formation is an iterative process that uses observations and information but does not follow directly or uniquely from them. Rather, it depends on how information is organized and structured, which can come about in any number of ways. The hypothesis also remains fluid, directing what information should be obtained to further test it but at the same time remaining open to change. Yet this still begs the more fundamental question of just where hypotheses come from. For this we must look to another element of diagnostic judgment.

INTUITION. In its diagnostic application, judgment identifies the problems to be addressed and defines the predictions to be made. Judgment determines what to look for. In this respect, the diagnostic role of judgment parallels the intuitive formulation of scientific theories that takes place even before they can be confirmed. And it testifies to the importance of intuition that the term is found at all in the lexicon of hard-nosed engineering. Focht (1994) unabashedly cites it several times in relation to judgment, particularly in last stage of the predictive process, where he accords it an overarching role:

> If a theoretical prediction … differs from the engineer's judgment as applied to historical data, or even if intuition alone raises questions, he or she should reexamine the numbers.

Terzaghi too cited the intuitive nature of hypothesis formation and its relationship to judgment in a way that presaged Focht's process. In so doing, he again contrasted these elements to the theoretical work of mathematicians with the same charitable kindliness that had long been his custom. In a 1944 letter to a friend, Terzaghi allowed that:

> ...mathematicians should be kept in cages and fed through the bars.... They are fit neither for properly formulating the problems—which requires intuition—nor for adapting their findings to practical conditions—which requires judgment. (Goodman 1999)

Terzaghi obviously had never met Poincaré. Still, the two had much in common. For Terzaghi, intuition was the "engineering instinct" he said had predated the development of soil mechanics theory and had guided earthwork engineering long before his *Erdbaumechanik* ever appeared. Later, it was intuition that guided how problems solved using his theories were formulated. Clearly for Terzaghi, intuition and judgment went hand in hand. Intuition was what diagnosed the problem to be solved, and judgment was what governed how its solution was applied. Others have also used the term "instinct" to describe intuition, with much the same relation to judgment. Legget (1979) contended that safety in geotechnical engineering ultimately rests on judgment, and judgment in turn depends on a kind of instinctive understanding of geology:

> No computer is ever going to decide when a suitable foundation bed has been reached, or when tunnel supports are necessary. In the final analysis it is human judgment that makes possible the safe uses of the earth. And judgment is based on sound experience that, whether so recognized or not, includes an instinctive appreciation of all geological factors.

Here Legget hits upon an important aspect of intuition or instinct: It operates without conscious recognition. Rather, intuition is a precognitive awareness that comprehends without need to think something through step by step. This can operate in several ways. First, an intuitive conclusion is one arrived at through an informal and unstructured mode of reasoning without the use of analysis or deliberate calculation. So an engineer might determine intuitively, that is, without calculation, where to locate borings to be drilled at a site. Second, a rule or simple fact of nature is said to be intuitive if it conforms to our tacit understanding of how things work. It is intuitively obvious that rock is stronger than soil and that water flows downhill. No training or instruction is involved in these determinations. Third, a rule or procedure is considered intuitive when it is applied during the course of ordinary activities so routinely as to require no formal proof or justification—one intuitively measures length along a straight line.

In each of these respects, intuition derives from expectations about how things should work. These expectations are embedded in experience, but they do not call upon it directly by way of specific instances. Intuition broadly applies experience as the general synthesis of many cases without drawing on any in particular. Moreover, intuition has the character of both foresight and hindsight. In its premonitory or diagnostic sense, embedded expectations anticipate possible outcomes and forewarn of potential difficulties. In its retrospective sense, intuition is interpretive. It measures conclusions against an internal standard of reasonableness—the standard of previous experience.

The conformity of judgment to intuition is no less than the embodiment of common sense, which can (and as Focht argued, should) prevail even in the face of analytical findings to the contrary. A solution is said to be "counterintuitive" if analysis and intuition are at odds. A counterintuitive solution is always suspect until the reasons are determined, which is why Focht's predictive framework calls for thoughtful review. By contrast, an "intuitively satisfying" solution is the signature of success. It conforms to intuition by simultaneously fulfilling the objective demands of analysis and the subjective requirements of common sense without conflict. Intuition then is far from the metaphysical property it is sometimes taken to be. It is central to judgment and to the predictive process itself. In the end, intuition, not calculation, is the ultimate test of engineering solutions.

A FEEL FOR THE PROBLEM. So far we have seen that the diagnostic aspect of judgment involves the formulation of hypotheses and intuitive expectations, where these hypotheses are generated by structuring information and observations within some organized format. But there is yet another element of diagnostic judgment. The structure in which information and observations are placed does more than classify them, it identifies their relationships and establishes connections. These relationships among otherwise disparate pieces of data and information are also an essential part of diagnosis. What they produce has sometimes been called a "feel" for the problem.

Peck (1962) allowed that the process of organizing information and developing its relationships is not something that comes naturally to civil engineers, perhaps speaking from his own training in structural engineering. Again, because it is set apart from formal instruction in the paradigm of theory, the diagnostic process is not well understood:

> The civil engineer, and particularly the structural engineer, is not trained in this fashion. He is not accustomed to collecting or being presented with a mass of detail which must be organized before conclusions can be drawn or before decisions regarding design and construction can be reached.

Using the example of a project in Cleveland adopting the observational approach for an ore storage yard, Peck went on to relate how masses of survey and movement data had to be sifted through, collated, and synthesized before any relationships could be established, or even before it was apparent which parameters to correlate:

> The work involved in this condensation could hardly be regarded as anything more than routine. It involved no great pleasure except the knowledge that it had to be accomplished before significant relationships would appear, and the knowledge that unless one did such work himself he would not be likely to get the real feel of the job.

Here, the "feel" of the job is arrived at by hands-on manipulation of raw data in its basic form, by direct inspection and not through review of any summary compilation. This is something too important to be left to subordinates, and the relationships "appear" only if one carries out such apparently menial tasks on one's own. It is interesting to observe that this process is directly opposite to what spreadsheets require. The spreadsheet must be formatted and the relationships established before any data are entered. This predetermines how the data are structured, rather than the data being used to discover the relationships, leaving one to wonder how much of the "feel" the data might otherwise provide may be routinely sacrificed in this way. Elsewhere, Peck (1969b) explains how this derives not only from data but also understanding of the underlying phenomena and mechanisms:

> The selection of proper quantities to observe and measure requires a feel for the significant physical phenomena governing the behaviour of the project during construction and after completion.

Having described how this "feel" comes about, Peck (1962) returned to its geologic origins and why it all seemed so foreign to engineers. They simply did not appreciate its importance:

> The routine work associated with the ore yard observations are not dissimilar to that carried out by the geologist in collecting his information concerning dips and strikes, entering it onto maps, and studying possible correlations.... Civil engineers do not like to do routine work of this sort, nor are they likely to see its significance. They are likely to feel that it belongs to the range of the subprofessional or technician. No geologist would feel this way.

So developing relationships among observations and other pieces of information is central to diagnostic judgment. But its culmination is to posit how these relationships might bring about some effect. Here we come to the final element of diagnostic judgment.

VISUALIZATION. We have seen how the observational approach is constructed around hypotheses. It requires formulating a suite of hypotheses that together describe every potential deviation from expected conditions, anticipating their possible effects, and developing courses of action to prevent those effects from occurring. Peck (1969b) couched these "hypothetical problems" in terms of modes of behavior, or failure modes, of the structure. The geotechnical engineer must, he said:

> ...take into account all possible modes of behaviour or failure, otherwise one of the underestimated, discounted, or unknown modes may develop without being noticed because no suitable observations have been devised for detecting it.

How one accounts for these failure modes is through the process of visualization.

Identifying potential failure modes comes about through a mental image of conditions and mental simulation of various failure processes. The judgmental exercise of visualization is what helps geologists sort through and organize information in forming hypotheses about how geologic deposits were formed in time and space and engineers about how adverse behavior might occur. In the earlier illustration of a structure on stiff clay, hypotheses related to compressibility, fissuring, and preexisting shear planes were posited as possible problems to be predictively addressed. Implicit in all three is the visualization of the various modes of failure they might bring about: excessive settlement in the first case, excavation instability in the second, and slope failure in the third. In the observational approach, Peck (1969b) went on to say:

> The essential ingredient, without which all the others may lead to nothing, is the visualization of all possible eventualities.

Neither is this visualization of failure processes limited to the observational approach. It underlies geotechnical predictions of all kinds:

> In most problems the most important step in design is the visualization of possible or probable modes of failure or deformation. Simple calculations based on sound conception of these modes are far more meaningful than elaborate calculations which ignore or which too readily overlook controlling factors. (Peck, 1973b)

Visualization, then, goes to the mental conception of a process that calculations by themselves do not provide. So again it is this element of diagnostic judgment that comes before analysis. Ferguson (1992) elaborates on visualization in design, contending that "visual thinking" is an intrinsic and inseparable part of engineering by which ideas are formed in mental images, not words or equations. Design is something that occurs in the engineer's head, in the "mind's eye." These mental images are communicated to others using drawings. And because these drawings are "neatly made and produced on large sheets of paper, they exude an air of great authority and definitive completeness." But the drawing does nothing more than transfer the mental image of the designer to a physical image, whereupon it is reconstituted back in the mind's eye of the person who reads it. The importance of visualization over any other form of representation is illustrated by an observation made of Albert Einstein by fellow physicist Richard Feynman: that Einstein was never able to consummate his unified field theory because he "stopped thinking in concrete physical images and became a manipulator of equations."[6]

So in its diagnostic respect, judgment has several key elements. It is intuitive, as opposed to derived from any rational inferential process, deductive or otherwise. This intuitive component does not stem from first principles, but precedes

---

[6]Agricola (1556) recognized what visualization and imagery would ultimately mean for the usefulness of his work, having included more than 250 illustrations in *De Re Metallica* like that in Figure 2-1 and saying in the Preface: "...with regard to veins, tools, vessels, sluices, machines, and furnaces, I have not only described them, but have also hired illustrators to delineate their forms, lest descriptions which are conveyed by words should either not be understood by men of our own times, or should cause difficulty to posterity." History has borne him out.

them. This is not to say that intuition should be considered irrational, rather that intuition lies beyond ordinary processes of reasoning as a nonrational (or better yet, "extra-rational") source of engineering insight into hypotheses concerning possible problems. Yet more than simply identifying the problem, judgment involves developing a "feel" for its significant elements and their interrelationships derived from direct manipulation of data and understanding of phenomena. And for diagnosis to be complete, these hypotheses must be visualized in mental simulation of their operation and effects. All of these diagnostic elements of judgment are, of course, highly personal and subjective, specific to the individual and varying in each. But together they provide a judgmental roadmap for problem solving that marks the diagnostic starting point and then directs where to look and what to look for in ultimately bringing the solution to its predictive destination. In its diagnostic application to problem solutions, as in the application of scientific method, judgment is that which comes before.

Induction

If this is what judgment does, at least diagnostically, then what are the underpinnings that make it work? As both a term and a concept, judgment is used as a universal shorthand for inductive methods of reasoning. In geotechnical practice, judgment is said to pertain almost by default to those things that cannot be deductively established from first principles. This is why inductive inferences, as judgment, are so easily taken as a defective form of deduction—a kind of deductive approximation that doesn't quite make the grade. But while the purpose of deduction and induction alike is to arrive at predictive conclusions, they have different functions and operate in different ways.

OBSERVATION AND EVIDENCE. Recall some of the characters we met in Chapter 2, starting with the 13th-century geotechnical engineer in the time of Thomas Aquinas and the slope he evaluated. Deriving inductive inferences about its stability was not possible for him. Induction first requires evidence in some form, but the concept of evidence was unknown. Predictive conclusions about stability could be drawn by deduction from first principles of mechanics or by *opinio* from approved doctrine of authorities, but not by observing the characteristics of the slope and relating them to other cases.

It took Agricola's prospecting for ore veins and his 16th-century colleagues to introduce the concept of evidence as "signs"—the signs of Nature—to be derived from observing nature itself. But signs could have different degrees of reliability as Fracastoro said, some to be trusted almost always and others merely often. A century later, Leibniz considered evidence in the form of judicial testimony, establishing that different kinds could carry different weight. So observational evidence could now become a basis for making conclusions, even if not deductive ones, by combining it in a way that accounted for its varying nature and predictive validity—in short, by inductive inference.

The basis for induction then rests on observational evidence, and Goodman (1999) shows that in both his personal and professional life, to the extent that the two could be separated, Terzaghi was consumed with observing nature, finding it to be a source of great personal pleasure and not incidentally the foundation of professional judgment. Of his work on a power project in the Croatian karst, Terzaghi wrote in 1909:

> All the beautiful and terrible things that nature could offer were assembled before me.... It is difficult to imagine times more rich and enjoyable.

Like Agricola before him, Terzaghi's signs of nature were the signs of geology, to be carefully observed and garnered. One's powers of observation, and the inclination to use them, were everything. What the engineering profession had lost when it turned so exclusively to theory in the 1880s was, he said:

> ...interest in observing and describing the whimsical manifestations of the forces of nature.

For as he later wrote in 1928:

> Keen observation is at least as necessary as penetrating analysis.

He added that in their powers of observation, most engineers he knew were "incurably blind." But the kinds of geologic observations and evidence so important to Terzaghi would be of little value if they were not synthesized in some way to constitute knowledge, and this is the role of induction.

Deduction starts with premises and works to conclusions from universal truths and principles. Geology, however, provides few universal truths, and its principles frequently change. Geologic observations are facts which are true by direct demonstration; if one doubts them, one can look for oneself. But a geologic conclusion is a matter of how these facts are interpreted, and the test of deduction—that it could not be otherwise—is one that is seldom met where alternative and ambiguous interpretations so often abound. So geologic inference must work by induction.

Terzaghi recognized that geology provided the vehicle for inductive inference of soil properties but never used the word as such. Instead, he called induction the "art" of soil mechanics, which he distinguished from deductive logic. Saying that soil mechanics by 1957 had "arrived at the borderline between science and art," he explained:

> I use the term 'art' to indicate mental processes leading to satisfactory results without the assistance of step-for-step logical reasoning. (Terzaghi, 1957)

Though the art of induction is indeed different from step-by-step reasoning, there are still some general approaches it characteristically adopts.

FORMS OF INDUCTIVE INFERENCE. An inductive argument draws conclusions from observations and other forms of evidence. There are no formalized rules of logic for selecting the evidence to be used (this, of course, is the diagnostic aspect of judgment) or for integrating it inductively, this being a matter of personal judgment as well. While deduction reasons from the general to the specific, from first principles to conclusions, induction works the other way around. Specific instances and observations are synthesized to a general rule, which may then be applied to some particular case. Induction can take several forms, and it also works by analogy. A large part of geotechnical reasoning is conducted apart from first principles, at least insofar as we understand how they actually operate in the field. Here we may look to similar cases and circumstances, reasoning by analogy from their similarities and differences to the conditions at hand. This involves things we've seen before that guide how we interpret those present. Our prediction is made by *inductive analogy*.

A related form of inductive reasoning is *induction by enumeration*, where we incorporate not only what we have seen but how often we've seen it, deriving generalizations from the number of instances observed. One might have encountered, say, a half-dozen highly plastic clays with low strength, leading to the inductive conclusion that all high-plasticity clays are weak. Induction in this sense is empirical, requiring no necessary substantiation in theory or other underlying knowledge. An inductive argument is never final and is always subject to revision. For all we know, the next high-plasticity clay we come across could just as easily be strong. But inductive inference makes no claim of absolute truth, as those who adopt it will readily concede. So it follows that neither can induction produce consistent answers. If people do not think alike, then if you ask different people you get different answers—an elementary fact of life established deductively since the conclusion, given the premise, could not be otherwise. Nevertheless, such a fundamental truth is easily overshadowed by the convenience of consistency. If induction does not yield consistent judgments or subjective probabilities, then there must be something wrong with it—and with them. But even if inductive conclusions can be inconvenient, that does not make them wrong.

THE PARADOX OF HUME. In his *Treatise on Human Nature*, Scotsman and 18th-century scientist and philosopher David Hume inquired into the nature of knowledge and with it induction. What he came up with is called Hume's "Problem of Induction" or "Hume's Paradox," cited by philosophers with such regularity as to give induction a bad name.[7] In it, he posed an innocent question: How do we acquire knowledge of things unobserved?

---

[7]Franklin (2001) claims that philosophy going back to the time of Plato has always been an "old enemy" of probability with its inductive roots. To distance themselves from rival Sophist proponents of rhetoric, most Greek philosophers aimed to establish knowledge beyond any doubt with proof, not plausibility.

He started out with a fairly restrictive definition of knowledge. To Hume, knowledge is what we know for sure, what we can objectively prove to be true by logical rules. Most geologists would say they know that sedimentary rocks were deposited in primordial seas, and most people would assert that they know the sun will come up tomorrow. But neither can be proven by any law of logic. No one alive today has ever seen these seas, and it cannot be proven that the features attributed to sedimentary deposition were not produced by something else. Neither can anyone prove by all past observations of sunrise that tonight the sun will not explode into a supernova by the time the alarm clock goes off. All we can do is argue inductively by past observations that we believe these things to be true with some fairly high degree of likelihood. But to Hume, knowledge did not encompass belief or likelihood. Knowledge dealt only with certitude, and certitude does not provide for degree.

Hume concluded that knowledge could not be gained by induction, but he committed some logical sins of his own in the process. In presuming that knowledge was only that which could be objectively proven, his conclusion had simply restated the premise—it was nonampliative. What Hume did succeed in demonstrating was that induction is not the same as deduction, with the former generating degrees of belief and the latter proof. Still, it follows from no rule of logic that if we can't know something for sure, then we know nothing at all. For most functional aspects of human existence, some sufficiently strong belief is enough to constitute knowledge whether we can prove it or not. Nor was Hume himself any stranger to belief. Recall from Chapter 2 that he was among those who sought to unify the frequency and belief concepts of probability inside the brain with neurosurgical precision. Yet by excluding belief, Hume's more restrictive notion of knowledge would relegate the entire field of geology, for one, to the realm of cartoon fantasy—to say nothing of intellectually incapacitating most of the human race. Indeed, it was from the pressing need to solve everyday problems that medieval physicians and miners first turned to observing the signs of nature, laying the foundations for inductive reasoning way back in the 16th century. So it might just as easily be concluded that Hume's Problem of Induction was less induction's problem than Hume's.

THE USES OF INDUCTION. In deductive logic, correctness is an all-or-nothing proposition; deductive inferences are either valid or invalid with nothing in between. But induction provides for degrees of correctness, and one inductive conclusion may be more strongly confirmed or supported than another. In admitting to degrees of confirmation, it becomes apparent how the inductive processes of judgment are connected so closely to degree-of-belief probability, which goes on to quantify this degree of confirmation. A predictive conclusion arrived at inductively may never contain absolute truth but may be probable even so, and probable truth manages to serve most human activities well enough. It further becomes apparent that those who might sniff at the empirical element of induction may be confusing apples with oranges. Deductive inferences with true premises must have true conclusions. But induction has a different function: to establish that conclusions with true premises are probable. Induction cannot be made into deduction, nor should we want to. An inductive conclusion cannot be shown to be true, but that it works. And that in the end to engineers is what matters most.

The staunch proponent of deduction would take issue with the well-known canard: "If it looks like a duck, and walks like a duck, and quacks like a duck, it's probably a duck." But for most of us, induction is why we think so. Provided we have seen enough ducks to recognize the characteristics of duckness, we reason from observations of the specific animal before us to the general category of ducks. Induction cannot guarantee it's a duck—which is why it only probably is—though a tissue sample could deductively prove the matter beyond any doubt from first principles of genetics. But this would be awkward, not to mention inconvenient to the duck. In everyday reasoning, we need something less cumbersome for establishing conclusions with a degree of certainty corresponding to the circumstances. So we call the evidence "circumstantial"—it does not offer proof. Nevertheless, induction from circumstantial evidence is entirely adequate for identifying the great majority of ducks, and we don't really require deductive proof from first principles to be sure. This is why induction is so closely related to what we call common sense and also why folklore (engineering and otherwise) is so full of anecdotes illustrating the absence of either one. Nor is this anything new. Aristotle recognized that the certainty of absolute proof is not always necessary, saying in the *Nichomachean Ethics* that "the same degree of precision is not to be sought for in all subjects … . It is the mark of an educated man to look for precision in each class of things just so far as the subject admits." And one cannot help but recall Savage's (1954) remarks in a similar vein, when he said that " … catastrophe is avoided primarily because in practical situations common sense saves all but the most pedantic of us from flagrant error."

This all has important practical implications. As shown by Fanelli's (1997) insightful treatise, safety is not an objective attribute that can be proven by deductive means. The safety of a structure is typically evaluated by determining if some calculated parameter from a geomechanical model, such as factor of safety, meets some established numerical criterion, such as 1.5. So when the factor of safety exceeds 1.5, we say that the structure is safe—that it will not fall down. But there is nothing to prove that it won't. The model is just that: some representation of reality such that if the structure conformed to some assumed law of behavior then it would react in a certain way. Here first principles are premises, and they cannot be proven to conform to the reality that exists in the field. Nor can it be deductively demonstrated that a factor of safety of 1.5 or any other such value requires a structure to stand of necessity. Instead, we observe over some sufficiently large number of cases that structures analyzed in certain ways and that meet certain criteria have a greater propensity to stand than those that do not. We reason from specific instances to the general case. This is simply induction by enumeration, and it is nothing less than the fundamental basis for engineering assessment of safety. We can never prove that a structure is safe; we can only believe that it is to some degree of likelihood by means of inductive inference.

Which brings us back to Aquinas' geotechnical engineer and his slope stability problem. The engineer observes that the slope contains highly plastic clay that could be weak according to induction by enumeration. But he observes no seepage, which theory tells could be a sign of detrimental pore pressures. His experience shows that other slopes of similar steepness, height, and in similar soils have remained stable, invoking induction by analogy. Integrating these pieces of evidence, assigning strength and weight to each, he predicts that the slope is likely to remain standing. The conclusion is inductive. It is a matter of judgment.

Interpretation

Any inquiry into judgment would be incomplete without considering it in Focht's third and final respect as a mechanism of critical review and assessment. Judgment determines whether things "make sense," and this of necessity means interpretation. For things to make sense or not is to determine whether and how they fit within some interpretive framework that does more than derive predictive conclusions from the evidence. By providing interpretation of the evidence and the conclusion, judgment gives order and meaning to both, a context in which they can be understood.

INTERPRETIVE CONTEXT. Any geotechnical practitioner knows that measured soil property data can seldom be taken at face value. Data always require some degree of interpretation before they can be used sensibly. Collections of laboratory data invariably contain outliers that are identified by inspection and sometimes discarded. They do not "make sense." But what does this actually mean?

First of all, induction by enumeration may be at work. In some number of previous cases the data might have fallen within some range and would thus be expected to do so in this case. Although induction by enumeration in itself is entirely empirical, the interpretive use of induction goes beyond the empirical by invoking other sources of underlying knowledge as well. One of these, of course, is theory, and its heuristic value as an aid to understanding comes from how it helps establish context. The theories of soil behavior might point to inconsistencies between the outlier data and other related soil properties. Or understanding of the procedures and processes of laboratory testing might identify some effect characteristic of test errors. The data are said to be "interpreted" according to factors like these. How? Using judgment to combine all of the sources of evidence in a way that gives meaning and context. Would one engineer interpret the data in exactly the same way as another? Probably not, and here again the personalistic and subjective character of judgment is evident. Many of these individual differences arise foremost from how patterns are (or are not) recognized and used.

PATTERN TEMPLATES. Establishing some framework for interpretation is done in many ways. However, pattern recognition is the element common to all of them. Pattern recognition is perhaps the single most important element of inductive reasoning, which could not operate without it. Induction by enumeration requires discerning patterns of sameness. Induction by analogy requires establishing patterns of similarity. But patterns of nonconformity can serve as well. Data outliers are not always discarded; indeed it may be foolish to do so. Taken together, outliers that appear in independently derived data sets can themselves provide a pattern—a pattern of anomalies—sometimes having greater evidential content than the mass of the data itself. In these ways, perceiving the patterns in previous experience is required for basing judgment on it. Without recognizing the patterns it contains, experience would be nothing more than an accumulation of random impressions. This what distinguishes judgment from experience alone.

Geotechnical practice uses various devices for establishing and recognizing the patterns of relationships in data and behavior. Recall Peck's analogy to how geologists assemble data in formulating geologic hypotheses. He called this a process of "organizing" such things as measured strikes and dips, and it requires some structure, here a spatial pattern, by which to organize. In this case the device for pattern recognition is as simple as a geologic map, where the act of constructing it lends order and meaning to the data. Indeed patterns are exactly what Poincaré discovered in the harmonious ordering of facts. And for Peck's Cleveland ore yard, the significant relationships in the data "appeared," as he said, but not by some act of magic. The hands-on manipulation of the data so important to deriving a judgmental "feel" was the vehicle for discerning their patterns of correlation.

Theory is yet another template for recognizing patterns, in this case the associative patterns of cause and effect. This provides interpretive context by organizing and structuring observational experience around the explanations that mechanics provides. Terzaghi recognized this clearly, as described in a 1919 diary entry (Goodman, 1999):

> Theory is the language by means of which lessons of experience can be clearly expressed. When there is no theory ... there is no collected wisdom, merely incomprehensible fragments.

The lessons of experience are contained in its patterns that theory helps reveal. Seemingly unrelated fragments of experience become comprehensible when hung on the branches of the theoretical tree of causality that relates them to each other. Reduced to this most basic element, pattern recognition is how theory works in assisting interpretive judgment and aiding understanding. Theory does not directly produce the prediction but instead provides an explanatory pattern of cause and effect that links its inductive elements. And here it is no small thing that Terzaghi used the word "wisdom" to express what such a synthesis of theory and experience was capable of producing.

Still, it was geology that Terzaghi took most as the framework for establishing spatial, causal, and correlative patterns in both the field observations and the theoretical calculations he made. According to Herbert Einstein (1991), geology never ceased to be "the context in which Terzaghi placed his work." Geology was so vitally important because the patterns it revealed gave interpretive order and meaning to everything else:

> The geological origin of a deposit determines both its pattern of stratification and the physical properties of its constituents...." (Terzaghi, 1955)

Einstein shows that for Terzaghi, this interpretive context of geology was the very essence of judgment. Geology and judgment were inseparable. Looking back on his career from the vantage point of 1957, Terzaghi reflected on:

> ...the array of useful knowledge that has filtered into my own system and crystallized into sound judgment. I find that it contains one ounce of geology for every pound of theory of structures and soil mechanics. (Terzaghi, 1957)

PATTERNS AND PERSONAL EXPERIENCE. A final anecdote concerns how two masters of engineering judgment, both recognizing the same patterns in the same data, interpreted them differently. Peck (1969b) relates the story of his work with Terzaghi around 1950 on a problem involving broad, areal subsidence beneath a chemical plant. It was underlain by a thick layer of normally-consolidated clay, beneath which was limestone bedrock, and at great depth a salt deposit previously mined by solution extraction. Peck concluded that settlement had been caused by consolidation of the clay layer, a hypothesis supported by calculations matching the measured values. On reporting these findings to Terzaghi, Peck recounts this dialectic exchange:

> "Now, about that report on subsidence..."
>
> "Yes?"
>
> "I think you have missed the boat..." He paused. "It is obvious that the settlement is in the bedrock."
>
> "But that is impossible. The bedrock is too thick," I protested.
>
> "How do you know it is impossible? You didn't establish any reference points at the surface of the bedrock, did you?"
>
> "No, but the general magnitude of the observed settlements agreed with the computed ones."
>
> "Didn't you notice that the real pattern of differential settlement is much more abrupt and erratic than the computed one?"

"Yes, but I think this difference is caused by the presence of erratic, compressible organic deposits near the ground surface."

"What is the evidence? You have forced the evidence to fit your preconceived notions."

Subsequent studies confirmed Terzaghi's suspicions that collapse of solution cavities, not clay compression, had caused the subsidence. A chastened Peck attributed his misdiagnosis to a "preoccupation with the wrong phenomenon," to his own "blind spot" that kept him from examining all the available evidence with an "open mind." But he also credited Terzaghi's experience. Terzaghi had, in fact, encountered similar collapse-induced subsidence situations before. Both recognized the pattern displayed by the settlement data, in this case its abrupt and erratic nature, but their interpretive judgment of its cause is what differed. Terzaghi was able to frame the external pattern of the data within the internal pattern of his personal experience elsewhere. Peck, lacking the same experience, was not. In this respect, it is not altogether clear that "preconceived notions" were at fault—Terzaghi's prior experience, after all, had produced his own—and readers can reach their own verdict on Terzaghi's indictment.

Call them what you will, differences in personal experience often lead to differences in the judgmental context of interpretation. Yet among the foremost achievements of the paradigm of geotechnical practice is to have documented and preserved its collective experience in the form of published field case histories. Access to experience today is not limited to the storehouse of individual knowledge, and with the published body of case histories continuing to grow year by year experience is no longer as personal as it once was. So too then should personal judgment become better correlated among those who take time to seek out these case histories and study them. Geology and the vagaries of soil behavior will always produce new and different surprises, but one is left to speculate whether Peck's and Terzaghi's interpretations, after a half-century's accumulation of this collective experience, would diverge so today.

The Elements of Judgment

Returning to Focht, we can now see why judgment was so essential to the entire "mental approach" that his predictive strategy embodied. Table 3-1 summarizes the key attributes of judgment organized around its diagnostic, inductive, and interpretive aspects, how they are used, and what they require. Of course, no such classification can be rigid or strictly categorized, and all of these elements of judgment interact with each other every step of the way. It would put too fine a point on the matter to insist that the intuitive expectations of experience should operate only in diagnosis and be restricted from interpretation, or that interpretive pattern recognition play no role in formulating diagnostic hypotheses. Neither can any such catalog be fully complete, and other important

TABLE 3-1. *Attributes of Judgment*

Aspect	*Use*	*Requirements*
Diagnostic	Hypothesis formation Problem definition Guiding what to look for Identifying predictions to be made	Intuition (precognitive awareness from expectations of experience) "Feel" for problem (organizing and structuring evidence, understanding of processes and mechanisms) Visualization, mental simulation
Inductive	Synthesis of evidence, information, and underlying knowledge from different sources Assessing probable truth of hypotheses Generalizing from specific cases	Recognition of evidence Awareness of "signs" Observational abilities Experience
Interpretive	Critical review Evaluation ("make sense") Establishing meaning and context	Pattern recognition: –Patterns of consistency and anomaly –Spatial relationships (conditions) –Correlative relationships (data, experience) –Causal relationships (theory, geology, mechanisms and processes)

attributes are bound to arise. Even so, the elements of Table 3-1 add substance to the meaning of judgment and go far toward explaining its central importance to the entire predictive approach in the paradigm of geotechnical practice.

A Sense of What Is Important

It is no oversight that defining judgment has been left to follow its description. Few have tried to do so, and even fewer with much success. Judgment, it seems, shares a certain indeterminate quality with obscenity: No one can define it, but everyone knows it when they see it.

Herbert Einstein (1991) related judgment to experience, calling it "the intelligent use of experience." This has a great deal to do with the kind of humility that geology engenders, because judgment is also:

> ...the recognition of one's limitations, of the limitations of the methods one uses, and of the limitations and uncertainties of the materials one works with; and this brings us back to geology.

This incorporates the inductive character of judgment, inasmuch as experience cannot be put to use without it. It also points to the interpretive aspect of judgment by framing experience in the context of limitations imposed by a variety of factors. Nor should it go unnoticed that for Einstein, judgment is linked to one's recognition of uncertainty.

Focht (1994) appeared to be stymied in his attempts at definition, although it was clear to him too that experience and judgment are closely linked. Asking rhetorically where good judgment comes from, he was left to cite this elliptical rejoinder he attributed to Mark Twain:

> Good judgment comes from experience. And where does experience come from? Experience comes from bad judgment.

Though one can certainly hope that bad judgment is not the only source of experience, this illustrates well how the words "experience" and "judgment" are used together so casually as to be almost synonymous. Yet experience and judgment are not one and the same. It is entirely possible, as sometimes demonstrated in certain individuals, to have accumulated a great deal of experience without deriving the least judgment from it.[8] Mechanical repetition of actions or observations does not alone give rise to the interpretive context that judgment provides. Experience does not become judgment until some useful form of generalization has been inductively inferred.

Then can judgment be had without experience? And if not, how much experience does judgment require? Terzaghi would surely have argued that there is no substitute for experience derived from one's personal skills of observation. This notwithstanding, the contemporary world of engineering is such that few today will ever have the wealth of opportunity for personal observation that Terzaghi did in his time, so they can afford even less to wait for these opportunities to "crystallize" into sound judgment as they did for him. Again, it should not be unreasonable that diligent study of published case histories can accelerate the acquisition of judgment, even if not replacing the need for experience personally obtained.[9]

Judgment and how it works have been taken up by entire fields of study in the behavioral sciences, and much more will be said on this topic. But even there the definition of judgment is taken to be so self-evident as to be left largely unstated. Perhaps the closest thing is the description of judgment as a scaling activity, one

---

[8]Almost everyone can cite a personal favorite.

[9]It appears that much of Terzaghi's experience was externally derived. The mimeographed notes from his early-1950s Harvard course on Engineering Geology compiled by student Walter Ferris contain some 400 case-history references (not including those he published) of every imaginable variety, many annotated and all of which he evidently studied himself. Interestingly, fully two dozen of these describe cases of subsidence from various mechanisms, including mined salt cavity collapses dating from 1904. Terzaghi's diagnosis of this mechanism for the chemical plant subsidence problem may not have been as clairvoyant as it might have seemed to Peck at the time.

that involves weighing and matching of things like data, hypotheses, arguments, and evidence (Smith, et al., 1991; Curley and Benson, 1994; Benson et al., 1995). Here we find allusions to the matching of patterns and the weighing by induction we have already seen demonstrated in geotechnical activities. But an important clue to defining judgment is in the concept of scaling, which means comparing in some way the relative importance of things with respect to each other.

For his part, Peck (1980) admitted that judgment is a vague attribute. He was not sure he could define it, and thought few would agree if he did, but went on anyway to define judgment as "a sense of proportion" (Peck, 1969a). By this he meant the proportion of dimension, of size and scale, of the relative magnitude of physically measurable quantities. One could not know if a quantity calculated or feature designed made sense unless its physical proportion had some meaning in relation to other things. By this definition one could achieve the interpretive context of judgment from the comparative scale of physical things and measurable quantities. Still, we have seen that judgment goes beyond physical objects to internal constructs. A more expansive definition, while retaining the concepts of sense and of proportion, might encompass such aspects.

Defining judgment as *a sense of what is important* seems to capture many of these ideas. In the same way that Peck used it, a "sense" is not something to be arrived at using any method of logic. It admits to intuition. To hold a sense is to perceive and be aware. This embodies the diagnostic aspect of judgment: the perception of what the problem is and the awareness of its significant elements, their relationships, their "feel," and how they guide its solution. At the same time, to have a sense is to observe and to comprehend, and this is interpretive: using the observations of field conditions and experience and comprehending their patterns within a context that gives them structure, order, and meaning. Judgment is surely all of these.

And what is its result? To establish the importance of things, not only physical things and quantities but also things conceptual and evidential. Again, it is diagnostic judgment that establishes which potential problems are important and how important they might be. It is interpretive judgment that puts problem solutions in context, evaluating their important aspects against an internal standard of reasonableness. Relative importance is what the patterns of evidence—its spatial, causal, and correlative relationships—produce, and this is where meaning comes from. A sense of what is important then seems to describe fairly well what these aspects of judgment, taken together, are truly all about.

This sense of importance, it turns out, operates at some fairly lofty levels and is by no means restricted to the geotechnical field. When Eric Cornell was asked to explain the success of his collaboration with Carl Wieman that culminated in the 2001 Nobel Prize for physics, he responded: "We agreed on what was important." Arlin (1990), citing other Nobel laureates, notes that they were uniformly convinced that what mattered most in their work was, in their words, "a sense of taste, of judgment, in seizing upon problems that are of fundamental importance." One said of his mentor, "He led me to look for important things, whenever possible, rather than to work on endless detail or to work just to improve accuracy rather than making a basic new contribution." Here judgment is more than mechanical, with its "sense of taste" harking back to Poincaré's esthetic feel. It evokes Kuhn as well, for these are the people whose sense of the important transcends the "mopping up" of ordinary science. And in relating judgment to what is important, Arlin (1990) takes it one step further. This, she claims, is a fundamental aspect of what we call wisdom.

It may be trite, but true nonetheless, to say that a working definition is one that is found to work. If so, the test is to replace the word "judgment" where one reads or uses it with "a sense of what is important," to see if added richness and content result. Readers are encouraged to try this and are free to judge for themselves. But whether by direct substitution or otherwise, viewing judgment in this way will almost always embody more than the vague attribute it has so often been taken to be.

Summary

Kuhn's concept of how paradigms guide a profession's activities is a most useful one for framing many matters that affect geotechnical engineering and its approaches to predictive conclusions. It has been shown that two operative paradigms exist within the geotechnical profession—the paradigms of theory and practice—that shape one's view of the activities in its domain. Each has its adherents, and probably most members of the profession are more comfortable with one or the other, though all are certainly acquainted with both. Fewer may be fully aware of just how and why they diverge, but the various accounts of Focht, Peck, and Terzaghi make it clear that their dichotomy has been present from the start. Within it is the explanation for a number of underlying things.

One is the duality between theory and judgment. The paradigm of theory holds to the deductive precept of reasoning from first principles. If the soil-property premise is true, then the predictive conclusion must be too, and theory and analysis are what make it so. Notwithstanding several troubling aspects of theoretical confirmation and application, the predictive conclusion is viewed as containing and expressing unqualified objective truth. This is the geotechnical embodiment of the scientific method as engineering science, where judgment, if invoked at all, is more a subjective contaminant to be eventually scrubbed away by the cleansing of scientific advancement.

While the paradigm of theory largely suppresses the theory/judgment duality in this way, the paradigm of practice actively embraces it. The predictive problem is seen in a larger context that begins with its diagnosis and ends with its interpretive assessment, where judgment defines both of these elements and guides everything in between. Here, theory becomes the companion of judgment. It provides but one of several ways for establishing a contextual framework for evidence, while direct deduction from theory falls short. The entire reasoning process of practice is different. It is inductive, and it has a different function. It does not seek to establish absolute truth of the predictive conclusion, only its probable truth. In this, the paradigm of practice seeks simply to do what works. The proof, as they say, is left to the reader.

The duality of theory and practice has everything to do with why the profession adopts probability in its own dual senses. We have seen how the Classical probabilists' original duality of frequency and belief was conquered by the dominance of statistical frequency that sought to eliminate all opinion. In the paradigm of geotechnical theory, this becomes the natural probabilistic outgrowth of the paradigm itself, which works toward similar ends. Not so for the paradigm of practice, where the equally logical extension of the inductive method is to quantify how probable the inductive conclusions might be. This could be through frequency-based enumeration, judgmentally derived belief, or some amalgam of both—the end result is much the same.

Through the discussions thus far, we have come to understand the dual nature of probability, its historic origins, and its relationship to the two views of itself that geotechnical engineering holds. If there is any lesson to be found in all this, be it for probability or for engineering, it is that the personal and subjective are not objectionable, they are indispensable. It remains to explore then how these concepts are manifested in the probabilistic methods the profession applies.

4

Reliability, Risk, and Probabilistic Methods

The uncertainties facing geotechnical engineering are legion, so much so that they are cited uniformly as the chief feature distinguishing it from its sister civil engineering specialties. They include the unknown presence of geologic defects, the uncertain values of soil properties, limited knowledge of mechanisms and processes, and human error in design and construction just to name a few. Numerical statements of probability are used to quantify at least some of these uncertainties according to a grab bag of procedures lumped under the heading of *probabilistic methods.* In various ways, they are intended to enhance the treatment of uncertainty by evaluating it systematically. What they mean is something else.

The use of probability in geotechnical applications has always been task specific, with different techniques and nuances for problems that outwardly might seem quite similar. A reliability analysis might calculate the probability of slope failure, and a risk analysis might do the same thing. Is one better than the other somehow, and if so which? Should a fault tree or an event tree be used? Or is Monte Carlo analysis the answer? Shouldn't hard data and analysis be preferred to something less rigorous? These are not simple questions where the meanings of probabilistic methods so often take a backseat to their calculated results. But they go to the heart of what these methods do, how they are interpreted, and how they are used together—all important matters when geotechnical uncertainties cover the ground they do.

The mechanics of probabilistic methods have come to be seen as an art form reserved for the few, and carrying them out a specialty within a specialty. But if probability is to have wider value, these methods must be understood by the many, in concept if not detail. Almost without exception, separate types of probabilistic analysis have grown up around different specialized areas of geotechnical problem application, from slope stability to offshore structures, from nuclear safety to dam safety, with techniques and protocols customary to each. This presents certain obstacles to those who would venture from one to another, and even more to the uninitiated in any of them. Navigating these probabilistic methods first requires mastering their different subtleties and practices, then mathematics arcane to the novice. But the final hurdle is perhaps the highest—to establish just what purposes these methods serve, what they mean, and what they don't. On this score the techniques themselves are silent, much like the probability calculus itself ever since Pascal and Fermat.

What has always been lacking is an overall framework, some broad, unifying treatment of how probabilistic methods fit together. Without this context, it is hard to select the right tools for the job, much less interpret their results. In short, it is hard to know how to use them sensibly.

We approach this task from two angles, both involving a broad distinction between reliability analysis and risk analysis as they are used in the profession. Although seldom clearly distinguished, with the terms often used interchangeably, they serve different functions and stem from some very different precepts. Reliability and risk approaches are the two umbrella concepts where probabilistic methods reside. If these are the tools for systematically addressing uncertainty, then they are the toolboxes. So where we start is to classify some of the more common probabilistic methods within the reliability/risk taxonomy to show how they relate to each other.

But even more important is to establish what the various probabilistic methods can realistically be expected to do, and what they can't. Here the reliability/risk distinction builds on what we have seen up to now. These two approaches are the procedural embodiment of the frequency/belief duality, and that of theory and judgment. We will see how reliability analysis is the natural outgrowth of the profession's paradigm of theory in preserving the concept of objective truth, while risk approaches reconcile this duality by acknowledging and accommodating both the objective and subjective, much like the paradigm of practice. And this is more than just armchair philosophizing. It is where these separate views of the profession's activities are manifested in how it goes about using probability, where those expecting one thing may get quite another and where bigger trouble still lies in store for those who may not know what to expect at all.

We go about all this in the spirit of overview. The purpose here is not to describe the mechanics of probabilistic methods since such tutorials can easily be found elsewhere, but rather to bring out their meaning and content. For even if engineers were never to perform a risk or reliability analysis themselves, this would not reduce the need to understand and interpret them when confronted with their results.

Reliability Analysis

Geotechnical reliability analysis is an offshoot of structural reliability and the two remain closely connected. It requires the adoption of some geomechanical model, typically the limiting-equilibrium variety, and starts with the precept that uncertainty in its results derives from uncertainty in its input parameters—mainly soil properties but sometimes loading as well. In this, geotechnical reliability analysis can be seen as simply a probabilistic adjunct to the drill–test–analyze strategy. What it addresses is uncertainty in soil properties, and what it produces is the probability that the factor of safety (FS) is less than unity, denoted here as p[FS < 1]. This is sometimes expressed as a *reliability index* by assuming some form of probability distribution on FS.

The reason, of course, that a factor of safety is applied at all in conventional deterministic analyses is to account for various uncertainties. But to the extent these uncertainties pertain to soil properties, then FS may not be a very good measure. Take, for instance, two sites that happen to have the same average value of undrained shear strength but different degrees of variability in the data sets, one tightly grouped and the other widely dispersed. The computed factors of safety based on mean strengths will be the same, but this does not speak to the greater degree of uncertainty for the more variable site. We might adopt a higher FS to compensate, but it would be difficult to know how much higher. Reliability analysis directly addresses this matter, and its fundamental justification is that it provides a more consistent measure of the effects of parameter uncertainty.

In concept, p[FS < 1] is found by performing the analysis for a variety of input values. The probability that FS falls below 1.0 is determined from the frequency with which the input values, from data statistics, would fall below that value which yields FS = 1.0 in the model. It is evident here that reliability approaches rely on the frequency interpretation of probability, but this is now taken one step further. Presuming that the geomechanical model is a unique and accurate descriptor of reality, then FS < 1 is taken to represent some failure condition, so that p[FS < 1] is said to be the probability of failure, or p_f. The reason this is not called a "probability of failure" analysis has much to do with the more reassuring tone of "reliability," which is related to failure probability by simply $(1 - p_f)$.

Lacasse and Nadim (1996) show how this works through the example in Figure 4-1, which provides two probability distributions of FS for pile capacity, one determined from the results of an initial field investigation and the other after additional sampling and testing. Here, FS is taken to conform to the familiar bell-shaped *Gaussian* or *normal* probability density function and hence is said to be *normally distributed.* In both cases, p[FS < 1] is found from the area beneath the respective distributions within the corresponding region.

Here the initial pile capacity analysis determined a comparatively high mean FS of 1.79, but with considerable variation in the data as indicated by the width

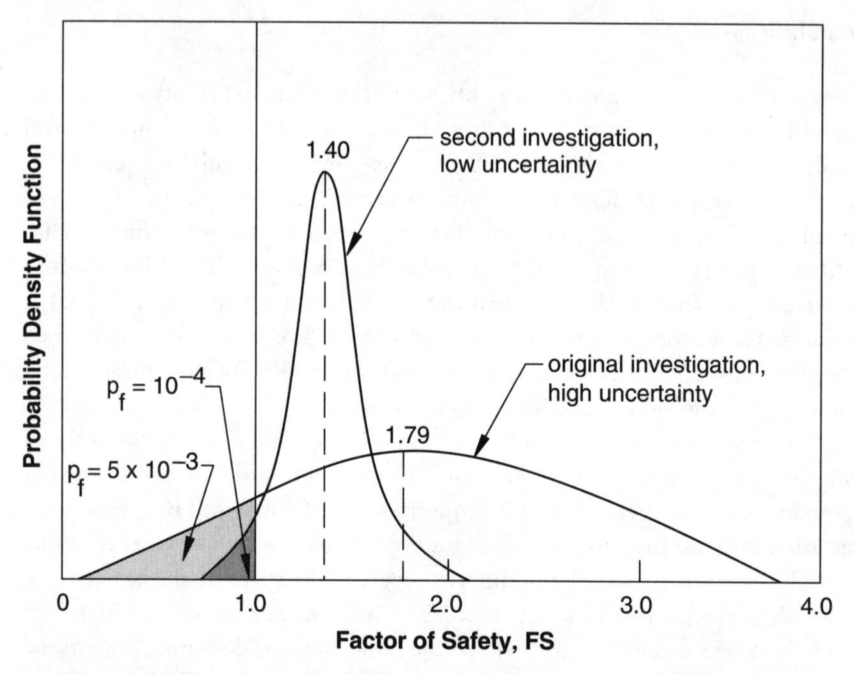

FIGURE 4-1. *Effect of soil property uncertainty in pile reliability analysis.*

of the distribution. A second analysis after the new soil investigation produced a lower mean FS of 1.40 because of average conditions less favorable than initially thought. At the same time, however, there was less variation in the input parameters from the additional information obtained, so the calculated failure probability was reduced. Thus, the reliability analysis of pile capacity showed in this case that a lower mean FS with reduced parameter uncertainty was less likely to fall below 1.0 than a higher FS with greater uncertainty. This also illustrates more generally that wider probability distributions, or those with greater *dispersion* about the mean, reflect greater uncertainty and vice versa.

Further to uncertainties in soil-property input parameters can be uncertainty in nongravity loadings imposed by waves, earthquakes, and the like. Geotechnical reliability analysis can be extended to these uncertainties by characterizing them probabilistically as well. Conceptually, this is shown in Figure 4-2 as the difference between the imposed load (or *demand*) and the structure's resistance to load (or *capacity*). In conventional deterministic analyses, FS is the ratio of resistance to load, and it would be equal for the two cases in Figure 4-2 where the respective mean values of load and resistance are the same. However, the degree of uncertainty—again, the dispersion about the mean—is different. Here $p[FS < 1]$ is schematically indicated by the size of the intersecting regions where load exceeds resistance, showing once more how the same FS can correspond to different

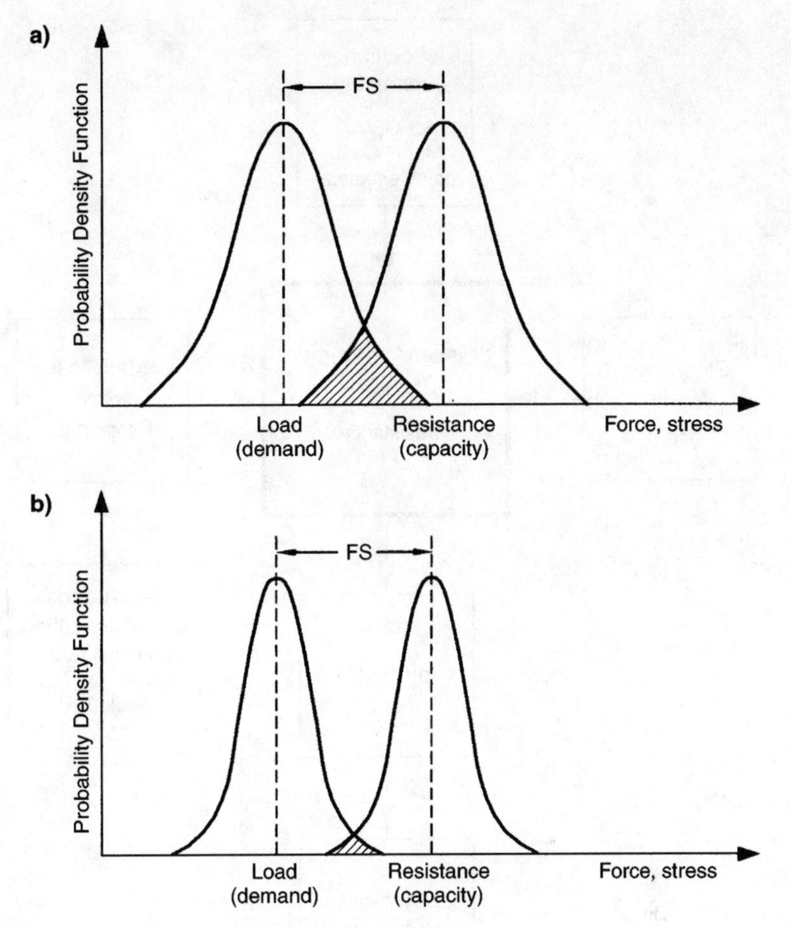

FIGURE 4-2. *Reliability analysis using load and resistance uncertainties: a, greater uncertainty, higher p_f; b, less uncertainty, lower p_f.*

imputed failure probabilities where both capacity and demand estimates contain uncertainty.

So far, these reliability principles are conceptually straightforward, but this can be less so for the array of related probabilistic methods that surround them. Closer examination of the various probabilistic components typically used in reliability analysis can help sort things out.

Probabilistic Methods in Reliability Analysis

Reliability analysis can incorporate a variety of probabilistic methods. Though these methods are usually described and developed as separate entities, few can actually be put to use without applying them within its framework. Figure 4-3

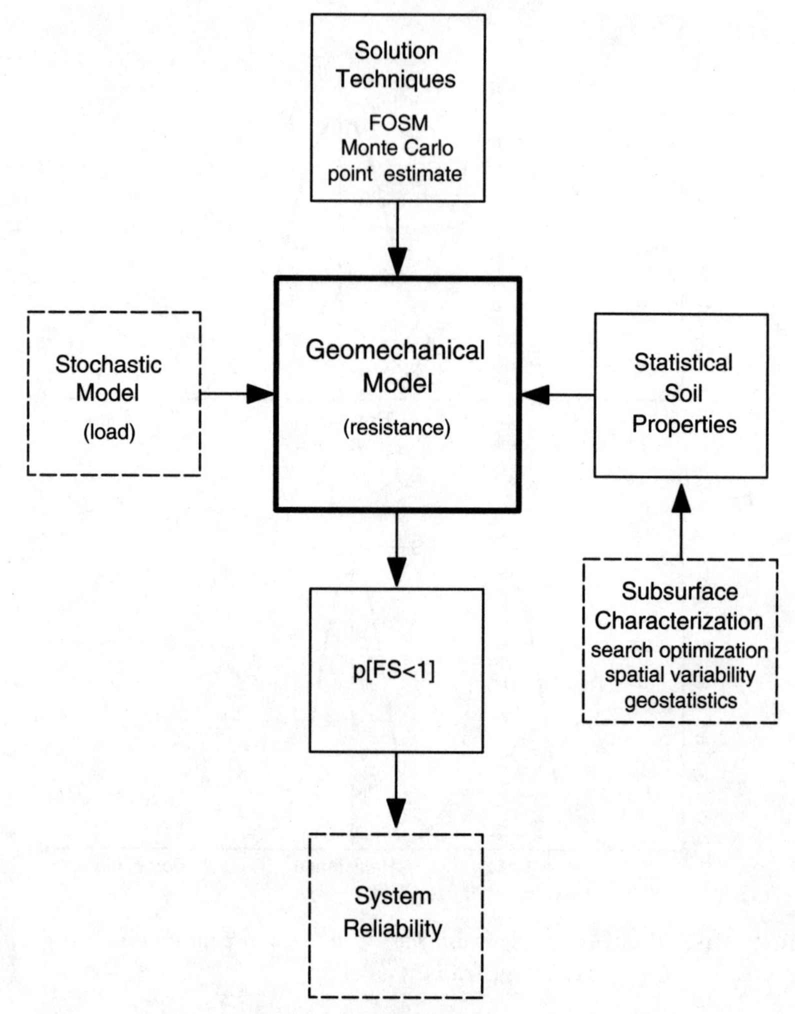

FIGURE 4-3. *Components of reliability analysis.*

provides a conceptual structure for relating various probabilistic methods used in geotechnical reliability approaches. Components shown in solid outline are usually mandatory, whereas dashed outlines indicate optional components depending on the circumstances. Each component is described separately below, with particular reference to the probability interpretations they adopt.

Geomechanical Model (Resistance)

Reliability analysis is centered around some geomechanical model. In principle this model need not be any different from that adopted in a conventional deterministic analysis of the same problem, but in practice those with closed-form computational solutions are most convenient. In fact, the entire reliability analysis can be thought of as occurring inside the geomechanical model, which is taken to be a faithful representation of the failure process. Since the model describes one distinct process, it incorporates one particular mode of failure. It also adopts one of sometimes many possible process representations or computational algorithms. So in this respect, there are as many kinds of reliability analysis as there are models. But how reliability analysis operates within them has much to do with its other components.

Stochastic Model (Loading)

As shown in Figure 4-3, the geomechanical model may also be accompanied by a *stochastic model* representing uncertainty in nongravity loading conditions. Unlike the geomechanical model, a stochastic model has no underlying deterministic algorithm and is purely probabilistic in nature. A stochastic process is one that exhibits long-run regularities in time or space over large numbers of repeated trials that can be characterized by a frequency-derived probability distribution. A number of such distributions exist, among the most common being the *binomial*, *Poisson*, and various forms of *extreme value* distributions as described by Benjamin and Cornell (1970) in engineering applications. Available data or measurements may not uniquely determine which should pertain, with selection often based on familiarity or mathematical convenience as illustrated by Le Mehaute (1986) for wave heights in reliability analyses of offshore structures.

Input Parameters

Statistical characterization of soil property input parameters to the geomechanical model is the core of reliability analysis, which in principle need consist of nothing more. Measured soil-property data are segregated into one or more statistically homogeneous populations, then characterized according to the *statistical properties* of the data (e.g., mean, variance, standard deviation, or coefficient of variation) along with a probability distribution that the data fit, typically normal or *lognormal* in form.[1] This is also often a matter of convenience, especially in the case of the lognormal distribution because of its inherent restriction on negative values. But the form of the distribution selected can have important effects because results are often sensitive to values farther out at their "tails," as Figures 4-1 and 4-2 show.

In practice, the same lack of soil-property data that produces input uncertainty in the first place can make it difficult or impossible to fully determine their statistical properties according to rigorous statistical protocols. Statistical properties and distributions must then be assumed, estimated from experience else-

[1]Alternatively, a distribution may be assigned directly to the factor of safety itself.

where, or judgmentally adjusted from those measured on more comprehensive data populations. As a result, input parameters to reliability analyses are always described by their statistical properties, but comparatively few of these properties may actually be derived from the data.

Sometimes various adjustments to data or distributions are made in the name of applying judgment to their interpretation. This represents a kind of normalized-frequency approach, which indeed has its place, but which as we saw in Chapter 1 is fundamentally subjective. A problem arises if this subjective exercise is not recognized and acknowledged as such, with the results ultimately put forward as objectively obtained from the data. This invites the kind of criticism made of the Rasmussen report described in Chapter 2, where subjective probability judgments were found to be embedded in assumptions, incorporated in other estimates, and generally not explicitly derived or adequately documented. It can also be an invitation to some of the heuristics and biases we will consider in the next chapter if conducted without awareness of their effects.

Subsurface Characterization

An extension to statistical characterization of soil properties is represented in Figure 4-3 as subsurface characterization, although the distinction is somewhat arbitrary. These probabilistic methods seek in various ways to statistically detect or describe systematic variations in conditions or properties. As demonstrated by Vanmarke (1977), spatial variation can be especially important in reliability analyses for long, linear structures such as dikes, levees, highways, or runways. The statistical property of interest is *autocovariance*, which too is usually estimated unless uncommonly large and areally extensive quantities of data are available. Related techniques include *geostatistical* methods, or *kriging*, for statistically estimating properties at unsampled locations, along with techniques for optimizing boring layout or sampling plans in searching for anomalies, defects, or other geologic targets. Probabilistic methods related to spatial variability and geostatistics are summarized by DeGroot (1996), while Halim and Tang (1993) describe search theory.

Solution Techniques

The result of probabilistic characterization of various soil properties, subsurface conditions, and loadings can be a considerable number of individual probability distributions that take the place of single-valued variables in the computational operations of the geomechanical model. These distributions must then be themselves manipulated and ultimately combined in some way to produce the end product. This is said to be to *propagate* these uncertainties or distributions through the analysis. Because closed-form operations on probability distributions are not simple, this requires any in an array of solution techniques that include *first-order second-moment (FOSM)* or *Taylor series expansion*, *point esti-*

mate, and *Hasofer-Lind* or *Monte Carlo simulation* (with or without *Latin hypercube* sampling). The choice can influence computational efficiency, precision, and other items of interest to mathematical probabilists. But they are simply agents, the algebra of reliability computations, and are mentioned here because they are so often confused with reliability analysis itself. It is not uncommon to hear calls for a "Monte Carlo analysis," or some such solution technique, for determining failure probability, which is something along the lines of asking a No. 2 Ticonderoga to produce the Gettysburg Address.

System Reliability

These components and their relationships in Figure 4-3 show how various probabilistic methods are integrated in geotechnical reliability analysis as it is commonly performed. The result is p_f as $p[FS < 1]$. Only one potential mechanism of failure has been addressed by the particular geomechanical model adopted but there may be others, some related and some not. To further extend the process to these areas and obtain a more comprehensive description of global p_f requires moving from reliability analysis into what is known as *system reliability*. As an adjunct to reliability techniques, system reliability is principally concerned with how physical components or features of complex systems, each having its own separate p_f, interact with each other, accounting for physical relationships and dependencies, redundancies, and external effects on system performance. A brief introduction to some simple system reliability concepts is provided in the Appendix, with more thorough treatments found in Ang and Tang (1984) and McCormick (1981). Few geotechnical reliability analyses, however, extend beyond the single failure mode that their geomechanical model addresses.

The Meaning of Reliability Techniques

Having outlined the components and structure of reliability techniques, we turn now to what their results can be taken to mean. This requires examining their conceptual underpinnings, which lie in two areas: the relative-frequency interpretation of probability and the paradigm of theory with its tenet of objective truthfulness. Here the probability of failure is often seen as a singular, unique, and correct value—in short, a "real" number derived from the statistics of hard data. However, some sticky practical problems intervene. While these do not diminish the usefulness of reliability analysis, they do qualify how its results are interpreted. There are two ways of looking at all this, and we start by examining the more prevalent.

The Conventional View

If we go back to the origin of the drill–test–analyze strategy of geotechnical prediction, we find its deductive foundations not far below the surface. The predictive conclusions follow from the soil-property premises, and the laws of theory and analysis are what make them true. But this deductive strategy has one important Achilles' heel: If the soil-property premises are uncertain, then they cannot be taken as true. And if the premises are not true for sure, then neither is the predictive conclusion.

Reliability analysis provides a tailor-made solution. By statistically characterizing input parameters, they contain qualified truth. The premise becomes true with some known probability, and therefore so does the predictive result. Moreover, since statistics is formal and rule based it too has an essentially deductive character—for any given data population, its statistical properties follow necessarily from statistical laws. This, of course, conforms to the relative-frequency interpretation, which holds that there exists in principle a singular and unique probability value that a sufficient number of samples, borings, or tests will provide. So in this way reliability analysis is seen to preserve the deductive truthfulness of the drill–test–analyze strategy while allowing it to extend to parameter uncertainty. In the conventional view, the result it provides is taken as the singularly and objectively correct probability of failure.

But this is achieved only by putting some important matters off to the side. Defining which data are relevant in the first place (the statistically homogeneous sample population) requires the interpretive aspects of geology and judgment, which are necessarily subjective. The form of the probability distribution to assign to the data must also be selected, and this is subjective as well. These factors can become especially important when there are too few data to characterize with any degree of statistical validity, and at least some of their statistical properties must be assumed or otherwise estimated. Yet it is exactly when data are lacking that the most serious parameter uncertainties arise, and there are those who have found this a source of discomfort. If there were enough data to characterize statistically, they protest, one wouldn't be uncertain about input parameters in the first place. And if the statistical properties and distributions do not come from the data themselves but are estimated—from judgment—then why not use judgment to begin with and dispense with the trouble? Others have noted that in ordinary practice, boring locations, sampling depths, and recovered soil specimens used for testing are not selected randomly as statistical protocols require. Instead, they are established by judgment to investigate the presence of hypothesized conditions, relating geology and past experience to possible modes of failure. Moreover, intentionally performing site investigations to actively seek out the worst conditions introduces statistical bias in the database, preventing it from being statistically representative from the start. None of this is to deny the value of attempts at statistical characterization of subsurface data, but simply to show that any such activity is inevitably influenced by a host of subjective factors—as well it should be if it is to incorporate all of the various kinds of relevant knowledge.

But the most difficult conceptual matter comes in relating p[FS < 1] to the probability of failure. This requires first that the model be a comprehensive descriptor of the failure process. The model must also be accurate. And it must be unique. Together these factors challenge the view that the product of reliability analysis can be taken as the probability of failure in any singularly correct or complete sense. But most of all, this has to do with model error and model uncertainty.

Model Error and Model Uncertainty

Geomechanical models are, of course, only idealizations of the processes they are taken to represent, and it is well recognized that the necessary simplifications and approximations can introduce error in model results. In principle, reliability analysis admits to the possibility of error in the model, though this is seldom addressed explicitly. Christian, et al. (1992) are among the few to directly incorporate the effects of model error in reliability analysis for embankment slopes, attributing it to such approximations in limiting equilibrium techniques as three-dimensional effects, convergence limits, and numerical approximations in the computer program. Such *model error*, or *prediction error* as it has also been called (National Research Council, 1994), encompasses the various simplifications inherent in a particular analysis procedure that may cause its predictions to depart from the true value. This is imprecision that may introduce "noise" or scatter to a greater or lesser degree for any given model. However, it still presumes that the model contains the true value somewhere within it and thus provides an objectively truthful, albeit imprecise, representation of actual behavior.

Figure 4-4a illustrates the effect of model error. A reliability analysis produces some probability distribution on the calculated factor of safety that derives from parameter uncertainty. Introducing the additional component of model error has the effect of widening the distribution, enlarging the region of the "tail" where FS falls below 1.0 and thus increasing p[FS < 1]. In this case, the combined distribution is symmetrical about the mean FS, implying that errors are equally likely to occur in either direction. Nevertheless, if p[FS < 1] is taken as p_f, Figure 4-4a shows that the result is to underestimate the probability of failure unless all sources of model error are accounted for and explicitly incorporated.

But further to model error is *model uncertainty* and the two are not the same. While model error refers to variability in the particular model adopted, model uncertainty concerns the degree to which the model faithfully represents field processes, where this includes its conceptualization of how the process takes place. There will always be different models for representing a physical process and how it operates. Some may invoke competing theories or alternative computational algorithms, while others are yet to be discovered. Model uncertainty is the extent to which any such model embodies a uniquely correct representation of the process it seeks to emulate. So while model error is a matter of precision, model uncertainty is a question of accuracy—of correctness in principle.

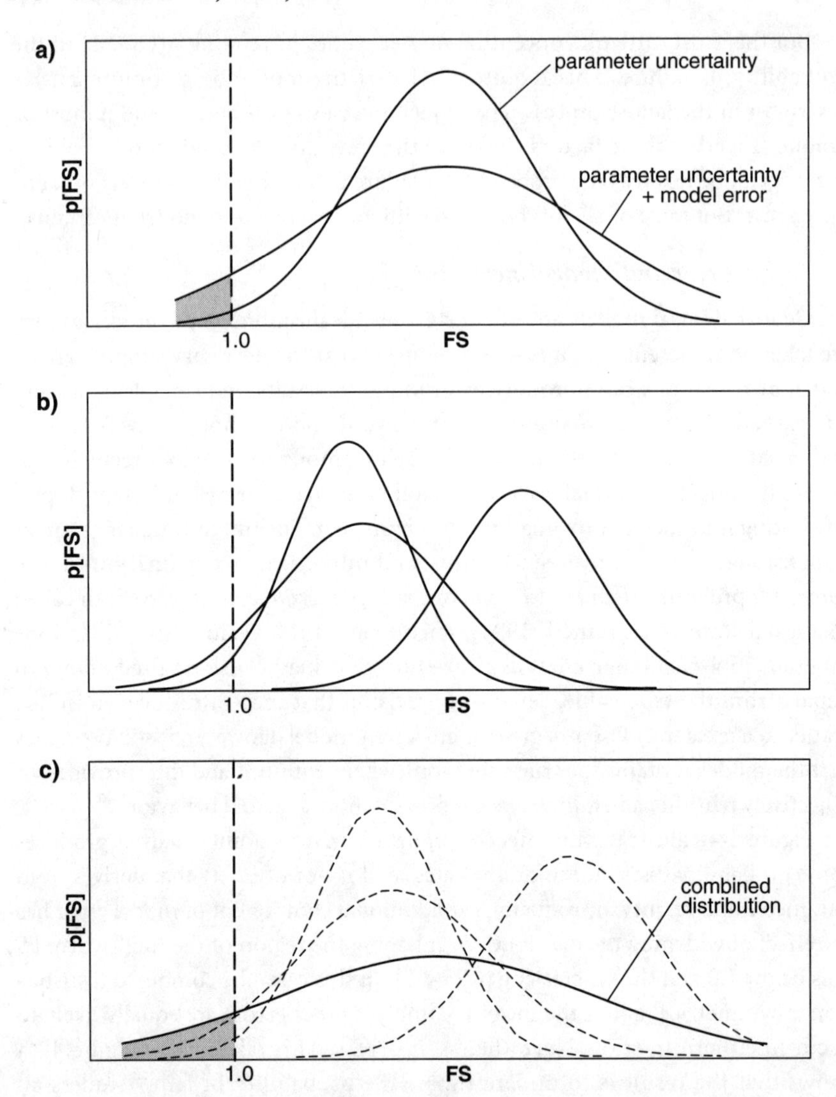

FIGURE 4-4. *Model error and model uncertainty: a, effect of model error on p[FS < 1]; b, p[FS < 1] from alternative models; c, effect of model uncertainty on p[FS < 1].*

In this, model uncertainty extends beyond analytical method, most importantly to how soil behavior is represented. In a larger sense, a geomechanical model is not simply a computational appliance but an internal construct that specifies the manner in which soil will react to loading. Even if there is only one generally agreed-on analytical approach, there are always a variety of alternative soil behavior interpretations it may adopt, and this is truly what distinguishes geotechnical reliability from its structural reliability counterpart. Model uncertainty is much more dominant because the mode of soil response must be assumed within the analytical framework. Here the most profound uncertainties are less in the variability of input parameter values than in which parameters are used to represent the process, and this has everything to do with our limited abilities to understand and reproduce in a model the behavior of soils in the field. While these matters were discussed in Chapter 3, it is well to repeat some of them here. Will the soils be contractive or dilative under the stresses imposed? Should the strength of contractive sands be represented by the peak or the steady-state strength, and for clays by peak or residual or something in between? Should liquefaction resistance be evaluated from laboratory approaches or field data, and how should both be interpreted? Will pore pressures on the failure plane at failure in the field be positive or negative? Will a failure occur slowly or rapidly with respect to the ability of these pore pressures to dissipate during shearing, and should a drained or undrained analysis correspondingly be performed? And what do we do about things like progressive failure, flow failure, or internal erosion where descriptive models may be lacking? Because every analytical model must incorporate a soil-behavior model, model uncertainty intrinsically includes soil-behavior uncertainty that goes beyond the imprecisions of model error.

Several alternative models or soil behavior interpretations are depicted in Figure 4-4b, each yielding its own $p[FS < 1]$. Indeed, there are as many failure probabilities as there are models to produce them, but how do we establish which, if any, is correct? This is a question that reliability analysis cannot answer, but it has major implications for what its results can be taken to mean. If the models in Figure 4-4b were aggregated into a single probability distribution that accounted for all of them together, the result would look something like that in Figure 4-4c. Here, the combined distribution is much wider than any one individually, and $p[FS < 1]$ is correspondingly greater.

To put this into perspective, consider the pseudostatic methods once used to represent the dynamic behavior of loose, cohesionless soils. A reliability analysis would incorporate parameter uncertainty by statistically characterizing measured values of strength, with the result a statistically legitimate and rigorously derived probability of failure. But without accounting for model uncertainty—here having to do with the potential for cyclically induced strength loss—it would also be recognized today as entirely wrong. The point is that any model and any soil behavior interpretation it uses are predicated on some state of knowledge. So the probability of failure produced by reliability analysis is not independent of this knowledge. Rather, it is a conditional probability given the particular state of knowledge its algorithms and interpretations reflect, and this reveals its inherently subjective underpinnings whether acknowledged or not.

Model uncertainty can sometimes be accounted for and included in reliability applications by means of statistical approaches. This requires that alternative models be tested by means of controlled experiments to determine their respective frequencies of correct predictions. One example is provided by Ronold and Bjerager (1992), who incorporated model uncertainty in determining offshore pile capacity in this way. It was by far the dominant contributor, greatly exceeding uncertainties associated with parameter values or wave loading, which were negligible in comparison.

But these circumstances are rare because of the difficulty in performing controlled field experiments whose results can be broadly generalized. While there is comparatively little discussion of model uncertainty in the reliability literature, Laskey (1996) describes how it is typically handled. Reliability analysis usually adopts some generally agreed-on model and set of assumptions. This does not mean that uncertainty is absent in either one, only that there is no controversy about (and usually no alternative to) the techniques given established convention at the time. So again we find what Poincaré found in the axioms of geometry. The uncertainties that accompany models are not really addressed but tacitly accepted, so that ultimately p_f is established by convention as well. Model uncertainty can be treated more explicitly by introducing subjective probability, something that orthodox reliability analysis does not usually condone. Here the p_f resulting from parameter uncertainty can be accompanied by a subjective probability statement that describes the likelihood of model correctness. If there are competing models or interpretations, the results can be weighted by the likelihood that one or another is correct, again a subjective judgment. But the probability axioms require that at least one be correct: Their probabilities of correctness must sum to 1.0, neglecting the possibility that all of them may be wrong. Kempton, et al. (2000) frame this in terms of geochemical modeling and the environmental policy decisions it influences. Here reliability approaches can statistically characterize the variability in some input parameters, but not all of the relevant uncertainties can be fully captured, so that the uncertainty band in model results and the policies they influence is ultimately underestimated.

But even if model uncertainty is not formally addressed, Laskey (1996) emphasizes the importance of presenting a range of reliability results that use different models. In the end, she argues that the inevitability of irreducible model uncertainty requires acknowledging that there is no single right answer to be obtained from models or the interpretations that go into them, and thus no correspondingly unique probability that reliability analysis can provide. Alternative predictive models can, however, be useful in directing attention to the different factors or uncertainties each might emphasize, producing a more complete picture overall (Einhorn and Hogarth, 1982).

Reliability analysis must always invoke subjective factors, be they in determination of statistical parameters, selection of probability distributions, or in specification of a model and the interpretations it adopts. This should not be surprising since it conforms to the necessary application of judgment in performing any analysis, probabilistic or not. But these subjective elements are seldom identified in reliability techniques, nor are their effects included in the failure probability it produces. Only by leaving them aside as externalities can its objective truthfulness be preserved. Still, there are other ways of interpreting reliability analysis that do not necessarily make this assertion.

An Alternative View

A different perspective sees reliability analysis as an aid to identifying the nature and sources of uncertainty affecting model results, as a supplement to conventional analysis. From this standpoint, the purpose of reliability analysis is to inform judgment about the effects of parameter uncertainty on model results to aid their interpretation.

Lacasse and Nadim (1996) note that by accounting for uncertainty in soil-property input, reliability analysis complements deterministic procedures the model embodies. In going back to its original justification, they argue that reliability analysis enhances these procedures by providing a more consistent measure of the effects of soil-property uncertainties. This is somewhat akin to a sensitivity study, where parameters are sequentially varied one by one within some arbitrary limits. But reliability analysis replaces these limits by varying parameter values jointly in proportion to data variability. The result is to establish not the relative influence of the various parameters per se but rather the influence of uncertainties as to their values.

In this perspective, reliability analysis is a means of enriching the content of the model's output; its calculated p_f is not an end in itself. Commenting on the meaning of reliability analysis, Christian, et al. (1992) amplify in relation to stability of slopes and embankments:

> ...there has been much confusion over precisely what is meant by reliability and probability of failure. The most effective applications of probabilistic methods are those in which the relative probabilities of failure are evaluated or in which the methodology is used to illuminate the effects of uncertainties in the parameters. Attempts to determine the absolute probability of failure are much less successful.

Several revealing things are found in this excerpt. First, it alludes to there being such a thing as an absolute probability of failure, and indeed this is a prime source of the confusion the authors describe. But these remarks call into question whether such a value can ever be found, with a nod to the subjectivist view. They suggest instead that failure probability calculated from reliability analysis be viewed in a comparative sense that allows effects of uncertainty in various parameters to be established proportionately. So here we find a meaning for reliability analysis more closely related to judgment—it instructs in relative importance of parameter uncertainties. From this standpoint, reliability analysis answers the following question: All other things being equal, what is the influence

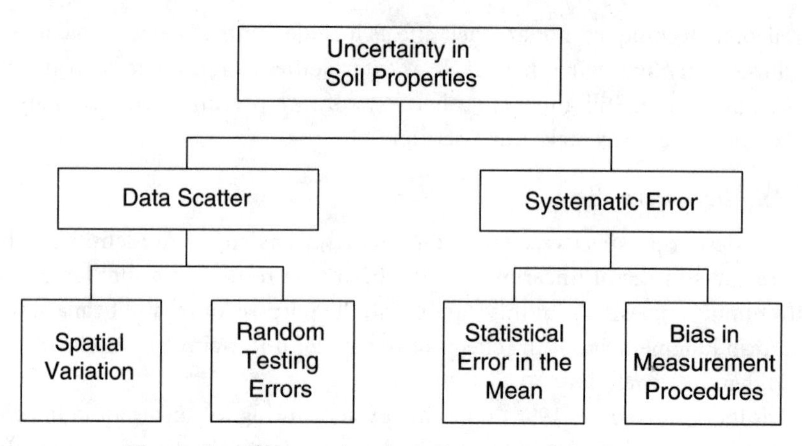

FIGURE 4-5. *Categories of uncertainty in soil property data.*

Source: Christian et al., 1994.

of parameter uncertainty on factor of safety? The calculated probability is the reference by which this comparison is made. Two things are important here. One is that this speaks to the relative comparison of one case to another, in the manner of Figures 4-1 and 4-2. The other bears repeating: all other things being equal—the model, the assumptions, and the interpretations inherent in both.

Christian, et al. (1992) demonstrate how this works on a different level having to do with the various internal contributors to soil-property uncertainty. They classify these sources according to the schema in Figure 4-5 and show how reliability analysis, properly structured, can separate and quantify their individual contributions. *Systematic error* can result from statistical error in mean values arising from limited numbers of measurements, as well as from bias in measurement procedures like that associated with field vane shear testing. *Data scatter*, on the other hand, includes both nonsystematic (random) testing errors and actual spatial variation in the soil profile. Depending on which contributions were most important, one might elect to obtain more samples, perform different types of tests, or perhaps simply accept the irreducible component of spatial variability and move on. But reliability analysis, suitably interpreted, is what enables informed decisions to be made in the face of soil-property uncertainty. Model uncertainty is much less at issue if p_f is not claimed to be the "true" value, but instead a reference value by which parameter uncertainties in a particular model are judged.

The Meaning of Failure Probability

It has already been noted that reliability techniques take place inside the analytical box to produce $p[FS < 1]$. This is called the probability of failure in a reliability context, but it does not pertain to a physical event. Analysis is predictive.

Solution component	Source of uncertainty	Probability imputed by reliability analysis
Diagnosis	Incorrect failure mode identified for analysis (wrong problem)	p = 0
Prediction	Incorrect prediction for identified failure mode:	
	• wrong parameter values	p[FS<1]
	• wrong model (model uncertainty)	p = 0
	• wrong result (model error)	p = 0
Implementation	Incorrect design/construction/ operation (human error)	p = 0
		Σ = p[incorrect solution]

FIGURE 4-6. *Reliability analysis in problem solution.*

And because it is, p[FS < 1] is the probability of an incorrect prediction—the probability that our prediction may turn out to be wrong when we think it is right. Moreover, reliability analysis deals with only one aspect of predictive uncertainty—that associated with input parameters. So the p[FS < 1] from reliability analysis is really the probability that we have chosen the wrong values in whatever model we have elected to use, without necessarily saying anything about the model itself.

Figure 4-6 puts the predictive process into the context of physical failure. There are three parts to the larger predictive problem: first diagnosis, then the prediction itself, then the implementation of those measures required for the predicted performance to occur in physical reality. When all three of these components work successfully, the solution to the predictive problem is correct and the structure behaves adequately. But if any one of them is faulty, the predictive solution, and with it the structure, will fail. So the probability of failure as a physical occurrence can be taken as the probability that any aspect of the problem solution is faulty—diagnosis, prediction, implementation, or any combination of the three.

Taking these aspects separately, Figure 4-6 describes the nature of the uncertainties associated with each. An incorrect diagnosis comes about if the wrong failure mode is analyzed. Unless extended to system reliability, reliability analysis treats only one failure mode with an imputed probability of zero that the diagnosis is incorrect. The prediction component is where reliability analysis resides, and here an incorrect prediction can come about in several ways. First, the analysis might use incorrect parameter values, as represented by p[FS < 1]. But the model might also give the wrong result, or it could be the wrong model. To the extent that reliability analysis neglects these contributions of model error and model uncertainty, it assigns zero probability to these occurrences. The prediction produces a numerical result that is translated into reality through the physical presence of some natural or constructed feature, and this is where human error comes in. A flawed subsurface investigation might cause some feature to be misidentified or misinterpreted; there might be errors in design, construction, or operation; or there might be miscommunication or erroneous documentation. But here too, usual practice in reliability analysis does not account for human error, with an implied zero probability of faulty implementation.

This has important implications for what p_f from reliability analysis can be taken to mean. Note again from Figure 4-6 that obtaining an inadequate solution—and thus physical failure—requires that only one of its components be incorrect. The respective probabilities are additive as the union of these events, so reliability analysis can provide only a partial probability of failure as a physical occurrence. Though it may be derived with statistical rigor, it is far from complete, which leads directly to a more fundamental implication. To the extent that any contributing source of uncertainty remains unaccounted for, a value put forward as an authentic probability of failure will always be too low. This goes far toward explaining why Christian, et al. (1992) might be leery of using reliability analysis to this end. More broadly, it holds for any probabilistic technique unless the full suite of failure-related uncertainties is addressed.

The upshot of all this is to place reliability analysis and its probabilistic methods into context. Residing within a particular geomechanical model, reliability analysis can help understand and accommodate uncertainties in its inputs, be they soil properties, loading conditions, or both. But because of constraints on knowledge about soil behavior (and therefore the models used to represent it), this is only part of the global uncertainty. If the calculated p_f is to be taken as the actual probability of failure in a true and literal sense, this requires going far beyond the confines of the model to how failures actually occur. And actual failures result far more often from incorrect diagnosis of the processes operating in the field than from incorrect parameters used in their analysis. So indeed there can be merit in deriving probabilities from the statistics of "hard data," but there can be some serious constraints as well. Reliability analysis has the ability to identify sources of data uncertainty in input parameters to a particular geomechanical model, but it cannot speak to uncertainty in the operation of the physical process the model seeks to emulate nor those of the emulation itself. In this respect, reliability analysis can be said to provide a statistically precise answer for a fairly small part of the problem.

The opposite side of the coin is represented by risk analysis. By taking a more encompassing view of the collective uncertainties influencing failure, a prime goal of risk analysis is to establish the relative importance of each, even—and especially—if this means that the probabilities used to do so are subjective and therefore nonunique. While reliability analysis is conducted around models, risk analysis is centered on processes. In conforming to the paradigm of practice, risk analysis incorporates both theory in the form of models, and judgment in the form of subjective probability. We can go on now to see how this comes about.

Risk Analysis

Risk analysis differs from reliability analysis in several respects, the first being that while it may make use of results from one or more geomechanical models, it does not necessarily rely on these models directly. Risk analysis provides the ability to address multiple failure modes—not just one—whether geomechanical models are available to represent them or not. It goes beyond where most models stop, to the full progression of failure processes and the factors affecting this progression—be they technical, human, or otherwise. As a probabilistic adjunct to design-based analyses, reliability techniques find their natural application in design of new structures. It is at this stage, before the actual performance of the structure has become manifest, that soil-property uncertainty is most often of greatest importance. Risk analysis, on the other hand, can be most useful for existing structures where it serves a diagnostic function in more comprehensively addressing all the components of uncertainty.

This follows directly from the meaning of risk. In engineering applications, risk is defined as the product of failure probability and consequences:

$$\text{Risk} = (\text{Probability of failure}) \times (\text{Consequences of failure}) = (p_f) \times (C_f) \quad (4\text{-}1)$$

So by definition, risk analysis must account for the physical consequences of failure, where these consequences are measured in any consistent metric one might choose. This is what makes risk analysis fundamentally different from reliability analysis, and it is the underlying reason why the two are distinguished here as distinct entities. In reliability analysis, $FS < 1$ is a surrogate for adverse consequences, a proxy for something bad. But bad things do not always, or even necessarily, happen as an outgrowth of this calculation. Taken at face value, $FS < 1$ in a limiting equilibrium analysis indicates that some movement will occur, but it does not say how much, how fast, or the extent to which it matters. There may be any number of intervening circumstances and conditions that might prevent any adverse consequences from occurring, but models are seldom available to represent them. Accordingly, risk analysis must rely to one extent or another on the application of judgment as subjective probability, another of its defining characteristics.

Take the prediction of $FS < 1$ in the slope of an earth dam. Accepting for the moment that the analysis is a perfectly accurate predictor, suppose a slide develops. The consequences of interest are whether loss of life would result from dam breach, and if so to what extent. Then to get from the condition of $FS < 1$ to these consequences requires that the remaining chain of events be carried through. Whether the slide allows overtopping to occur depends on the elevation of the reservoir at the time, whether it fully engages the dam crest, and how much movement it undergoes. If the slide occurs during the daytime or if monitoring is in place, it might be detected. It might also develop slowly enough for repairs to be made in time—provided that people react and respond the way they should. In the worst case, rapid and undetected slide movements could allow overtopping, but this might not proceed to full breach if its depth and duration were limited and the materials resistant to erosion. Otherwise, all of these things would affect the characteristics of the dam-breach floodwave, and whether lives might be lost would depend on the inundation depth, number of people within the inundated area, their distance from the dam, time of day, and other factors that govern warning time and evacuation effectiveness.

The list could go on, but the point is that none of these events or their ultimate physical consequences are captured by an analysis that produces a factor of safety less than 1.0—or by p_f representing uncertainty in its input parameters. Nor were they ever intended to be, since geomechanical models are almost always developed in the implied context of design of new structures and proposed facilities. For design, it is entirely sufficient to establish that some indicator of possible failure like FS < 1, or some failure precursor like the development of a slide, will not occur. The purpose of design, after all, is to prevent failure, and preventing its precursors or incipient development guarantees that no failure process can progress. The designer might be justifiably reluctant to rely on a whole string of "mights" in avoiding the consequences of failure for a structure yet to be completed. This is exactly why design assumptions are customarily conservative, or at least they're intended to be at the time.

But uncertainties interfere with a host of design assumptions, especially in structures already built. For most existing structures, soil-property uncertainty is largely obviated after their successful completion. If the uncertain event was once whether material properties would be adequate to maintain the structure's stability, then this event has come and gone once it stands barring some change in physical properties or new loading conditions not yet experienced. The outcome is known and therefore no longer uncertain. Instead, a variety of processes or conditions never anticipated in design may become evident as its performance is revealed.

When a structure is conceived, it is a gleam in the designer's eye reflecting the hope that design assumptions and all other good intentions will come to pass. But once born, it develops a personality with idiosyncrasies all its own, and as it grows older, the creaks and groans of aging will eventually settle in. Geotechnical engineering brings its fair share of problem children into the world, some who become cranky and cantankerous. It is for identifying and learning how best to cope with these offspring that risk analysis finds its most successful and useful applications. So risk analysis must take the structure as it is, not as the designer intended it or as the analyst would idealize it. It must go beyond design to processes, events, and conditions that design either cannot or has no need to anticipate. In short, design analysis and risk analysis are not the same. They do different things, and the latter can supplement but never replace the former. Because risk analysis must extend all the way to physical consequences, it cannot stop where design ends. It must continue beyond p[FS < 1] where the design-based models of reliability analysis leave off.

Risk analysis resides outside models. Although it may incorporate their results where it can, it must also incorporate uncertain events, conditions, or processes that no model, geomechanical or otherwise, is able to describe. This requires judgment, and judgment must be quantified as subjective probability. With this, subjective probability becomes a virtually mandatory component of risk analysis, which can seldom arrive at physical failure consequences without incorporating it to one degree or another. The application of judgment becomes explicit, unlike its covert operation in the background of reliability techniques, as we will see in several of the following probabilistic methods.

Probabilistic Methods in Risk Analysis

Risk analysis encompasses the entire process of how adverse effects are incurred. Any such process has three parts: (1) an *initiator* that begins it, (2) the *response* of the structure to the initiator, and (3) the *consequences* if inadequate response results in failure. Just as for reliability techniques, a number of probabilistic methods can be adapted to each of these parts with Figure 4-7 showing how risk analysis structures them. The mandatory components are shown in solid outline, with the three basic elements highlighted in bold and optional components enclosed in dashed outlines.

Initiator

The initiator is a hypothesized occurrence or condition that sets a potential failure process in motion. It is often associated with imposed loadings such as the occurrence of an earthquake, where it describes all possible levels of loading such an event could produce. In this respect, initiator uncertainty is treated the same way as loading uncertainty in reliability analysis, and often using the same kinds of stochastic models. The initiator, however, is not limited to measurable quantities. Geologic conditions can result in an unknown state of nature, such as the existence of boulders that might affect pile driving or tunneling, which could initiate some sequence of undesired occurrences. These initiators can be represented as using statistically based stochastic models as well.

In risk analysis, an initiator may also have a nonstochastic origin in hypotheses about process causation, and this allows multiple failure hypotheses to be accommodated. As we have seen, a clay deposit might possess compressibility characteristics, or fissuring, or pre-existing shear planes that could affect the structure in different ways. The degree of validity or confidence in such hypothe-

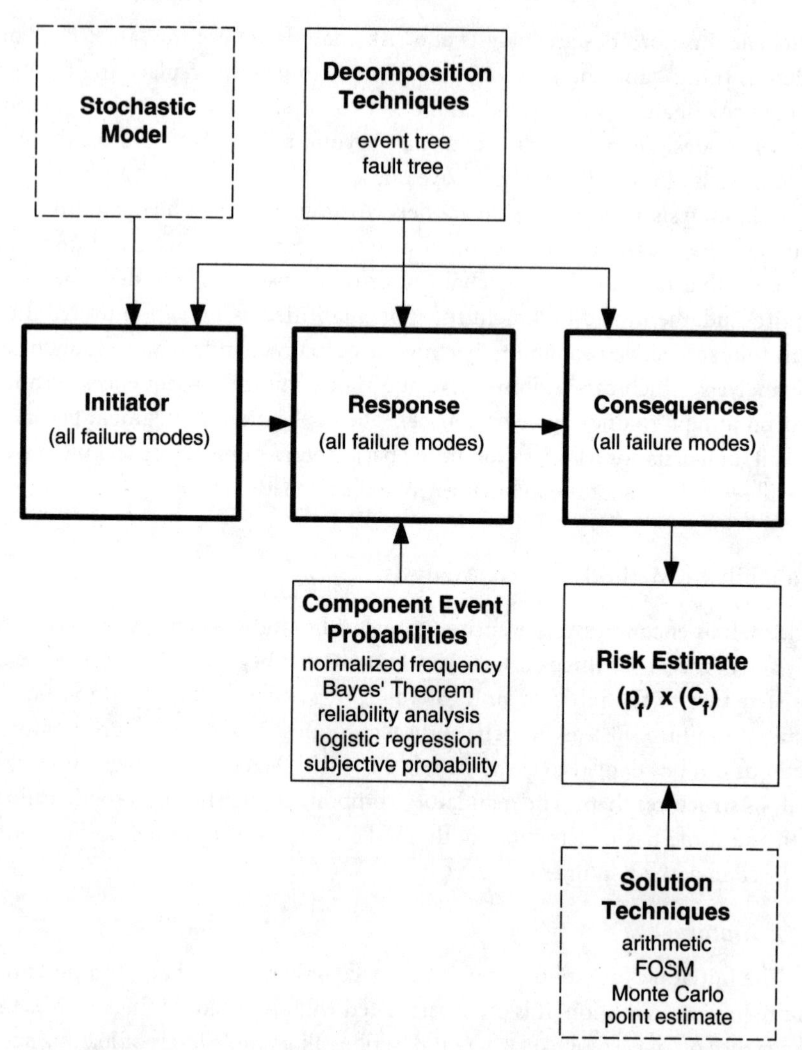

FIGURE 4-7. *Components of risk analysis.*

sized geologic conditions is expressed subjectively. And a hypothesis can extend beyond conditions produced by nature to any unknown feature or condition the structure might contain, like construction defects or hidden damage. Any of these conditions or states of nature could produce different types or sequences of failure development, even should they result in the same consequences. This again is one of the principal differences between risk analysis and reliability techniques: It is not necessary to preselect any one particular mode of failure described by a specified model. The diagnostic aspect of the problem is explicitly incorporated rather than being externalized. Indeed, as we will see, the entire process of risk analysis can be considered an exercise in diagnosis.

Response

The various failure modes identified in connection with initiation of each failure process may affect the structure in different ways, and this too is almost always uncertain. The occurrence of the initiator will lead to failure only if the structure is unable to sustain its effects. This is the *response* of the structure to the initiator, and the associated uncertainties can be evaluated using several probabilistic methods and interpretations (Figure 4-7) that are described as follows.

FAILURE STATISTICS. Many risk analysis applications deal with manufactured electrical, mechanical, or structural components such as pumps, valves, switches, or structural members. Their manufacture is identical and so are they, except for deviations in production or materials making some more failure-prone than others. The corresponding failure frequencies can be determined statistically from service records over large numbers of identical applications—a direct, frequency-based approach for response probability. But geology is not in the manufacturing business, and it does not produce identical parts. Although frequency-based statistics are used in other aspects of geotechnical risk analysis, it is uncommon for component failure statistics to be applied directly as response probabilities.

NORMALIZED FREQUENCY. Sometimes data statistics are available for data populations that are outwardly similar but not identical to the case of interest. It can be possible to subjectively adjust, or normalize, the resulting frequency value to better conform to the circumstances at hand. For example, historic data are available for geotechnical structures such as earth dams that provide their failure frequency per dam-year of operation. Figure 4-8 illustrates how one such compilation (Von Thun, 1985) might be used to estimate a normalized-frequency response probability for internal erosion, where the initiator is filling of the reservoir and its subsequent annual fluctuation.

Figure 4-8 shows internal erosion failure frequencies for all U.S. dams; for those in the western United States; and for those greater and less than five years of age to reflect first filling when internal erosion failures tend to occur disproportionately. This shows that different subpopulations of the data have different failure frequencies according to age and location. These are called a *base-rate frequency* or simply the *base rate.* However, every dam and its foundation are unique, and the probability of some particular dam failing by internal erosion must be normalized by adjusting upward or downward from the observed base-rate frequency according to conditions it displays and features it contains such as those listed.

Determining the degree of departure from the base-rate frequency is a subjective matter of judgment. It is complicated, however, by physical and resulting statistical nonuniformities within any given subpopulation. Neither dams nor foundations are identical widgets, and each subpopulation contains a vast but

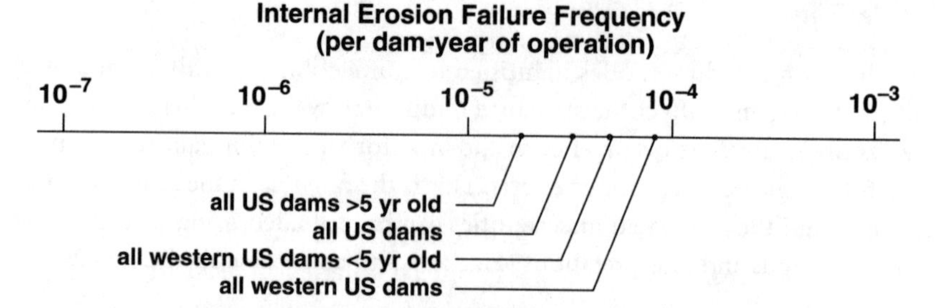

Factors Evaluated in Frequency Normalizing

Decreased Probability	Increased Probability
processed filters	uncontrolled seepage exit
surface treatment of foundation	no processed filters
impervious foundation	open untreated foundation joints
foundation grouting	openwork granular foundation soils
long operation at maximum pool	history of seepage carrying fines
shaping of rock foundation	dispersive soils in dam or foundation
well-graded core	structures penetrating embankment
plastic core	irregular or steep foundation profile
positive cutoff into rock	fine-grained erodible core
low gradients	high gradients
minor seepage	high seepage quantity
good monitoring	poor monitoring

FIGURE 4-8. *Normalized frequency example for dam failure by internal erosion.*

unspecified variety of features and conditions unique to each member. In this respect there is no "typical" or "average" dam in the database, no truly representative member of the subpopulation, so the importance of the characteristics of any particular dam can be difficult to gauge in comparison and the normalized departure from the base-rate frequency correspondingly hard to establish. Nevertheless, as we will go on to see, base-rate effects can have an important influence on subjective probabilities, where they often tend to be neglected. The normalized-frequency approach can serve to good purpose in confirming that response probabilities are in the "right ballpark."

BAYES' THEOREM. Bayes' theorem, which we first encountered in Chapter 2 in its historical context, reappears here as a formal way to adjust frequency data for factors specific to a particular case. To the extent it might make failure more, or less, likely to occur, such a factor can be considered an indicator of failure, with those listed in Figure 4-8 again serving as potential examples. Recall from Eq. 2-3 that Bayes' theorem can be expressed as:

$$p[\text{failure} \mid \text{indicator}] = \frac{p[\text{indicator} \mid \text{failure}] \times p[\text{failure}]}{\{p[\text{indicator} \mid \text{failure}] \times p[\text{failure}]\} + \{p[\text{indicator} \mid \text{no failure}] \times p[\text{no failure}]\}}$$

Here the prior probability p[failure] is adjusted by one or more indicators of failure, which might include various observed conditions or analysis results. The reliability of the indicator in predicting failure is p[indicator | failure], its false-positive rate is p[indicator | no failure], and both of these indicator properties must be supplied. Frequency-based approaches for doing so can be limited because postfailure investigations usually provide insufficient statistics on the prefailure presence of the indicator, and because of multiple indicators whose individual influences are hard to separate (the statistical problem of so-called *confounding effects*). Alternatively, subjective estimates of indicator properties can be made if there is some sense of the relative importance of the indicator in both past failure occurrences and nonfailure conditions.

Typically there are several indicators and counter-indicators of failure that each influence adjustment of the prior to different degrees. Provided that the properties of all such indicators can be specified, the advantage of Bayes' theorem over the normalized-frequency approach lies in its ability to separate the effects of each one, allowing them to be considered individually and combined sequentially where the posterior probability calculated from one becomes the prior for the next. But this advantage is often diluted by limited quantitative understanding of indicator significance. It would be difficult, for example, to find any dam or collection of dams that would allow the effects of the indicators and counter-indicators in Figure 4-8 to be evaluated individually because they often if not always occur in various combinations. Still, there has been little systematic collection of data for common geotechnical indicators in the context of Bayes' theorem, and its potential for deriving inductive inferences from observational evidence remains largely untapped.

RELIABILITY ANALYSIS. Much has already been said about reliability analysis as a stand-alone technique, but Figure 4-7 shows that it can also be used in the response component of risk analysis. Reliability analysis can be a prime candidate for those aspects of response where applicable geomechanical models exist and where soil-property uncertainty may dominate, especially for types or magnitudes of loading not yet experienced by an existing structure. In a risk analysis framework, however, the calculated p[FS < 1] is this and nothing more—the probability that uncertainty in input parameters to a geomechanical model could cause its results to fall below a desirable threshold. Here p[FS < 1] is no longer accepted at face value as the probability of failure, which incorporates additional aspects of the failure process and other information about it.

LOGISTIC REGRESSION. While a geomechanical model produces the magnitude of some quantity such as factor of safety, most processes associated with response have a binary outcome with only two possible states: Either something occurs or it doesn't. Geotechnical engineering abounds with empirical correlations for such processes that relate observed occurrences to various parameters. The familiar result is a two-dimensional space in these parameters, where the observational data are separated into regions of occurrence and nonoccurrence by a boundary typically established by "eyeball." The data almost always overlap, though, so no one boundary separates them perfectly. *Logistic regression* or *binary regression* is a statistical procedure that replaces a single boundary with a family of boundaries, each with the probability of occurrence for any associated [x,y] data pair.

Figure 4-9 provides an example for level-ground liquefaction of clean sands under M7.5 earthquake shaking. Figure 4-9a shows the classic empirical correlation of Seed, et al. (1985), whereby observed indicators of liquefaction like sand boils, and the absence of these indicators, are characterized by two parameters: the normalized cyclic shear stress ratio produced by earthquake shaking τ/σ_o and corrected Standard Penetration Test (SPT) blowcount as $(N_1)_{60}$. The liquefaction boundary is constructed to separate most of the liquefaction/nonliquefaction occurrences, but it is not entirely precise, with a few misclassified data. The logistic regression of Figure 4-9b (Liao, et al., 1988) shows much the same thing with an expanded database, except that the deterministic boundary is replaced by isochrones to express liquefaction probabilities.

In logistic regression, the empirical correlation replaces the geomechanical model of reliability analysis, and there is no model uncertainty as such because there is no underlying algorithm. However, the parameters themselves may be affected by uncertainty, and regression probabilities can be further affected by how much data there are and how they are interpreted, so that different logistic regressions—and different probabilities—can be obtained for the same or similar database (Youd and Noble, 1997). These matters notwithstanding, logistic regression techniques hold great potential for characterizing uncertainty in a wide variety of empirical correlations, with Honjo and Veneziano (1989) providing another example for internal erosion and filter criteria.

SUBJECTIVE PROBABILITY. In risk analysis as elsewhere, subjective probability is the quantified expression of engineering judgment about the likelihood of occurrence of an uncertain event, the existence of an unknown condition, or the confidence in the truth of a proposition. It can be used for response probability estimation in basically two ways. One is as a conditional value given a probability derived by some other method, where this subjective conditional serves to adjust the initial value in similar fashion to normalized frequency. For example, a subjective conditional might be appended to $p[FS < 1]$ from a reliability technique to account for uncertainty in the model—a judgment of its being correct. Or such a value might be assigned to the level-ground liquefaction probability derived from logistic regression to express uncertainty in how it might pertain beneath a slope.

But for response-related events with binary outcomes, subjective probability provides a stand-alone measure of outcome likelihood. And where response involves some quantity, like movement, the range of possible values can be partitioned into discrete increments and a subjective probability assigned to each.

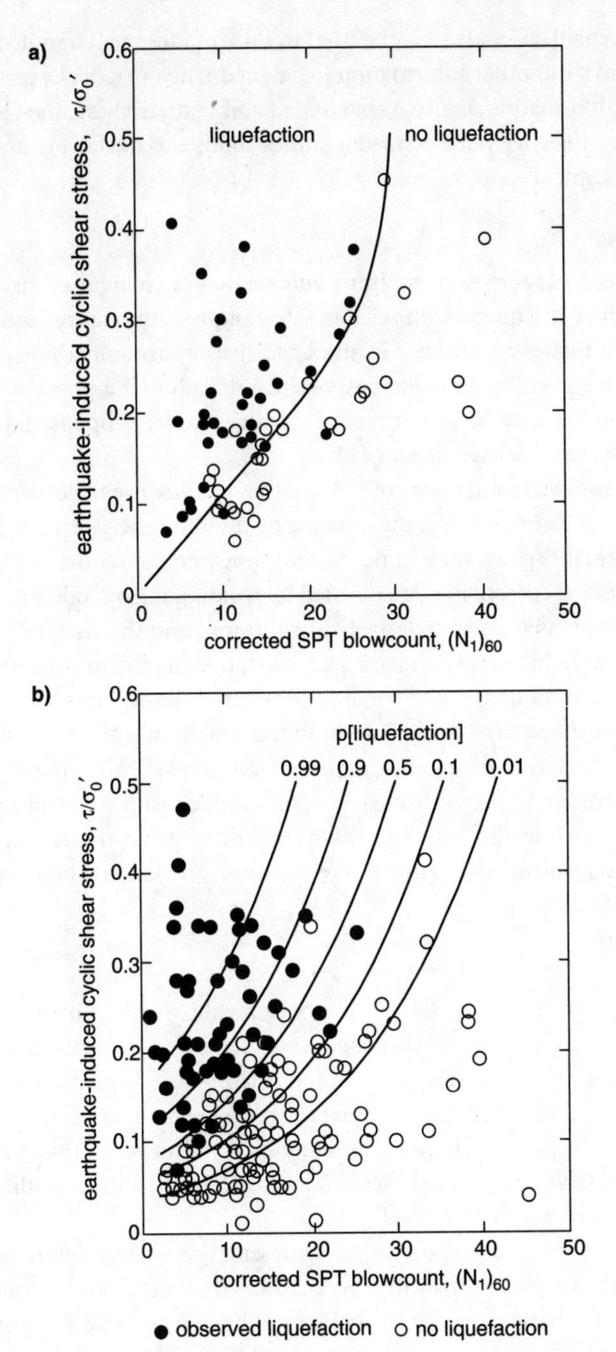

FIGURE 4-9. *Level-ground liquefaction correlations: a, deterministic liquefaction boundary; b, liquefaction probability by logistic (binary) regression.*

Sources: a, Seed et al., 1985; b, Liao et al., 1988.

Here, geomechanical models can be used to inform judgment about the occurrence of events, but other information like data quality, personal experience, or case-history information, is also incorporated and inductively synthesized in formulating the subjective value. How this comes about and the factors involved are topics of subsequent chapters.

Consequences

Returning to Figure 4-7, the third and final core element of risk analysis involves failure consequences. In reliability techniques, adverse consequences are inferred from model output, but in risk analysis they are something physically measurable like lives lost or dollar costs incurred. Failure is a physically observable condition like dam breach or retaining wall collapse, as opposed to an indirect indicator like calculated factor of safety.

The consequence component of risk analysis can involve uncertainties itself, often as much if not more than the initiator or response elements. But engineers do not ordinarily spend their time determining the consequences of failure because most of their activities are devoted to avoiding it. So models for predicting failure consequences are relatively uncommon, and this usually relies to a large extent on failure case histories and subjective judgment for interpreting how their effects might apply to conditions at hand. Beyond this, consequences often involve the potential for intervention that might prevent them. People do not stand idly by and watch a failure happen if they are aware it is in progress and able to intercede, and the case-history literature is filled with "near misses" where failure was averted in this way. So consequences can incorporate a human element, and this involves uncertainty of the most perverse kind that only judgment can address.

Decomposition Techniques

Figure 4-7 shows that the three primary elements of risk analysis considered so far—initiator, response, and consequences—are linked by methods of *decomposition.* This is simply the disaggregation of each failure sequence into its constituent events, all of which must happen for failure consequences to be realized. Decomposition applies to uncertainties contained within each of the three basic elements, and having separated these elements to begin with is nothing more than decomposition on a gross level.

Event trees and *fault trees* provide the representation of how failure sequences are envisioned, and they portray the decomposed elements each one contains. As such, they are the basic tools of risk analysis. Still, these are only devices for portraying its organizing structure and not the analysis itself. Although risk analysis is based on processes, not models, the event tree or fault tree documents how the failure sequence is visualized to occur in the same way that an algorithm describes the underlying operation of a model. And here, visualization is undoubtedly key. This, as we have seen, is central to diagnostic judgment, and it is ultimately what event tree or fault tree construction requires: conceptualizing the sequential and logical organization of conditions and events in each failure process from start to finish. Because visualization is judgment based, there can be no unique way to specify it or the decomposition that results, and failure processes may be conceptualized and organized differently by different individuals or decomposed to different levels of detail.

Conventions used in event tree and fault tree construction are treated by Ang and Tang (1984). Both methods work through decomposition, but they operate in different ways that adapt them best to different classes of problems (Paté-Cornell, 1984). Consider the generic failure process portrayed in Figure 4-10 in both formats. Figure 4-10a shows an event tree containing both initiator and response events that together could lead to failure, with each having some associated probability of occurrence. The event tree starts with initiator event I, which may or may not occur. If it does, the next thing to happen in order to progress to failure would be response event R_1, followed by R_2. Either response event R_3 or R_4 could happen next. In this illustration, the *branch pathways* $IR_1R_2R_3$ and $IR_1R_2R_4$ constitute separate failure modes, each containing a unique set of decomposed events. The failure probability associated with each is the product of their event probabilities (or distributions), and this is multiplied by the consequence value at the end of the branch pathway to derive failure mode risk. These products are summed to find the total risk, but their separate derivation allows the relative contribution of each failure mode to be easily identified.

The fault tree in Figure 4-10b portrays the same failure process. It starts with a *top event,* which ordinarily represents failure. Failure could be proximately caused by the occurrence of either failure mode previously represented by the two branch pathways containing event sets $IR_1R_2R_3$ and $IR_1R_2R_4$. These individually become known as the *basic events.* In general, a fault tree implicitly attributes the same consequences to each failure mode it contains, without portraying them directly. Computing the probability of the failure top event would adopt the same arithmetic operations as before, but if the consequences of the two failure modes were different then separate fault trees would be needed to reflect their respective top events.

Comparing their development shows that event trees use a "bottom up" or "forward logic" construction beginning with the initiator and taking it to consequences. Each succeeding event in the decomposed failure sequence is identified by asking what must happen next for the failure process to progress. This captures how failure sequences are internally visualized by expressing the logical order, progression in time, or physical dependence of their events. This is a natural adjunct to judgment and can be seen as merely its formal representation. By contrast, the activity of fault tree construction is more one of categorization and classification using a "top down" or "backward logic" process. Starting from the top event, each

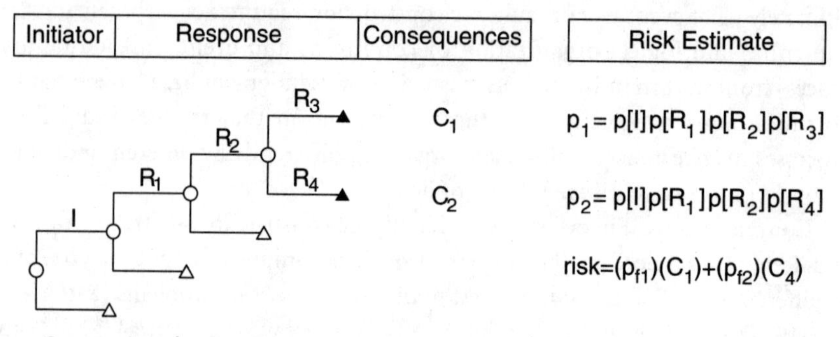

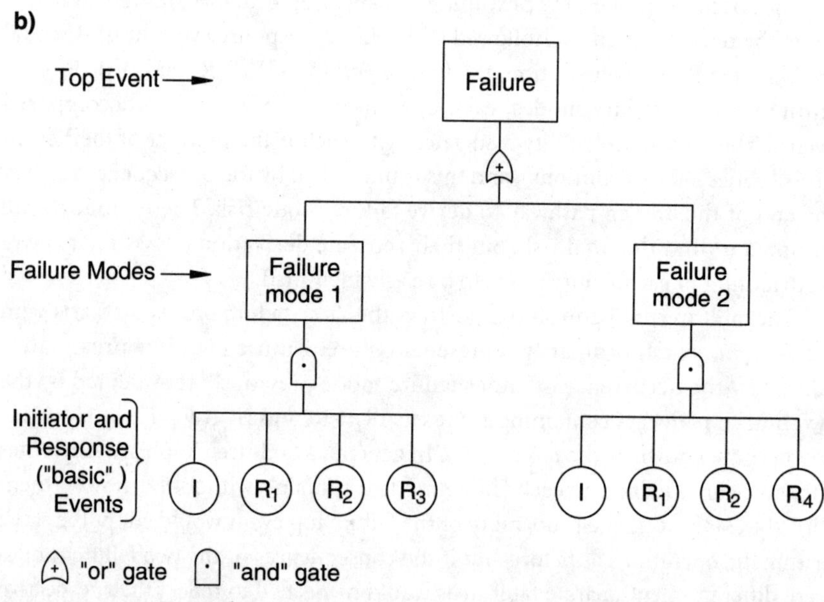

FIGURE 4-10. *Example event tree and fault tree formats: a, event tree; b, fault tree.*

set of underlying events is identified through a search for possible causes. This can be better suited to evaluating complex systems with large numbers of physical components whose interconnections and interactions make system operation intrinsically hard to understand. The great strength of fault trees lies in the symbolic logic of their *gates* (the "and" and "or" symbols in Figure 4-10b). This allows computations to be made using *Boolean algebra* for failure processes far more complex and interactive than any expanding thicket of event tree branches could ever handle, particularly for complex mechanical or electrical systems.

Because they mimic the visualization of failure progression embodied in diagnostic judgment, event trees are often the method of choice for geotechnical risk analysis applications. Some use fault trees, mostly for reasons of familiarity or convenience where event trees might otherwise become unmanageably large. Either way, fault trees and event trees provide the fundamental tools for portraying how failure processes are conceptualized and failure sequences decomposed.

Solution Techniques

Equally as in reliability analysis, the component probabilities in risk analysis must be mathematically manipulated and combined. Because many of the component events in risk analysis usually have binary outcomes, *discrete* or single-valued probabilities rather than continuous probability distributions on quantities often dominate. Certainly there is nothing to prevent the use of continuous distributions for parameters that may take multiple values, which would require the same solution techniques previously reviewed for reliability analysis. It is usually more convenient and sufficiently accurate, however, to separate the range of possible values into increments and assign discrete probabilities to each that require only simple arithmetic to propagate.

Decision Criteria

To this point we have reviewed the probabilistic components of risk analysis and how they fit together in its framework. The result is a probability of failure that reflects various failure processes and the variety of uncertainties affecting them. So presumably this value can be taken as some reasonably comprehensive, if approximate, descriptor of failure likelihood, which when multiplied by consequences becomes a quantified measure of risk. The question is now what to do with this number, which is not nearly so obvious as it might appear.

There once was a day, now lost in the fog of time, when some engineers got together—nobody really knows who—and decided what the minimum factor of safety should be for certain structures and analysis methods. These minimum values, now having become embedded in various codes and standards of practice, accounted for both uncertainty and failure consequences, albeit in an informal and unstructured way. It would seem then a most natural outgrowth of quantifying risk probabilistically to do exactly the same thing by establishing some limiting value of risk or some maximum failure probability. A recurring theme has been that the subjective and objective views of probability are mutually complementary. So it is ironic indeed that such an innocuous exercise should put them on a collision course.

There is a view which holds that the calculation of risk is inseparable from some decision regarding what, if anything, to do about it. This may be self-evident by simple inspection. We might be reluctant to live with a 10% chance of a major dam failing, where one-in-a-million might be more to our liking, so we would decide to reduce this somehow. But the same probability might be entirely suitable for a temporary excavation with very different failure consequences, so here we might decide to do nothing more. Probabilities and consequences are often weighed informally, along with other factors like the feasibility or cost of mitigation. But this requires judgment, it is not rigorous, and it leaves the door open for inconsistency among decisions and decisionmakers. So some have looked to the establishment of a uniform *decision rule*. This might consist of some maximum value for failure probability, with consequences implied or inferred, or it might adopt the criterion of risk. Various such criteria have been proposed to distinguish adequate safety from inadequate. Such a discriminant has been variously called an "accepted," "acceptable," or "tolerable" risk magnitude, each with its own nuances.

Figure 4-11a shows one example in the form of a plot containing the two dimensions of risk—probability of failure and failure consequences—on logarithmically scaled axes. Consequences could be of any type, say dollar costs incurred, black-footed ferrets endangered, or lives lost, but the latter is of greatest interest for failures that could affect public safety. The number of lives lost is commonly designated "N" and probability as frequency "f," thus the designation "f-N" plot. From the computed failure probability and quantified failure consequences a point results, with the structure determined to be adequately safe or not depending on which side of the boundary it lies.

Just how such a boundary should be established, precisely where it should be, and for that matter whether there can realistically be one at all have been the topics of much discussion, with undoubtedly more having been written on this and the related matter of risk perception than on the conduct of risk analysis itself. Suffice it to say that if public safety is at issue, then somehow this determination must come from society. But society has always been reluctant to oblige in any overt way. It is not the intent here to enter broadly into this area, except to note that there have been a variety of attempts through various means to infer such a risk boundary indirectly by looking to the risk magnitudes that society manifestly accepts, or *revealed preferences*. Some have proposed a straight-line boundary with constant slope, in effect a constant value of maximum risk as the product of probability and consequences. Others contend that societal attitudes toward risk depend not only on its magnitude but also on the level of consequences involved in individual events—the example often cited of the plane crash with 300 victims that receives newspaper headlines, with the same 300 fatalities in separate automobile accidents scarcely mentioned.[2] In some f-N plots, this inferred *societal risk*

[2]This implication of media coverage has not gone unquestioned, with some studies (Combs and Slovik, 1979; Greenberg et al., 1989; Koné and Mullet, 1994) indicating that media reports are instrumental in producing these responses rather than passively mirroring them.

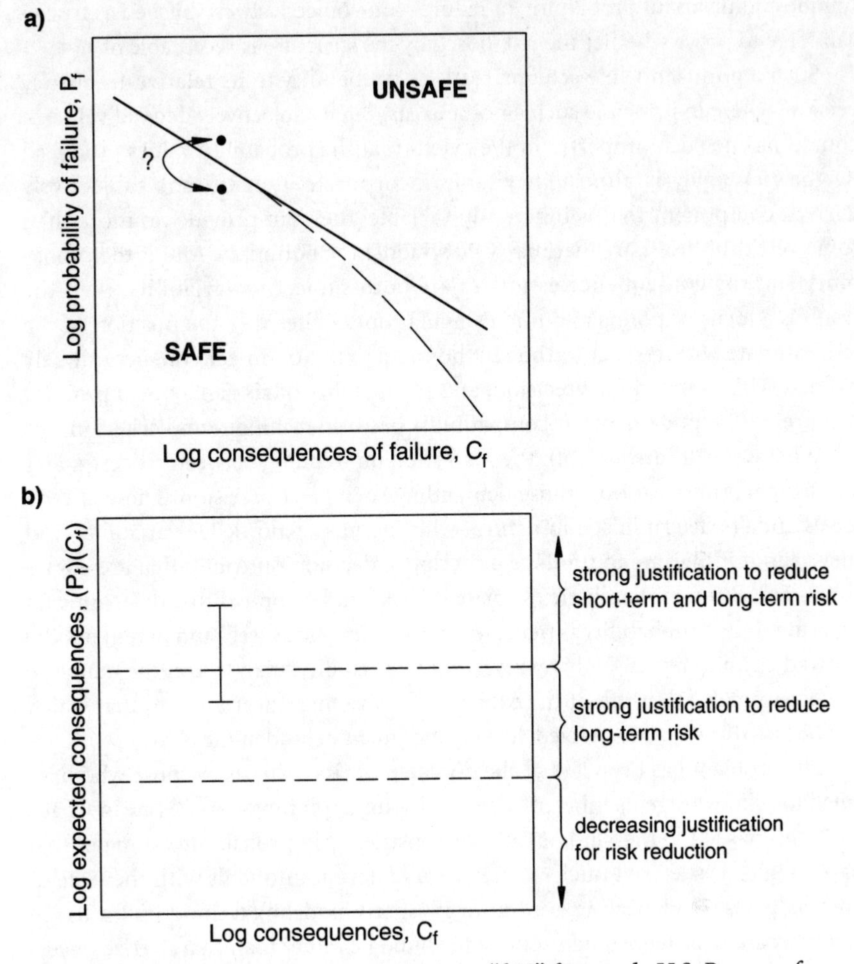

FIGURE 4-11. *Risk-based decision criteria: a, "f-N" format; b, U.S. Bureau of Reclamation format.*

aversion is reflected by a downward-sloping boundary that requires ever-lower probabilities of occurrence for higher-consequence events.

The marriage of risk analysis and these kinds of decision criteria has always seemed such a match made in heaven that few have seen through to the irreconcilable differences beneath. The underlying problem can be appreciated in terms of simple geometry. Wherever the risk boundary might lie and whatever its form, a line is always of infinitely narrow width. This means that an infinitely small point for risk must also be found in order for it to lie unequivocally on one side or the other, as an unambiguous determination of safety requires. In other words, the constituents of risk, both probability and consequences, must be determined precisely and uniquely if the boundary is to have prescriptive value. So only some unique and invariant probability of failure—one objectively established as true—can serve to prove whether the risk posed by the structure is acceptable or not.

Such a point-and-line schema restricts probability to its relative-frequency version, where in principle such a point exists. But a subjectively derived value, of course, has no such property. To the extent that the probability values produced by the risk analysis almost inevitably incorporate a significant subjectively derived component that is inherently variable, they can provide no more than some vertical "error bar" that may well straddle the boundary. And if the probability and the consequence estimates are both subject to variability, then the result is a locus of points and not a singular one. Either way, the position of the risk estimate with respect to the risk boundary cannot, in general, be uniquely defined. This demand for precision, and the inability of risk analysis to produce it, represents a fundamental incompatibility between technique and criterion.

What we really are left with then is a shotgun wedding between risk estimates and formal criteria, at least those demanding such great precision. These criteria are essentially deterministic in nature—they do not acknowledge variability, and they cannot if they are to produce prescriptive decision outcomes that are always consistent. But a probabilistic estimate of risk can be applied to a deterministic criterion only if probability is made to be deterministic as well, and herein lies the contradiction in terms. Only by reverting to reliability analysis can the marriage be consummated, but this at the expense of restricting uncertainty to that which can be statistically characterized, leaving judgment behind at the altar.

The problem has been laid at the doorstep of risk analysis by those who have failed to acknowledge its inherently subjective underpinnings. As we saw in Chapter 2, this has led to the kind of calls for consistency in probability estimates that have long characterized nuclear safety. Some have found fault with the state of knowledge itself, advocating ever more advanced modeling techniques for use in reliability analysis that would presume to eliminate rather than characterize uncertainty. Others have devoted considerable effort to performing risk analyses, only to abandon the technique altogether after belatedly realizing what it can and cannot do. But this need not be the case, and there are those who have taken a more insightful approach in reconciling the capabilities of risk analysis with the requirements of decision criteria. One such example is shown in Figure 4-11b, developed by the U.S. Bureau of Reclamation (1997) for use in dam safety decisionmaking.

In this format, the expected consequences in terms of loss of life, or risk, are plotted against the absolute value of lives lost in order to convey the element of societal risk aversion.[3] What is most important here is not what this criterion

[3]"Expected" consequences and "expected" loss of life use the term in its connotation of mathematical expectation, or expected value. This can convey an unintended impression that these effects are anticipated or known in advance, which has led some to categorically reject the use of the term in reference to fatalities.

contains but what it does not. There is no unique line, no constant boundary, that rigidly separates "safe" from "unsafe" dams, but instead a relative scaling. The plot is divided into broad regions with purposefully indeterminate delineations measured in graduated degree, in this case degrees of justification for taking risk-reduction measures.

At the same time it avoids the incompatibility between criterion and technique, this format does something else. When risk analysis is coupled with hard-and-fast decision rules like that shown in Figure 4-11a, for all intents and purposes the risk analysis becomes the decision. The decision follows automatically and prescriptively from the risk analysis results. But risk analysis cannot make a decision, only a decisionmaker can, and it is only intended to assist the person or persons doing so. The great promise—but at the same time the great danger—of coupling risk estimates with risk criteria is to make the decision process mechanical. But the use of risk analysis, like anything else, requires interpretation and knowing what went into it. It requires determining whether the results make sense in the broader context of all the considerations involved. In short, it requires the exercise of judgment. The format of Figure 4-11b discourages using risk analysis alone as the singular source of objective truth and helps keep it from dictating the decision without considering other factors alongside it. And chief among these factors are various alternative measures available for reducing risk.

What acceptable-risk criteria themselves do not account for is that decisions about risk are made in many dimensions. No well-informed decision can come about without considering not just risk magnitude but at the same time what can be done about it, how feasible this is, and what it might entail. We all accept a certain degree of risk to our safety every time we drive a car. But there is an alternative for reducing this risk: a seat belt. It is readily available, it costs us little, and there is no doubt of its effectiveness. Some would not consider driving without it. It is simply a sensible decision to reduce risk where we can easily do so. So risk is not evaluated in a vacuum regardless of what its magnitude might be, and we do not decide to accept it or not before more fully exploring our options. Thus, risk-based decisionmaking requires considering risk-reduction alternatives in tandem with the magnitude of the risk itself. And this, it turns out, is exactly what decision analysis is intended to do.

Decision Analysis

Decision analysis is a probabilistic technique that bundles risk analysis and evaluation of risk-reduction alternatives. It internalizes the decision rule and compares risk-reduction measures through a modest extension to the risk analysis techniques already described. Decision analysis adopts no absolute standard for risk. What it does do is compare the effectiveness of various risk-reduction measures, with the result being to identify that option which achieves the greatest risk

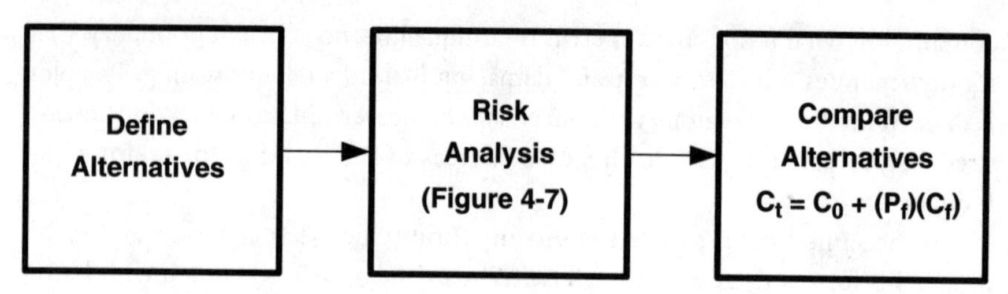

FIGURE 4-12. *Components of decision analysis.*

reduction at least cost. Correspondingly, decision analysis is customarily used for business decisions where both consequences and the costs of implementing the alternatives have the common denominator of economic value.

These extended components of decision analysis are depicted in Figure 4-12. First, a suite of alternative measures for risk reduction is defined, any of which could reduce either the probability of failure or its consequences in some way. These alternatives will be effective to varying degrees and may incorporate uncertainties of their own, and none of them need eliminate risk entirely. They might include performing additional studies that could reduce uncertainties in the initiator or in soil properties; or structural measures that could improve the response to imposed loadings; or monitoring that might allow intervention before failure or its consequences could occur. For each such alternative, the elements of conventional risk analysis, including any of its related probabilistic methods, are then applied. Where consequences can be expressed as economic losses, the risk associated with each alternative is its failure probability multiplied by failure cost, or $(p_f)(C_f)$ in Figure 4-12. This is the *expected cost* or *risk cost*, which is then added to the cost C_0 incurred in implementing the alternative to find its total cost C_t. The embedded decision rule is to minimize total cost, allowing the lowest total-cost alternative to be identified, including among them the "do-nothing" or *null alternative.*

A *decision tree* is a slightly modified version of the event tree used for portraying event sequences in risk analysis. As shown in Figure 4-13, it begins with a *decision node* that represents the choice between alternatives A_1 and A_2, with the remaining uncertainties represented in the same fashion as the event tree of Figure 4-10a but with component event probabilities and consequence values that reflect the risk-reduction measures they incorporate. The result is essentially two subset event trees, one for each alternative.

With this, it can be seen how decision analysis adapts risk analysis techniques to bottom-line business decisions. Still, this has much to do with how their economic consequences are valued. In Figure 4-13, assume that A_1 represents the null alternative and A_2 some added design feature that improves response. Then A_2 can be viewed as an insurance policy and the cost of constructing it the premium. Suppose that the failure costs in both cases were in the hundreds of dol-

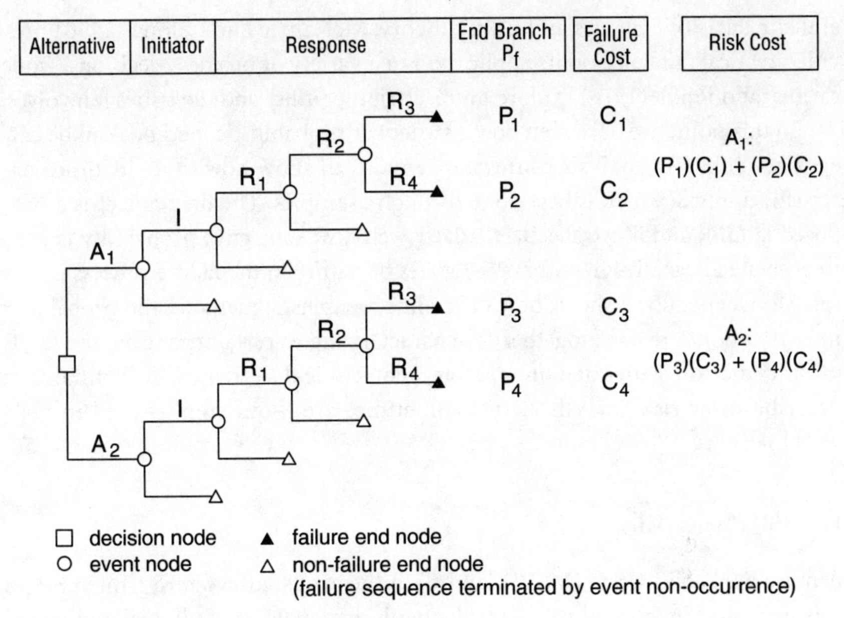

FIGURE 4-13. *Example decision tree format.*

lars. Many people would not insure against losses this small and would pay little for the added protection that A_2 affords, choosing to accept A_1's risk cost instead. However, if failure costs were measured in the millions the worthwhile premium for A_2 over A_1 might be proportionately much greater. This is called *monetary risk aversion*, which expresses the notion that most people are more averse to losing large amounts than small ones under equivalent probabilities of loss, and it can affect the use of the *expected monetary value* (or EMV) criterion in decision analysis accordingly. To account for this effect, EMV can be replaced by a *utility function* that expresses monetary risk aversion. It is derived from the decision-maker's preferences with gambles involving various sums, a concept that goes all the way back to Daniel and Nicolaus Bernoulli. This, of course, assumes that a single decisionmaker can be found, which may not be easy considering how decisions actually come about within the structure of many organizations.

Beyond monetary risk aversion, nonmonetary consequences can result from failures—things like damage to reputation, environmental effects, or loss of life—that cannot readily be expressed in dollars. These problems involve *multi-attribute* utility, and these utility functions can, in theory, be assessed by evaluating a decisionmaker's or even society's preferences and value tradeoffs. However attractive this might be, there has always been widespread reluctance to associate public-safety consequences with dollar value, so decision analysis is usually reserved for situations that do not involve safety-related consequences. Benjamin and Cornell (1970) present comprehensive descriptions of the workings of probabilistic decision analysis and utility theory, McNamee and Celona (1990) provide practical guidance on its application to a variety of business decisions, while Keeney and Raiffa (1976) explore multi-attribute utility and the issues it involves.

To this point, we have seen how a variety of probabilistic methods can be categorized in a risk analysis context. It remains to show how these methods are actually applied, which is best done through examples. The first concerns a proposed landfill and illustrates particularly well how subjective probability is used to evaluate alternative hypotheses or states of nature in the failure process initiator. More generally, it shows how different probabilistic methods and probability interpretations are used together for characterizing uncertainties using the kinds of information generated in routine geotechnical activities. And finally, it describes how risk analysis results sometimes are—and aren't—used in real-world settings.

Landfill Case History

A municipal solid waste (MSW) landfill in the arid southwestern United States was proposed in a broad valley infilled with over 3000 m of alluvial and lacustrine sediments. The alluvial aquifer had until recently supplied agricultural irrigation for many years, resulting in considerable drawdown of the water table. In such regions of groundwater overdraft, broad areal subsidence can result from compression of basin sediments and even small vertical strains can produce considerable total compression over the great sediment depths. This settlement produces horizontal tensile strains at the ground surface, especially over abrupt bedrock irregularities. The net result can be what are called *earth fissures*. They begin as subsurface tensile inclusions that produce a buried vertical crack in the soil. Over time, this appears at the surface as a thin opening about a centimeter wide that can extend for hundreds of meters. As infiltration occurs, the crack walls begin to soften, slake, and collapse at depth, and the surface opening widens as the surficial soils drop to fill the void. If the enlarged crack captures surface drainage, erosion ensues to form lineations visible on airphotos.

Such was the case in the valley of the proposed landfill, where earth fissures had occurred near the site and elsewhere. Naturally, this was cause for some concern. The landfill was to be developed sequentially in modules, each underlain by a geomembrane liner. Each module would be filled with MSW over a brief period, then covered with an impermeable cap. But should an earth fissure develop beneath an operating module and rupture its liner, leachate resulting from direct precipitation on the exposed MSW surface could find its way to groundwater and from there offsite. New earth fissures were considered unlikely to form because irrigation had ceased in recent years and groundwater declines were diminishing. But the possibility could not be altogether ruled out, especially in light of another peculiar feature.

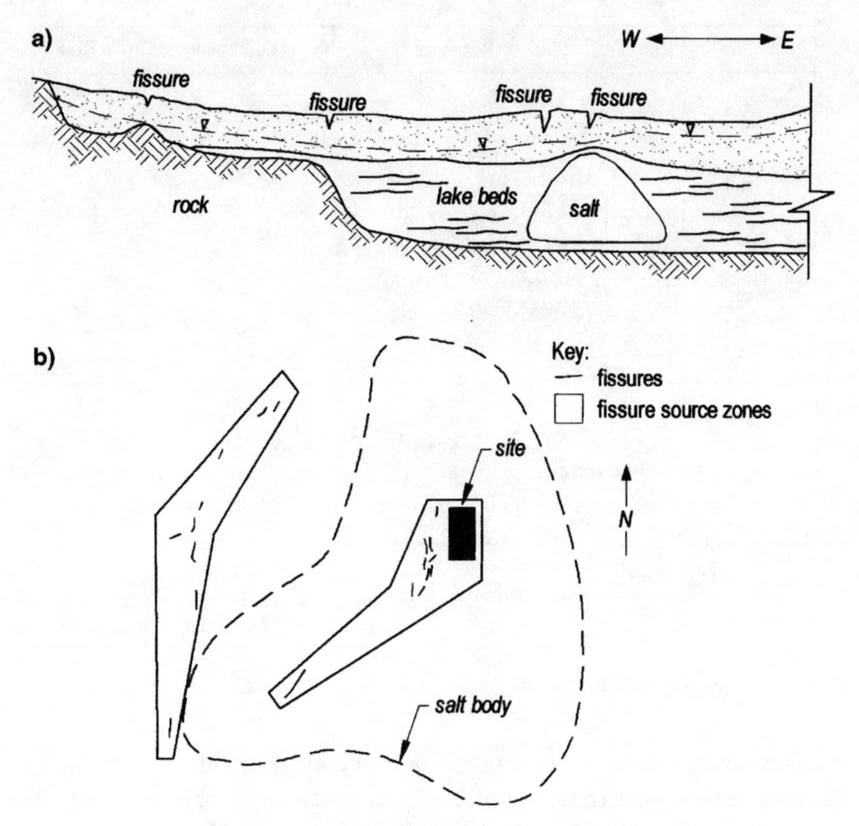

FIGURE 4-14. *Earth fissure example: a, schematic valley profile; b, plan of site area.*

Figure 4-14 shows that the site was situated over the flank of a dome-like salt body originally derived from evaporite deposition in an ancient saline lake. Earth fissures had developed over bedrock irregularities to the west (Fig. 4-14a), where pumping had reduced groundwater levels by nearly 100 m. To the east, the reduction had been only 60 m, and no fissures had ever been observed on this side. The salt body therefore might be acting like any other bedrock nonuniformity in promoting earth fissures from the induced tensile strains, and the group of fissures just west of the salt dome (Fig. 4-14b) had almost certainly formed in this way. But salt is not mechanically stable and it creeps. What once started out as a tabular deposit of evaporites had pushed its way upward in the classic mechanism of salt dome formation, which too has been known to cause fissure-like features. Neither is it all that slow, and active creep of the salt body beneath the site was evidenced by broken well casings.

So groundwater withdrawal and salt creep were two competing geologic hypotheses for how fissures might form beneath the site. Groundwater reduction was the leading candidate, even though it was unlikely to continue and had pro-

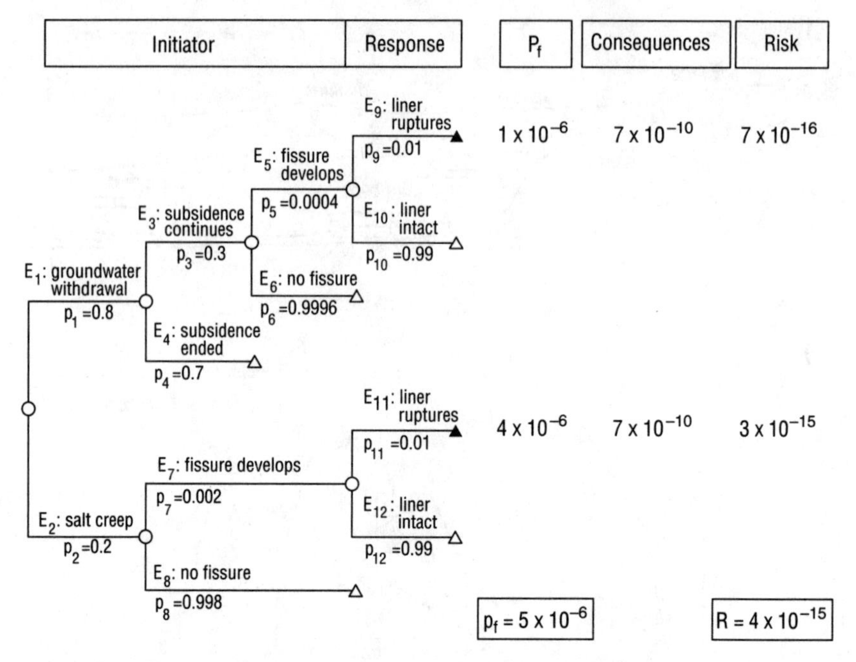

FIGURE 4-15. *Landfill event tree.*

duced no evident fissures on the side of the salt dome where the site was located. But geologic evidence could not preclude the possibility in the future, nor that salt creep would not produce similar effects. If geologic evidence was inconclusive, then it boiled down to a matter of likelihood, and quantifying it became the task of a risk analysis.

Figure 4-15 provides its structure in event-tree format, with the familiar initiator, response, and consequence components. The failure process initiates with the development of an earth fissure beneath one of the operational landfill modules by either geologic mechanism. The response component reflects whether the geomembrane liner would rupture in the event of fissure occurrence, allowing leachate to escape. Here, failure is defined as the occurrence of liner leakage, and its consequences are expressed in terms of the potential health effects on someone who might consume leachate-contaminated groundwater. Reviewing the information and evidence step-by-step for each component event will help illustrate how the associated probabilities are formulated and the probabilistic methods behind them.

Initiator

Events E_1 and E_2 in Figure 4-15 are the two mechanisms that might produce fissures beneath the site. Evidence favoring the groundwater withdrawal hypothesis was more persuasive and consistent with similar circumstances of fissure formation at other locations in the valley. Accordingly, the geologists subjectively assigned $p_1 = 0.8$ to reflect their degree of confidence (in other words, they were 80% sure) in groundwater withdrawal as the potential fissure cause beneath the site (event E_1), with the complementary $p_2 = 0.2$ for salt creep mechanisms (event E_2).

Continuing along the upper branch pathway for groundwater withdrawal, future fissures may form under this hypothesis only if subsidence continues to occur. Even though pumping had been discontinued, continuing primary or secondary compression might still produce some future subsidence. The geologists were reasonably sure that significant subsidence beneath the site vicinity had ended (event E_4), but slightly less so than for event E_1, so they assigned a subjective probability $p_4 = 0.7$ to it, and correspondingly $p_3 = 0.3$ to the complementary event E_3.

Damage from continuing subsidence would require that fissures form beneath a landfill module during its limited operational life. Assuming that fissure development occurs in the future at the same rate it has in the past, it can be modeled as a stationary stochastic process using the binomial theorem as described in the Appendix. The probability (P_m) that at least one fissure develops beneath a landfill module during its operating life t is:

$$P_m = 1 - (1 - p)^t \tag{4-2}$$

where p is the probability of fissure occurrence within the module area per unit time.

The fissures in Figure 4-14b were mapped on aerial photographs first taken when their development began, then again some 30 years later. To determine their occurrence rate per unit area and time, two regions encompassing them were constructed using their spatial distribution, salt body geometry, and depths of underlying alluvium and groundwater, much like delineation of earthquake *source zones* in probabilistic seismic hazard analysis. Fissure occurrence rates per unit area for the two regions were found to be nearly identical and after adjusting for landfill module area are easily used to find the probability of fissure development during the active module life. For the salt creep hypothesis, this value is used directly for event E_7: The past fissure occurrences are taken to reflect what will happen beneath the site in the future. The corresponding p_7 of 0.002 is purely frequency based and comes from this stochastic model of fissure occurrence.

For the groundwater withdrawal hypothesis, even if subsidence beneath the site should continue in the future, its rate and magnitude would be less than in the past because of the termination of groundwater pumping and declining consolidation rates of basin sediments. In this case, the same stochastic process rate of the past would not directly pertain to future occurrences, and with these factors in mind the probability p_5 for event E_5 was estimated to be five times lower than p_7, or 0.0004. This is the normalized-frequency approach, applied here by subjectively adjusting fissure frequency from the stochastic model.

Response

The response element in Figure 4-15 consists of the liner-rupture component event. Its probability was estimated with the aid of laboratory testing on liner materials and a geomechanical model of liner behavior over an unsupported opening. Some 40 uniaxial tension tests were performed on geomembrane materials from several manufacturers to evaluate yield tension, tensile stiffness, and anisotropy, with values adjusted for viscoelastic effects used in a soil-arching/tensioned-membrane model. For an unsupported liner spanning a 1-cm-wide fissure, the calculated factor of safety against long-term rupture was in excess of 100. Alternatively, the liner was predicted to be capable of sustaining a theoretical fissure width of some 40 cm.

In principle, input parameter variability in liner properties could have been statistically characterized and used in a reliability analysis based on the arching model. However, its results were not especially sensitive to these values and other uncertainties in the analysis like geomembrane creep were more important. So instead, a subjective probability $p_9 = 0.01$ was assigned to event E_9 for liner rupture from fissure occurrence due to groundwater withdrawal, taking into account the results of the model and the factors affecting it. For a similar fissure width formed by salt creep the same value was assigned to p_{11} for event E_{11}.

Consequences

The result of liner rupture beneath an operating landfill module would be to allow some amount of leachate to enter the alluvium, travel through the vadose zone, and reach the groundwater. This could produce adverse health effects if someone were to drink enough of this groundwater over a sufficient period of time. Therefore, the consequences of failure were defined here as an excess cancer (over and above that which might occur from unrelated causes) to an exposed individual consuming groundwater from just outside the site boundary. This is one of the rare cases where a model for consequence prediction exists, here an algorithm produced by the U.S. Environmental Protection Agency (EPA). This algorithm is probabilistic, so consequences are expressed in Figure 4-15 as the probability of adverse health effects. But before proceeding further, we must digress.

Risk analysis has another meaning from how it is presented here. This pertains to the fields of toxicology, health physics, and industrial hygiene, where the entire approach is different. As applied in these areas, risk analysis (termed "risk assessment") treats only the adverse consequences on human health of chemical substances, mainly their carcinogenicity, mutagenicity, or neurotoxic effects. Thus within the framework of Figure 4-7, such a *health risk assessment* resides entirely within the consequence component. It does not deal with the initiator or response aspects of any failure process, and it begins by presupposing that some chemical compound has been released. Van Zyl (1987) further explains these differences and the most unfortunate confusion in terminology that results, a matter of some importance since health risk assessment figures so prominently in contaminated site cleanup activities in which engineers participate.

EPA has rigorous protocols and regulatory requirements for health risk assessment with a vernacular all their own. Probability enters only in specifically circumscribed and generally limited ways (Barry, 1989; U.S. EPA, 1992). In this respect, health risk assessment contains the following stepwise elements:

1. *Hazard identification* establishes that a released substance has the potential to cause various toxicity effects.
2. *Exposure assessment* identifies the pathways by which individuals or various human populations (including children or others especially susceptible) might be exposed to a toxic substance. One such pathway can be groundwater, and here geochemical and geohydrologic modeling is used in support of exposure assessment. Ultimately, it predicts the dose that might be ingested and by whom.
3. *Dose–response assessment* describes the incidence of adverse health effects as a function of dose received. It is derived from toxicology studies on laboratory animals where, for example, lc_{50} is the concentration lethal to 50% of them. A frequency distribution is established accordingly and extrapolated to human health effects.
4. *Risk characterization* combines the dose–response relationship and exposure data to produce the probability of adverse health effects for an individual or their incidence in a population.

Even to the engineer, it is plain that substantial uncertainties are involved in every step. Their probabilistic quantification is acknowledged but not particularly encouraged, and they are seldom comprehensively or systematically propagated through the analysis. Health risk assessment is fundamentally a deterministic exercise. It conforms to the paradigm of science, where "good science" is based only on "hard data." Thus judgment, if applied at all, is not explicitly incorporated, global uncertainty is seldom quantified, and subjective probability hardly ever used. The only unavoidably probabilistic feature of health risk assessment is the frequency-based dose–response relationship derived from the animal experiments.

Returning then to the landfill case, the empirically based health risk algorithm specified by EPA was used to estimate failure consequences. It uses as input the leachate release rate and groundwater flux as a measure of contaminant dilution, dispersion, and degradation during travel to the site boundary. There, a person is presumed to consume this groundwater over a lifetime, and this exposure is combined with a dose–response relationship for constituents typically found in

MSW leachate. The result in Figure 4-15 is the consequence probability of 7×10^{-10} that an excess cancer could result.

Outcome

Figure 4-15 displays the end-branch failure probabilities for groundwater withdrawal and salt creep fissure mechanisms. The sum of the two is p_f for fissure-induced liner rupture, or 5×10^{-6}. The consequence probability is multiplied by the end-branch failure probabilities and the two products summed to find the estimated risk R of 4×10^{-15} of an excess cancer to an exposed individual as a result of fissure occurrence. A closer look at the numbers shows that while salt creep had been considered the less likely initiator, its risk contribution actually exceeds that from groundwater withdrawal. This is because of the reduced subsidence and fissure occurrence rates assigned to the groundwater hypothesis, illustrating how the risk analysis better distinguished their relative importance.

The landfill example also illustrates how the risk analysis framework accommodates a variety of probabilistic methods. Subjective probability was used for evaluating alternative geologic hypotheses, and later for the response element where a geomechanical model was used. For one of these hypotheses, a frequency-based stochastic model was used directly to characterize the initiator, while normalized frequency was adopted for the other. Finally, the health risk algorithm used for consequence evaluation is essentially a reliability approach that statistically characterizes one of its input parameters. Each of these methods was adapted to the kinds and levels of information typically available in many geologic and geotechnical studies without any especially sophisticated statistical data-gathering activities. The landfill risk analysis is a simple case, and many refinements can be easily imagined. And while the earth-fissure failure mode was of particular interest, it is one of many ways a liner system might fail, and other failure modes could be incorporated in an expanded event tree.

It is interesting to learn what came of all this. The calculated risk of adverse health effects was minuscule by any measure, and the consequence probability—much more than failure probability—is what drives this outcome. In fact, EPA adopts a broad probabilistic criterion, a "health risk target," for evaluating the health effects of groundwater contamination by landfill leachate, which in this case would be in the neighborhood of a 10^{-4} to 10^{-7} probability of an excess cancer. This criterion would be easily satisfied by the consequence probability alone even if the probability of liner failure were 1.0, or for that matter if the liner were not present at all. The reason for this is the minimal infiltration of precipitation in the arid site climate that could produce so little leachate from the exposed MSW in the first place. The landfill designers, of course, had recognized this, and the geomembrane had only been provided as a gratuitous redundancy. What could it matter then if it did rupture? The exceedingly low numbers calculated in the risk analysis, it seemed, had made the entire question of earth fissures go away.

The regulator, though, saw things a bit differently, this being a part of the job description. He dutifully sat through a presentation of the risk analysis, and he earnestly examined the results. However, the calculated risk numbers were so incomprehensibly small, with so much of them within the black box of the consequence algorithm, that the regulator remained unpersuaded that the earth fissure matter was really so benign, much to the risk analyst's chagrin.[4] He would be prepared to issue the necessary permit, but only if a geogrid were added beneath the liner for supplemental reinforcement. No calculations showed the need for this, and certainly the risk analysis didn't, but the matter was settled and that was that.

Although the regulator's decision might seem somehow illogical, this is not necessarily so. He was evidently assigning some credence to the risk analysis by virtue of having allowed the project to proceed at all. And even if its numbers meant little to him, it did demonstrate that the earth-fissure question had been systematically considered. But he was accounting for additional factors and weighing them in a different light. In his eyes, the liner had been provided for a reason, whatever that reason might be. Most landfills have liner systems of some kind, and it was only reasonable to expect that one subject to earth fissures should be something more. Besides, an alternative—the geogrid—was readily available for reducing the fissure-induced risk, no matter how small. Here there was nothing wrong with the risk analysis or with the regulator's decision. It's just that decisions are about more than risk analysis, and risk analysis is about more than the calculated results. Risk analysis above all is about judgment, not just incorporating it but enhancing it as well. This comes about because risk analysis, properly structured, yields insight into the relative importance of the things that contribute to risk. Which brings us to the next example.

Duncan Dam Case History

As dam foundations go, Duncan Dam's was difficult from the very beginning. With sands, gravels, silts, and clays to depths of up to 380 m, it was both pervious and compressible, and as would later be found, susceptible to seismic liquefaction. As described by Gordon and Duguid (1970) and shown in Figure 4-16a, the design incorporated a partially penetrating slurry trench cutoff and downstream relief wells to accommodate the anticipated foundation underseepage through the pervious sand and gravel units. Exceptionally flat embankment slopes would be needed to maintain stability during construction, and extraordinary foundation settlement in excess of 4 m was predicted from consolidation of the soft foundation silt and clay layers shown in Figure 4-16b.

---

[4]Actually, he said it all seemed like "smoke and mirrors" to him.

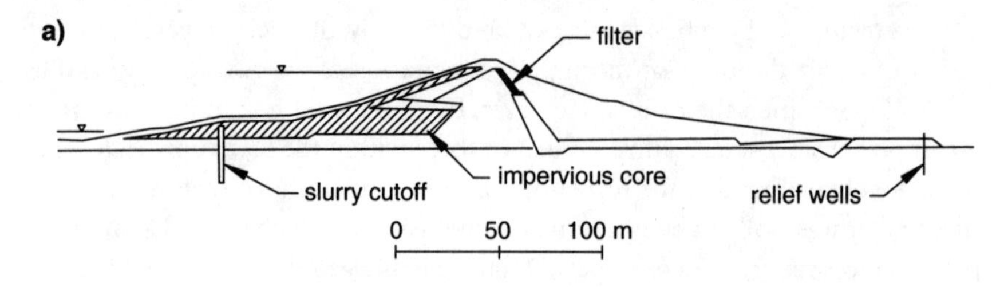

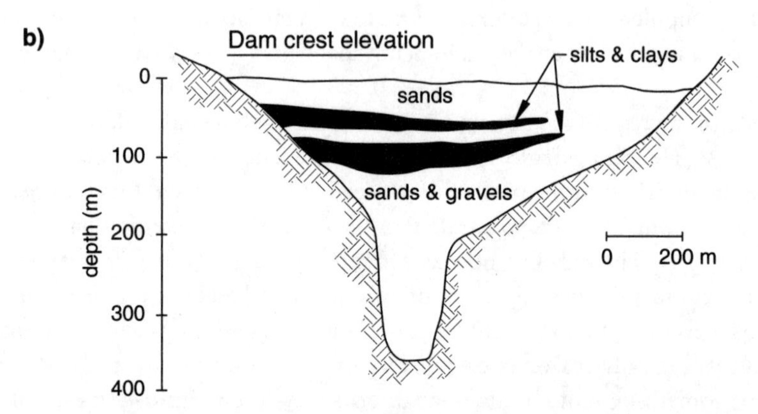

FIGURE 4-16. *Duncan Dam: a, embankment cross-section; b, foundation profile.*

What was not expected was the nonuniformity of this settlement that developed even while construction was still under way. It was concentrated at one abutment, reaching 4.4 m over a distance of only 120 m, to produce magnitudes of differential settlement unprecedented for any dam at the time. This produced some serious transverse cracking of the core, requiring redesign in the field. The result was complex internal zoning with a thin upstream core section (Fig. 4-16a). Still vulnerable, however, to renewed cracking from continuing settlement, the core had to be well protected against internal erosion and particle transport. The filter zone would therefore be the critical element, although with an intricate geometry it would be as little as 1.0 m wide in places. Within it would be a complicated system of finger drains and culverts able to carry off large quantities of seepage that cracking of the thin core could allow.

Seismic Analyses

In the 1990s, attention turned to the potential for liquefaction of the loose to medium-dense foundation sands, particularly those beneath the downstream shell. The seismicity of the site was not especially high, with peak horizontal ground acceleration (PGA) of 0.12g and magnitude (M) of 6.5 assigned to the Maximum Credible Earthquake (MCE), but simplified analyses indicated that liquefaction triggering and downstream flowsliding could occur under these ground motions. Due to the great depths involved, however, these procedures required extrapolation beyond their empirical database. Supplementing them required innovative subsurface sampling techniques and an uncommon degree of analytical refinement.

As summarized by Byrne, et al. (1993), this advanced seismic assessment included elaborate ground-freezing techniques for obtaining undisturbed samples of the cohesionless foundation sands. Sophisticated laboratory testing was used to measure the effects of confining stress and static stress anisotropy (K_σ and K_α) on cyclic shear resistance. Post-liquefaction residual strength and dynamic soil properties as a function of in situ void ratio were also determined, then used in two-dimensional dynamic response and deformation analyses. The end result was that the potential for liquefaction-induced flowsliding was judged to be "negligible" (Salgado and Pillai, 1993).

Detailed as they were, the analyses had accounted for only one possible mode of seismic failure—downstream flowsliding. Another was a related possibility for the upstream portion of the dam. Also, the analyses had predicted that seismic deformation of the dam crest would not be sufficient to allow overtopping by the reservoir, but the internal effects had not been clearly established. Especially the filter and associated drainage features, narrow to begin with and possibly further affected by postconstruction settlement, might become more distorted or even disrupted by any additional seismic displacements. This raised the possibility of internal erosion should their integrity be compromised. The slurry cutoff, relief wells, and portions of the core were also sensitive to displacement, and resulting increase in seepage might affect stability of the downstream slope.

Yet another potential failure mode had to do with volume change and settlement (sometimes called "seismic compaction") of the cohesionless foundation soils that would accompany earthquake shaking, whether they liquefied or not (Tokimatsu and Seed, 1987). The deep and irregular foundation profile meant that even small strains could produce substantial differential settlement. Indeed this was the same problem that had occurred during construction, and it too could affect the thin core, filter, and drainage elements, thus promoting internal erosion. None of these additional failure modes had been addressed in the previous seismic analyses, and they became the basis for a risk analysis to evaluate their effects.

Risk Analysis

Procedural aspects of the Duncan Dam risk analysis as described by Vick and Stewart (1996) included initiator and response components in event-tree format. Failure consequences were not required for relative comparison of failure modes because they would be comparable for any manner of seismic failure, allowing failure probability to stand as a proxy for risk.

The initiator, here the occurrence of various levels of earthquake ground motion, is described by the seismic hazard curve in Figure 4-17 showing

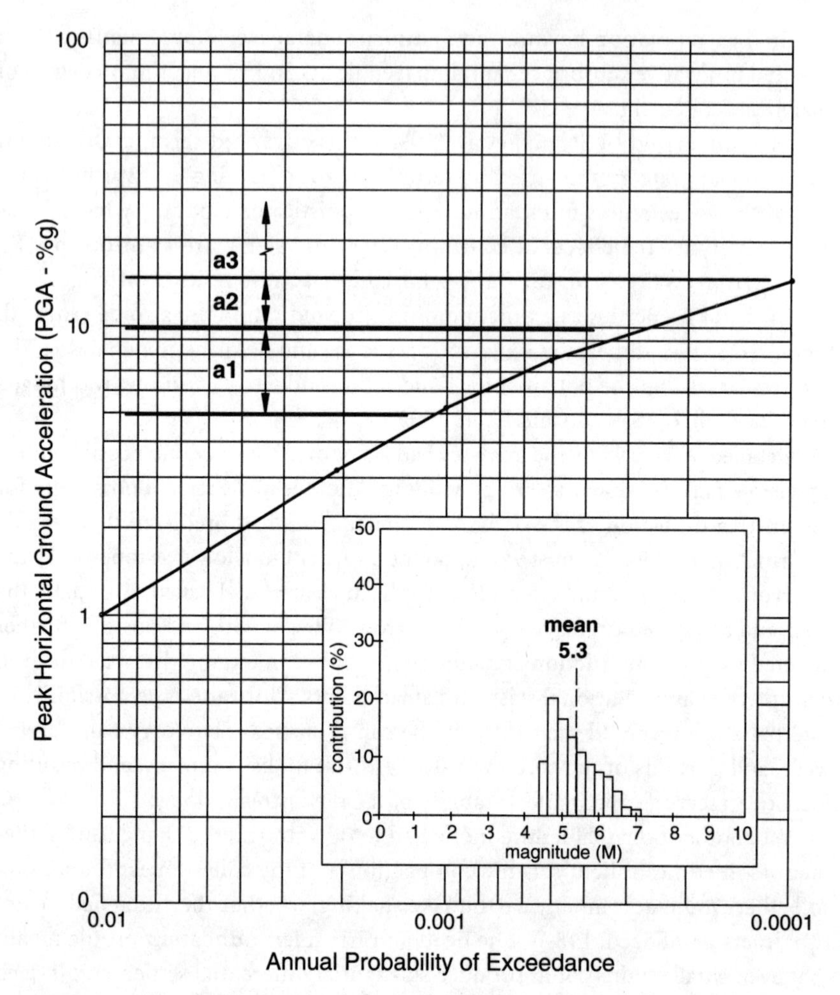

FIGURE 4-17. *Duncan Dam seismic hazard curve.*

exceedance probabilities for various PGA levels. Earthquake magnitude contributions to PGA are also shown, which were essentially the same for all PGA levels. For risk analysis purposes, the continuous seismic hazard curve was discretized into three PGA increments: a_1 (0.05–0.10g), a_2 (0.10–0.15g), and a_3 (> 0.15g). The hazard curve was truncated at 0.05g, with PGA less than this being judged incapable of initiating any failure sequence.[5] The response of the dam to these

[5]Figure 4-17 shows that PGA can be greater than the MCE value of 0.12g at low exceedance probabilities. This occurs when seismic hazard is controlled by so-called "floating" earthquakes whose location is not constrained by known faults. Deterministic MCE procedures place some arbitrary limit on the nearest site–source distance, and therefore PGA, while probabilistic procedures do not.

TABLE 4-1. *Duncan Dam Seismic Failure Modes*

Designation	*Description*
m_1	Foundation liquefaction beneath the downstream shell, flowsliding, and overtopping
m_2	Foundation liquefaction beneath the upstream shell, flowsliding, and overtopping
m_3	Embankment deformations, disruption of filter integrity, and internal erosion
m_4	Foundation settlement, disruption of filter integrity, and internal erosion
m_5	Deformation and loss of core, cutoff, or relief well integrity; excessive seepage or piezometric pressures; and instability of the downstream slope

initiator levels centered around the five primary seismic failure modes identified (Table 4-1).

These failure modes were each decomposed into their respective failure sequences according to the event tree structure shown in Figure 4-18. Reservoir level plays an important role because only at or near the maximum level (Fig. 4-16a) can any of them propagate to breach of the dam. Operating records showed full-reservoir conditions only during about half of each annual operating cycle; otherwise it was effectively empty. Also, discharge capacity of the outlet facilities was capable of reducing the water level by up to almost 2 m/day in response to seismic damage. Both of these factors are incorporated in the event tree. Probabilities were assigned to each component event using the various probabilistic methods previously described, with many adopting subjective approaches.

Outcome

Table 4-2 provides probabilities calculated from the event tree, with explanations of the columns provided in the accompanying notes. The contributions of the three PGA increments to total seismic failure probability are identified separately, with those for each failure mode at each PGA level also shown. As expected, the conditional probability of failure increases with higher PGA (column 5). Also, each of the PGA increments represents a roughly similar proportion of the total seismic p_f (column 6), with about half of this total attributable to PGA greater than the MCE value used in the previous liquefaction flowslide analyses (0.12g) and half to lesser values.

But the most informative insights from the risk analysis, hence its diagnostic value, are contained in column 4 of Table 4-2 showing the individual failure mode contributions. Recall the conclusion from the earlier seismic analyses that there was negligible risk of downstream flowsliding. This corresponds to failure mode m_1, and by multiplying $p[m_1 \mid a_i]$ by $p[a_i]$ and summing the three products, the total probability of downstream flowsliding can be found to be 1.4 ×

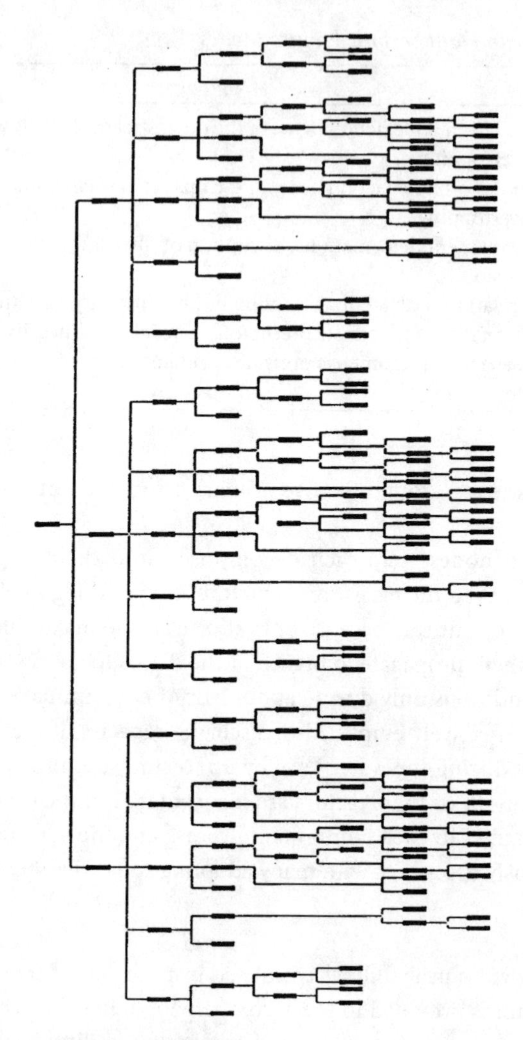

FIGURE 4-18. *Duncan Dam event tree.*

10^{-5}, a value not inconsistent with this conclusion. However, the risk analysis arrived at this conclusion for different reasons. Whereas the previous seismic analyses had been performed using an M6.5 earthquake corresponding to the MCE, Figure 4-17 shows that the actual magnitude contributions predicted by the seismic hazard analysis are generally much lower, with the mean M5.3 corresponding roughly to the minimum threshold for observed liquefaction occurrence (Ambraseys, 1988). So the risk analysis conclusions concerning downstream flowsliding are mostly attributable to the low corresponding likelihood of liquefaction triggering (i.e., a low initiator probability), an observation that

TABLE 4-2. *Duncan Dam Seismic Failure Probabilities*

(1) a_i (PGA)	(2) $p[a_i]$	(3) Failure mode, m_i	(4) $p[m_i \mid a_i]$	(5) $p[f \mid a_i]$	(6) $p[a_i]p[f \mid a_i]$
a_1 0.05–0.10g	0.00075	m_1: Downstream flowslide	0.0025	0.0362	0.000027
		m_2: Upstream flowslide	0.0050		
		m_3: Embankment deformation/ internal erosion	0.0020		
		m_4: Foundation settlement/internal erosion	0.0257		
		m_5: Deformation/seepage/slope instability	0.0010		
a_2 0.10–0.15g	0.00015	m_1: Downstream flowslide	0.0300	0.1078	0.000016
		m_2: Upstream flowslide	0.0150		
		m_3: Embankment deformation/internal erosion	0.0188		
		m_4: Foundation settlement/internal erosion	0.0416		
		m_5: Deformation/seepage/slope instability	0.0024		
a_3 >0.15g	0.00010	m_1: Downstream flowslide	0.0800	0.2381	0.000024
		m_2: Upstream flowslide	0.0400		
		m_3: Embankment deformation/internal erosion	0.0736		
		m_4: Foundation settlement/internal erosion	0.0408		
		m_5: Deformation/seepage/slope instability	0.0037		
p_f for all a_i and m_i					$\Sigma = 0.000067$/yr $= 7 \times 10^{-5}$/yr

Notes for table columns:

(1) For PGA increments given in Figure 4-17.
(2) Probability of occurrence of PGA within increment a_i, calculated as the difference of end-member exceedance probabilities from Figure 4-17.
(3) Failure modes as defined in Table 4-1.
(4) Conditional probability of failure mode m_i given PGA increment a_i, calculated from event tree component probabilities.
(5) Conditional probability of seismic failure given PGA increment a_i, calculated as the sum of $p[m_i \mid a_i]$ assuming probabilistic independence of failure modes.
(6) Incremental seismic failure probability for a_i, as the product of columns 2 and 5.

TABLE 4-3. *Seismic Failure Mode Contributions*

Failure mode	*Proportion of total p_f*
m_1: Downstream flowslide	21%
m_2: Upstream flowslide	15%
m_3: Embankment deformation/internal erosion	18%
m_4: Foundation settlement/internal erosion	44%
m_5: Deformation/seepage/slope instability	2%

might have modified the nature or emphasis of the detailed seismic analyses had the risk analysis preceded them. Instead of downstream flowsliding, the risk analysis indicated the principal seismic vulnerabilities of the dam to lie in other areas. This can be seen in Table 4-3, which sums the products of columns 2 and 4 in Table 4-2 separately for each failure mode, expressing their contribution to total seismic failure probability p_f as a percentage.

As shown in Table 4-3, foundation settlement (m_4) is the dominant failure mode, contributing nearly half of the total p_f and more than twice as much as any other, because seismic shaking could produce differential settlements of the same order as the filter dimensions in key areas. Related effects on the filter could arise from deformations induced in the embankment directly, as reflected by failure mode m_3 constituting 18% of the total. Taken together, these filter and deformation-related failure modes comprise nearly two thirds of the seismic risk.

By contrast, deformation-induced core cracking or related effects on the slurry cutoff and relief wells (failure mode m_5) is of minor importance, principally because the flat embankment slopes originally made necessary by the soft foundation clays—and the long seepage flowpaths that result—would reduce the effects of elevated seepage and pore pressures on downstream slope stability. Here especially, the effects of reservoir drawdown capability come into play, with good prospects that seepage effects could be mitigated well in advance of any instability.

In the end, the most important outgrowth of the Duncan Dam risk analysis was to show that it was the potential for internal erosion, not liquefaction-induced flowslides, that constituted the most serious seismic vulnerability of the dam. This pointed the way to further investigations of the condition of the filter, as well as to the seismic deformation characteristics of the dam and foundation. It also emphasized the importance of prompt action to reduce reservoir level in the event of seismic damage, which in turn highlighted the potential for rapid drawdown effects on stability of the upstream slope with its weak foundation clays. By showing the relative importance of issues like these, and identifying new ones that might not have been considered, risk analysis becomes not only an aid to judgment but an aid to analysis as well. It directs attention to those things which most need analysis and which don't, which kinds of information are most needed and where. In this respect, the risk analysis becomes a blueprint for engineering analysis, and a roadmap for directing all sorts of investigations and actions to the major contributors of risk.

The Duncan Dam case has been developed here in some detail to show how risk analysis, used and applied in a diagnostic sense, aids judgment by providing insight into the relative importance of failure modes and the underlying conditions that produce them. Sometimes a test of insight can be whether in retrospect the things it reveals seem obvious, and possibly to those having uncommonly mature judgment—an exceptionally well-developed sense of what is important—all of these things might have been self-evident. But for others, the insights the risk analysis provided could probably not have been obtained in any other way. Without them, the availability of geomechanical models, especially those of greatest complexity, all but invariably tends to dictate which conditions and failure modes receive greatest attention, at the expense of those lacking models of the same intricacy, or for that matter those lacking models at all. Failure modes are often accorded importance in proportion to the complexity of corresponding analytical techniques for reasons much as Mallory said of Everest: because they're there. Diagnostic judgment, and its embodiment in risk analysis, shows otherwise.

Qualitative Procedures

Risk analysis techniques like those used for Duncan Dam are quantitative inasmuch as they assign numerical probability values to decomposed uncertainties. But it is not always practical—or possible—to perform a full-blown quantitative assessment without first having narrowed down those elements of uncertainty most important to the problem at hand. As for any other engineering technique, risk analysis is best performed iteratively, starting with simple screening techniques then proceeding to quantitative procedures only for those aspects that require more detailed assessment. This is especially so for complex systems.

There exists a whole family of qualitative procedures for evaluating system safety not discussed so far that are best explained by illustration. They deal with complex systems—those having many components that may interact with each other. In various ways, these qualitative techniques aim to bring about a better understanding of system functioning and vulnerabilities, to enhance system reliability, and to improve overall system safety. In this respect they can all be classed as system reliability techniques, but they are not the same as those discussed in relation to reliability analysis. Recall that reliability analysis addresses some particular failure mode or component on a detailed analytical level. If enough failure modes are addressed in this way, then formal system reliability procedures can be used to combine these assessments to produce a quantified system failure probability along the lines of some of the simple illustrations in the Appendix. This quantitative approach works from the bottom up. It starts with detailed assessments of individual component failure probabilities, then combines their interactions as a system. The qualitative procedures discussed here work the other way around. They start at the top with a broad overview of the system, from which quantified probabilistic methods can then be used to focus on key features as needed. This makes them particularly useful during design, where the emerging understanding of how a complex system functions is equally important to system safety as the operation of its individual components.

This family of qualitative techniques originates largely in the chemical process industry but also in nuclear safety and aerospace, where a variety of safety effects can arise from complex component interactions. These effects must first be understood on a broad level, then managed individually. As reviewed by Greenberg and Cramer (1991), the procedures fall under such headings as: Preliminary Hazards Analysis (PHA); WHAT-IF Analysis; Hazard and Operability Studies (HAZOP); and a variety of similar techniques. Another considered here in more detail is Failure Modes and Effects Analysis, or FMEA.

The techniques are simple, intuitive, and above all pragmatic, focused on identifying and improving system safety vulnerabilities. As their names imply, risk—as a function of both likelihood and consequences—is not the only metric they use, though all are directed toward risk management in one way or another. Some focus mainly on identifying failure modes: what could go wrong, and what system responses would occur if it did. Others go a step further to the "hazards" (in this context, the consequences) associated with them. But what they all have in common, and their great strength, is their systematic approach. Each component and each failure mode is evaluated within an organized framework which allows a broad overview to be assembled. This requires little more than a tabular format designed to best suit the particular system characteristics, with no quantification or mathematical treatment. Thus these procedures are highly versatile, and they rely extensively if not exclusively on professional judgment. For only by capturing this judgment systematically can broad system functioning—and therefore malfunctioning—be understood.

Failure Modes and Effects Analysis

Failure Modes and Effects Analysis (FMEA) is a qualitative technique for understanding the behavior of physical components of an engineered system, the influence their failure would have on each other, and effects on the system as a whole. It can be used in two ways, one as a precursor to probabilistically quantified reliability, risk, or decision techniques. The other is as a stand-alone procedure for relative ranking of failure modes that screens them according to risk. As a risk evaluation technique, FMEA treats risk in its true sense as the combination of likelihood and consequences. It is not, strictly speaking, a probabilistic method because it does not generally use quantified probability statements. It does, however, require qualitative likelihood judgments on failure mode occurrence, even if they are not taken the final step to numerical quantification.

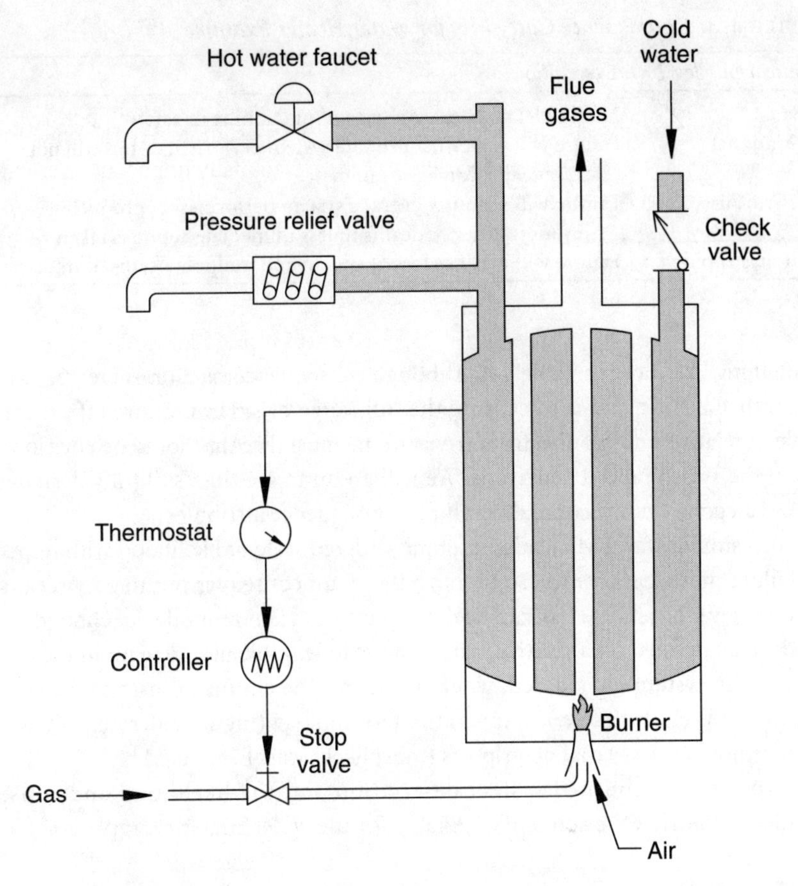

FIGURE 4-19. *Water heater system description.*

There is no better way to describe the operation of FMEA than the water heater example provided by McCormick (1981). This household device is familiar to everyone, as is its potential for serious, though rare, risks to safety. FMEA starts with describing the system and the components included in the assessment. This involves "drawing a box" around the system to define the components considered and those not included. In the water heater system description of Figure 4-19, for example, plugging of the rooftop chimney and carbon monoxide effects would fall outside the system description, as would rupture of the gas line buried outside. The system description must also make note of other external effects not addressed, like earthquakes causing toppling and fires.

The next element in designing the FMEA is to define various kinds and levels of consequences that can be envisioned to occur from system malfunction. This is highly specific to each system, its particular failure effects, and its surrounding environment. Consequence categories for the water heater system might look

TABLE 4-4. *Consequence Categories for Water Heater Example*

Consequence level	*Description*
I. Safe	Negligible; no effects on system functioning or safety
II. Marginal	Failure will somewhat degrade system performance but will not cause major damage or injury
III. Critical	Failure will seriously degrade system performance, producing serious injuries or deaths unless immediate action is taken
IV. Catastrophic	Failure will produce loss of system and multiple deaths or injuries

something like those in Table 4-4. Although consequence categories are not easily generalized, their aim is to capture the full range of system failure effects on a scale of relative catastrophe, from trivial to the most dire that could be envisioned given the system and its environs. We will go on to see that, skillfully designed, these categories can encompass multiple consequence attributes.

In a similar way, FMEA adopts a rank-ordered scale of likelihood with respect to failure mode occurrence, so that together with consequence category, a rank-ordered level of relative risk can be derived for each failure mode. Likelihood categories are defined on a relative scale of most to least likely and again in the context of the system considered, where these may be expressed using numerical designations (I, II, III, etc.), comparative terminology (high, moderate, low, etc.), or commonly used verbal descriptors (most likely, least likely, etc.).

From the established system description and the likelihood and consequence categories, the actual FMEA table for the water heater example is shown in Table 4-5.

The FMEA table is constructed by sequentially considering each component in turn, with those that could be subject to any potential failure mode listed in the first column, and the corresponding description and effects in the following three. Moving on through the columns, consequence and likelihood categories are assigned to each failure mode. This now makes it possible to establish the relative risk associated with that failure mode.

Perhaps the most important part of any FMEA are the two right-most columns in Table 4-5. They capture means for detection and intervention, as well as measures that could be taken to mitigate the risk by reducing either the likelihood of failure mode occurrence or its consequences. Used at the design stage, this is where inspection or monitoring requirements are identified, along with risk-reducing features or modifications.

For the water heater, inspection of columns 5 and 6 in Table 4-5 provides an indication of system safety. It can be seen that no high-risk failure modes have been identified which would produce "catastrophic" consequences (category IV) in combination with the highest or "probable" likelihood of occurrence. Most of the failure modes have consequences classifying as "safe" (category I). In this case

TABLE 4-5. *FMEA for Water Heater Example*

(1) *Component*	(2) *Failure mode*	(3) *Effects on other components*	(4) *Effects on whole system*	(5) *Consequence category*	(6) *Failure likelihood*	(7) *Detection methods*	(8) *Compensating provisions*
Pressure relief valve	Jammed open	Increased gas flow and thermostat operation	Loss of hot water, more cold water input and gas flow	I	Reasonably probable	Observe at pressure relief valve	Shut off water supply, reseal or replace relief valve
	Jammed closed	None	None	I	Probable	Manual testing	No consequences unless combined with other failure modes
Gas valve	Jammed open	Burner continues to operate, pressure relief valve opens	Water temperature and pressure increase; water turns to steam	III	Reasonably probable	Water at faucet too hot; pressure relief valve open (observation)	Open hot water faucet to relieve pressure, shut off gas supply, pressure relief valve compensates
	Jammed closed	Burner ceases to operate	System fails to produce hot water	I	Remote	Observe at faucet (cold water)	
Thermostat	Fails to react to increased temperature	Burner continues to operate, pressure relief valve opens	Water temperature rises; water turns to steam	III	Remote	Water at faucet too hot	Open hot water faucet to relieve pressure, pressure relief valve compensates
	Fails to react to decreased temperature	Burner fails to function	Water temperature too low	I	Remote	Observe at faucet	

the greatest system risk is posed by the gas valve jamming open ("reasonably probable" with "critical" category III consequences). This would emphasize the importance of corresponding detection methods and compensating provisions during normal operation. Given this proviso and the limits of its system definition, the FMEA would imply that the water heater is a comparatively safe and reliable device. In this case, the risk ranking can be done informally by inspection. For systems with greater numbers of failure modes, risk ranking can be conducted in a more structured format through a process called *binning* described subsequently.

The water heater example illustrates all of the essential elements of FMEA as a stand-alone procedure for risk evaluation, and it is applied in a similar way to most mechanical and electrical systems. It can also be adapted to other kinds of complex systems whose performance includes not only engineering functionality but effects on the natural environment as well.

Windy Craggy Case History

Mining produces all of the basic materials and much of the energy society depends on and without which it would cease to function. At the same time mining cannot be conducted without risk: not only financial risk and risk to personnel safety, but also risk to the surrounding environment. Environmental risks make mining one of the most contentious activities in society, and are also among the hardest to understand. Various stakeholders bring with them values that interpret these risks from different perspectives and in different contexts. This can produce a plethora of perceived environmental risks but without any global sense of which are most important, the sources from which they derive, and whether or not they have been accounted for and reduced. And public policy decisions regarding land use are where society's need for minerals and its aversion to environmental risk meet head-on.

FMEA has been adopted by some segments of the mining industry and related regulatory authorities in an attempt to systematically identify, structure, and reduce risk to environmental attributes. A mine contains a variety of processing facilities, waste deposits, and supporting infrastructure, as well as the orebody itself and natural features of its setting. These can be treated just like any other complex system, with components including both constructed and natural features whose failure to function as expected could bring about undesirable environmental consequences.

The proposed Windy Craggy copper mine was an extreme case in every respect. It would confront geomorphological processes active to a degree rarely found elsewhere in the world. Its transportation corridors would parallel the Tatsenshini River, a mecca for whitewater rafting and host to all five species of Pacific salmon plus arctic char. Located high in the St. Elias range on the Alaska/Yukon/British Columbia border, the project land use decision would have international implications. And if this weren't enough, the area is surrounded by

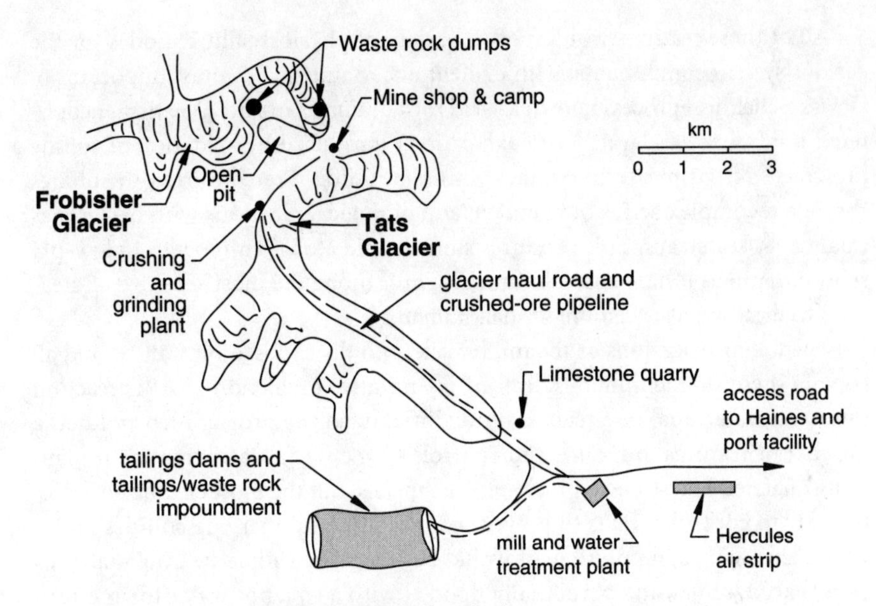

FIGURE 4-20. *Windy Craggy system description.*

three U.S. and Canadian national parks. Accordingly, risk-mitigation measures had been adopted to an unprecedented degree. But FMEA, or something close to it, would be the only hope for comprehensively evaluating the project's risks to the surrounding environment in any systematic and unified way.

The system description in Figure 4-20 includes features typical of many surface mines—the mill processing facilities and water treatment plant, an access road, two tailings dams to enclose the impoundment where tailings and waste rock would be deposited, and other waste rock dumps. What was far from typical were the geologic hazards and potential failure modes they produced. First, the site is in one of the most seismically active parts of the world, with the nearby Fairweather and Denali fault systems producing M8.0 to M8.8 earthquakes with an average recurrence interval on the order of 1000 years and M>7.0 every 100 or so. Thus, components including the tailings dams, although of the most inherently stable rockfill type, would be subject to strong seismic shaking.

But the dynamic glacial geomorphology is what made the site truly unique. The pit would require digging through 100 m of glacial ice to expose the orebody. Its waste rock would be placed on tributaries to the active Frobisher glacier, one that occasionally exhibits a curious acceleration known as "surge" behavior. The haul road and crushed-ore pipeline would be directly on the Tats glacier, active as well though having receded hundreds of meters in the last century to leave extensive deposits of ice-cored ablation till. The tailings dams would contain these materials in their abutments, plus active rock glaciers, solifluction lobes, and debris flows on the mountainsides above.

All of these features would pose a variety of physical failure modes for the various system components, with evident uncertainties. But most important of all was a chemical process known as acid rock drainage, or ARD. We first encountered it in Agricola's "warm, dry exhalations" produced by oxidation of sulfide minerals such as pyrite in contact with air. Sulfides are converted to sulfates through a complex series of chemical and biological reactions with bacteria as catalysts at key stages. This produces sulfuric acid along with metals in low-pH solution. And as it happens, salmonid fish are among the most sensitive of creatures to these metals, far more so than humans.

The tailings, portions of the mine waste, and the exposed pit walls would all contain these sulfide minerals. But the quantitative prediction of ARD reaction rates and water quality effects is quite difficult, so the project incorporated a number of features to reduce and control it. One was a water treatment plant using quarried limestone to precipitate the metals. But the most chemically effective way to control ARD is to prevent oxidation by submerging sulfides under water. The tailings impoundment would be where the sulfide-bearing materials were placed, remaining perpetually flooded with a cover of water to, in effect, smother the oxidation reactions.

It was clear from the outset that water quality effects on aquatic life would be the consequence of chief importance in this setting, and the FMEA was structured accordingly. Here, the salmonid fish represent a "sentinel" species whose sensitivity makes them a surrogate for effects on others, but this may also be represented by organisms at lower trophic levels where things like bioaccumulation or biomagnification of metals can affect the entire food chain. On a relative scale of catastrophe, the worst thing conceivable at Windy Craggy would be a tailings–dam breach, with widespread distribution of tailings and all but irreversible effects from physical habitat destruction and uncontrolled ARD generation. On the other end of the scale were things like treatment-plant upsets or a truck accident that might dump fuel oil into a stream. The consequence categories designed to reflect this continuum are shown in Table 4-6. Few engineers know much about aquatic biology, but such FMEA applications require that they become conversant with biological effects. Table 4-6 incorporates three main factors. First is toxicity, which ranges from chronic to lethal as a function of constituent, concentration, and species as they affect individuals or larger populations. Next is the duration of the effect from temporary to permanent, with corresponding implications for population recovery. And the area affected is also important because it pertains to dilution effects and the ability of more mobile species to avoid affected areas.

This also highlights the multidisciplinary nature of FMEA. It requires experts in different technical specialties for evaluating various physical components and failure effects. In the case of Windy Craggy, these experts included a menagerie of specialists in mining engineering, wastewater treatment and geochemistry, engi-

TABLE 4-6. *Consequence Rankings for Windy Craggy FMEA*

Consequence level	*Effects on aquatic life*
Low	Avoidance likely or localized effects Short-term effects on population Recovery likely
Moderate	Temporary and recoverable effects Some lethality to individuals or chronic effects on population
High	Permanent or large-scale habitat destruction Lethal to significant proportion of population Avoidance not possible

neering geology and geotechnical engineering, glaciology, and aquatic biology. Each brings their own contribution, but it is their interaction that yields great benefits as each gains broader appreciation of how their individual activities and specialties affect risk as viewed from the perspective of the others. In this, it is as much the process of performing the FMEA as its result that provides understanding of risks and how to reduce them. This pertains especially to failure modes with unappreciated significance or unanticipated consequences made apparent through these multidisciplinary interactions.

Table 4-7 shows a typical FMEA sheet for the Windy Craggy assessment from among the 40 or so produced, which happens to be for the tailings and waste rock impoundment. It is provided to show mainly how its structure corresponds to that of the water heater example in Table 4-5, but with two main differences. Additional columns are added to reflect the confidence associated with the assigned consequence and likelihood categories. This often aids these judgments and can become important later on when the resulting failure mode risks are binned. Another added aspect pertains to the several discrete stages of the project when the failure mode or particular effects may occur, ranging here from construction through abandonment. Further details of this and similar FMEA applications are provided by Dushnisky and Vick (1996).

The comprehensive risk ranking for all of the components and failure modes considered in the Windy Craggy FMEA is not reviewed here. Suffice it to say that many of the highest risk failure modes concerned long-term stability of the tailings dams, as Table 4-7 suggests. Recall that a primary means for controlling ARD from the sulfide-bearing tailings and waste rock was to permanently submerge them in the impoundment. This meant that the tailings dams would have to serve as water-retaining structures in perpetuity to prevent release of these materials at any point in the future. However, they would be subject to repeated episodes of seismic shaking and exposed to all of the other geologic hazards at the site for an indefinite period of time.

TABLE 4-7. *FMEA Format for Windy Craggy Example: Tailings/Waste Rock Impoundment*

Failure mode	*Effects*	*Project Stage**	*Consequence Category*	*Consequence Conf.*	*Likelihood Category*	*Likelihood Conf.*	*Compensating Factors Current*	*Compensating Factors Potential*
Seismic dam cracking and internal erosion leading to dam breach	Release of tailings and dissolved metals to Alsek River, with habitat destruction and oxidation	A	Severe	High to med.	Moderate	Med.		Removal of surface water from impoundment after closure
Fault offset of dam leading to breach	Release of tailings and dissolved metals to Alsek River, with habitat destruction and oxidation	O,CL,A	Severe	High to med.	Low	Low	Assumes further tectonic studies	Removal of surface water after closure
Overtopping and dam breach by displacement of tailings/water from landslide, avalanche, debris flow	Release of tailings and dissolved metals to Alsek River, with habitat destruction and oxidation	O,CL,A	Severe	High to med.	Moderate to significant	Low		Removal of surface water after closure; further engineering geology studies
Loss of water cover due to excessive seepage	Oxidation of tailings and acid-generating waste rock, causing low pH and metals in Tats Creek and Alsek River with chronic habitat degradation	A	Moderate	Med.	Moderate to significant	Low		Assess effectiveness of seepage control during operation; obtain improved hydrologic and water balance data during operation
		O,CL	Moderate	Med.	Very low to low	Med.	Control of water input, treatment of discharge	

* C = construction; O = operation; CL = closure; A = abandonment

Significant in this respect is that the system the FMEA evaluated was not the first to have been considered. Instead of permanent submergence for ARD control, a previous plan would have adopted a more conventional cap and cover at the end of mine operation, whereby the tailings would be allowed to drain, then covered with low-permeability materials to reduce oxygen and water ingress. While perhaps less chemically effective than permanent submergence, this would have been more reliable from a physical stability standpoint in all but eliminating any prospect of tailings dam failure in the long term. Thus, in having changed the system to enhance ARD control, a risk tradeoff had occurred. This was a tradeoff of effectiveness for reliability in the ARD control measures, of long-term acid-generation risks for long-term dam stability risks.

FMEA can be well-suited for evaluating these kinds of risk tradeoffs, especially during design when different alternatives are being considered. This is shown in Figure 4-21, which also illustrates the binning process. This simply takes the form of a two-dimensional matrix of likelihood and consequences, where each failure mode is "binned" according to its {likelihood, consequence} pair as indicated by an "x" in Figure 4-21. Risk increases to the upper left and decreases to the lower right, providing a convenient visual portrayal of system risks as a whole.

Figure 4-21 shows binned risks for two systems A and B that might represent different design alternatives. Two factors can be used for comparison. One is the total number of failure modes: The more things that can go wrong, the greater the potential for at least one to do so, and their cumulative effects can add up. The other is the position of the binned failure modes and the proportion that fall into the higher-risk regions of the matrix. The two systems A and B have similar numbers of failure modes. Although these will pertain to different components and system features, their overall distribution within the matrix is comparable. Likewise, the proportion of the failure modes that fall within the higher-risk region outlined in Figure 4-21 is similar, although again with minor differences in how they are distributed. So in this case, neither system can be said to be clearly superior in terms of risk. They simply pose risks of a different nature, as in the case of Windy Craggy a tradeoff of one kind for another.

To complete the Windy Craggy story, in the end a land-use decision was made to set aside the project area from all future mining activity. The social, economic, and political fallout had far-reaching implications, and both the project and the decision remain controversial to this day. It might be inspiring to think that the FMEA was able to carve out some small part of this process where systematic treatment of risk was able to inform the policy decision. But it became known only after the FMEA was completed that the decision had already been made before it was ever started.

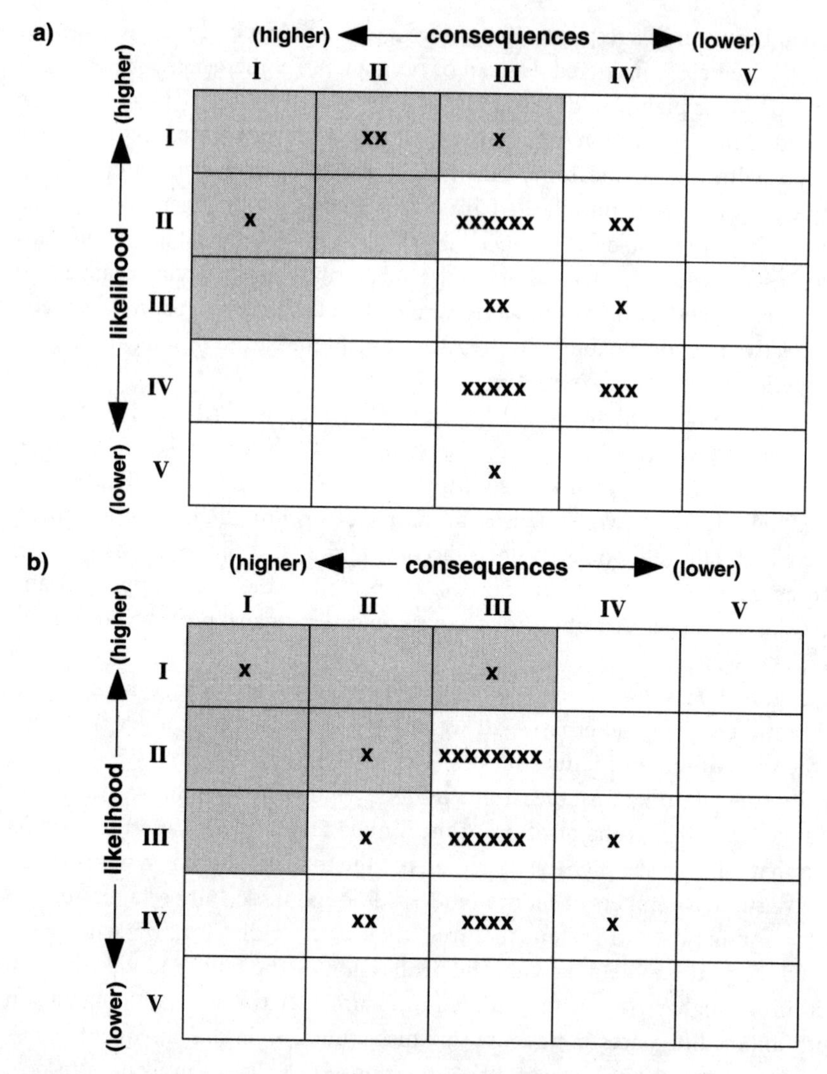

FIGURE 4-21. *Binned risks and risk tradeoffs: a, system A; b, system B.*

This once again underscores that no process for risk evaluation, FMEA or otherwise, takes place in a vacuum, nor is systematic evaluation of risk by itself able to rationalize decisionmaking processes conducted in the political sphere. The risk evaluation process seeks to better define where risks lie and what they are. The political process decides who bears them and how benefits are distributed. They simply do different things and they work in different ways, and this is as it should be, though often to the engineer's consternation. The responsibility of engineers is to evaluate risk in their domain as fairly and honestly as possible and to reduce it wherever they reasonably can. This duty has been fulfilled when their product is turned over to the messy, unsystematic, and unavoidably nonrational world beyond. In the end, it is engineers and scientists who produce technology but society which determines whether and how to use it.

Ground Improvement Case History

We turn now to a final example with a more down-to-earth geotechnical flavor. Instead of a stand-alone technique for risk evaluation, FMEA is used here as a precursor to quantitative techniques.

Most if not all structures subject to geotechnical failure modes are part of some larger project. It would not be practical to conduct risk or reliability analyses for each one separately, nor desirable because of their interactions. What is needed is some way to first evaluate them more broadly with the aim of grouping them according to similar behavior and system functioning. Following this, quantitative reliability, risk, or decision analysis can be applied more effectively to those components or groupings of components found to be key. FMEA can be ideally suited to such a screening purpose, though as a precursor to quantitative methods this need not necessarily be according to risk. Here it may be sufficient for categorization of components to look mainly to their structural characteristics and vulnerabilities. Used to this end, FMEA can be viewed as a "preprocessor" for complex systems in preparation for quantified techniques to follow. The Annacis wastewater treatment facility was such a system, and the risk was due to seismic liquefaction.

The Annacis treatment plant was to be constructed on an island in a distributary of the Fraser River near Vancouver, British Columbia. The soils were alluvial sands and silts susceptible to liquefaction. The plant would cost some $400 million, no small investment to put at risk from seismic damage. But the other thing was—you guessed it—the salmon. The Fraser hosts some of the most prolific runs in all the Pacific Northwest, with a commercial value alone close to $100 million annually. Treated wastewater would be discharged to the Fraser through a diffuser, but in the event of an earthquake raw sewage could enter the river throughout the duration of repairs. So as far as the salmon were concerned, cost would be directly related to time, and time to ease of repair.

All this could be avoided by improving the ground with vibroflotation and stone columns to reduce the potential for liquefaction triggering. But that would cost $30 million, almost 10% of the plant's cost. Nor was it clear what treatment depth should be used. Since the Fraser parallels the site, lateral spreading enhanced by the "free face" of the riverbank would produce horizontal as well as vertical movements. Partial protection could be provided by treatment to 10 m depth, but full protection would mean treating all 22 m. Such an effort would constitute the largest ground improvement project ever undertaken in Canada at the time.

Given these uncertainties, on top of that posed by earthquake occurrence, the ground improvement decision was tailor-made for decision analysis techniques. But first a better understanding was needed of how the plant worked, the process flow stream, its components and their structural vulnerability to ground movements—and, of course, the salmon. Examining these interactions became the job of FMEA, with decision analysis to follow.

Geotechnical engineers are not ordinarily well-versed in seismic behavior of wastewater treatment plants. So the first step was to review case-histories of their performance in earthquakes, of which several exist in Japan. This showed that many components are tank-like structures, and liquefaction induces the kind of behavior expected of tanks—floating, tilting, and breaking of connections. Their performance might be enhanced by things like slab reinforcement and nonstructural concrete for added weight, but only up to a point. Then too, there would be costly equipment such as pumps and centrifuges for sludge dewatering. Some structures would contain methane or chlorine gas whose release would produce widespread safety concerns both on and off the site, while the administration building would have to remain functional as a designated center for post-earthquake disaster coordination.

Moreover, some key components would be buried beneath others or so difficult to repair that the plant could be down for weeks if not months—a great deal to ask of a salmon headed upstream during any of the five annual runs. But for other structures repair would be easy and quick, since tanks can be made to work with temporary or improvised connections and cracks can be readily patched. For their part, the structural engineers were not accustomed to dealing with such matters. Their job was to provide movement tolerances, and the geotechnicals' to make it so. For sure, there were structural modifications that could reduce movement vulnerability, and things like flexible or fusible connections. But they never had reason to think about how structures would behave under movements of tens of centimeters or even meters, or what it would take to repair them, mostly because they had never been asked. Nor had they considered the salmon.

In light of all this, a group of geotechnical engineers, structural engineers, treatment plant operators, and aquatic biologists were sequestered in a room to produce an FMEA. Figure 4-22 shows the system as they defined it, and Table 4-8 shows what they came up with.

Table 4-8 is a typical FMEA sheet listing the components in the plant's liquid effluent stream. It can be seen that it is of different design from those presented previously, foremost by its absence of any likelihood ranking. Here, all of the failure modes are conditional, given the occurrence of an initiating earthquake and liquefaction triggering. Their consequences are grouped in two categories: effects on effluent quality and those on human safety. Remaining columns capture the engineers' judgments about general movement sensitivity and magnitudes for each structure that together would mark the onset of damage and gross damage

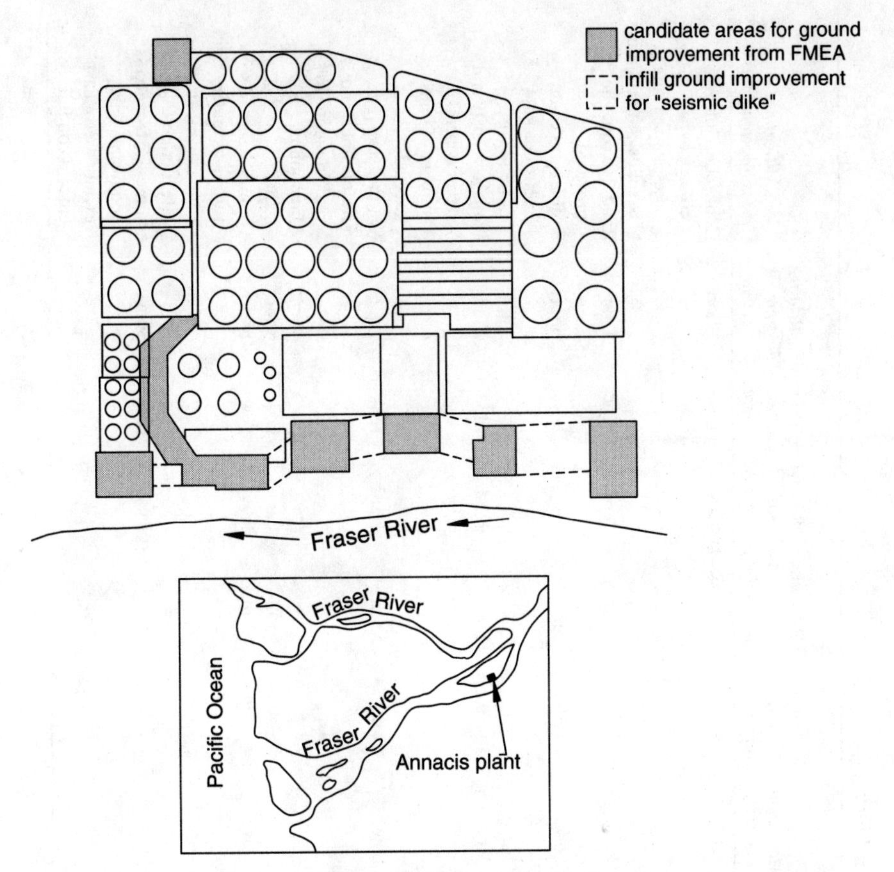

FIGURE 4-22. *Annacis treatment plant system description.*

along with ease, cost, and time for repair. As well, the effluent quality and repair time columns allowed the biologists to estimate effects on salmon.

In this case, failure modes were not binned according to risk because at this point risk had yet to be evaluated. Instead, structures were binned according to the following consequence attributes: those with high life-safety effects or critical effects on water quality, those requiring extensive time for repair, and those with the highest repair costs (including equipment) as a ratio of footprint area. These structures were designated as candidates for full-depth ground improvement without further evaluation, as portrayed in Figure 4-22. But this still left the ground improvement decision outstanding for all the rest. The FMEA showed that the remaining structures all had similar behavior, failure mode effects, and vulnerabilities to movement. Hence, these components could be considered together for ground improvement taken as a group. This could now provide the basis for a probabilistic decision analysis to evaluate the depth of ground improvement.

TABLE 4-8. *FMEA for the Annacis Wastewater Treatment Plant*

Structure/ component	*Failure mode due to ground movement*	*Failure effects on effluent quality*	*Failure effects on safety*	*Movement sensitivity*	*Movement damage threshold*	*Destructive movement threshold*	*Relative ease of repair*	*Relative cost of repair*	*Relative time for repair*	*Mitigating measures/comments*
Influent pump station	Rotation	Critical	Low	Low	>1 m	>10 m	Easy	Low	Short	Connection integrity critical, bypass necessary to route flow to solids contact basin
Trickling filters	Structural collapse	Mod.	Low	Mod.	>10 cm	>1 m	Mod.	Mod.	Long	Rigid slab could be provided
Solids contact basin	Cracking or horizontal separation	Low	Low	Low	10–30 cm	1 m	Easy	Low	Short	Fusible links could isolate cracking for easier repair
Secondary clarifiers	Cracking or horizontal separation	Mod.	Low	Mod.	10–30 cm	1 m	Mod.	Mod.	Mod.	Damage severity and repair depend on number of units affected
Secondary effluent conduits	Shearing, buckling, or tensile rupture	Low	Low	Low	>1 m	>1 m	Hard	Mod.	Long	Assumes steel pipe, fusible links, and remaining process functioning
Disinfection	Cracking	Low	Low	Low	10–30 cm	>1 m	Easy	Low	Short	—
Chlorine/SO_2 reagent storage	Cracking	Low	High (gas release)	Low	3–5 cm	>1 m	Easy	Low	Short	Assumes design of structure as secondary rigid containment and fail-safe suction feed lines
Miscellaneous at-grade channels	Cracking or horizontal separation	Mod.	Low	Mod.	3–5 cm	>1 m	Easy	Low	Short	Assumes surface access and fusible links, excludes secondary effluent conduits

The first branch set of the decision tree in Figure 4-23 shows the three options, including the null alternative of no treatment. The second set portrays liquefaction triggering in untreated soil, with probabilities derived from a stochastic model for ground motion occurrence coupled with deterministic Seed liquefaction criteria (Atkinson, et al., 1984). Here the full-treatment option retains a nonzero probability of liquefaction that reflects the degree of improvement.

The next branch set portrays various movement states, each corresponding to specified horizontal and vertical deformations, that together cover all possible movement magnitudes. The associated probabilities are subjective and incorporate the results of vertical settlement analyses (Tokimatsu and Seed, 1987) and assessments for lateral spreading (Hamada, et al., 1987; Byrne, 1991). A movement state may produce any of four damage states reflecting repair costs and fisheries losses, as portrayed by the final branch set. Damage state probabilities are subjective as well, incorporating the Japanese field performance experience to relate damage to liquefaction-induced movement magnitudes.

The decision tree incorporates a cost matrix showing plant damage and fisheries losses for each end branch. These costs are multiplied by the end-branch probabilities and summed for each treatment depth to find its expected or risk cost. Also shown for each option is the cost of ground improvement on an annualized basis. The final column shows the total cost for each alternative as the sum of the risk cost and the cost for implementing it. From this, the no-treatment option can be seen to have the lowest total cost, a result found to be insensitive to fairly wide variations in probabilities and costs assigned. That no treatment should be the preferred alternative for the great majority of the site area came as somewhat of a surprise, so this was revisited to find the underlying explanations.

One is that the soils' potential for liquefaction is not all that great, with corrected blowcount $(N_1)_{60}$ averaging about 12. Another is that most of the structures affected by cracking and tilting could be repaired and returned to service at comparatively small cost and delay. But Figure 4-23 also shows that fishery damage contributes very little to the outcome. It turns out that while salmon are indeed sensitive to reduction in dissolved oxygen in a plume of untreated effluent, this plume would be diluted within short distances by the prodigious flows of the Fraser. Nor, it seems, are they stupid. When encountering a region of low-oxygenated water, these most persistent of creatures in the animal kingdom are not to be denied. They can simply swim around it, or if worse comes to worst, retreat down the south arm of the Fraser and swim up the north as shown on the inset to Figure 4-22.

So not only did the combined FMEA and decision analysis make sense in the end, they brought to light some serendipitous opportunities. Figure 4-22 shows that nearly all of the structures that binned out from the FMEA as requiring ground improvement are aligned along the riverbank where liquefaction-induced lateral movements would otherwise be most severe due to proximity to the free face. For only a modest additional cost, the areas between them could also be

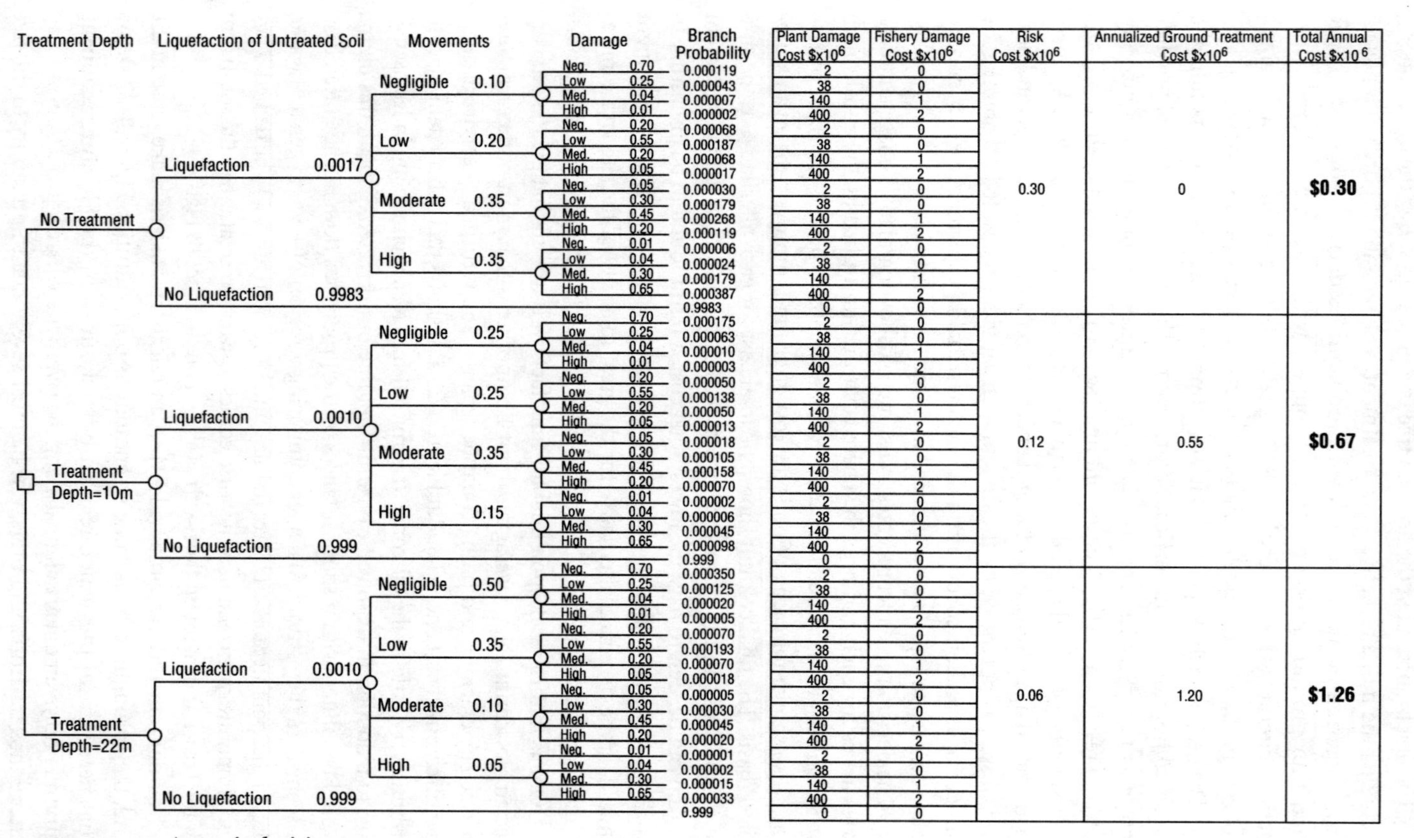

FIGURE 4-23. *Annacis decision tree.*

infilled with ground treatment to provide a continuous "seismic dike." By acting to restrain liquefied ground across much of the site, this would provide a substantial degree of lateral movement protection for interior structures even without treatment directly beneath them.

Though obvious in retrospect, such an intuitively satisfying engineering solution was not at all obvious before. The ground improvement measures could be carried out for $9 million, much less than the $30 million for the whole site. In the end, therefore, the combined FMEA and decision analysis were able to enhance judgment in a most unexpected and beneficial way. Note too that neither technique by itself directly produced the final ground improvement decision. What they did was reveal in a structured way the factors that made a sound decision possible. Similar uses of these combined techniques for airfields and port facilities are described by Vick and Watts (1994), with their application to other complex systems limited only by imagination and ingenuity.

The Uses of Qualitative Techniques

Qualitative methods—FMEA and others like it—are the most versatile and user-friendly in the spectrum of risk-based techniques. Either as stand-alone procedures or as precursors to quantitative methods, they provide the ability to understand complex systems and how these systems function. Widely adaptable to constructed and natural systems, qualitative procedures are well-suited to those having a large number of interacting components and those with multiple consequence or risk attributes.

As a stand-alone tool for risk screening, the FMEA process serves several purposes. The first is to better inform decisions that must necessarily be made in the face of imperfect knowledge, and therefore risk. In adopting risk as the metric for evaluating project features, FMEA differs from techniques like environmental, safety, or regulatory-compliance audits that base their assessments on prescribed criteria. Instead, risks are based on physical effects, making it adaptable to any set of criteria or project objectives that might be applied. FMEA will not guarantee that system risks are suitably low according to some standard or absolute measure. What it will do is provide some assurance that risks which exist have been identified and thought out in a systematic and organized way, and that means to reduce them have been highlighted in a structured and thorough process. Much like its quantitative cousins, FMEA seeks to direct investigations toward the greatest risk contributors. This is especially useful during design, where the system description represents a discrete "snapshot" in time of the emerging design features and elements. As these change during design development, the FMEA can be updated or supplemented accordingly. This can distinguish design features that truly do reduce risk from those which may simply shift it to other areas with ensuing risk tradeoffs.

But a separate purpose of qualitative techniques derives from their multidisciplinary character. To understand the reliability and functioning of complex systems usually takes a variety of specialists in diverse technical fields. The workshop format of the procedures brings these people together, with opportunities for interaction that seldom occur when each works independently. This is important because individual components that achieve optimal performance in their own domain do not necessarily yield optimal system performance. It is not unreasonable to expect that engineers should appreciate the workings of biologists, and vice versa, when proper system functioning depends on both, and this extends to any of the other specialists involved in multidisciplinary projects and systems. FMEA brings this about using risk as the common denominator, the focal point for interactions. In this larger sense it serves as a means to the end of not just identifying vulnerabilities, but also enhancing system functioning.

If all this is so, then qualitative techniques might seem to provide the magic answer, for system reliability at least. But this is far from the case, and their versatility comes at a price. There is a tendency to ask of these techniques more than they can do, and this first of all goes to the basic construct of risk.

While all such qualitative methods are billed as risk-management techniques, not all of them address risk as such. This can easily mislead the unwary. Either by design—as was the case for the ground improvement FMEA—or by oversight, some procedures address consequences alone. Consequences can produce risk only in combination with their likelihood, however, so those who think risk is being evaluated may be getting something quite different. And as we will see in the next chapter, something so simple can have profound implications. Evaluating risk qualitatively means that likelihood judgments are not quantified, but this does not mean they can be neglected.

This points to another limitation of FMEA. As a broad screening tool for complex systems, it can isolate the major elements of risk, but it cannot explore at a detailed level the specific conditions and events that produce it. To put it another way, complex failure modes are not the same thing as the kind of complex systems that FMEA is best suited for addressing. Complex failure modes are failure sequences with many component events, as distinct from many system components, and to evaluate them requires breaking these sequences down by decomposition. FMEA cannot do this and must lump together component events for purposes of likelihood assessment. Not only can this make likelihood judgments more difficult, it also makes them more prone to cognitive bias as subsequent chapters will show.

What all this amounts to then is simply that there are different tools for different jobs. And as with anything else, knowing which one to use and use skillfully with understanding of its capabilities is always the first requirement for producing a well-crafted product.

TABLE 4-9. *Comparison of Reliability and Risk/Decision Analysis Methods*

	Reliability analysis	*Risk/decision analysis*
Does	Account for parameter uncertainty in a geomechanical model	Account for broad range of uncertainties
	Use model output as surrogate for adverse consequences	Account for adverse consequences explicitly
	Adopt model as uniquely truthful	Allow for alternative hypotheses and models
	Address single failure mode that model incorporates	Address multiple failure modes and processes
		Account for processes that cannot be modeled
		Explicitly incorporate judgment using subjective probability
	Adopt relative frequency interpretation to provide precise answer for part of the problem	Adopt subjective and frequency interpretations to provide approximate answer for broader problem
Does not	Produce probability results that explicitly incorporate judgment	Produce probability results that are precise and invariant
	Address model uncertainty	Directly produce a decision or action

Summary

Geotechnical engineers encounter probabilistic methods in a variety of settings, where their use is often highly specific to their context of application. Inasmuch as emphasis is usually placed on the mathematics of their operation, their purpose can be easily obscured and their underlying meaning difficult to discern. One way to categorize them is by the relative-frequency and degree-of-belief interpretations, and various probabilistic methods usually adopt one or the other but sometimes both in combination. These two interpretations in turn give rise to the respective reliability and risk analysis frameworks.

As vehicles for the application of probabilistic methods, risk and reliability analysis operate under different precepts, as summarized in Table 4-9. Reliability analysis is essentially a probabilistic overlay to an ordinary deterministic method of analysis that, with rare exceptions, takes its geomechanical model as a given without explicitly incorporating model uncertainty. The input parameters to this model are characterized probabilistically using frequency-based statistical methods. Risk analysis, by contrast, is process-based, not model-based. It seeks to address every failure process that could occur, relying on separate assessment of the decomposed events contained in each one. If analytical models are available for some of these events, so much the better, and if uncertainty in analysis input is a primary concern then reliability techniques may be incorporated as one of the methods employed. But because risk analysis must go beyond the predictive capability of ordinary models, it always requires judgmental assessment for many elements of the failure process, and hence subjective probability.

One product of either a reliability or a risk analysis is a calculated probability value, and the question is always what it means. Ideally, one would like it to tell the whole truth about uncertainty and nothing but. Nevertheless, in the realm of probability, as in the courtroom, idealism often must bow to pragmatism, and this ultimately requires a choice between having one or the other.

The calculated result of a reliability analysis is typically the probability that the factor of safety is less than 1.0 according to the particular model adopted, the particular failure mode it addresses, and the particular database that parameter statistics incorporate. If one believes the model to be objectively truthful, and if statistical rules are faithfully followed, then $p[FS < 1]$ is equated with p_f. What now becomes the calculated probability of failure likewise achieves the status of objective truth—provided that one is prepared to neglect the necessarily subjective uncertainties in problem formulation and model interpretation as external to the analysis and therefore to p_f as well. So in this restrictive sense, p_f may be nothing but the truth, but doesn't contain the whole of it.

Risk analysis, on the other hand, puts forward its calculated result squarely as a probability of the physical failure event, but it can make no claim as to the objective truthfulness of this or any other such number. Its component probabilities will be substantially subjective, expressing a quantified belief in outcomes that varies according to personal judgment. So too then will the resulting p_f be subject to variation, and whether any such value is the "true" probability of failure becomes, like Poincaré's rhetorical query about the metric system, a question that has no meaning. Risk analysis does, however, accommodate alternative causal hypotheses and multiple failure modes, so its conception of failure probability is a more comprehensive one that better encompasses uncertainty in its global sense. Its p_f aims at expressing the whole truth, without representing it to be nothing but.

Ultimately then, the probability of failure is what you believe it is, whether this belief rests implicitly in the singular veracity of a modeling procedure or is expressed outright through subjective probability. With many parallels in other engineering activities, either a precise probability can be calculated for a small part of the problem or an approximate value can be obtained for the whole, but there is no escaping that both precision and comprehensiveness cannot be had at the same time. This is why no calculated failure probability can be advanced for communicating uncertainty without some accompanying statement about which aspects of failure-related uncertainty it addresses—and perhaps more to the point which it does not. This is also why unwarranted expectations sometimes lead to skepticism and even disillusionment about probabilistic methods that diminish their deserved place in the full suite of engineering tools.

Uncertainty will always remain the defining characteristic of the geotechnical field that distinguishes it from others, where this encompasses not just uncertainty in soil properties but even more in soil behavior as it operates in the field. What probabilistic methods in their reliability and risk manifestations have always promised to do is to evaluate uncertainty systematically. In large measure they have made good on this promise, with reliability offering a systematic treatment of parameter uncertainty, and risk analysis providing a systematic approach to uncertainty in failure modes and processes. It was never really part of the bargain that these methods would provide both the precision and the comprehensiveness demanded simultaneously by prescriptive decision criteria, yet the mismatch produced by this expectation has always been a kind of fatal attraction. By making the square peg of probabilistic estimates conform to the round hole of deterministic criteria, decisions could be made mechanically, with uncertainty all but banished in practice if not principle. But when you pull up a seat at uncertainty's table, probability offers no free lunch, just a better description of what's on the menu.

And unfortunately or not, depending on one's point of view, these decision criteria have yet to rush forward and make themselves broadly known to the profession. So with their implied promise remaining unfulfilled, much of engineering practice has in the past put probabilistic methods on the shelf. But some have come to see their promise differently. For a decision to remain a decision, it cannot be instructed. Instead it must be informed, and the role of systematic treatment of uncertainty is to do just that. And if probability is to inform the decision, it too must be informed by judgment. This perspective holds that the real promise of probabilistic methods is in their ability to lend a sense of what is important to uncertainty, whether this lies with data or with mechanisms and processes. This view of probability is ultimately diagnostic, as opposed to prescriptive. Its purpose is to aid understanding first and thereby decisions second. And to make it work requires that judgment be quantified as subjective probability.

The case studies presented here have been selected for their range of probabilistic methods and outcomes. From quantitative reliability, risk, and decision analysis to qualitative procedures like FMEA, their success can be judged on two levels. When used within the engineering domain, there is no doubt that in each and every case, a much improved understanding of risk and uncertainty was obtained. The various techniques were able to shed light on the nature and source of uncertainties, with discovery of things otherwise unseen. They were able to inform and aid engineering decisions, and as engineering tools for the systematic treatment of uncertainty their power is unparalleled. But a fairly distinct line is crossed when these results are presented to others for decisions outside the engineering sphere, and here their outcomes were decidedly more mixed. In matters of public policy, systematic treatment of uncertainty can only go so far and many other factors enter into the risk equation at this level. Most of all these have to do with how risk is perceived through the lens of one's values, and how the cost of risk and its benefits are distributed among society's competing factions. This is what the political process has always been for, and passing acquaintance with the writings of Jefferson and Madison and Hamilton suffices to show that even the wisest among us haven't yet figured out how to do this systematically while preserving a democratic society. So one is left to conclude that probabilistic methods simply may not work very well in the cauldron of advocacy where political decisions take place by design. But then again, the same could be said of any other engineering technique.

There remains a great deal more to learn about subjective probability and where it comes from. While the laws of statistics govern its frequency counterpart, subjective probability is contained in laws of human nature. These are the cognitive processes by which people internally conceptualize uncertainty, and we can now go on to explore them more fully.

5 Subjective Probability and Cognitive Processes

The discussions on subjective probability thus far have been largely devoted to establishing its conceptual underpinnings. As a measure of confidence or belief, the subjective interpretation is as old as the notion of probability itself, and its inductive roots go back to the medieval signs that gave birth to the very concept of evidence. In its engineering context, subjective probability is inseparable from the judgment it serves to quantify. And judgment, in turn, goes to the heart of geotechnical practice. So it is that the suite of probabilistic methods used in the profession today must include subjective probability as one of the essential tools for evaluating, expressing, and communicating uncertainty.

This is all well and good, as far as it goes. But to actually use subjective probability, to put these conceptual foundations into practice, is something else, and here is where the waters have always turned murky. There is no guidance in the engineering literature, and probability texts do not instruct in its use. Subjective probability lies outside the purview of these areas, but this does not mean it is without investigation or study. It is people, not engineering principles or probabilistic mathematics, who formulate subjective probabilities, and understanding how this comes about requires understanding how people think. This is the job of the behavioral sciences, and fields like experimental and cognitive psychology are where research on subjective probability is found.

Engineers by their nature shrink with horror at the mention of psychology. It is for good reason that engineers have chosen to become engineers, and psychology is everything that engineering is not. It deals with vague attributes and ill-defined properties, with how people behave and how things are perceived—in short, with the subjective. But by now we have come too far to deny the subjective elements that constitute the core of engineering judgment or the subjective influences that underlie almost every aspect of engineering practice. One need not become a psychologist to glean what the field has to offer, and it is in this true spirit of engineering inquiry that these areas are now explored.

That said, behavioral science operates very differently from engineering in how it selects the topics it investigates and the protocols it uses to address them. Unlike engineering, it is not fundamentally a problem solving enterprise. Rather, identifying and investigating such "problems" as peculiarities and inconsistencies in human behavior can be sufficient in itself to better explain this behavior and why it comes about. Some trait that for an engineer would be a flaw to correct and remedy to the behavioralist becomes an opportunity for understanding the cognitive processes that produce it.

Much of the research related to subjective probability has been conducted in this vein. Recall that there are two requirements for the validity of a subjective probability value: first that it reflect the assessor's actual beliefs about the uncertainty at hand and second that it be coherent, or consistent with the basic probability axioms. The premise of a considerable body of related behavioral research is that the cognitive processes people actually use formulating subjective probability values are not always well suited to these requirements. In their natural course of affairs, people do not internally perceive or externally express uncertainty using mathematical rules. Instead they adopt various *heuristics*, simple mental strategies or rules of thumb, shortcuts of probabilistic reasoning that are usually successful tools for coping with the uncertainties we all confront every day. But sometimes these heuristics can lead to *bias,* or inconsistencies with mathematical probability principles. There has been so much work along these lines that it has come to be known as the "heuristics and biases" school, and it constitutes one of the primary topics of this chapter. Understanding the pitfalls for subjective probability assessment can then set the stage for dealing with them in Chapter 6 to follow.

The specification of a subjective probability value requires some internal operation by which it is developed and expressed. Ideas about how uncertainty judgments are formulated have evolved since the contemporary foundations of subjective probability were first developed by Savage (1954) and others of that era. The initial concept saw the mind as a kind of "intuitive statistician" that adopted mathematical reasoning automatically and subconsciously to produce probability values conforming to mathematical rules.

The Intuitive Statistician

From the very start, Savage's techniques for extracting personal probabilities from lottery preferences evoked a kind of subliminal calculation. Recall from Chapter 2 how a gamble involving the occurrence of some uncertain event would be compared with a reference gamble with specified probabilities and rewards. Presumably the mind was subconsciously calculating and equating the expected values of the two bets. In other guises the intuitive statistician had already been around for some time. As early as the mid-1700s, the associationist theories of Locke, Hartley, and Hume had seen the mind as a kind of mechanical device for mapping observed frequencies onto beliefs, with a physiological etching of grooves into the brain so unerringly precise as to discriminate among probability values at the 10^{-4} level.

But subjective probability requires integrating and synthesizing different kinds of information from different sources. This takes inductive reasoning, and here Bayes' theorem emerges once again, this time as the intuitive statistician's principal tool. On registering the image of a duck-like object, the mind would presumably formulate a prior probability of it being a duck. With sensory inputs like observations of waddling or quacking, the properties of these Bayesian indicators of duckness would be retrieved from past experience and used to adjust the prior in arriving at the posterior duck probability. Bayes' theorem provided a mathematical rule for updating evidence, and even if the mind did not operate in exactly this way it could be supposed to provide at least some close approximation (Peterson and Beach, 1967). So Bayes' theorem became the gold standard for how the mind should work, and many of the subsequent studies on heuristics and biases adopted it as the reference against which various cognitive deviations could be measured.

By the 1970s cognitive psychologists had developed some serious doubts about the intuitive statistician. First of all, they argued, inferring subjective probabilities from reference gambles had it backward. People don't back-calculate their uncertainty judgments from their betting preferences, instead they first have to formulate these judgments before being able to decide how to bet. But it was the idea that people somehow combine information using a mental calculator programmed with Bayes' theorem that became the main target of many cognitive researchers (Slovic and Lichtenstein, 1971). Their methods were experimental. In controlled settings, groups of research subjects would be asked to specify subjective probabilities or answer related questions designed to reveal departure from the values prescribed by mathematical rules. The results were sometimes less than encouraging.

Summarizing this work, Hogarth (1975) concluded that people have only a limited capacity for processing information. This means they must be selective in the information they perceive and must proceed sequentially from one piece to the next instead of integrating it simultaneously. It also means that various shortcuts and simplifications must serve in place of subliminal calculations. These are the heuristics which, together with the biases they could produce, served to focus much of the cognitive psychology research over the next decade on dispatching the intuitive statistician, even though experimental findings remain almost evenly divided on this score (Beach, et al., 1986). Still, almost as a by-product of these efforts have come the beginnings of an understanding of the mental processes used in conceptualizing and expressing uncertainty judgments.

The Cognitive Framework

While the intuitive statistician described how people ought to reason about uncertainty, behavioral approaches are more concerned with how they actually do. Humans are above all remarkably adaptive organisms, and even if their capacity for information processing is limited they compensate by molding their strategies to the kinds of tasks they face. In this respect, people use different kinds of reasoning depending on the content and the context of uncertainty judgments.

Reasoning Strategies

The problems presented to subjects in an experimental setting usually contain cues or hints about the nature of the uncertainty they contain, and these cues give rise to two characteristic reasoning strategies. *Statistical* reasoning is brought to bear for judging membership of some element to a larger reference class. Here the elements are substitutable as opposed to unique, and their membership in the set does not depend on their individual characteristics. Urn-type problems, for example, would be a contextual cue for adopting a statistical strategy: All red balls or all white balls in the urn are presumed to be identical to each other, and their individual characteristics are of no significance.

It is important to note that in the context of reasoning strategies, the word "statistical" carries no connotation of statistical methods of data manipulation in any formal sense. The pummeling received by the intuitive statistician has largely disabused any such notion. Instead, statistical reasoning can be best understood as deriving from a sense of "oftenness" of things observed or experienced, from how one thing conforms to the pattern of some larger class of things that have happened before. So what we would call reasoning from past experience is a statistical strategy. Recall from Chapter 3 how Peck described retaining walls, the vast majority of which are designed not by soil mechanicians but by landscaping contractors, and most of them turn out to perform quite adequately. The contractor knows simply what worked the last time and the time before that, and also what collapsed and had to be repaired out-of-pocket. These are repeated trials of retaining wall design from the contractor's experience, and they require no causal knowledge of earth pressures. In judging the likelihood of success of the next retaining wall, the contractor compares it to the reference class of retaining walls that usually work.

The other reasoning strategy is knowledge based and termed *causal.* Here, the uncertainty judgment derives from one's knowledge about the unique characteristics of a specific feature or event and how these characteristics fit within some framework of cause and effect. In a causal setting, specific knowledge about an element cannot be ignored in assessing its membership in some general class. In the urn problem, it might be known that some red balls are larger or perhaps that some have a textured surface—specific characteristics that could affect their likelihood of being drawn. A geotechnical engineer would account for properties of the particular backfill soils in causally establishing their effects on retaining wall stability from earth-pressure theory. In both cases, there are specific factors influencing the outcome that make the conditions unique in some way. These are single-event occurrences that have not been, or cannot be, replicated by previous trials or experience, and such circumstances provide contextual cues for causal strategies. Statistical and causal approaches have also been called aleatory and epistemic reasoning strategies (Beach, et al., 1986) from their correspondence with these concepts.

Beach and Braun (1994) highlight the difference by posing the following experiment. Suppose you stand outside a church where a wedding is being held. When the Best Man emerges you ask: "What is the probability that any randomly selected couple being married today will still be married to each other 10 years from now?" If the Best Man has any familiarity with divorce rates, he would account for this in his answer. But now suppose you ask: "What is the probability that this happy couple, whose marriage you have just witnessed, will still be married to each other 10 years from now?" The answer might be quite different. While both probabilities are subjective, they rely on different reasoning strategies according to the different cues contained in their problem statements—statistical reasoning in the former and knowledge-based causal reasoning in the latter.

According to the so-called *contingency model*, people tend to adopt one or the other of these reasoning approaches, or sometimes both in combination, depending on the circumstances of the problem and the cues it presents (Beach, et al., 1986). So the retaining wall contractor might rely on previous experience in familiar cases, consult a geotechnical engineer in other circumstances, or rely on experience together with the consultant's recommendations both at the same time. Indeed, Peck was describing the contingency model when he related in Chapter 3 how he and Terzaghi realized that the causal strategy that soil mechanics provided could be best used to extend the statistical strategy of experience in adapting empirical retaining wall design practices to unique or different conditions.

When applied to assessments of likelihood, these contingent strategies result in probability statements, although their likelihood judgments are arrived at in different ways and do not necessarily mean the same thing. When assessing the probability of precipitation, a weather forecaster might use statistical reasoning in evaluating the number of times that current conditions have produced precipitation within the set of previous such observations. Or, the forecaster adopting causal reasoning might use satellite photos and barometric pressure data, mentally projecting forward in time and space how these patterns might progress according to knowledge about how they can produce rain. These two approaches are, of course, simply embodiments of the frequency and belief interpretations of probability, and behavioral research shows that these concepts are internalized in the cognitive strategies people adopt in assessing likelihood. Although they are equally important, causal strategies are of particular interest because subjective probabilities are so often associated with nonrepeatable, single-event occurrences in a geotechnical setting.

Causal Reasoning

That people crave and seek out certainty is such a basic aspect of human nature that it is not surprising for this to be manifested in cognitive research. As a means for coping with the uncertainty that surrounds us, we use causality to cast it in the form of what is known whether this is really the case or not. Effects are attributed to causes we can understand. Causality is what underlies deductive logic, and the drill–test–analyze strategy incorporates it through soil properties that cause a predicted effect to occur. Causality is also the basis for formulating hypotheses in diagnostic judgment, and ultimately for establishing the interpretive framework that judgment provides. People cling to causal reasoning so tightly because it is such a natural adjunct to a comfortably deterministic world where causes produce known and predictable effects.

The perception of causality has deep cognitive roots. Beach (1992) reviews several classic studies where simple moving geometric forms were projected onto a screen. The observers uniformly interpreted the motions of the objects in causal terms. When one object contacted another, it was reported to have "pushed" or "hit" the stationary one, with a quasi-Newtonian inference of cause and effect. In another experiment, subjects attributed the motions of a triangle, a square, and a circle to the shapes having fallen in love with a storylike causal scenario. Here, as elsewhere more broadly, the perception of causality is motivated by the compelling need to structure observations in a way that gives them coherence.

Assigning causal attributions to events is also fundamental to inductive reasoning. Through past observations, people posit inductive "if–then" relationships by enumeration. These are not hard-and-fast rules but loose relations expressed in terms of likelihood: If A occurs, then you can usually expect B because A usually causes B. When such a causal association is established by means of statistical reasoning from observed outcomes, we say it is empirically derived. Similar causal propositions can also be arrived at by training and instruction, with theoretical relationships in engineering a case in point. Either way, they indicate what to expect if certain conditions are met.

Causal reasoning often involves complex sets or sequences of events that involve many such if–then propositions. These are linked in the form of a structured narrative—in effect, a "story." The story relates important events or characteristics in an internally consistent and plausible way that enables the outcome to be anticipated. While behavioralists use the term metaphorically, others adopt it more literally. "The story" is what geologists call a hypothesis for the causal sequence by which things got put together, whether the arrangement of the continents or a grain of oolitic sand, a kind of shorthand for some chronological narrative related in geologic time (McPhee, 1980).

Jurors too have been shown to use evidence presented by the prosecution and defense to construct a storylike narrative. The precise form of the narrative can be influenced by such factors as the order in which the information is received, an effect well known to attorneys who bury their weakest arguments in the opening of their summation and finish with the smoking gun. While the narrative may vary from one juror to another, it allows relationships among events and circumstances to be inferred when no direct evidence is available—in other words, to fill in the blanks. Verdicts tend to follow from these narratives, and the ease with which one can be constructed determines its plausibility and hence the jurors' confidence in their verdict (Pennington and Hastie, 1986). The formulation of a narrative representation is closely related to the visualization of failure processes that we have already encountered in connection with diagnostic judgment. Likewise, decomposition of failure sequences in risk analysis is essentially a formalized representation of some causal narrative used to describe the failure process.

The importance of the causal narrative can hardly be overstated. To construct a causal narrative is what it means to understand something. It is what makes the light bulb go on. The narrative gives form to the relationships and patterns that made Poincaré's facts come together. And until there is a causal narrative, there can be no comprehension of how things work. It is entirely possible to perform a calculation without the slightest idea of what it represents. An article of faith in engineering education is that the student instructed in theory will eventually be able to penetrate its mathematical abstractions to a conceptualization in physical reality of why its agents interact the way they do. It is this causal narrative that gives the problem deeper structure and coherence extending beyond the case at hand. Not only does this make it possible to draw inferences where specific information is not available (to "read between the lines"), it is also used to identify small but critical details whose importance would not be obvious without its contextual framework. We have already seen how such contextual frameworks constitute the basis for interpretive judgment. The framework of the causal narrative is what provides its ability to anticipate the outcome. This is ultimately how inferences about past events and predictions of future ones are constructed from causal propositions.

But where does the narrative come from? Beach (1992) proposes that memory contains a library of story plots or "prototypic narratives." These, for example, are what enable people to recognize so quickly the themes and plots in novels they read. For our purposes, prototypic narratives can come from knowledge about what theory and analysis predict or from past experience of causes having produced certain effects. Here the role of case histories becomes especially evident. There are always a few particularly well-documented case histories that readily come to mind, which constitute the "type cases" for a particular mode of failure. The Lower San Fernando Dam would be one for seismic liquefaction, the Tower of Pisa for differential settlement, and Teton Dam for internal erosion through jointed rock foundations. Many geotechnical engineers would be familiar with how these failures or near-failures progressed and the factors responsible. They are prototypes that best illustrate the necessary conditions and the development of a failure process, and what we mean when we say they are "well-documented" is that all of the pieces are in place and they fit within a coherent whole. So memory draws on all of these sources of prototype narratives—theory and analysis, personal experience, and case histories—to formulate at least separate episodes of the story. These are then linked together to tailor a unique narrative for the circumstances at hand in forming what we might otherwise call a hypothesis, yet another central aspect of judgment. Recall from Chapter 3 how preparing a geologic map provided the framework necessary for hypotheses about geologic origin. This was simply the vehicle for assembling a causal narrative. With this, we can begin to see the wheels of cognitive machinery turning within all of engineering judgment.

Causal reasoning therefore has three components: (1) the perception of causality, (2) the inductive formulation of causal if-then propositions, and (3) the construction of a structured narrative that assembles these propositions. Causal strategies provide the means by which people conceptualize knowledge-related uncertainty. The associated cognitive processes contain a great number of parallels to various aspects of judgment that engineers employ, and they go far toward illuminating some of its inner workings. In this, engineering and psychology share more elements than we might suspect.

Cognitive Models

Behavioral researchers have proposed a variety of conceptual models for the cognitive processes people use in formulating and expressing subjective probability judgments. We have already seen one such model, the contingency model of Beach, et al. (1986), and McClelland and Bolger (1994) review and compare a number of others on a more detailed level. Of these, the *stage model* (Koriat, et al., 1980; Smith, et al., 1991) has particular appeal. As its name implies, the stage model holds that subjective probabilities are formulated in a series of more-or-less distinct operations, although these may be interactive or iterative and not strictly sequential. Three such stages involved in formulating and expressing a likelihood judgment can be characterized as follows:

1. Determination of an anticipated outcome which appears most plausible or expected,

2. Assessment of uncertainty associated with this outcome, and
3. Quantification of this uncertainty in the numerical probability metric.

In the first stage, the predictive problem is framed, memory is searched for relevant information, and a predictive answer is chosen. Initially, this selection of some preferred "best" answer comes about in a deterministic way, without at first specifically considering what its degree of support or plausibility might be. In the second stage, the relevant evidence is evaluated to arrive at a sense of certainty about the answer. In assessing belief or confidence, one interrogates their "feelings of doubt" about it (Koriat, et al., 1980). This involves a process of weighing arguments and evidence for and against, and it is where the internal conception of uncertainty takes shape. This process may call into question the initial answer, and if so, some alternative answer or outcome may be examined by going back to the first stage. At this point the likelihood or uncertainty judgment has been formulated but has not yet been expressed. Only in the third and final stage is this feeling of doubt, uncertainty, or likelihood translated into a numerical probability value.

This helps demonstrate that a probability value is not something swimming around in peoples' heads just waiting to be fished out. People process information sequentially to circumvent their capacity limitations and the part that usually gets most attention—the third stage of quantification—merely floats most visibly on the surface. So the process itself is much more than trolling for numbers. But we have also seen from the contingency model that different problem cues give rise to different reasoning strategies, and these strategies can be combined with the stage model.

The contingency and stage models are integrated to develop the generalized framework for the cognitive process shown in Figure 5-1. First, problem cues involving the kinds of information available are used to establish whether the problem is to be represented as a repeatable event or a single-event occurrence and hence whether the corresponding statistical or causal reasoning strategy best pertains. When there are cues both for similarity to some reference class of like situations and for uniqueness as a nonrepeatable circumstance, the two strategies may be applied jointly or the process may cycle back and forth between them. In an experimental setting cues are contained in the problem presentation, but in the real world much depends on which cues are actively sought out. Recognizing and selecting the best approach can be difficult, and many heuristic and bias effects can be traced back to strategies that prove to be inappropriate to the task. In particular, statistical strategies are often underemphasized, sometimes being applied only after all causal options have been exhausted. Brehmer (1980) suggests that this is because people are so strongly attuned to the structured, cause-and-effect relationships in the world around them, making the apparent disorder of inconsistent outcomes much harder to evaluate.

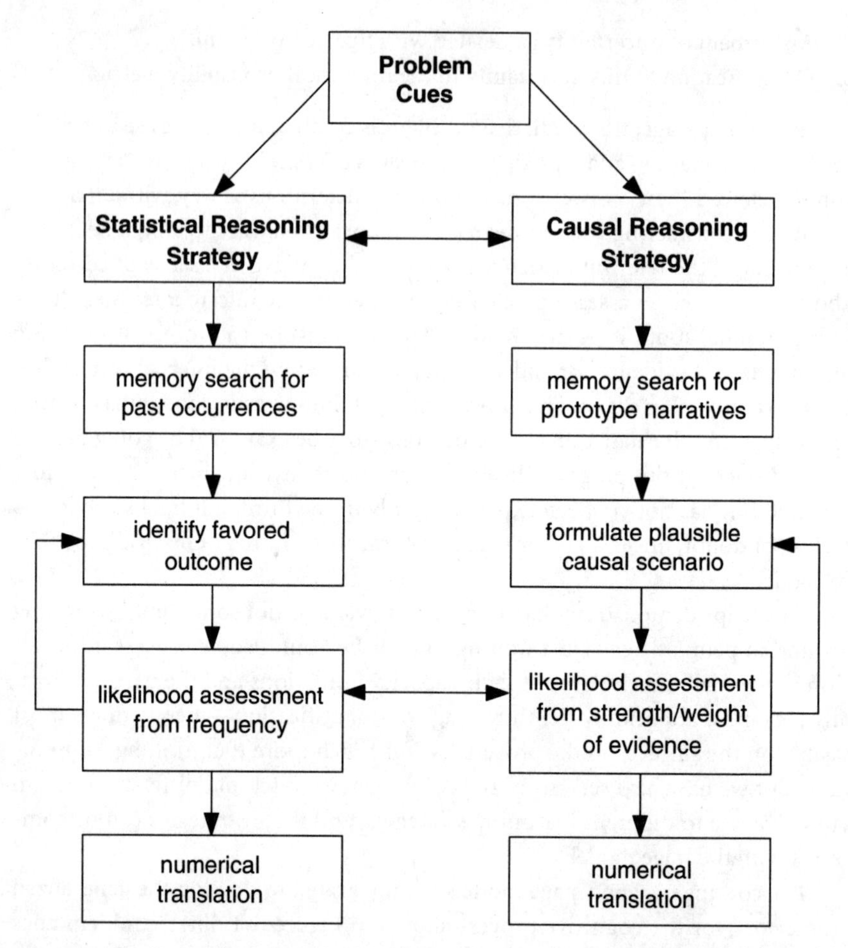

FIGURE 5-1. *Reasoning strategies in subjective probability assessment.*

In Figure 5-1, the first step in either strategy involves search of memory, either for past occurrences or for prototype narratives relevant to the problem. These are then used to formulate some outcome or causal scenario seen to be most tenable or plausible. The next step is assessing just how likely this outcome or scenario is felt to be. For statistical reasoning this involves assessment of recalled frequency of past outcomes, while for causal strategies scenario likelihood is judged according to both strength and weight of supporting and disconfirming evidence (Griffin and Tversky, 1992). The assessment may take on an iterative character similar in some ways to hypothesis testing, such that if the evidence or recalled frequency fails to provide adequate support, then a new outcome or scenario may be posited and tested again. Once this process is completed, the final step is to scale the resulting likelihood judgment as numerical probability.

The significance of this or any other such cognitive model is to illustrate several important things. One is that specifying a subjective probability value involves much more than simply picking a number from 0 to 1. Even the generalized framework in Figure 5-1 shows that complex cognitive processes for formulating likelihood and frequency judgments must precede their numerical expression. Moreover, there are definite reasoning strategies at work. One's degree of belief may depend on how these strategies are selected and applied, but it is not something pulled from thin air. Beyond this, the cognitive process also provides a context for understanding how and why heuristics and biases come about, as we will go on to see.

Cognitive Strategies and Engineering Judgment

Every geomechanical model contains within it a causal narrative. We call this an algorithm, and it specifies the underlying cause-and-effect relationships the model adopts. The algorithm of limit equilibrium, for instance, is so deeply embedded as a causal narrative in how we conceptualize real processes that some geotechnical engineers might assert that soils really do behave as rigid bodies—though nobody has bothered to tell this to the soils. The algorithm thus becomes reality. And if the analyst's results don't conform to reality, then it is reality that must be wrong.

At the other extreme lies experience and its application through purely statistical strategies. But we would view as something short of an engineer someone who would rely on experience alone without any idea of the causal factors that have produced it. The reason why we would reject both of these approaches, and why they sometimes seem at war, has much to do with the contingency model contained in Figure 5-1. Neither a statistical strategy as experience, nor a causal strategy as theory, can always provide the answer. It is judgment that determines which to apply, and when.

Judgment operates at the top of Figure 5-1, where problem cues are first identified then used to select the reasoning strategy or strategies best suited to what the problem presents. This is where the cognitive process begins, and it may well be the most important part. But how this operates in real-world activities stands in sharp contrast to behavioral experiments. In real problems, these cues are not pre-packaged but instead must be coaxed out from an almost infinite variety of facts the setting contains. The reasoning strategy selected depends largely on which of these facts are observed and which are determined to be most important. The problem must be interrogated closely and thoroughly to determine how it lends itself both to the causal aspects of theory and the statistical elements of experience. This is what we have previously described as developing a "feel" for the problem, one of the diagnostic components of judgment.

But more than this, judgment determines how these two strategies are combined as the problem solution proceeds, as they often must be. This is where judgment unites the pure theorist and the complete empiricist in melding their two strategies together. The interpretive aspect of judgment sets the statistical element of experience in the context of a causal framework. And in the opposing respect, judgment invokes the patterns of experience so that the causal explanation becomes plausible and "makes sense." The exercise of judgment can be viewed from a cognitive standpoint as the ability to move freely back and forth between these reasoning strategies, in much the same way as the Classical probabilists were able to do so fluidly with their two probability interpretations. Here too, then, the essence of judgment is to synthesize the causality of theory with the repeated occurrences of experience.

Clinical Diagnosis

So much of engineering is focused on prediction—with the drill–test–analyze strategy and Focht's predictive model of Chapter 3 as cases in point—that diagnosis is often seen as a "black art." While prediction projects future effects from causes, diagnosis seeks past causes from observed effects. So diagnosis is just prediction in reverse. But this does not mean that diagnosis is the predictive process run backward. The diagnostic process has its own approaches that employ some of the cognitive reasoning strategies we have seen.

Diagnosis constitutes one of the many similarities between geotechnical engineering and medical science: for engineering the diagnosis of failure modes and for clinical medicine the diagnosis of disease. A closer look at clinical diagnosis shows how causal and statistical strategies work in practice, insights central to the role of diagnostic thinking in engineering judgment.

Clinical diagnosis is the classical exercise of medical judgment. Given the patient's symptoms, medical history, and laboratory or other test results, the causative disease must be identified. This is purely an inductive process: Specific information unique to the patient must be generalized to the disease. In reviewing some 50 diagnostic exercises published over a period of years in the *New England Journal of Medicine*, Eddy and Clanton (1982) found that physicians use a diagnostic strategy consisting of the following six steps:

1. *Aggregation of information about the case.* The first step involves learning about the patient's presenting symptoms and history of the illness. Because there is often a large amount of information, it is summarized by aggregating it into more limited sets of correlated symptoms.
2. *Selection of a "pivot" finding from this information.* Perhaps one or two of these aggregated symptoms called "pivot findings" are then identified as particularly significant, while the rest are temporarily put aside. This pivotal finding is often something known to be a generally important diagnostic indicator and emphasized in medical education as suggestive of a certain disease or class of diseases. As a cognitive expedient, the selection of a pivot find-

ing reduces information-processing demands by emphasizing a select few aggregated symptoms felt to be most significant.

3. *Use of the pivot finding to generate a cause list of possible diseases.* This involves generating a cause list of diseases that could be responsible for the pivotal symptom. Note here that the direction of reasoning is changing. Before, it was forward from symptom to disease; from here on it will be backward from disease to symptom. At this point, the process shifts from hypothesis forming to hypothesis testing (Clancey, 1988).
4. *Pruning of the disease list.* Here the cause list is pruned or screened by comparing each disease with all of the aggregated symptoms. Each incompatibility reduces the plausibility of the candidate disease as the cause of the illness. This continues until some minimum threshold of plausibility results in a hypothesis being culled from the list.
5. *Selection of the diagnosis.* Now the actual diagnosis is formulated from the remaining hypotheses. If multiple candidate diseases survived the pruning process, these are compared to see which best matches the symptoms.
6. *Confirmation of the diagnosis.* In the final step, the diagnosis is reviewed to determine its adequacy in a larger context. This involves evaluating the extent to which the selected diagnosis could comfortably fit all the symptoms, a retrospective reassessment of the fourth step. Any insufficiency here results in repeating the entire process using the unexplained symptoms as the new pivot finding.

In many ways, this process reveals a more structured version of the elements of judgment described in Chapter 3. The first step of aggregating information makes observations and organizes information in some framework to establish correlations and develop a "feel" for the problem. The pivot findings of the second step identify its most important aspects from associative patterns. These give rise to hypotheses which are then tested, again by matching patterns of causes and effects. And the final step of review invokes interpretive judgment to see if it all "makes sense."

This process emphasizes causal reasoning, viewing the patient as a unique, single-event occurrence and establishing the causative disease from its symptoms. But these same symptoms have also been observed in many other patients before—which, after all, is how they come to be associated with the disease to begin with. So statistical reasoning should also have its place in accounting for how often the disease produces the symptoms. A modest extension would ask the physicians to provide a subjective probability to express their confidence in the diagnosis. Such an exercise illustrates the effects of neglecting statistical reasoning, at least according to Bayes' theorem.

Eddy (1982) describes an experiment in which physicians were asked to estimate the probability that a patient with a lesion had cancer. The subjects were informed that an initial examination indicated a 99% probability that the lesion was benign, but also that a subsequent X-ray test had given a positive result for malignancy. With regard to the X-ray procedure itself, the subjects were told that biopsies had determined that the test correctly identified 79.2% of all malignancies and 90.4% of benign lesions. These are typical cues for statistical reasoning strategies.

What is desired is the probability of cancer in light of the apparently contradictory information from the negative examination and the positive test. This answer is expressed as p[cancer | positive], where p[cancer] is the prior probability from the examination. As a Bayesian indicator, the test is 79.2% reliable, so p[positive | cancer] = 0.792. But since 90.4% of benign lesions were correctly identified, 9.6% were not, so the false-positive rate p[positive | no cancer] is 0.096. Bayes' theorem would then specify that:

$$p[\text{cancer} \mid \text{positive}] = \frac{p[\text{positive} \mid \text{cancer}] \times p[\text{cancer}]}{\{p[\text{positive} \mid \text{cancer}] \times p[\text{cancer}]\} + \{p[\text{positive} \mid \text{no cancer}] \times p[\text{no cancer}]\}}$$

$$= \frac{(0.792)(0.01)}{(0.792)(0.01) + (0.096)(0.99)}$$

$$= \mathbf{0.077}$$

Despite this result, most physicians (over 95% of them) provided a subjective probability of about 0.75 that the patient had cancer, demonstrating bias of roughly an order of magnitude in relation to Bayes' theorem. In so greatly increasing the prior probability of cancer from 0.01 to the adjusted posterior value of 0.75, they were vastly overestimating the predictive validity of the test. The explanation for why Bayes' theorem gives a much smaller value is because there are many more people who do not have cancer than who do, so there are many more benign lesions than malignant ones in the population as a whole. While the group of people with positive test results will include many (79.2%) within the tiny minority (1%) who actually have the disease, it will also include a significant proportion (9.6%) of the overwhelming majority (99%) who do not. Thus, Bayes' theorem views the test results not in isolation but in the context of their broad applicability.

Despite the statistical nature of the cues given in this diagnostic problem, the physicians were either misinterpreting these cues or were failing to adopt a statistical strategy altogether. Their probability of 0.75 for p[cancer | positive] was remarkably similar to the specified probability of 0.792 for p[positive | cancer] for the test, and their heuristic was to substitute one for the other (as in fact many admitted having done). The retrospective accuracy of the test was being mistaken for its predictive validity. Simply put, p[B | A] and p[A | B] are different things and cannot be taken as equivalent. All of this might mean merely that the physicians were not trained or instructed in the use of Bayes' theorem, but Ayton (1992) points to other reasons as well.

For one, medical education is organized around diseases rather than symptoms. Medical texts present information as the likelihood that patients will present certain symptoms when they already have the disease, or p[symptom | disease]. But diagnosis requires determining the disease for the patient who presents certain symptoms, or p[disease | symptom]. This encourages the two to be confused when they are clearly not the same. Another reason is that physicians receive clinical information in a way that reinforces unwarranted perceptions of validity. On ward rounds, patients with certain diseases are examined and the corresponding symptoms noted, while those having the same symptoms but without the disease are not examined, never having been admitted to the hospital in the first place.

Here as elsewhere, the engineering parallels are evident. Our training is organized around solving various types of problems, not the symptoms that would identify them. And we tend to devote more attention to our deceased patients—our failures—and their retrospective indicators than we do to the everyday successes that vastly outnumber them, even though the same indicators may be present routinely. In a causal framework this is no shortcoming. Indeed, it is how the understanding of failure causes is achieved. But an exclusive emphasis on failures can be less helpful for gauging the diagnostic value of the indicators when they are brought to bear on probability estimation tasks.

An empirical correlation is a kind of diagnosis—a causal association established by statistical reasoning. But in interpreting these correlations, some of the same factors can come into play in mistakenly reversing the direction of inference between cause and effect. Figure 5-2 provides data compiled for flowslides in end-dumped coal waste material in a region of the Rocky Mountains, showing that all of them occurred on slopes steeper than 17 degrees. Because flowsliding is so much more prevalent on steeper slopes, p[flowsliding | steep slopes] appears to be quite high. So high, in fact, that statistical reasoning could easily suggest that steep slopes are what cause flowslides.

But the data really portray the opposite—that p[steep slopes | flowsliding] is high—such that when you have flowslides you can expect to find steep slopes. Here again we encounter the physicians' problem that p[B | A] does not equal p[A | B], or affirming the consequent from Chapter 3. The reason that flowsliding seems to occur so infrequently on flat slopes might simply be that there aren't any, level ground being notoriously hard to come by in mountainous terrain. And we know nothing about the false-positive rate of slope steepness, since the data do not include waste dumps on steep slopes that may well have remained stable.

These data do tell us something, however, and here it takes a combination of causal and statistical reasoning to determine what. Separately, we know from laboratory studies that wetting-induced collapse resulting from rainfall, seepage, or snowmelt can cause severe loss of strength and fluid-like behavior in materials like these. In this context, the data would indicate that steep slopes are associated

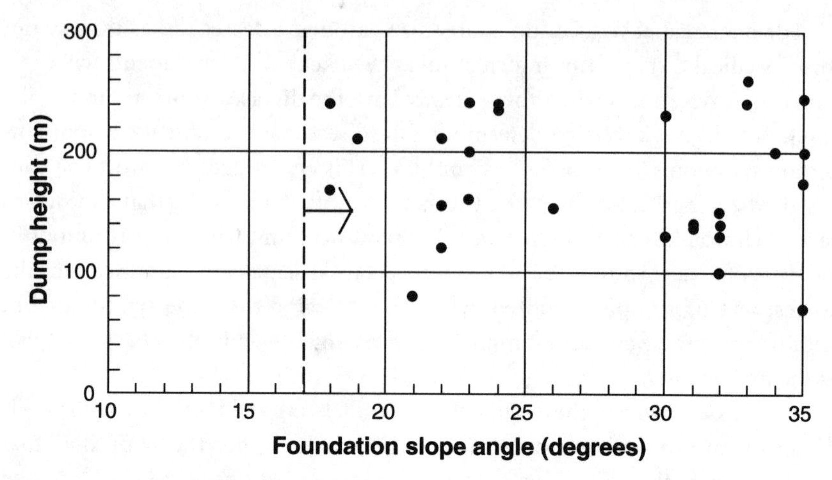

FIGURE 5-2. *Coal waste flowslide data.*

with flowslides because they enhance the mobility of such material. Integrating these causal and statistical elements allows us to say that steep slopes aren't the cause of flowsliding; what they do is magnify its effects once strength loss occurs. On another level, the application of these two reasoning strategies can be seen as the melding of theory and judgment.

Diagnostic Strategies

Rasmussen (1993) frames diagnostic strategies more broadly in a variety of technical and managerial settings. He starts by noting that the Greek root of the word *diagnosis* translates to "knowing the difference." Diagnosis, then, requires distinguishing one condition from another, picking out some important aspect from its general context. In this, diagnosis is closely related to the categorization or "scaling activity" that Smith, et al. (1991) used to describe judgment. Indeed, Rasmussen calls these activities "diagnostic judgment," so it follows that diagnosis, like judgment, cannot be governed by rule-based criteria that apply to every case. Instead, diagnosis is personal, subjective, and context dependent. A geologist will classify and categorize a rock outcrop using an entirely different taxonomy of petrographic characteristics than the geotechnical engineer according to engineering properties of the rockmass. Their respective contexts of what is important are different, and for good reason. Likewise, diagnosis depends on the situation, who is performing it, and what they want to achieve. This is why diagnosis can take so many different forms.

Even so, all diagnostic procedures have certain elements in common. One way or another, diagnosis involves "causal backtracking," first from effect back to cause, sometimes then from cause forward to effect, and sometimes alternating back and forth (Rasmussen, 1993). But first the key features or events must be found, and this involves searching for abnormalities and anomalies. Possible explanations for what happened are then developed by constructing causal narratives around these factors by linking them into temporal sequences or causal chains in the same way as event trees, breaking them down into their more primitive constituents much like failure mode decomposition. The test for significance of these links in the chain could be things like spatial proximity, the relation of events in time, the tendency for certain events to occur in association, and different explanations or causal chains that can be found (Einhorn and Hogarth, 1982).

But these diagnostic activities do not go on forever, and they continue only until some criterion of adequacy or *stop rule* is invoked. In arriving at a causal explanation, Rasmussen (1993) contends that diagnosis proceeds until the narrative of what happened and the relationships among its elements match intuition, until the "story" becomes sufficiently similar to the pattern of other cases. This is what makes the explanation, and the outcome of the diagnosis, plausible or believable to some degree of certainty—it "makes sense" by conforming to the expectations of previous experience.

The way in which these elements are integrated into a particular diagnostic strategy depends on the direction of inductive inference and the perspective of those who use them. Rasmussen (1993) describes three such strategies:

1. The *variationist* strategy: from the intended condition to the actual case;
2. The *empiricist* strategy: from the expected condition to the actual case; and
3. The *generalist* strategy: from the actual case to a larger generality.

We can go on to examine these cases more closely from the perspectives of those who might be involved.

VARIATIONIST. The variationist strategy uses inferences from intended system behavior to abnormalities. This is deductive in character because it derives from principles embedded in the intended or design system function. For most engineered systems, there is a reasonably well-formed knowledge of how they should work in ordinary circumstances that derives not so much from past performance but mainly from how they are designed to work. Only for the most resoundingly complex systems like nuclear reactors or spacecraft is this ever in doubt, and even then mostly for performance after a failure sequence is initiated. Knowing how the system ought to work forms the basis for determining why it didn't. From the engineer's standpoint, the departure of what actually happened from what should have happened will be traced backward through the event sequence to determine in which system component the deviation initiated. This constitutes the "root cause" explanation for system failure.

An operator of the system may have a more urgent perspective even while adopting the same strategy. An immediate diagnosis must be made by a chemical plant operator to prevent the situation from further deteriorating or by an emergency medical technician to first stabilize the patient's condition. This involves causal backtracking too, but to identify some compensating action that can bring the system back under control whether the root cause is found or not. As Rasmussen (1993) puts it, when you see a fire you look first for a bucket of water, not the match that started it.

An extension of this perspective is that of the repairman. The diagnostic target is still to find out what is broken, but this can also be a question of where. A weir at the toe of a dam might be registering high seepage, but the problem cannot be located until the seepage pathway in the embankment or foundation is found. *Topographic diagnosis* is where the spatial location of the system fault is of central importance. Alternatively, in *search by hypothesis and test*, if the source of the seepage is X, then seepage quantity would be Y. Multiple hypotheses are evaluated deductively from cause to effect, proceeding by trial and error until an adequate match is found.

The attorney's approach is flexible, adopting any or all of the others' perspectives and diagnosing from normal to abnormal in similar ways. But the stop rule is different. Diagnosis extends beyond what happened to who did it, continuing on until the guilty party is found. This is someone whose actions were connected with the system fault, who had enough responsibility and control to have prevented it—and, of course, whose pockets are sufficiently deep to pay.

EMPIRICIST. The empiricist strategy also adopts a normal-to-abnormal approach, and it too is employed by engineer, operator, repairman, and attorney alike. But here causation is established according to some expectation of system performance developed from prior cases. This is an inductive inference about how the system ought to work from how it has worked before, rather than deduction from its design as the variationist strategy adopts. Once the general case is established in this way, the empiricist derives inferences about the actual case by matching it to patterns of behavior associated with various causes. These patterns provide the *search template* for identifying a matching cause, and once it is found the diagnosis is made by inductive analogy. This is the same strategy physicians use in clinical diagnosis, deriving the cause of the patient's disease from the similarity of the symptoms presented to those associated with the disease as noted from other cases.

GENERALIST. The final diagnostic strategy builds on the previous two but reverses the direction of inference, going from the diagnosis of a system fault in the particular case to generalizations about its wider implications. As such, this strategy is purely inductive in character. From the engineer's perspective, such a broader view of failure causation goes to diagnosis of more general design flaws that may apply to the same type of system across the board. So this strategy generalizes from the particular system fault to changes in design that could prevent its recurrence. Indeed, the historical course of engineering has always been driven by this diagnostic strategy, and much of its progress occurs in this way. We call it learning from failures, and through its specific-to-general character, empirical

TABLE 5-1. *Strategies and Perspectives in Diagnosis*

Approach	*Basis of inference*	*Type of reasoning*	*Perspective and purpose*
Variationist	Intended → anomaly	Deduction from design principles	Engineer: Determine root cause Operator: Control system Repairman: Locate anomaly Attorney: Assign responsibility
Empiricist	Expected → anomaly	Induction by analogy	
Generalist	Anomaly → general	Generalization by induction	Engineer: Modify design procedures Operator: Modify operating procedures Manager: Modify organizational procedures

knowledge becomes embedded in engineering practice or encoded in procedural standards. Again, though, other actors will have different perspectives on the same technique. Managers are concerned with organizational practices and policies, and if a failure occurs their interest is in generalizing to the deficiencies in organizational procedures or procedural violations responsible. The same holds true for operators. They generalize to a cause grounded in operational process—where the operating procedure fell short and how it might be changed to prevent recurrences.

Table 5-1 summarizes these various descriptions and perspectives of the diagnostic process, and through them we see that diagnosis is no monolithic procedure. Different strategies may pertain, and their targets depend considerably on who uses them and why. If this weren't enough, Rasmussen (1993) shows that people shift back and forth in both strategy and perspective as situations and circumstances change. Diagnostic inference and diagnostic judgment are above all pragmatic. The great strength of cognitive processes is their adaptability in how people use them, and perhaps in diagnosis even more so than elsewhere they are used innovatively and opportunistically in finding an approach that works.

Regardless of strategy, diagnosis can be applied in two different temporal settings. A diagnosis performed after the fact is a retrospective search for causation, and the ordinary postfailure diagnosis is of this type. Conversely, risk analysis is simply a prefailure diagnostic exercise—a postmortem before the fact. This is what links risk analysis to diagnosis so closely, and they employ many of the same underlying cognitive processes and strategies. There is really only one thing that makes them different. The diagnostic process seeks to find the cause of a single failure mechanism after it has already occurred. But risk analysis does not know which, if any, of many potential failures may take place in the future or how they will occur. So risk analysis must consider the whole spectrum of failure modes, diagnostic hypotheses, and causal chains in their future prospective sense. It does this by using probability to reflect the likelihood of future occurrences.

Heuristics and Biases

Having reviewed a variety of cognitive strategies, it remains to extend them to subjective probability and to examine how they operate there. Here we return to heuristics and biases and how they affect they way people formulate and express uncertainty judgments. Cognitive research has compiled an extensive catalog of these heuristics and biases, and we start by considering them in turn.

The Availability Heuristic

The availability heuristic is one of the common simplifying shortcuts or rules of thumb people use in forming likelihood judgments. Figure 5-1 shows that an important component of both statistical and causal reasoning strategies is retrieval of information from memory, either past occurrences or prototypic narratives, relating to the situation at hand. The ease with which these instances are recalled is termed their *availability*, and those that come to mind most readily tend to be judged most likely (Tversky and Kahneman, 1974). In a great many circumstances this heuristic serves well. The ease of recalling certain events or scenarios can be a useful and oftentimes reliable proxy for their frequency or likelihood. Things can be difficult to imagine or hard to conceive because they are indeed rare or implausible. But availability can also mislead, producing bias in judged probabilities.

AVAILABILITY BIAS. Tversky and Kahneman (1982) describe a simple experiment involving the position of the letter "k" in words that contain it. Subjects were asked whether "k" was more likely to be encountered as the first or the third letter in words selected at random. Almost 70% judged the initial position to be more likely, whereas it actually appears most often in the third. Here the availability heuristic was at work—people perceived words that begin with "k" to be more common because these words are more readily retrieved, an effect well known to any aficionado of crossword puzzles. Availability bias can be understood by analogy to a statistical sampling problem. Memory retrieves a sample consisting of the most readily recalled subpopulation, which does not fully reflect the population as a whole.

Along with this is *salience,* the vividness of the instances or scenarios retrieved. Here again, the availability heuristic can be a useful one. We learn early in life that a hot stove can burn, and such a searing impression is not easily forgotten. But here too salience can mislead. A person having just passed the scene of an auto crash may overestimate the probability of a traffic accident from the salience of this experience. Likewise, a geotechnical engineer having just conducted a failure investigation, or having read an especially well-documented and hence salient account of one, might well adopt the availability heuristic in judging that particular mode of failure to be more common, and hence more probable, than it really is. The operative cognitive proposition is that if examples come to mind quickly there must be a lot of them, or that if an association is made easily it must be accurate.

HINDSIGHT BIAS. A related effect is called *hindsight bias*, the tendency to exaggerate in retrospect what was known before the outcome was determined (Fischhoff, 1982). After the event has transpired, the reasons why seem obvious because they are now especially salient, whereas all of the reasons it might not have occurred are by then long forgotten. It is perhaps one of the most common traits of human nature to say in hindsight that we knew it all along. The salience of things revealed after the fact allows them to overshadow how one's state of knowledge has changed.

CONFIRMATION BIAS. If availability bias were simply a problem of faulty memory, then those with good recall would be immune. Instead it derives from selective retrieval of information, and apparently its selective encoding in the first place. In particular, disconfirming evidence—that which would run counter to or contradict a projected outcome—is less readily encoded and retrieved (Ross and Sicoly, 1982). This leads to *confirmation bias*, and it follows from the everyday observation that people are prone to ignoring or discounting information inconsistent with preconceived views. Confirmation bias is our tendency to see what we want to see and hear what we want to hear, suppressing or ignoring the rest.

The Anchoring and Adjustment Heuristic

Once having formed a likelihood judgment, in many situations people estimate the corresponding subjective probability value by starting with some initial value then modifying it to produce the final answer. The heuristic is to anchor on the initial value then adjust it for other factors that influence it (Tversky and Kahneman, 1974). By allowing these factors to be evaluated sequentially and one at a time, the *anchoring and adjustment* heuristic helps overcome cognitive limitations in information-processing capacity. This heuristic, for example, is directly employed in the normalized-frequency approach to probability estimation, where an initial base-rate frequency value is then subjectively adjusted.

INSUFFICIENT ADJUSTMENT. One tendency from this heuristic is for *insufficient adjustment* in accounting for the effects of additional information or evidence according to how Bayes' theorem says it ought to be weighted. This produces an adjusted probability biased toward the initial value (Slovic and Lichtenstein, 1971). Such situations can arise in risk analysis when, for example, a conditional probability for earthquake damage initially estimated for one level of seismic shaking is then adjusted when considering the next level. This could also apply to the normalized-frequency example in Figure 4-8 where the initial estimate anchored to failure frequency might be insufficiently adjusted for any or all of the specific factors listed.

CONJUNCTIVE DISTORTION. Another form of anchoring and adjustment bias is known as *conjunctive distortion*, which occurs when the joint probability of combined events is overestimated compared to the aggregation of their separately estimated values or, alternatively, when the separate values are underestimated relative to the combined value. Again, the estimate is anchored on the probability of one event and insufficiently adjusted for the others. This is illustrated by the famous "Linda problem," so consternating as to have become a minor classic. In it, Tversky and Kahneman (1983) put the following question to their experimental subjects:

> Linda is 31 years old, single, outspoken, and very bright. She majored in philosophy. As a student, she was deeply concerned with issues of discrimination and social justice, and also participated in antinuclear demonstrations. Which of these two alternatives is more probable?
>
> a) Linda is a bank teller
>
> b) Linda is a bank teller and active in the feminist movement

Which answer would you choose? If (b), you would be guilty of conjunctive distortion (as were 80% to 90% of the experimental subjects) and in violation of the basic probability axioms. Answer (b) cannot be probabilistically correct because bank tellers who happen to be feminists constitute some smaller subset of bank tellers as a whole. This invokes the "oftenness" character of statistical reasoning—unless all bank tellers are feminists, there must always be more bank tellers than feminist bank tellers. Still, Linda's description virtually screams that (b) simply must be right, and to choose otherwise would be to ignore all manner of relevant information.[1]

Linda may indeed be a feminist. It's just that this can't make it any more likely for her to work in a bank. Although the Linda problem was designed to illustrate conjunctive distortion, it shows even better how hard it can be to integrate causal and statistical information. It is the strong perception of causality from the problem cues in Linda's description that makes knowledge-based causal reasoning so dominant. It is easy to see this if we include the third possibility missing from the set—answer (c), that Linda is a feminist. Now causal reasoning from her description would lead us directly to this answer, and rightly so because the statistical aspect of her occupation becomes irrelevant. In the Linda problem, the probability of the alternatives is judged almost exclusively according to one type of information—her description—which brings us to yet another important heuristic.

The Representativeness Heuristic

People often judge the probability of an event by the extent to which it resembles something else or its similarity to related circumstances. This is the *representa-*

[1]Naturalist Stephen Gould (1992) said he was particularly fond of the Linda puzzle because even though he knew the conjunction was less probable, "a little homunculus in my head continues to jump up and down, shouting at me 'but she just can't be a bank teller; read the description!'"

tiveness heuristic, and like others it informs the probability judgment in generally useful ways. Resemblances are convenient heuristics that allow us to go through everyday life without deep analysis of every uncertainty we come across. We might evaluate the probability of large settlement for an embankment on peat according to its similarity to the general class of other embankments on such materials that have settled excessively. This might be an entirely valid and useful generalization. But the representativeness heuristic can also mislead because similarity is one, but not necessarily the only important factor in the probability judgment.

REPRESENTATIVENESS BIAS. Representativeness bias results when similarities to some reference class are taken as the sole indicator of likelihood, ignoring or overlooking other relevant information, especially that of a simpler or more intuitive nature (Tversky and Kahneman, 1974; Bar-Hillel, 1982). Not just the presence of peat but also its thickness is an important factor in settlement that the representative heuristic might encourage us to neglect.

This overreliance on similarities has been described as a kind of stereotyping, as in Linda's most convincing resemblance to the reference class of those with feminist inclinations. As we have seen before, it goes like this: If A usually indicates B, then B is likely to indicate A. So if Linda's description relates closely to that of a feminist, then Linda is likely to be one, neglecting the other information that would say she must be more likely to be a bank teller. Likewise, if embankments on peat usually settle a lot, then an embankment with large settlement is likely to be underlain by peat. This, of course, is not necessarily so because it could be underlain by any compressible soil.

The problem here again is that the reference class of similar cases is just a subset of a much larger body of information. The representativeness heuristic judges probability for the case at hand by comparing its resemblance to the subset of similar cases, not to the entire population. In this way, representativeness is closely related to availability. Whereas availability refers to the selective encoding and retrieval of information, representativeness refers to the selective use of information derived from similarities.

INSENSITIVITY TO PREDICTABILITY. Inasmuch as the representativeness heuristic has been characterized as the tendency to confuse $p[A \mid B]$ with $p[B \mid A]$ (Sage and White, 1980), it is what was at work in the physicians' assessment of the probability of cancer. The reason why their assessed probability of the disease given the test results was so nearly equivalent to the probability of test results given the disease is that they put too much reliance on the resemblance between patients who test positive and those who have cancer. In other words, they overemphasized the predictive validity of the test. What they neglected was the simpler yet meaningful counterindication provided by the examination of the patient, and a bias in their assessed probability by almost a factor of 10 was the result. This effect has been called *insensitivity to predictability*, which occurs due to overreliance on some predictive approach with insufficient consideration for the uncertainties that affect it or the information it incorporates (Tversky and Kahneman, 1974). For the physicians, we also saw how this misperception of predictive validity was embedded in medical training.

The representativeness heuristic can operate in related ways for engineering problems where a subjective probability incorporates analytical findings. The tendency again is for overreliance on their predictive validity that can overshadow other significant factors. We even use the same terminology, a telling similarity in itself. The analysis is said to incorporate "representative" soil properties, and the model is said to be "representative" of field behavior. These judgments of similarity are, of course, assumptions and sometimes made for little more than convenience, but assumptions can be forgotten when the representativeness heuristic encourages overemphasis on analysis results. Forgotten too can be simpler kinds of evidence. The geotechnical literature abounds with calculated failure probabilities for existing slopes that neglect to incorporate the simple but crucial observation that any such slope must, in fact, still be standing, and other kinds of information from field observations can be equally important diagnostic indicators. As with medical training, the representativeness heuristic operates as often as it does in relation to the predictive validity of analysis at least partly because the theoretical paradigm in engineering instruction reinforces it.

BASE-RATE NEGLECT. Yet another form of the representativeness heuristic is termed *base-rate neglect*, where overreliance on descriptive similarities encourages pertinent frequency information to be ignored. Again as embodied in the normalized-frequency technique, such statistical information might otherwise constitute a starting point, or base rate, from which a probability judgment might be adjusted to account for specific factors. But base-rate neglect allows these specific factors to dominate, with how often things tend to occur going unrecognized. Here the classic experiment is the "Jack problem" which, as it happens, involves engineers and their natural predators. The subjects were divided into two groups. Each was given the following description and asked to provide the probability that Jack is either an engineer or a lawyer (Kahneman and Tversky, 1973):

> Jack is a 45-year old man. He is married and has four children. He is generally conservative, careful, and ambitious. He shows no interest in political and social issues and spends most of his free time on his many hobbies which include home carpentry, sailing, and mathematical puzzles.

The first group of subjects was told that Jack had been selected at random from a sample of 70 engineers and 30 lawyers. For the second group the proportions were reversed, with the sample consisting of 30 engineers and 70 lawyers. The base rates obviously vary, but mean estimates of the two groups were virtually the same (55% versus 50% mean probabilities of Jack being an engineer), showing that they almost entirely neglected the base-rate frequency information provided.[2] In Jack's case—as with Linda—the perception of causality must not have been far below the surface since the representativeness heuristic overshadowed what statistical reasoning should have contributed. This effect failed to occur when the subjects were given no description of Jack and statistical reasoning operated unhindered, demonstrating once again that the difficulty lies in combining causal and statistical information.

INSENSITIVITY TO SAMPLE SIZE. The representativeness heuristic finds still another incarnation as *insensitivity to sample size*, where it operates at variance with simple statistical notions. The principle of "regression to the mean" says that trials should converge more closely to a stable frequency as their number increases, and vice versa.[3] So if we were to actually measure the proportion of times a coin comes up heads, we would expect it to approach 50% after some reasonably large number of tosses, perhaps 20 or 30 but not necessarily two or three. Yet this is not accounted for in experiments where subjects assign the same probability to there being more boys than girls born in a hospital with 1000 infants as in one with only 100 (Kahneman and Tversky, 1982). So while base-rate neglect ignores frequency information, insensitivity to sample size accords it disproportionate significance—the sample is taken as more representative of the population than its size warrants.

In engineering, insensitivity to sample size can be encountered in the evaluation of precedent: If something has worked before the presumption is that it will work again without fail. That is, its probability of future success conditional on past success is taken as 1.0. Accordingly, a structure that has survived an earthquake would be assumed capable of surviving a recurrence of the same magnitude and distance, with the underlying presumption being that the operative causal factors must be the same. But seismic ground motions are quite variable in their frequency content, attenuation characteristics, and many other factors, so that precedent for a single earthquake represents a very small sample size.

In general, where x successes have occurred in n trials, the mean probability of success p_s can be taken as:

$$p_s = \frac{(x+1)}{(n+2)} \tag{5-1}$$

If there were no precedent at all, and supposing the absence of any other information, p_s would then be 0.5. A single success would increase the chance of sur-

[2]There is little doubt that any engineer would assign a much higher probability to such an uncannily accurate stereotype, and some might even suspect Jack of being a colleague. Although this was not part of the experiment, it still shows how persuasive representativeness can be.

[3]Recall that Bernoulli said even the "stupidest man" should know this instinctively.

vival to two out of three—an improvement to be sure but far from certainty—and eight successful occurrences would be required to have 90% confidence in a favorable future outcome. In short, precedent and a proven track record are not necessarily the same thing. But here again the perception of causality eclipses its statistical counterpart in relation to sample size effects.

In geotechnical circumstances, it is easy to become insensitive to sample size in a more direct way. Seldom do soil samples taken represent more than a small fraction of a percent of the total volume of the materials. While depositional processes and other geologic information may warrant inferences about spatial variability from relatively small numbers of samples, this is rarely so when geologic anomalies are involved. Much is sometimes made, for instance, of rock cores retrieved from known karst terrain showing little more than minor vugginess, where this is taken as a counterindication of more serious solutioning and the presence of cavities. But absence of evidence is not evidence of absence when the volume sampled is statistically insignificant.

Insensitivity to sample size appears to operate in more than just these instances. Chapter 3 described the widespread confidence once placed in the inherent seismic resistance of earthfill dams, largely because few of them had failed. Then again, few had ever been subjected to strong shaking, and it took the Lower San Fernando incident to show how insensitive to such a small number of trials this confidence had been. The former and pervasive confidence was revealed to have really been overconfidence, which brings us to perhaps the most important cognitive bias yet.

Overconfidence

Overconfidence is the tendency for people to be more sure about uncertain occurrences than they should be, more confident than the evidence really warrants. They exaggerate the extent to which what they know is correct. The resulting *overconfidence bias* is manifested by assigning subjective probabilities for single-event occurrences that are too extreme at either end of the probability scale—too close to 0 or 1, often with either too many zeros or too many nines behind the decimal point. Similarly for continuous variables or quantities, overconfidence promotes assigned probability distributions that are too narrow, with insufficient dispersion about the mean—in effect, a tendency to discount outliers too much. Overconfidence is by far the most persistent and tenacious form of bias in subjective probability estimation, and it seldom fails to be demonstrated in almost every study across widely varying types of experiments, subjects, and questions. It is reasonable to expect that as one's knowledge or information increases about an uncertain event or condition, so too should one's confidence increase as reflected by subjective probabilities that are closer to one end or the other of the probability scale. But it appears that this confidence increases far too fast, with a doubling of confidence being associated with perhaps a 50% increase in knowledge (Fischhoff and MacGregor, 1982). This all is produced by an unwarranted perception of how much we really know.

Terzaghi recognized overconfidence in how engineers handled uncertainty, though he never used the term as such. But there is no mistaking the concept in what he had to say about dam engineers and the importance of geologic details. And while he was no cognitive psychologist, his views about where overconfidence came from were remarkably prescient against the background of today's behavioral research. This had to do with precedent and the "dangerous tendency to generalize" it brought about (Terzaghi, 1929). Design practice, he said, was based on generalizing from successful precedent in one case to some other, which he couched in terms of statistical correlation. But among all of the variables on which safety depends, only a few at most were ever considered, with others—especially geologic ones—disregarded. As a result, dam engineers were overconfident, taking insufficient account of the possibility of departures from expected performance; in effect, their probability distributions were too narrow. Depending too much on textbook rules, they failed to appreciate what they didn't know:

> Present practice of design is based on precedent, whereby not only minor but also certain major geologic details are disregarded. Such practice means blindly trusting in purely statistical relations with an extraordinarily wide range of deviation to both sides from the average. As most of the textbooks fail to call the attention of the readers to the great uncertainty associated with rules of design based on this practice, many engineers engaged in dam design have an exaggerated conception of the degree of reliability of their methods of procedure... (Terzaghi, 1929)

Certainly dam engineers are not the only ones who could be charged with exaggerated confidence. Although the cognitive basis for this unwarranted certainty is complex, a key element seems to be people's lack of awareness that their knowledge is based on assumptions which may be quite tenuous. Ultimately, this springs from the underlying need for certainty in an uncertain world, and this applies to engineers perhaps even more than to people in general. Engineers' very reason for being is to provide answers and provide them confidently, and this is encouraged not only by the theoretical paradigm of our training but also by our deeply held view of ourselves and our profession. So overconfidence bias in engineers' subjective probability estimates should come as no surprise. It is where proof-seeking deductive reasoning meets the inductive, hence uncertain, synthesis of evidence. Focht (1994) spoke of overconfidence in this broad sense, though again without using the term. Referring to the entire predictive process, he said:

> It is very often more important to know what we do not know than what we do know.... It is better to be approximately correct than to be precise and wrong.

Calibration

One measure of overconfidence bias is the relationship of subjectively assigned probabilities to measured long-run frequencies of the same occurrence, otherwise known as *calibration.* For example, one might be asked to provide the answer to a question along with the degree of confidence, expressed as a subjective probability, that this answer is correct. If many such questions were asked, the probabilities provided could be compared to the actual proportion of correct answers overall. Where the judged probability of correct answers closely matches their frequency, the probability assessor would be said to be *well calibrated.* So if a perfectly calibrated assessor were to assign probabilities of, say, 0.7 to various events, these same events would actually occur 70% of the time when observed. Conversely, overconfidence bias would be determined as the departure from these conditions.

Fischhoff, et al. (1977) describe one of many such experiments in which three groups of subjects were asked to answer different kinds of questions and to provide their subjective probabilities that each answer was correct. One set of questions involved judging which of two lethal events occurred more frequently (for example, drowning or bee stings), and these were given to two of the groups. The remaining group received general-knowledge questions (e.g., whether potatoes are native to Ireland or Peru). Each group included 40 to 60 subjects, all volunteer graduate students, and in total some 13,000 answers and subjective probabilities were provided.

The aggregated results shown in Figure 5-3 plot the corresponding subjective probabilities of error against actual error frequencies, and they are illuminating in several key respects. One is that the type of question seemed to make little difference, and the three groups' responses were fairly tightly bounded. But more significantly, they were well calibrated only within a rather small probability range from values of about 0.2 to 0.5 (estimates higher than 0.5 were not provided because the alternative answer would have been chosen instead). Below this range, overconfidence bias was demonstrated by actual error frequencies that were much higher than estimated error probabilities. In all three groups, the subjects thought they knew more than they really did.

Cognitive Discrimination

Another important feature of Figure 5-3 is revealed by the slope of the three group–response curves. For subjective probability values less than roughly 0.01, the curves are almost flat with very little reduction in error frequency even for estimated error probabilities that were negligibly small. This shows that the subjects had little ability for *cognitive discrimination* among subjective probability values below this threshold. Once they had made the determination that their likelihood of error was very low, they were all but unable to distinguish 10^{-6} from 10^{-2} or much of anything else in between. Whether they thought there was a one-

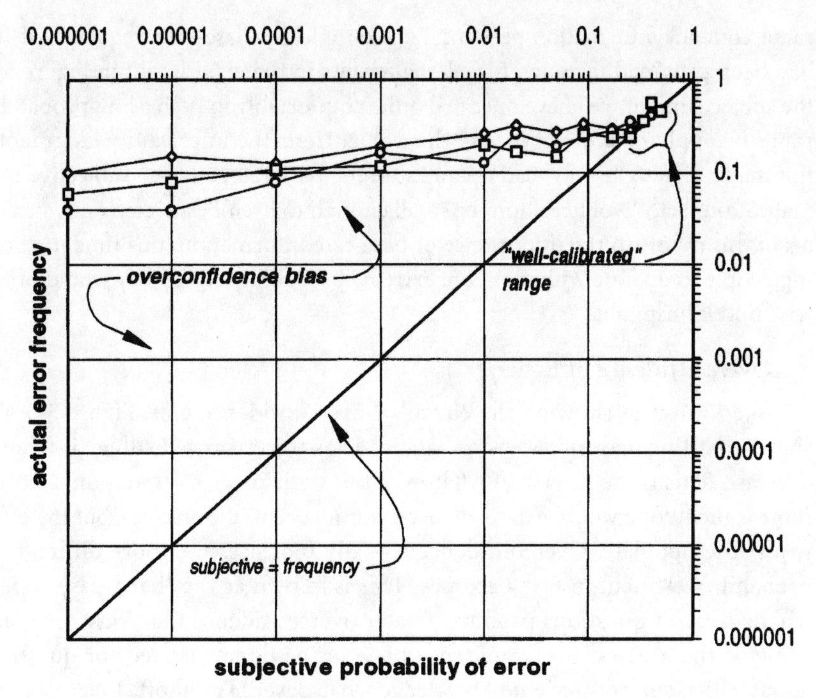

FIGURE 5-3. *Overconfidence bias and calibration.*

Source: Experimental data from Fischhoff, et al., 1977.

in-a-million, 1-in-10,000, or 1-in-100 chance of being wrong, their error frequency was virtually constant across the board. Within this range, no matter what odds were assessed they all turned out to represent essentially the same thing.

From this and related research, it appears that people have little ability to distinguish among very small degrees of likelihood, and their cognitive abilities in making subjective probability judgments do not extend very far into the range of extreme probability values (Hogarth, 1975). People simply do not operate very well with extremely small probability numbers because instances of such rare frequency or limited plausibility do not easily come to mind, and even when they do it can be exceptionally difficult to distinguish one minuscule degree of likelihood from another. And similarly at the other end of the probability scale, the number of 9s following the decimal point may not have much real significance. In this respect, cognitive discrimination can be viewed in terms of resolution. If probability were an instrument to measure degrees of uncertainty, it would operate with any precision only within some well-calibrated range. But this would dramatically diminish outside it, with differing degrees of uncertainty at some point becoming indistinguishable as the quantity measured becomes very large or very small.

There is, however, an important exception. The problem of cognitive discrimination may not apply to probability estimation tasks where some relevant base-rate frequency information pertains. For example, if a base-rate frequency of 10^{-5} had been provided for one of the lethal events in the experiment in Figure 5-3, the subjects might well have operated quite comfortably in such a low-probability range by applying adjustments to this value. Here, the uncertainty judgment is not made directly but instead operates on the base rate, and the subjective normalization factor would seldom be small enough to itself be adversely affected. So again this points to the importance of base-rate information, this time in allowing people to operate within a more extreme probability range they would otherwise find inhospitable.

Overconfidence of Experts

In addition to showing the effects of overconfidence bias, Figure 5-3 also shows that this bias increases in magnitude as the estimated subjective values decrease. This is the reverse of what might be anticipated. As questions become more difficult one would expect reduced confidence in the answers, but the effect is just the opposite: Overconfidence actually increases the more difficult the probability estimation task becomes. This is known as the "hard–easy" effect, where difficult questions produce greater overconfidence than easy ones, and some of the greatest overconfidence bias can be demonstrated for questions about which subjects have no knowledge whatsoever (Fischhoff, 1982; Lichtenstein, et al., 1982). It would appear that in general, the less people know about the subject matter, the less they recognize this lack of knowledge.

If so, then it might follow that those most accomplished in their knowledge domain should be less overconfident in their area of subject-matter expertise where questions for them are easier. It is often noted that posing general-knowledge questions to the graduate students who most often participate in these experiments does not necessarily mean that experts operate the same way. This too has been put to the test, with no less than geotechnical engineers the unwitting subjects in one example of particular interest. Slovik, et al. (1982) cite the case of predictions made by a group of internationally recognized geotechnical experts for the stability of a test-fill embankment. Focht (1994) describes the same case in more detail in the context of "lessons learned from missed predictions" in his Terzaghi lecture of the same title.

An embankment had been initially constructed in 1969 on soft clay in connection with Interstate 95 near Boston. Several years later, a symposium was held at the Massachusetts Institute of Technology to predict the height of additional fill that would cause the embankment to fail. The embankment crest was to be raised over a length of some 300 ft. A berm already existed on the west side (Fig. 5-4a), so the east side would be steepened to the angle of repose to induce failure there. The experts were furnished with detailed geotechnical data, including piezometer records, slope-indicator deflections, and comprehensive soil-property

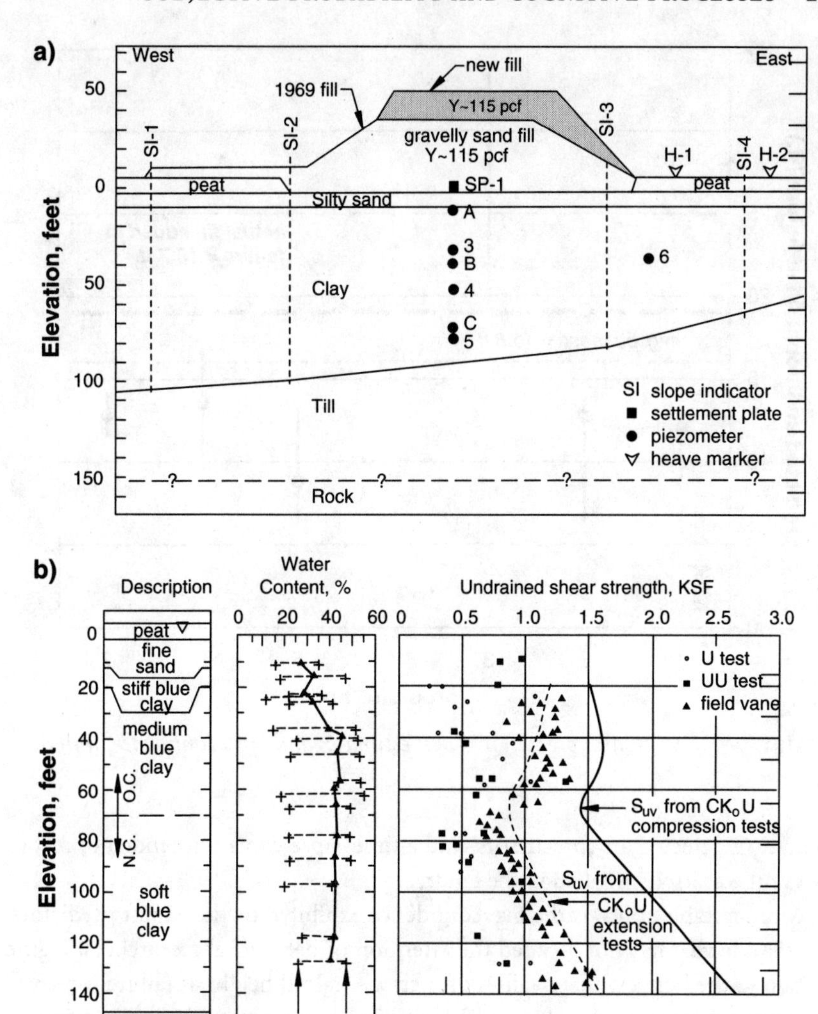

FIGURE 5-4. *I-95 test fill: a, cross section; b, soil properties.*

Source: Focht, 1994.

data (Fig. 5-4b). They were then to predict the incremental fill height at failure using whatever method of analysis they might choose.

The first hint of overconfidence came on the part of the symposium organizers themselves. When failure of the embankment occurred, it was of entirely unanticipated dimension and "extraordinary proportions." The crest dropped some 30 ft, more than half the total embankment height, and the failure extended an astonishing 450 ft beyond the heightened section in either direction. Not only did the east side fail as planned, but the west side did as well. None of the instru-

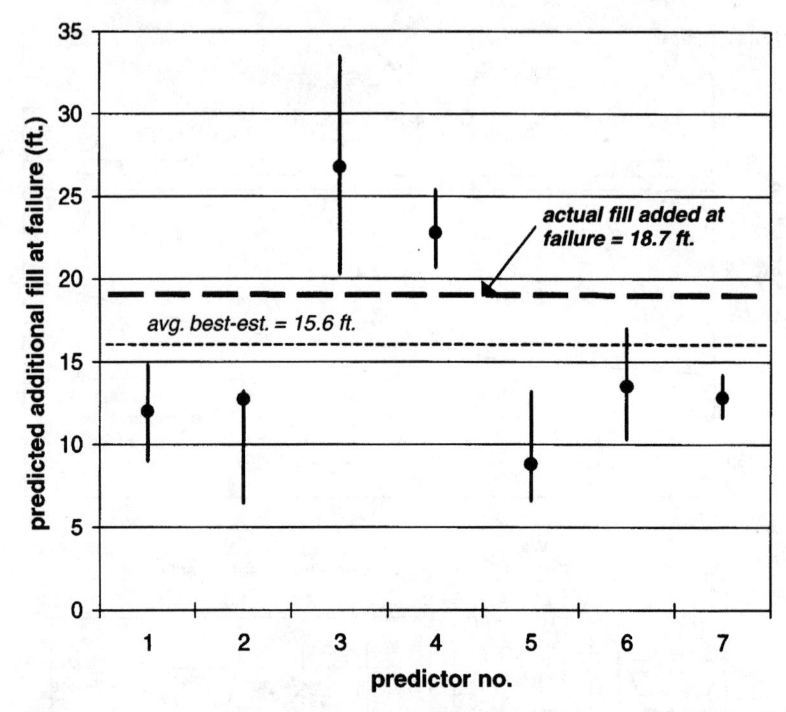

FIGURE 5-5. *I-95 test fill predictions (best estimates and 50% confidence limits).*

mentation detected any precursors, and a time-lapse camera intended to record the event was itself engulfed in the failure.

More notable still was the overconfidence exhibited by the expert predictors, an example that has not escaped the attention of behavioral researchers. Figure 5-5 shows the best-estimate values of incremental fill height at failure for seven predictors along with the 50% confidence limits they were asked to provide on these values (Hynes and Vanmarke, 1976). The best-estimate points were individually no closer than 5 to 9 ft to the actual result, although interestingly the group mean was within 3 ft, showing that the group as a whole significantly outperformed any one predictor.[4] But their overconfidence was complete. Had they been well calibrated as a group, half of their 50% confidence ranges would have encompassed the actual result. None did. Evidently, and as cognitive research would predict, neither the organizers nor the participants fully recognized how tenuous their assumptions and methods really were.

[4]Clients receiving different opinions from consultants are sometimes heard to complain that they should just split the difference and be done with it. The group performance in this case shows that there may be more to this than they suspect.

In terms of subjective probability assessment, an expert is one who has both *substantive* ability within their specialized knowledge domain and *normative* ability for providing unbiased uncertainty judgments. This example and other studies like it show that substantive expertise is no guarantee of normative expertise and that subject-matter experts can be just as affected by overconfidence as people in general (Hogarth, 1975).

Overconfidence in Subjective Probability Distributions

Overconfidence bias to this point has been described in relation to discrete or single-valued subjective probability estimates. It can also affect subjectively assigned probability distributions on continuous variables like strength, permeability, or other engineering properties. Here again, geotechnical engineers oblige in providing a case in point.

Folayan, et al. (1970) describe the reclamation of a former marshland of San Francisco Bay that required evaluating settlement of a fill placed on a uniform deposit of bay mud, the very soft and organic silty clay common to the area. As a first estimate of compressibility, and before any laboratory testing was performed, four engineers with 3 to 17 years' local experience and a single graduate student were asked to subjectively assess the mean and variance of compression ratio for the bay mud underlying the site. To express the confidence in their estimates, the participants were also asked how much they felt these estimates were "worth" in terms of an equivalent number of tested samples. After their estimates were provided, actual data were obtained from 39 one-dimensional compression tests.

Baecher (1972) analyzed the engineers' estimates, assuming compression ratio to be normally distributed. Figure 5-6 compares the subjectively derived distributions and those obtained from the subsequent data. The first thing to be seen is that all of the engineers underestimated the mean of the data, but Baecher notes that this site was indeed atypically compressible judging from typical bay mud compression ratios cited in the literature. Another is, not surprisingly, that the first three more-experienced engineers had greater confidence in their estimates judging from their equivalent numbers of tests.

These effects were examined more closely using this number and the associated mean and variance estimates to infer the participants' 95% confidence limits on the average compression ratio. The regions within which the average value should lie with 95% probability are shown in Figure 5-6 for the subjectively assigned distributions along with those for the measured data. If the participants were perfectly calibrated, these regions would correspond. Comparing them provides interesting insight into the overconfidence bias and how it varied according to the estimators' levels of experience and presumably their substantive expertise.

For the three most-experienced engineers the regions do not overlap at all, and their distributions were too narrow about the mean to encompass the major-

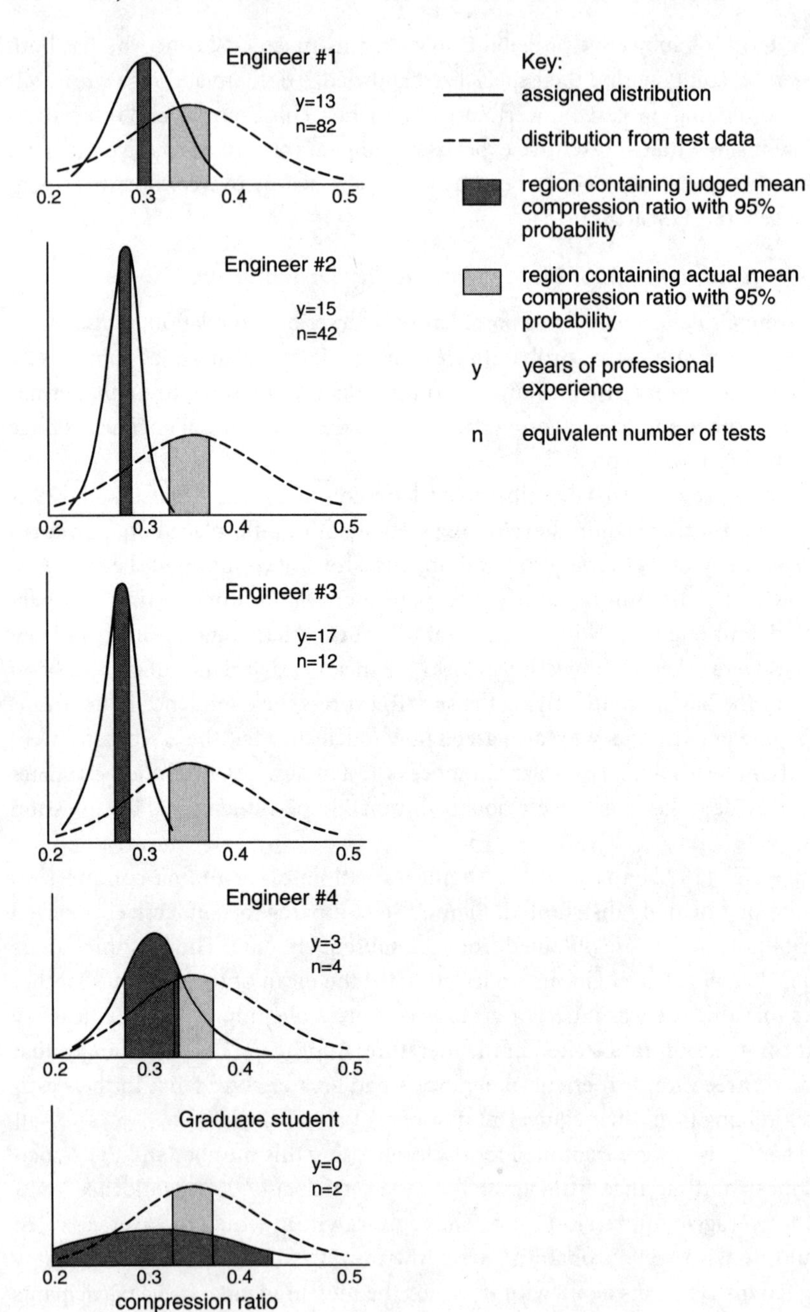

FIGURE 5-6. *Subjective probability distributions for bay mud compression ratio.*

Source: Baecher, 1972.

ity of the measured data. Their experience elsewhere apparently led them to misjudge how well it applied here to what turned out to be an unusually compressible site, and they were overly confident in their estimating abilities as a result. The novice engineer's distribution (no. 4) was more widely dispersed, and its 95% confidence region for mean compression ratio just reached the corresponding limit from the data.

But it was only the graduate student who was comparatively free from overconfidence bias, where the two regions for mean compression ratio in Figure 5-6 do overlap. This person, and to a more limited extent the novice engineer, were apparently able to overcome the "hard–easy" effect that afflicted their graduate-student counterparts of Figure 5-3 where specialized knowledge was not involved. This suggests it can be possible to reduce overconfidence bias when one's limitations within their specialized knowledge domain, along with the inherent limitations of this knowledge itself, are more fully comprehended. With training still in progress or only recently completed, the young engineer and the graduate student seem to have been better able to appreciate just how much they didn't know.

In this case overconfidence bias was inversely related to experience, showing again that substantive expertise in subject-matter knowledge does not necessarily give rise to normative expertise in subjective probability judgments. This is not to say that the inexperienced are the better settlement predictors, but that those more keenly aware of their limitations may be better probability estimators. Those with greater specialized knowledge can come to trust it more completely and thus less prepared to expect surprises. And while all of this has been couched in the vernaculars of cognitive psychology, it is really no different from what some in the geotechnical field have said all along. Here we need only return to Mitchell's (1986) admonishment:

> It is essential to expect the unexpected and to deal with soils as they are, not as we might wish them to be.

Weighing the Evidence

One factor common to the heuristics and biases described here is that nearly all of them in some fashion involve integrating various kinds of evidence. We usually call this "weighing the evidence." It lies at the very core of inductive reasoning and belief formation as a basic element of human thinking processes, though most often recognized in jurisprudence. We speak of a convicted criminal as being "proven guilty," but such a verdict is really nothing of the sort, at least in any deductive sense. The jury has delivered its finding with a statutory qualification—"beyond a reasonable doubt"—and this makes it a belief statement as Leibniz long ago recognized. We have seen from the stage model of the cognitive process that feelings of doubt are the essence of an uncertainty judgment. So an equivalent translation of the jury's criterion for conviction would be that "with a reasonable degree of certainty" they "hold belief" in the defendant's guilt. They have made this judgment of reasonable certainty by weighing the evidence even though it has been conflicting by the very design of the adversarial judicial process, the duty of both prosecution and defense to make it so. There are no rules for weighing the evidence, otherwise juries would be unnecessary and verdicts could be rendered by some formula. But there are cognitive strategies that juries and other people use, and these help reveal connections between the diverse collection of heuristics and biases.

Griffin and Tversky (1992) isolate two cognitive properties of evidence, its *strength* and its *weight*. By strength of evidence they mean its forcefulness or extremeness—its intrinsic persuasiveness in arguing for a proposition. Weight, by contrast, reflects the quality and quantity of the evidence, how much of it and how reliable it is, things like sample size and predictive validity. The strength of the evidence reflects its significance, while weight goes to its goodness. So a report of water gushing from a dam could be strong evidence of imminent failure, but if it comes from a schoolyard rendition of the Little Dutch Boy story it might not deserve much weight.

The reason why the strength/weight distinction is so important is that the two attributes cannot be considered simultaneously. With limitations on information-processing capacity being what they are, people instead treat strength and weight one at a time in arriving at their feelings of doubt and hence their uncertainty judgment. It is the sequence of doing this—first judging the strength of the evidence, then adjusting for its weight—which can explain overconfidence through the agency of two other biases: representativeness and anchoring-and-adjustment.

In initially evaluating the strength of the evidence, representativeness often comes into play. As we saw in the Linda and Jack problems, this reflects the extent to which something resembles a certain class. So representativeness bias almost automatically acts to emphasize the strength of the evidence, since it wouldn't be strong in the first place without being considered representative. When it comes time to adjust for weight, however, the anchoring and adjustment heuristic causes the necessary compensation to be insufficient. So here the strength of the evidence dominates, and overconfidence results from discounting its weight.

This natural tendency will be reinforced and uncertainty judgments will be particularly overconfident when strength of the evidence is high but its weight is low. So when judging a slope's propensity for failure, a calculated factor of safety of 3.0 might promote unwarranted confidence in its stability if this were based on, say, only two laboratory tests. But the reverse situation can produce the opposite effect. Should there be many indicators of slope instability (perhaps seepage and bulging or other such signs) but none of them entirely persuasive—no "smoking gun" as it were—then the strength of the evidence would be low but its collective weight high: There is a lot of it pointing in the same direction. Here the tendency would be for *underconfidence* in underestimating the potential for slope failure, with the evidence as a whole discounted as "circumstantial."

More generally, people assess their confidence in one of several competing hypotheses according to the balance of their arguments for and against this hypothesis. The arguments are judged primarily according to the nature and degree of the supporting and conflicting information, with insufficient attention to its quantity and quality. This gives rise to overconfidence when people form a strong impression with limited information and underconfidence when they form a moderate impression on the basis of more extensive information (Griffin and Tversky, 1992).

All of which takes us back to our jury and in particular to death penalty cases. Prosecutors are more apt to win such convictions on the basis of evidence with high strength than high weight. Their job is to increase the jurors' perception of its strength, while the job of the defense is to counter this strength by challenging its weight and validity. But in a recent review of such convictions in Illinois since 1977 it was found that 12 men had been executed but 13 freed after DNA testing established their innocence. With convictions on strength of evidence apparently no better than flipping a coin, one's philosophical views on capital punishment need not be at issue to question such verdicts, as did the state's governor in imposing a moratorium on executions. It is interesting to note that according to Griffin and Tversky's interpretation, the accuracy of verdicts might be increased by greater reliance on evidence considered circumstantial; the "smoking gun" would appear to be substantially overrated. Indeed, confession provides the strongest evidence possible. But an Illinois commission found negligible weight in the evidence provided by supposed confessions to jailhouse snitches in recommending that such testimony be restricted. Of course, none of this speaks very well of the judicial process, but the point here is not to indict it. The lesson for engineers is that uncertainty judgments are vulnerable to grave distortion when conducted in an adversarial or advocacy setting that promotes biases by design and encourages attempts to manipulate them.

To sum up this discussion on heuristics, we might observe that the engineering folklore contains some of its own, and while doubtless untutored from a cognitive point of view, they're not at all bad ones. These are the simple rules of thumb which serve as guideposts to our daily activities, and they have as much as anything to do with how we weigh evidence. One might be Murphy's Law: If anything can go wrong, it will. This serves as an ever-present reminder that overconfidence can reap unwanted rewards. Another might be "garbage in, garbage out," and though it usually refers to the representativeness of input for analyses it could just as well apply to representativeness bias more broadly. When judging the strength of evidence provided by analysis, or for that matter any other indicator, we must look equally to its weight. So it's not just having information that matters but also how much and whether it's any good.

Sound rules to go by these, in conducting everyday engineering activities and likewise in making uncertainty judgments. But before leaving the topic of bias, there is one more that remains. It too has to do with another basic human quality and perhaps the most fundamental tenet of engineering—honesty.

Motivational Bias

Recall that a valid subjective probability value must meet two criteria: It must be coherent with respect to the probability axioms, and it must reflect the assessor's actual beliefs. Up to now, we have considered only the cognitive biases that produce deviations from mathematical rules. These stem from the underlying conceptualizations that take place whether one is aware of them or not. But distinct from cognitive bias is another kind that operates on a much different level of awareness in violating the second criterion. *Motivational bias* occurs when a probability assessor has some stake, something to gain or lose, in the outcome of the assessment, where this includes the outcome of the risk analysis, decision, or any other purpose to which the probability value may ultimately be put. Here the assessed probabilities do not reflect the assessor's actual beliefs but are intentionally manipulated to reflect some form of direct or indirect punishment or reward.

As might be expected, cognitive research has little to say about motivational bias and seldom addresses it at all. While much can be learned about cognitive processes from peculiarities and inconsistencies in human behavior, most garden-variety intentional distortions are not especially enlightening. Yet motivational bias is the constant companion of subjective probability judgments in practice, though it receives disproportionately small and infrequent attention in relation to its power to influence them, with Otway and von Winterfeldt (1992), Roberds (1990), and McNamee and Celona (1990) among the few to even mention it.

It is useful to consider motivational bias as it operates in two spheres, one personal and the other organizational. To illustrate how it can occur on a personal level in distorting a value to an assessor's own self-interest, we turn to several fictitious engineers.

- *The designer.* Designers want to see some proposed structure or remedial measure go forward when the task of designing it would fall to them. Probabilities are accordingly underestimated or overestimated as the situation requires. Others may have similar interests in promoting activity or funding in their particular area.
- *The builder.* Builders know most everything there is to know about their structures—most importantly, that they built them to last. They aren't about to admit to anything that might suggest otherwise, and their probabilities could just as well be for the Rock of Ages.

- *The opponent.* Opponents don't like the project no matter what, and any roadblock in its path will do. They're not fussy, and if providing probabilities that show it is doomed is what it takes, then so be it.
- *The investigator.* Investigators slant a probability judgment to emphasize some topic from whose study they would stand to benefit in stature or funding. The investigator holds that certain things should be researched because good science requires it. Good science inevitably is that in the investigators' own field of interest, and of course they stand ready to assist.
- *The loyalist.* Loyalists are keenly attuned to the expectations of organizational culture. Loyalty is always rewarded, but disagreeing with the boss is not. Some structures can simply be too important to even contemplate anything going wrong, and the loyalist aims to please. But loyalists are above all flexible, and if some other result should be favored, well that's all right too.
- *The captive consultant.* Captive consultants are a subspecies of loyalists who know where their bread is buttered. Mostly dependent on a single client or industry, they arrive at a risk analysis session only to announce that they are there to ensure that it doesn't kill the project. The probabilities they provide are sure to guarantee nothing does.

As crudely drawn as these caricatures might seem, they have each made their appearance at one time or another. But engineers are jealously protective of their ethical standards and have always been quick to defend them—one reason why the public in every survey holds engineers in such consistently high regard. Still, Layton (1986) argues that ever since the "corporatization" of engineering became essentially complete in the early 1900s, turning what once were individual practitioners into organizational employees, there have remained built-in tensions between the bureaucratic loyalty demanded by organizations and the autonomy implicit in professionalism. Safeguarding the integrity of the practicing professional is a primary reason why the "founding societies" of engineering, among them the American Society of Civil Engineers, were established 150 years ago. This makes it all the more incredible that for engineers who could never conceive of fudging data or fabricating analyses, subjective probability assessment can be an invitation for party time at an ethical debauch, simply because there is no objective test of their subjective judgments. It does not excuse motivational bias on this personal level, however, to note that it can occur on the sponsor's level as well.

In risk analysis, a principal purpose of subjective probability is to communicate uncertainty judgments. But the sponsor of a risk analysis can sometimes be inclined to use it to an altogether different end—to persuade—where the public or regulators are the target, and in either case to establish that the project's risks are low enough to win their approval. Subjective probability can only express belief, and advocacy is something it was never intended to do. It can inform as to the sponsor's belief about risks and explain the reasons why, though others may or may not concur for perhaps equally legitimate reasons of their own. But when the chief purpose becomes to promote some desired result, an all but inevitable outgrowth is to condone or encourage likelihood assessments slanted to this end, even if only with a wink and a nod.

Motivational bias can be illustrated using the concept of belief sets, similar in principle to Venn diagrams. A belief set contains all possible beliefs an individual might hold about uncertainty, where these could result from changing states of knowledge, differing interpretations, vagueness in likelihood judgments, or even cognitive biases. The corresponding belief space contains all of the individual belief sets for those persons evaluating the uncertainty. Figure 5-7a shows the frequentist's belief space. To the frequentist, a belief set contains only a single point to which all rational assessments of uncertainty must necessarily converge. The belief space extends beyond this point only insofar as approximations are unavoidable, and beliefs outside it are rationally inadmissible.

To the subjectivist, the belief sets of equally rational persons are not required to coincide, just as Savage (1954) asserted. Like those in Figure 5-7b, individual belief sets can contain areas of overlap or intersection, although nothing demands that they do. But a mutually held understanding requires identifying these relationships, be they similarities or divergences. Risk analysis and probabilistic methods can help explore belief-set relationships and the factors responsible, ultimately aiding construction of a belief space encompassing them all as their union. Such a belief space represents a consensus, and it is possible for the very act of constructing it to make it shrink as areas of agreement are identified or areas of disagreement clarified.

But motivational bias precludes these insights by distorting the belief space as shown in Figure 5-7c. It is no longer possible to identify the relationships among belief sets or their causes when motivational bias intercedes. The situation is made even worse should the biased belief set be put forward as objectively derived and exclusively true, in the manner of Figure 5-7a.

Certainly it is a fine line that separates informing from persuading, achieving consensus from gaining consent. But when probability ceases to be a vehicle for communication and instead becomes and instrument of persuasion, the door is thrown open to motivational bias on every level. The point of the exercise is no longer to communicate insights about uncertainty but to achieve an outcome by purposefully misrepresenting it. In this, motivational bias becomes a particularly insidious poison. It pervades the process and does its damage by giving tacit approval to the characters we have met. And this damage is hidden if probability as an expression of belief is allowed or encouraged to be taken as a statement of objective fact. The art of persuasion is called advertising, and advocating a client's position is what attorneys are pledged to do. Engineers are in neither of these businesses, and subjective probability is not a marketing or public relations tool.

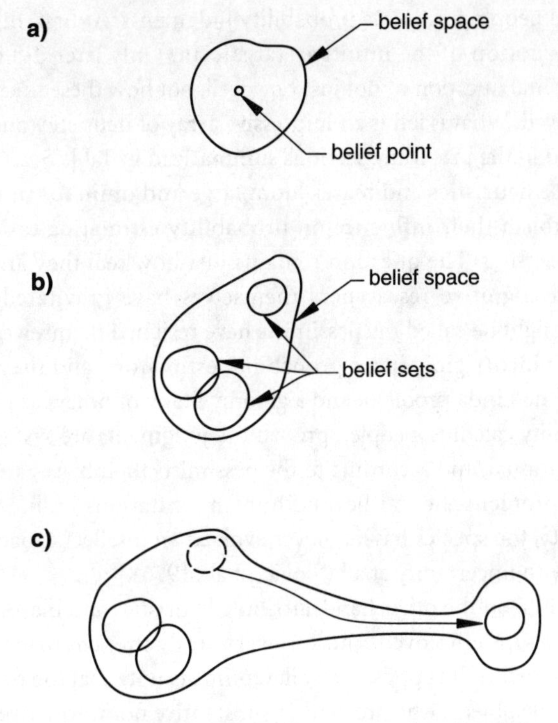

FIGURE 5-7. *Uncertainty belief spaces: a, frequentist; b, subjectivist; c, motivational bias.*

Suffice it to say that if subjective probability is to be used responsibly, then motivational bias is to be avoided at all costs. The assessor must act as a disinterested party to the outcome in providing probability judgments as fairly and honestly as possible. Motivational bias impairs judgment and is the antithesis of using it wisely. But if motivational bias is consciously induced, then it can also be consciously curtailed. The tradition and integrity of engineering judgment, and with it engineering itself, demand nothing less.

Some called it a scandal, others a collosal blunder. Whatever it was, the accounting profession until then had been trusted without question, held in the same public esteem as engineering. Time will tell if they can regain it. But it would be equally devastating to our profession's integrity if the very same motivational bias were ever allowed to become engineering's own Enron.

Heuristics and Biases in Perspective

A look back at their development provides a useful context for the heuristics and biases described here. Investigation of heuristics and biases was initially to determine how well people formulate probability judgments. At first, this was mostly to debunk the notion of the intuitive statistician. Only later did it turn to the more fundamental question of not just how well, but how these uncertainty judgments are derived. What is left is an impressive array of heuristics and biases, with those discussed in the preceding sections summarized in Table 5-2.

While these heuristics and biases loom large and ominous in research, it is harder to establish their influence on probability estimating tasks outside an experimental setting. The question remains just how real they are in everyday activities, and cognitive researchers themselves have gravitated to different camps. What might be called the pessimists have resigned themselves to people as hopelessly and incorrigibly inept probability estimators, and they offer up the Jack problem, the Linda problem, and a gloomy litany of others as proof. Judging by the probability calculus, people's probability judgments are systematically and stubbornly irrational, and according to the pessimists their biases are indelible. In this view, the problem goes far beyond human limitations in information processing capacity, the species having never evolved an intellect capable of dealing conceptually with uncertainty at all (Slovic, et al., 1976).

The optimists, on the other hand, attribute heuristics and biases partly to the experimental setups employed, and for every study that confirms one of them they cite another with the opposite result. Optimists note that the performance of graduate student subjects who are neither substantive nor normative experts cannot be definitively extrapolated to those who are. In their view, experiments can present a distorted picture of how people actually function. In identifying cues for reasoning strategies, people must seek out and determine for themselves what information is relevant, and no mathematical rule can prescribe this. So the "correct" answers narrowly specified by experimenters are not necessarily the only or even the best ones within a less restrictive interpretation of experimental problems (Beach, et al., 1986). Some optimists note that certain heuristics and biases can vanish entirely if experimenters' questions are formulated differently, so it is more the question than the subject whose rationality is at issue (Gigerenzer, 1994).

But there is also a middle ground. While acknowledging bias, this view holds that it does not necessarily result from reasoning that is faulty in some way, but from reasoning strategies that are not well matched to the probability estimation task (Fischhoff, 1982; Curley and Benson, 1994). By pointing no fingers at either experiments or subjects, a probability judgment can become more tractable if it allows people to use their cognitive skills to best advantage. This perhaps is the most useful perspective for understanding the role of heuristics and biases in the real world because it admits to the potential for human adaptability and learning to change their effects. For if people are not very good intuitive statisticians, this may simply be because (with all due apologies to Bernoulli) statistics for most

TABLE 5-2. *Summary of Heuristics and Biases*

Type	*Heuristic or bias*	*Description*
Cognitive	Availability heuristic	Judging likelihood by instances most easily or vividly recalled
	Availability bias	Overemphasizing available or salient instances
	Confirmation bias	Selectively retrieving supporting evidence while neglecting disconfirming evidence
	Hindsight bias	Exaggerating in retrospect what was known in advance
	Anchoring and adjustment heuristic	Adjusting an initial probability to a final value
	Insufficient adjustment	Insufficiently modifying the initial value
	Conjunctive distortion	Misjudging the probability of combined events relative to their individual values
	Representativeness heuristic	Judging likelihood by similarity to some reference class
	Representativeness bias	Overemphasizing similarities and neglecting other information; confusing $p[A \mid B]$ and $p[B \mid A]$
	Insensitivity to predictability	Exaggerating the predictive validity of some method or indicator
	Base-rate neglect	Overlooking frequency information
	Insensitivity to sample size	Overemphasing significance of limited data
	Overconfidence bias	Greater confidence than warranted, with probabilities that are too extreme or distributions too narrow about the mean
	Underconfidence bias	Less confidence than warranted in evidence with high weight but low strength
Motivational	Personal bias	Intentional distortion of assessed probabilities to advance an assessor's self-interest
	Organizational bias	Intentional distortion of assessed probabilities to advance a sponsor's interest in achieving an outcome

people is not intuitive. If so, there should really be no more reason for expecting people to mentally arrive at mathematically correct answers than to solve differential equations in their heads. Edwards (1975) maintains that people are not given the probability tools they would need or the opportunity to apply them, and to this extent it is surprising that they approximate mathematically governed answers as much as they do.

So just how real are heuristics and biases, and what is their practical significance? It would be hard to deny their existence in engineering activities, and we have seen too many examples involving geotechnical engineers to think otherwise. But this does not mean that they operate under ordinary circumstances exactly as they do in experimental settings. People are like soil specimens in the triaxial cell in certain ways. Both are inherently variable, and there is always a difference between their laboratory and field behavior. They are subjected to differing states of stress and can occasionally be somewhat disturbed. But experimental results nonetheless provide useful guidance about what to expect and a general indicator of field behavior, in people as in soils. Yet for all the heuristics and biases that affect them, people do manage to cope with uncertainties of everyday life, the probability calculus notwithstanding. So heuristics can best be seen as a double-edged sword: They serve well in many situations, handled with proper knowledge and due care, while bias may otherwise injure their user. This knowledge and care, then, are the keys to using heuristics effectively and using them wisely.

All of the discussions so far have viewed cognitive processes and assessments of uncertainty as occurring in isolation. But reality is much more complex. Decisions and actions under uncertainty often occur in chains and sequences, one dependent and building on another, rather than neatly separable in time. Each of us too operates in a framework of others, most often within the structure of organizations. Such things do not lend themselves to the experimental settings that researchers prefer, so understanding them requires looking elsewhere. One place has to do with engineers, science, and technology of the highest order and the organization where they all resided. It has to do with probability and risk and the diagnostic process, and above all with judgment. And it has to do with one of the most tragic technological failures of our time.

The Challenger Disaster

No one alive on January 28, 1986 will ever forget that casually bifurcating plume as it etched into the sky the place of the space shuttle Challenger's demise. So engraved in our minds is its image that we might be surprised to find it missing there now. It could seem almost irreverent, then, to mar such a monument in memory with the minutiae of technical detail. But what happened is this. At 58.788 seconds after ignition, a small flame emerged from the right Solid Rocket Booster (SRB) in the area of the aft field joint. As this enlarged to a continuous and well-defined jet, its eccentric thrust tore out the struts attaching the SRB to the External Fuel Tank, the massive unit that dwarfs the Orbiter and contains the liquid hydrogen and oxygen to fuel its main engine during later stages of ascent. The SRB now began rotating, first smashing into the Orbiter's right wing then back again into the External Tank. The torch-like flame from the SRB became increasingly deflected by the aerodynamic slipstream, impinging on the External

Tank and breaching it at 64.660 seconds into the flight. Challenger erupted into a ball of flame nine seconds later. The crew compartment, detaching from the Orbiter as the vehicle disintegrated, plummeted to the sea over the next 2.5 minutes and hit the water at 200 miles per hour with its seven crew members aboard.

In the postfailure technical diagnosis, recovery of debris and review of telemetry showed that the O-rings sealing the aft field joint of the right Solid Rocket Booster had been badly burned, with failure delayed from occurring on the launch pad only because charred material had temporarily sealed the joint. Other factors may have contributed as well. The charred residue might have remained intact if not for the severe wind-shear turbulence during launch that dislodged it, allowing hot propellant gases to penetrate the joint. Ice could have formed in the joint in the subfreezing cold of the pre-dawn hours preceding the launch then sublimated after ignition. But the root technical cause remained the O-rings themselves, and the decision to launch at an ambient temperature of 36°F that had not accounted for their decreased resiliency and inhibited ability to properly seal. Failure diagnostics from an organizational perspective would remain for later.

Manned space flight is an enterprise with inherent risks, as the National Aeronautics and Space Administration (NASA) would be the first to admit. But it could not be conducted without accepting certain risks, and NASA had in place at the time an extensive program to formally evaluate them. Clearly the Challenger mission failed, but did NASA's risk assessment process fail as well? If so, why and what can we learn? The launch decision and the entire shuttle design process that had preceded it were conducted under uncertainty of the most difficult kind. What cognitive processes might have affected these decisions? And why was the problem not diagnosed earlier and disaster averted?

In exploring these questions, we turn to Vaughan's (1996) detailed account of the actions and words of the shuttle program managers and engineers during the years, and ultimately the moments, leading up to the Challenger launch. A flurry of investigations followed in the wake of the accident, among them the Presidential Commission on the Space Shuttle Challenger Accident, or Rogers Commission (perhaps best known for physicist and member Richard Feynman's dramatic demonstration of O-ring resiliency by dipping one in a glass of ice water), another by the U.S. House of Representatives Committee on Science and Technology, and still others spawned by the recommendations of these two. Through the thousands of pages of interview transcripts now housed in the National Archives, Vaughan (1996) allows us to hear those involved tell their own stories in their own words. She argues that the Challenger disaster was not a failure on the part of the engineers, nor was it due to any bureaucratic incompetence or wrongdoing by NASA, but instead was attributable to the organizational norms and expectations, the organizational culture, that pervaded the space program. We will further see here that just when it mattered the most, the judgment of the engineers directly involved was subsumed by this organizational culture and the paradigm of science in which it operated. At its most fundamental level, one of the most celebrated technological failures of our time was itself a failure in judgment.

Probability, Risk, and Technical Culture in NASA

To fully understand NASA's conception of risk requires going back to its forerunners in the aerospace industry who pioneered quantitative reliability techniques. Among the first applications of these methods was in Germany in the 1940s during the development of the V-1 rocket (led by the legendary Wernher Von Braun), where the first 10 prototypes all blew up or fell into the English Channel. Given Hitler's great hopes for the V-1 in bringing England to its knees, German mathematician Robert Lusser developed some of the first reliability principles in response. Later, the U.S. Department of Defense adopted and refined these procedures, in the 1960s originating the first fault tree methods and using them in development of the Minuteman missile, with NASA's first forays into this area during the Mercury and Gemini programs for manned spaceflight (Henley and Kumamoto, 1992). By the time NASA embarked on the Apollo program to land a man on the moon, it was deeply involved in the use of quantitative probabilistic methods to address questions of safety, reliability, and risk (Garrick, 1989). Then something happened.

According to Cooke (1991), initial estimates of catastrophic failure probabilities for Apollo moon missions were so high that their release would have threatened the political viability of the entire space program, with one estimate of the likelihood of a successful manned lunar landing putting it at less than 1 in 20. John F. Kennedy's famously public pronouncement had, of course, already mandated that the United States would land a man on the moon and return him safely, so NASA, forced to choose between a presidential directive on one hand and probability on the other, perhaps understandably elected to abandon the latter. NASA never again used numerical probability measures to quantify risk, at least openly or officially, and continued to resist doing so even after post-Challenger recommendations by a National Research Council committee to this effect (Garrick, 1989).

What NASA adopted instead were qualitative, nonnumerical methods, and understanding how they worked requires some patience in penetrating several acronyms and protocols. Two such techniques were in place during development of the Space Shuttle and were applied sequentially: Failure Modes and Effects Analysis (FMEA) which we encountered in Chapter 4, along with Hazard Analysis (HA). NASA's use of both was prescriptive and detailed, with formalized procedures for making a determination of "accepted risk" based on their results.

The purpose of FMEA was to identify hardware items critical to the performance and safety of the vehicle, in particular those not meeting mission requirements. NASA applied it in the following way (Garrick, 1989):

1. Define the system and its performance requirements;
2. Specify the assumptions and conditions to be applied in the FMEA;
3. Develop reliability block diagrams or simple representations of system component interactions and dependencies;
4. Devise an FMEA worksheet and complete it for every identified failure mode, identifying "worst case" effects; and
5. Recommend and evaluate corrective actions, including design improvements or redesign.

In NASA's procedures, the upshot of the FMEA process was to identify those components not meeting certain design, operational, or fail-safe requirements. From these, a Critical Items List, or CIL, was constructed. Hazard Analysis took a complementary but slightly different tack. Using the failure modes and associated information identified from the FMEA, the HA developed accident scenarios, essentially working backwards to the critical components that might participate in these scenarios. HA took the further step of ranking the critical items according to their failure consequences using the criteria in Table 5-3. The importance of component redundancy is underscored here. Many design specifications required redundancy for safety purposes, and all but the lowest criticality category distinguished between failure of a single key component (i.e., C-1) and failure of a backup along with it (C-1R). Perhaps more important is that this scheme accounted for the consequences of failure but not its likelihood. Unlike the standalone FMEA procedures used for risk ranking in Chapter 4, nowhere in these procedures does likelihood appear in any form, qualitative or otherwise. In abandoning numerical probability, NASA, it seems, had abandoned considerations of occurrence likelihood altogether.

These results were used by NASA in what it called its Acceptable Risk Process (Vaughan, 1996). This was essentially a formal signoff procedure meant to do two

TABLE 5-3. *Criticality Ranking*

Criticality category	*Potential consequences of failure*
C-1	Loss of life or vehicle
C-1R	Redundant hardware element failure that could cause loss of life or vehicle
C-2	Loss of mission
C-2R	Redundant hardware element failure that could cause loss of mission
C-3	All others

things: first, to ensure that critical items not meeting fail-safe requirements received extra attention—in essence a diagnostic function—and second, to communicate this within the organization so that everyone would be fully appraised of the implications. Before the shuttle could fly, critical items had to be subjected to design modifications or corrective actions to meet redundancy requirements. If corrective actions were not feasible (and many weren't) a waiver request had to be submitted to NASA management that specified a "rationale for retention," a technical justification for retaining the critical item in the design. This might invoke such things as testing, inspection, operational experience, or measures to reduce failure potential. If the waiver was approved, the item was termed an "accepted risk" (Garrick, 1989). The important thing here is that the Acceptable Risk Process required justifying why a critical item should remain as part of the design, with a presumptive remedy of redesign. At this stage the burden of proof lay with the design, but this would later change in the case of the Solid Rocket Booster field joints and the O-rings they contained.

Matters involving accepted risk were more than just a paperwork exercise. They were central considerations in another formal process called Flight Readiness Review (FRR). Each shuttle flight was preceded by 15 months of activity on the part of thousands of people and countless hours of engineering work. The purpose of the FRR before each flight was to establish that the shuttle was ready to fly and fly safely. The FRR for the Solid Rocket Boosters was conducted at NASA's Marshall Space Flight Center in Huntsville, Alabama, where it took on a character all its own with an uncompromising ambiance that has much to say about NASA's organizational norms.

According to Vaughan (1996), NASA's highly respected technical culture was born at Marshall from a heritage of scientific achievement and military discipline. Wernher Von Braun himself had been its first director, where he would shepherd it through the spectacular successes of the Apollo program. Marshall was the citadel of NASA's scientific reputation, and as rightful heir to Von Braun's accomplishments it would harbor no fools.

The FRR process at Marshall was conducted before a review board, which Vaughan (1996) calls the "quintessential embodiment of Marshall culture." Chaired by its Director William Lucas, the proceedings were orchestrated in a cavernous auditorium before upwards of a hundred engineers with drama worthy of any impresario. To say that the FRR process was rigorous would be an understatement—adversarial would be more to the point. One engineer maintained that such large audiences were drawn by masochism, the engineering equivalent of Christians being thrown to the lions as they watched their colleagues perform under the spotlights then be eaten alive. Suffice it to say that Lucas demanded—and received—data, charts, and rigorous engineering analyses that left no stone unturned. No intuitive argument would survive the purposefully contentious challenges of Marshall's review board, and few engineers ever saw fit to throw themselves on its tender mercies in advancing one. Scientific proof and only proof would do.

The Flight Readiness Review was where performance experience and information gained from each successive shuttle mission were evaluated prior to launching the next. Any previous flight anomalies received close scrutiny, along with determinations of accepted risk they might affect. The O-rings in the booster rocket field joint received their share of this scrutiny and discussion. However, even as evidence of O-ring damage mounted, this damage eventually came to be seen as expected, not anomalous, performance.

The Solid Rocket Booster

Serious design work on the space shuttle began in the early 1970s, and the two Solid Rocket Boosters were among its chief components. Figure 5-8 illustrates the launch configuration, showing the familiar Orbiter vehicle, the External Tank to which it is attached, and the twin SRBs on either side. NASA awarded the contract for SRB design and fabrication to Morton Thiokol of Utah in 1973. As with many other shuttle components intended for reuse, Thiokol's segmented design offered certain advantages in transport and refurbishment. Its disadvantage was that the segments had to be connected, the joints constituting a point of vulnerability to leakage of the searingly hot gases produced by propellant combustion. So redundancy of the system used to seal the joints would be a crucial factor. At the business end of the SRBs were the rocket nozzles that directed the thrust, and these required joints as well. Although of somewhat different design from the field joints connecting the SRB segments, these "nozzle joints" used a similar kind of sealing system, both relying principally on two rubber O-rings.

Fabrication would require that the SRB segments be assembled at Cape Kennedy using the field joints shown in Figure 5-9. Consisting of a "clevis and tang" arrangement, the joint was to be sealed by two 1/4-inch-diameter O-rings, primary and secondary, made of rubberlike Viton material encircling the 12-ft diameter of the SRB. Also present, and shown schematically, was an annulus of flexible and heat-resistant putty of asbestos-filled zinc chromate to seal the thin gap at the top of the clevis and prevent direct gas impingement on the primary O-ring. The inner space to the right of the joints in Figure 5-9 contains the solid propellant, with the entire system serving to prevent the escape of hot gases which provide the rocket's thrust. So the primary O-ring was the first line of defense, the secondary O-ring its backup, and the putty was there to protect the primary.

Even before Thiokol began producing the rocket motors, predicting field-joint behavior was not simple, and this led to different interpretations between Thiokol's engineers and those with oversight responsibility at Marshall. Almost

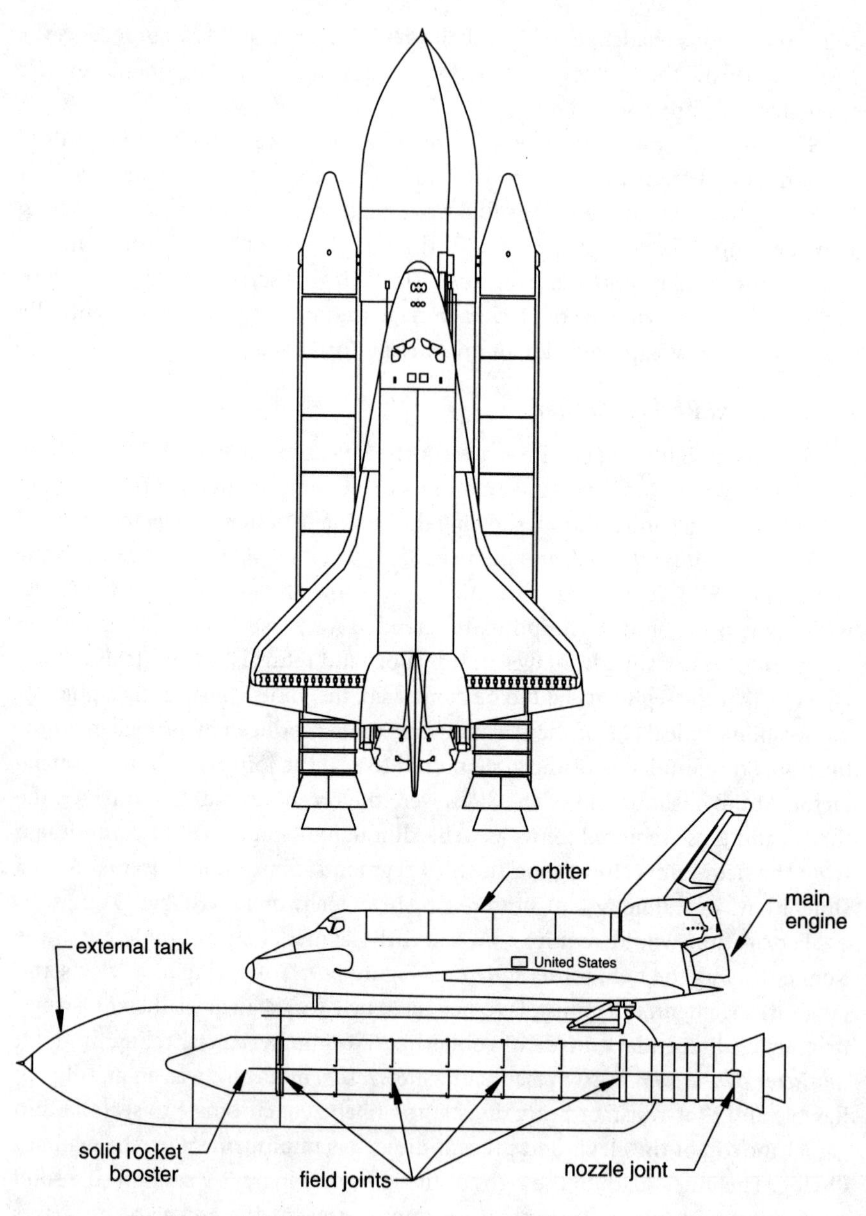

FIGURE 5-8. *Space shuttle system.*

immediately on ignition, internal pressure within the SRB would cause it to bulge ever so slightly, producing a minute but temporary eccentricity between the cylindrical walls of adjoining segments. This phenomenon of "joint rotation" between the tang and clevis was well-known in the industry, and it is why flexible sealing materials were needed throughout. But while Thiokol indicated that the

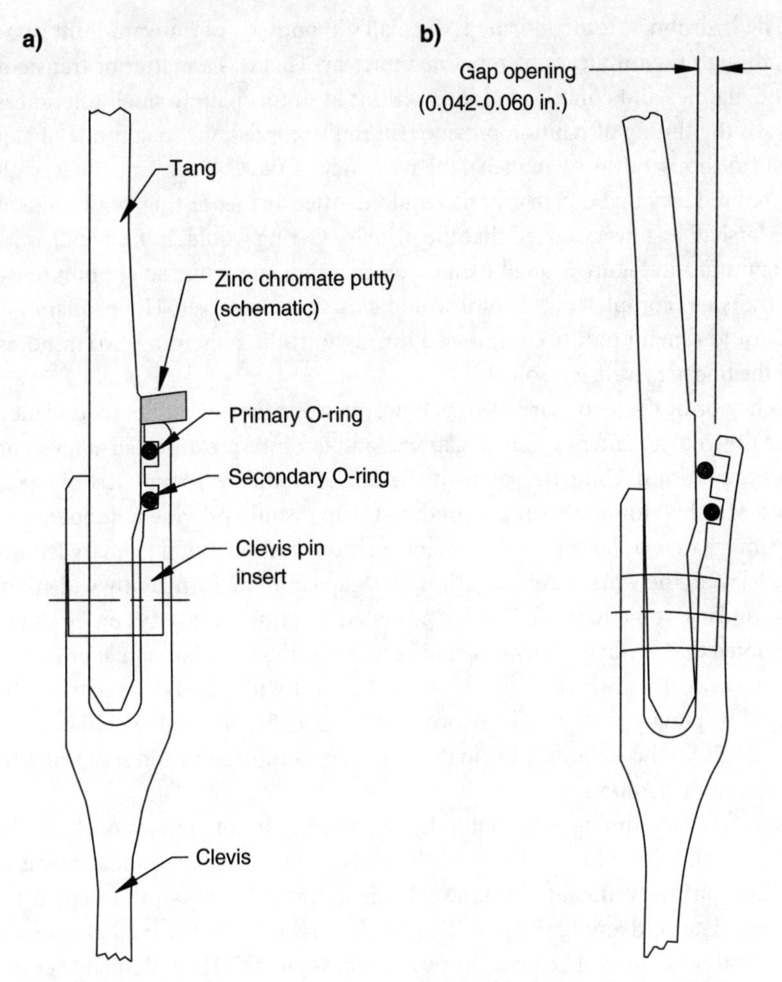

FIGURE 5-9. *Solid rocket booster field joint: a, unpressurized joint with no rotation; b, pressurized joint showing rotation effect (exaggerated).*

joint would close on the O-rings, compressing them and enhancing the seal, Marshall's calculations indicated the opposite, whereby ignition pressure would force the tang outward, forming a gap for a brief instant after ignition. The rotation controversy was to be put to rest in September 1977 by performing a "hydroburst" simulation using water pressurization. If the joint sealed as intended, the primary and secondary O-rings would both remain in compression, as the unpressurized conditions in Figure 5-9a depict. But if not, outward rotation of the tang (Fig. 5-9b) could allow the primary O-ring to slip out of its seat and jam the gap, delaying its closure. Should the primary fail at this precise instant, the secondary O-ring would be uncompressed and unable to seal.

The hydroburst test confirmed Marshall's hypothesis of outward joint rotation, though the implications remained unclear. This was a matter of transient joint dynamics and complicated interactions at unforgivingly small tolerances between the timing of ignition pressure (in milliseconds), the magnitude of gap width (in thousandths of an inch), the resiliency of the O-rings, and most of all their redundancy that design criteria required. After further testing, both Thiokol and Marshall engineers agreed that the primary O-ring would, in fact, seal, as its configuration in Figure 5-9b illustrates. Once it did, and with the ignition pressure transient completed, the joint would close back on itself. The primary O-ring would remain tightly compressed throughout the remaining two minutes until the boosters were jettisoned.

Behavior of the secondary O-ring, however, was problematic. Its redundancy boiled down to a matter of timing, gap size, and test interpretation, all at levels of exquisite precision. Using the gap width measured in the hydroburst test of 0.060 inches, Marshall found that the secondary O-ring would provide redundancy if the primary were to fail almost instantaneously on ignition. But if primary failure were delayed, the worst-case condition, this gap could be sufficiently wide that the secondary would not seal. Thiokol protested that this was overly conservative. They noted that the hydroburst test had been performed in a horizontal position, not the vertical launch configuration, so the joint width had been artificially enlarged by gravity ovalling. Their corrected 0.042-inch gap width would be narrow enough for the secondary O-ring to remain in compression and seal properly throughout the ignition cycle.

In view of the timing issue, redundancy was present but perhaps only conditionally so, and the matter was never fully resolved. Provisional redundancy was a situation that the criticality ranking scheme of Table 5-3 was not designed to accommodate, and ambiguity in criticality class assigned to the field joints was the natural outgrowth. The field joint was first assigned C-1R (redundant) criticality status and later changed to C-1. But all of the engineers unequivocally agreed at the time that one way or another the joint would seal, whether by the primary O-ring, the secondary, or both. A waiver was granted, a determination of accepted risk was made, and neither was ever again questioned.

Flight Experience

The developmental period involved four initial shuttle test flights for field verification of design performance. The first of these was designated STS-1, the launch of the shuttle Columbia on April 12, 1981. Postflight inspection of the SRB field joints showed that they had performed according to expectations in all respects. The field-joint design had received its field baptism.

Things were different, however, for STS-2 which flew in November 1981. When Thiokol engineers disassembled the retrieved boosters and inspected the field-joint O-rings, at first they did not realize what they were seeing. It took laboratory analysis to show that hot motor gases had eroded 0.053 inches of the primary O-ring in the right SRB's aft field joint, the same one that would later haunt Challenger. But since none of the other 16 O-rings in the two boosters had shown any damage, the problem seemed to be isolated and related to the putty. By further laboratory experiments they found that an O-ring with 0.095 inches of material removed would still seal. They had now established that the factor of safety with respect to the eroded dimensions had been at least 1.79, a prodigious margin as many aspects of space flight go. This was taken as further evidence for the security of the field-joint design. Besides, the problem was in the putty, not the O-rings. The putty could be, and was, reformulated.

Neither of the next two developmental flights had any O-ring anomalies, and the first operational flight, STS-5 in November 1982, also performed without incident. The putty modifications had apparently done the trick. But on STS-6 in April 1983, heat again penetrated the putty, reaching though not eroding primary O-rings in the left and right SRB nozzle joints. Still, these nozzle joints were not identical in design to the field joints, and the remaining three flights in 1983 experienced no anomalies. This demonstrated again that there was no design problem with the field joint—if there were, it would appear consistently, not intermittently as the idiosyncratic flight experience was showing.

But in 1984, flight STS-41B sustained O-ring erosion that Vaughan (1996) contends fundamentally changed the engineers' perception of risk. Again, as in the only previous case of damage for STS-2, minuscule bubbles and imperfections in the putty appeared to be the problem, but one thing had been notably different. During booster assembly, air pressure testing of the field joints was required to verify that the O-rings were properly seated and not pinched or otherwise misaligned. While all of the previous testing had been done at 50 to 100 psi, this had been increased to 200 psi for STS-41B in the interest of better assuring the seating. Even though the increased pressure had apparently been responsible for damaging the putty, verifying that the O-rings were properly positioned was the most important thing, and this could not be compromised. The root cause of the O-ring damage—the pressure testing itself—was now thought to be understood. The accepted risk it entailed now involved a tradeoff of increased O-ring erosion for assured O-ring seating. This brought about another fundamental change in thinking that invoked a causal reasoning strategy. Thenceforward, O-ring erosion was no longer considered anomalous but an expected condition within the operational performance parameters. It became taken for granted because now it could be causally explained, and more than that even rationalized as necessary. Thus it was no surprise when erosion again occurred for flight STS-41C in April 1984: O-ring erosion had become a predicted, and hence normal, aspect of joint performance and a risk to be routinely accepted on future missions.

But something new still happened during shuttle launch STS 41-D in August 1984. This time, erosion was accompanied by "blow-by" of propellant gases. This was evidenced by soot behind the primary O-ring of the nozzle joint, indicating that hot gases had reached it before it sealed. Blow-by was more serious than erosion because it demonstrated that the initial O-ring seal had been delayed. This was the very circumstance that had led Marshall engineers to qualify the redundancy capabilities of the SRB joint during design. Even so, the secondary O-ring had done its job. Now, both erosion and blow-by were predictable and expected phenomena, and both an accepted risk. Seven of the nine launches in 1985 were to have erosion and blow-by, but such incidents were now anticipated. What was not was the severity of two in particular.

For the first time, flight STS 51-B experienced complete burn-through of a primary O-ring along with serious erosion to the secondary O-ring, although in a nozzle joint not a field joint. Nevertheless, this was the first instance where the primary seal had been violated entirely, calling on the secondary to carry the burden alone. Equally disturbing was flight STS 51-C. Blow-by and erosion had affected the primary O-ring, and here the secondary had experienced blow-by as well showing that it too had been compromised if only instantaneously. But this experience now called into question for the first time the governing assumption of redundancy. What is more, the question of cold temperature had been raised for the first time. Shuttle components for the most part had been designed to withstand extreme heat, not cold, and this was something that had not been fully considered. The launch of STS 51-C in January 1985 had been preceded by three consecutive nights of record low Florida temperatures down to 18°F. Although by launch time the ambient temperature had risen to 66°F, allowing for the effects of thermal lag, it was calculated that the O-rings had been at a temperature of 53°F at liftoff. While it was apparent to everyone that cold, in principle, would stiffen the putty and harden the O-rings, the engineers were left without any quantitative data on the influence of cold temperature on O-ring resiliency. The issue simply had not come up before.

Concern now escalated, and internal memos at Thiokol began to speak of "red flags" and even "catastrophe." Thiokol engineers had personally inspected the O-rings from the recovered STS 51-C boosters. Seeing firsthand the unprecedented and alarming jet-black soot behind the secondary, they drew the seemingly plausible conclusion that the cold had decreased O-ring resiliency, delaying their sealing and producing the damage. But others at Marshall would not hear of such a thing and sharply questioned the basis for any such causal relationship. O-ring damage was known to result from a whole variety of different causes, and if cold had been responsible then why had some of the O-rings on STS 51-C been damaged but not others? Thiokol had no hard data, no quantified evidence, to demonstrate such a conclusion. And in true Marshall tradition, field observations and intuitive arguments were mere personal impressions too subjective to qualify as evidence in its scientific arena.

Thiokol's suspicions therefore remained just that and were never presented at the subsequent Flight Readiness Review, the cause of the problem identified yet again as minute gas paths through the putty. Nevertheless, testing on temperature effects was commenced, albeit without any real sense of urgency. Yet underneath this complacency, for the very first time, a nascent statistical reasoning strategy was beginning to emerge in the background and with it the first allusion to likelihood, at least for the weather-related part of the problem. If the effects of cold temperature could not be causally discounted, their occurrence was still so statistically improbable as to be of little concern. Speaking of the likelihood of experiencing a repeat of the cold temperatures of January 1985 again in 1986, one Thiokol engineer related it to the occurrence of back-to-back 100-year events, saying:

> Nobody had any idea that you would have like a 100-year storm two years in a row, and that's what that amounts to, tantamount to a 100-year storm. That statistically is so improbable that it won't exist.... It was nobody's expectation we would ever experience any cold weather to that degree before we had a chance to fix it again.... (Vaughan, 1996)

The assumption of negligible likelihood for ever again experiencing such cold conditions was also an underlying, if unstated, rationale for maintaining the acceptable risk status. This assumption continued to hold for flight STS 51-D in April 1985, but despite its moderate temperature this launch sustained the greatest depth of O-ring erosion to date: 0.068 inches, putting it outside the previous experience base of 0.053 inches encountered for STS-2 early on. Other than that, the only thing about STS 51-D considered somewhat out of the ordinary, but not especially ominous, had been its crew payload specialist, an astronaut by the name of Jake Garn—who also happened to be a United States Senator from Utah.

Meanwhile, by July 1985 some of the results of the temperature testing were beginning to come in from the lab. Not surprisingly, they showed that as the rubber got harder, the O-rings took longer to seal, although the data were just preliminary and for a limited temperature range. At the same time, an analytical model coupling thermal effects and joint dynamics had been devised to quantify the conditions under which the secondary O-ring would fail if the primary burned through, as it had already on STS 51-B. This showed that the secondary O-ring should remain intact, confirming redundancy in theory. Both the temperature testing and analytical findings were presented at Marshall, and they affirmed accepted risk yet again. Nevertheless, in view of the seriousness of the primary burn-through on STS 51-B, an engineering task force was established to solve the O-ring problem once and for all, even if this meant changes in field-joint design. And as it turned out, the deadline for completion of this work was set for January 1986, the scheduled month of STS 51-L—the fatal Challenger flight.

The Launch Decision

As Florida temperatures go, the forecast overnight lows at Cape Kennedy for January 27, 1986 were cold. Almost as cold as they had been prior to the launch of STS 51-C a year earlier that experienced such unprecedented erosion and blow-by of both primary and secondary O-rings. But unlike before, temperatures were predicted to rise only nominally by Challenger's scheduled liftoff. It was calculated that the temperature of the O-rings would be 29°F, some 24 degrees colder than their 53°F temperature for STS 51-C. The freakish conditions thought so unlikely to recur just had, and with thermal effects that would be even more severe. Moreover, what were supposed to be the definitive O-ring studies remained in progress.

At 8:45 PM EST on the evening before Challenger's liftoff, a three-way teleconference was held with participants from Kennedy, Marshall, and Thiokol after having transmitted various data, calculations, and notes by fax. For the next two and a half hours the final launch recommendation would be debated by some 34 engineers and managers with engineering backgrounds. The topic would be the field booster joints, the predicted launch temperature, and the effects on the O-rings.

As the discussions unfolded, several Thiokol engineers presented their concerns. Their argument came down to this. There had been two documented instances of blow-by in field joints: STS 51-C and STS 61-A in January and October of the previous year. Temperature data for both as presented in the meeting is shown in Table 5-4. Thiokol engineers had inspected the field joints in both cases, and there was no doubt in their minds that the damage to STS 51-C had been much worse. They reasoned again that temperature is what had made the difference between this and the warm launch of STS 61-A, since the effects of temperature on resiliency were at least qualitatively understood. After further roundtable discussion of the data and what they might indicate, Marshall's project manager in due course asked Thiokol for their launch recommendation. Thiokol responded that they could not recommend launch at an O-ring temperature outside the available precedent, that is, any lower than the calculated 53°F experienced for STS 51-C. The launch should wait until O-ring temperatures could sufficiently rise later in the day.

Immediately Marshall managers began contesting this interpretation just as they had before. Since field blow-by had been experienced for both warm and cold

TABLE 5-4. *Field Joint O-Ring Blow-By Occurrences and Temperature*

Flight	*Calculated O-ring temperature*
STS 51-C (January 1985)	53°F
STS 61-A (October 1985)	75°F

launches, there was no conclusive evidence, no hard data, to support Thiokol's claim. As Thiokol's engineers themselves were aware, what evidence they had was merely circumstantial and really little more. Besides, even their own presentation had included some results of smaller-scale static testing showing no blow-by for either warm or cold simulations, contradicting their own position. Thiokol, it appeared, was talking from both sides of their mouth. Marshall managers continued to hammer away, noting that no definitive Launch Commit Criteria for booster joint temperature had ever been established and that Thiokol was attempting to do so now on the very eve of the launch. They were not being objective. Since the data themselves were inconclusive, Thiokol's recommendations and observations could be nothing more than their subjective interpretations.

Then a Marshall manager attending at Kennedy piped in on the conversation, asking for his reaction to all this from George Hardy, Marshall's well-known and universally respected Deputy Director of Science and Engineering. Shirtsleeves rolled up and taking careful notes on a pad in concentration, Hardy looked up at the speakerphone, responding sharply and without hesitation that he was "appalled" at Thiokol's recommendation, the first and only time in the entire shuttle program that Thiokol had asked for a launch delay, and this from data that were inconclusive.

It is reported that after Hardy spoke there was dead silence on the line for several seconds. Many of the 34 listening were stunned. This was to be the defining remark, the turning point of the meeting and, what is more, a turning in point of view. Up to now, the system had always been treated as redundant, erosion and blow-by had always been anticipated by everyone, and everyone had concurred that this was an accepted risk. After all this time Thiokol was changing its tune, but the burden of proof had suddenly changed as well. The precedent for accepted risk was entrenched, and whereas its initial determination from the Hazard Analysis had placed the burden of proof on the field-joint design, it was now on Thiokol to show the design faulty. They now were required to prove that the launch should not go forward as scheduled. And they didn't have the data to do it.

At this point in the teleconference Thiokol asked for a break, and a 30-minute recess was held with all three sites on mute. Present in Utah were four Thiokol managers at the vice president level, one of whom later recalled that "we had to make a management decision" apart from the engineers. When there were engineering disagreements that could not be resolved by the data, as it was clear by now there were, management had to break the deadlock. In what ended up as an informal poll among themselves, the managers caucused. The initial tally was three in favor of launch. The lone holdout was reminded that as an engineer in a management position, "you've got to put on your management hat, not your engineering hat." Acquiescing, he voted with the rest.

The teleconference now reconvened. Marshall was informed that Thiokol had reconsidered, and their revised flight-readiness recommendation was read off.

Although temperature effects were admittedly a concern, the data predicting blow-by were inconclusive and therefore insufficient to justify launch delay. With this, the determination of accepted risk had been preserved for the last time, and the meeting ended at 11:15 PM EST. One of the final comments heard on the line before the parties hung up, addressed to whom is unclear, is said to have come from a Kennedy attendee: "Don't forget what George [Hardy] said about the secondary, you know there's always the secondary."

Some two hours later at 1:30 AM EST on January 28, 1986, the Ice/Frost Inspection Team at Kennedy alerted Rockwell International in California, the Orbiter's prime contractor, that despite de-icing, solid ice as much as 3 inches thick remained in places on the shuttle's multilevel 235-ft. high service tower along with icicles up to 18 inches long. The walkway used for emergency flight crew evacuation was also covered with ice. Rockwell expressed concern that acoustic effects at ignition would create ice debris that might damage the Orbiter's delicate heat tiles or be aspirated into the SRBs at liftoff. Rockwell's position was that they had not launched before in such conditions, so they could not be sure it was safe to fly. Their data were inconclusive, and this justified launch delay.

Launch was moved back slightly until later in the morning to allow ice to be cleared and the sun to melt it. Challenger's crew were awakened by 6:00 AM, and after a leisurely breakfast were given a weather briefing. The ambient temperature was then 24°F. At 7:00 AM the ice team made another inspection, and at 8:30 AM the crew were strapped into their seats. By 10:55 AM the ice was melting in areas of direct sun and icicles were falling from the service tower. Ambient temperature was 36°F when the final ice inspection and cleanup were completed, with an O-ring temperature later calculated to be 29°F.

Shortly after 11:25 AM EST the terminal countdown began. Liftoff of flight STS 51-L occurred at 11:38 AM, and 73 seconds later Challenger went down in history.

What Went Wrong

How could all this have happened? The effects of temperature on O-ring resiliency were not particularly difficult to grasp, and it did not take rocket science to comprehend the implications of icicles on the launch pad. Even to the geotechnical engineer, the Challenger story is perhaps as close as anything one might find to the observational approach run amok. Field experience was progressively documenting more and more observational evidence of problems. These problems did not occur consistently on every flight, but their progressively escalating severity was unmistakable even at the time. What was missing was interpretation of what was being observed, a trigger point for implementing some basic change in the design or at least for postponing the launch. Without this, observations without actions led directly over the cliff.

The Challenger disaster was a failure on many levels, and it holds probably as many lessons as there are those who care to learn them. But from our perspective there are three that stand out. There was first of all a failure of diagnosis. The entire joint sealing mechanism had been suspect from the start, but these suspicions were in the end ignored. There was a failure of NASA's risk analysis process. It had been relied on to perform a diagnostic function in identifying high-risk shuttle components and bringing about changes in their design, but this it did not do. And most of all, the Challenger failure was a failure in judgment, not just a failure of judgment to be applied but an outright rejection of judgment in principle. We can go on to examine these failures, in the process using some key post-failure recollections of the engineers and managers involved, as provided in the investigative hearings.

Recall from the cognitive model in Figure 5-1 that there are two kinds of reasoning strategies, causal and statistical, where one or the other is adopted according to problem cues. NASA had always tried to diagnose the O-ring problem using the causal strategy of cause and effect. Testing and analysis were counted on to find the cause of the problem and fix it. The hydroburst testing said there shouldn't be a problem, but when it appeared anyway it was first attributed to the putty and then to the field assembly pressure testing. Analyses of increasing sophistication were relied on to find the answer, with the thermal analysis of field-joint dynamics ostensibly confirming O-ring redundancy. And throughout all this, more general kinds of information took a back seat to data and analytical complexity. That cold makes rubber stiffer, an observation of childlike simplicity, was simply ignored without hard data to prove it.

But flight experience kept undercutting these causal strategies. The problem cue it provided pointed increasingly to the statistical nature of the problem, as continuing flight experience failed to confirm each explanation in turn. Problems kept occurring despite each fix adopted. The field performance of the O-rings was not reliably predictable. It was idiosyncratic, occurring on some flights but not others, in some field joints but not all of them, sometimes in field joints and sometimes in nozzle joints and sometimes not at all, and none of this with any consistency that causal strategies were able to explain. NASA did not ignore this flight experience, but they never saw the O-ring problem in the statistical light that its unpredictable nature would engender. This would have required recognizing the patterns that flight experience contained, and it is here that NASA's diagnostic process broke down.

Precedent stood in the way of pattern recognition because of how precedent came to be viewed. As instances of anomalous O-ring performance kept accumulating, they were incorporated into the body of what was seen as successful flight experience. Despite increasing evidence of O-ring charring, blow-by, and even complete burn-through, the field joints still had not caused a mission failure up until that time. So instead of seeing a pattern of anomalies, the anomalies became

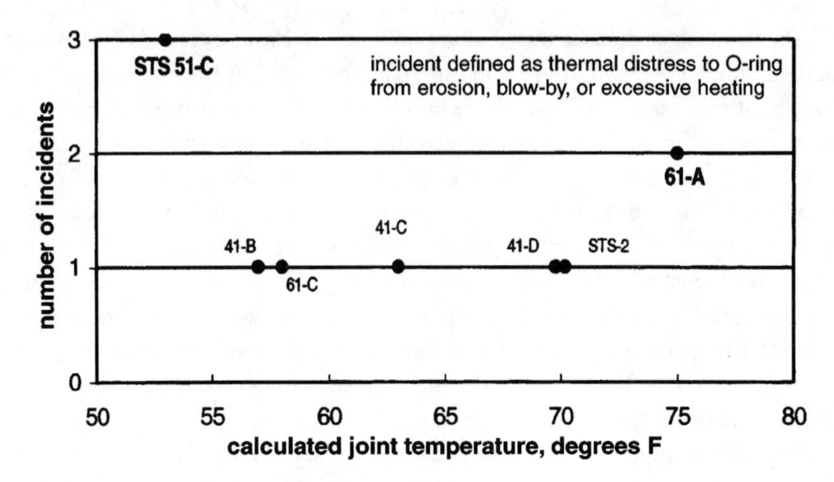

FIGURE 5-10. *Field joint O-ring incidents.*

the pattern—the pattern of successful precedent. The patterns of anomaly did not come forward to announce themselves, but they were there to be found if one were prepared to tease them out.

Things become crystal clear in hindsight, and the patterns of temperature data are no exception. Two investigators from the Rogers Commission, one an attorney and neither of them engineers, compiled and interpreted a more comprehensive database than was ever done at the time (Vaughan, 1996). Figure 5-10 shows all of the incidents involving field-joint O-rings that had occurred up to the Challenger launch as a function of calculated O-ring temperature, defining incidents broadly as occurrences of erosion, blow-by, or excessive heating and highlighting the two incidents raised in the prelaunch teleconference. Here the number of O-rings affected per launch is also shown, recalling that the two SRBs contained a total of 16, so the weight of the evidence as well as its strength becomes more apparent. In this format, perhaps some temperature-dependent trend might have been discerned from the highlighted STS 51-C and STS 61-A, the warmest, the coldest, and the only two cases the teleconference discussed. Yet considering the rest of the data, this is far from clear: Both strength and weight of the opposing arguments appear to cancel out.

A different picture emerges, however, when not just the incidents of thermal distress, but also flights with no distress are included as in Figure 5-11. Here, two separate subpopulations of data are identified, those above 65°F and those below. While less than 20% of the warm launches had problems, every launch colder than 65°F experienced at least one. It now becomes apparent that temperature was indeed not the only cause of the O-ring problems, with other factors at work as their occurrence for some warm launches demonstrates. But one would have needed to know nothing at all about what caused O-ring problems to observe from

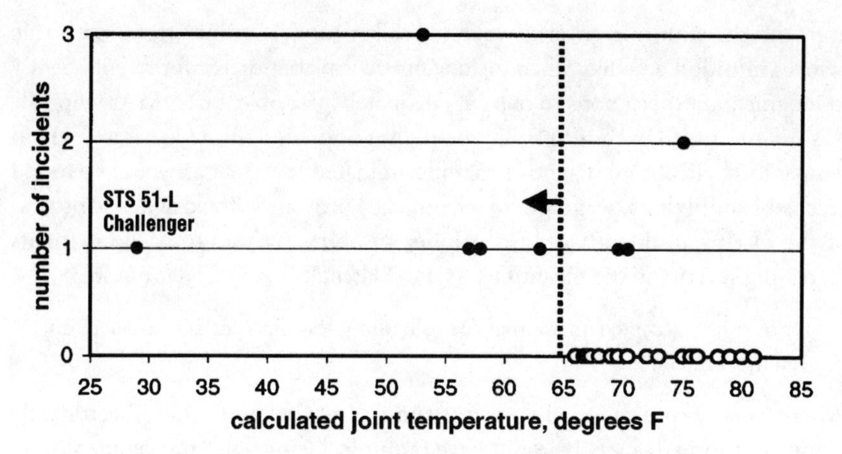

FIGURE 5-11. *Field joint O-ring incidents and successful performance.*

Figure 5-11 that the conditional probability of experiencing a problem given a temperature less than 65°F was fast approaching 1.0 (just before Challenger it had reached 0.83 from Eq. 5-1) to say nothing of just how far beyond past precedent Challenger's calculated 29°F O-ring temperature actually was. Whether this might have revealed the true nature of problem is something we will never know. But it speaks volumes that it took two nonengineers to develop and apply a statistical strategy to what NASA had always treated as a causal reasoning situation. For not just a pattern of empirical correlation, but causal proof was what NASA required. And by the time of the Challenger launch decision the burden of proof had shifted dramatically. Even at the same time that Rockwell in the pre-launch hours was accorded a launch delay because data on launchpad icing were insufficient to assure flight safety, much the same data insufficiency with regard to O-ring effects was not enough to overturn the established precedent in favor of launch.

These problems can also be viewed in the context of NASA's diagnostic approach, where both variationist and empiricist strategies went awry. Review of Table 5-1 shows that the variationist strategy works from the intended design functioning to the anomaly. The intended function of the entire joint-sealing system had always been to assure the integrity of the joint, but the pressure testing pushed this into the background. Even though the increased pressure was damaging the putty, the O-rings were doing exactly what the design said they had to do—seat properly during assembly. So from the variationist standpoint, there was no anomaly. And correspondingly from Table 5-1, the empiricist strategy works from the expected system functioning to the anomaly. But so many flights had performed successfully despite O-ring damage that this damage itself became expected. So the empiricist strategy said there was no anomaly either.

But the problem went deeper. Recall from earlier discussions that every diagnosis has some kind of stop rule. The diagnostic process does not continue forever, and the stop rule governs when to "call it good." Ordinarily, the stop rule invokes intuition, in effect when the diagnostic explanation conforms sufficiently to judgmental expectations to make it adequately plausible. But NASA's stop rule was different, and even if to some the flight anomalies pointed increasingly to temperature effects, no diagnosis that incriminated the O-rings would be found acceptable until this assertion were scientifically proven. So the diagnostic process of the O-ring problem would continue, without any way to account for its increasing severity in the meantime. As one Marshall engineer later put it:

> Once you've accepted an anomaly or something less than perfect ... where do you draw the line?

Where exactly to draw the line is not something that could have been neatly planned out in advance. It would have required seeing what was happening in flight experience, an awareness of the developing situation that comes from discerning its patterns and projecting them forward, and doing this all by using judgment. Without it NASA was blindfolded, proceeding steadily closer to the edge of the cliff without ever sensing when to stop.

Vaughan (1996) adopts an organizational perspective on NASA's diagnostic process. She calls the continued acceptance of the flight anomalies and the risks they posed the "normalization of deviance." What she means is that risk, to the extent that it represents a deviation from desirable safety conditions or design specifications, kept being tolerated and justified by all manner of rationalizations despite the manifold factors arguing otherwise. Risk acceptance was "normalized" within the organization in that it became a normal state of affairs. But there is more to it than this. For NASA to have accepted certain risks was not deviant per se; it was part and parcel of their cost of doing business. The problem was not that they accepted risk but that they never knew how much they were accepting, or where. They could not recognize that field-joint risk was steadily escalating in the face of the accumulating evidence because, as the window through which they viewed risk, their risk assessment tools could not alert them to this. And this all goes back to NASA's mistrust of probability that began with the Apollo program.

To appreciate the deficiencies in NASA's assessment of risk, we must start from the fundamental precept that risk equals a function of two things: the likelihood of an adverse occurrence and its consequences. NASA's techniques, in the way they were applied, incorporated the latter but never addressed the former. In ignoring likelihood, they were neglecting fully half of the risk equation. With each successive flight, the state of knowledge was changing. The likelihood of something very bad happening with the O-rings was increasing with time as more and more of their vulnerabilities were revealed, from erosion to blow-by then burn-through. And if likelihood was increasing, therefore risk had to be incrementally increasing too. But to NASA, without formally acknowledging the building likelihood of failure or having any way to incorporate it into risk estimates, the accepted risk remained static. This has everything to do with why risk became "normalized" in Vaughan's sense. Something that is not perceived to change becomes the norm by default, and things that are normal and ordinary are accepted as a matter of course.

And why it was that NASA chose not to address likelihood could only have been the outgrowth of their longstanding aversion to probability. In the end the cost was high. Without probability, or some expression of likelihood, their tools misled them into thinking that what they were getting was a measure of risk, when in fact it was not. The Critical Items List constructed according to failure consequence ranking did not provide the needed information. At the time of the Challenger launch, there were some 748 "Criticality 1" items for the space shuttle whose risk was formally accepted and fully 114 of these were for the Solid Rocket Boosters alone (Vaughan, 1996). While all of these items may have had the same consequences of failure, they could not all have produced the same mission failure probability. So risk necessarily had to vary among them. But with no way to define the magnitudes of these risks, NASA had no way to prioritize critical items according to their relative importance, so there was nothing to distinguish the SRB field-joint risk from any other in the same criticality category. In its attempt to cleanse itself of potentially embarrassing probability estimates, NASA had thrown the baby out with the bathwater. They were mistakenly equating consequences with risk.

There was another significant aspect to this. Recall that it was after flight STS 41-B in 1984 when engineers first recognized that the increased 200-psi leak check pressure was contributing to O-ring erosion. But it was more important that the O-ring seating be verified at this pressure, even as it became clear that microscopic holes in the putty would unavoidably result. As one Thiokol engineer recalled:

> We thought it was of the utmost importance to have a verified primary O-ring and so we increased the leak check pressure to 200 psi, ... realizing that blow holes are not desirable either, but yet it is more important to know that you have a good O-ring and have some [more] putty blow-through than otherwise.

In continuing to accept the risk of O-ring damage, they were knowingly engaging in a risk tradeoff, one they felt to be warranted. Perhaps they were right given what they knew at the time, but being unable to quantify the risks involved, to compare the relative risks with and without implementing the high-pressure procedures, they really had no good way to tell. This again would have required some measure of relative O-ring failure likelihoods associated with the original and the increased testing pressures that incorporated both seating and putty effects. But something still more was at work.

The crucial factor throughout, and one that continued to support faith in the field-joint design, had been the redundancy of the O-ring system—if the primary failed, the secondary would be there for backup—even though Marshall had reckoned early on that this redundancy might only be provisional on failure of the primary O-ring early in the pressure transient. For redundancy to be maintained would require that each of these components behave independently: The failure of one could not be affected by failure of the other, nor could they both be affected simultaneously by the same thing. What was never considered in a redundancy context is that cold temperatures would equally affect both O-rings. This results in what are called *common cause* or *common mode* failures, where components presumed to behave independently are mutually subject to the same external influence, and even with sophisticated fault-tree techniques these kinds of failures in complex systems can be especially difficult to identify (McCormick, 1981). But NASA's Hazard Analysis procedures, although based on failure scenarios, could not incorporate this kind of effect (Garrick, 1989). To do so, these scenarios would had to have been decomposed in far greater detail and a more formal event or fault tree representation adopted.

If one were to trace the key events in the field-joint failure sequence in highly simplified form, they might have looked something like the representations in Figure 5-12 at various points in time as the situation unfolded. Recall that shuttle launch STS 51-C after the unprecedented Florida cold snap of January 1985 had been the first to really alarm the Thiokol engineers and alert them to the possible effects of cold temperatures. With this incident, the field-joint failure pathway shown in Figure 5-12a had become apparent. In fact, it was a near miss, with the conditions just experienced having incorporated all but its final branch. But recall too how improbable they felt a recurrence of these temperatures to be. This was the initiator of the failure sequence, and for future launches they were relying in effect on some very low probability of the first branch occurring. As for the second branch in the failure pathway, the engineers never directly said what probability they might have assigned to it, but in July 1985 a Thiokol memo would rate as a "jump ball" the likelihood that the secondary would seal should the primary fail—a 50/50 chance of the third and final branch occurring.

By the eve of the Challenger launch, at the time of the fateful teleconference, the situation looked like Figure 5-12b. The unimaginable had just transpired. With a recurrence of the cold conditions having now come to pass, the initial branch set, the occurrence of cold temperatures, was no longer germane. The neglibibly small initiator probability (the "100-year storm two years in a row") counted on to prevent the entire failure sequence had just become 1.0. All that remained were two events: that the primary would burn through, then the "jump ball" secondary too.

This, however, still assumed that these two features were redundant and hence that their potential failures would constitute two independent events. Had common-cause effects of temperature been recognized, the situation now would have looked more like Figure 5-12c. From this standpoint, the remaining two

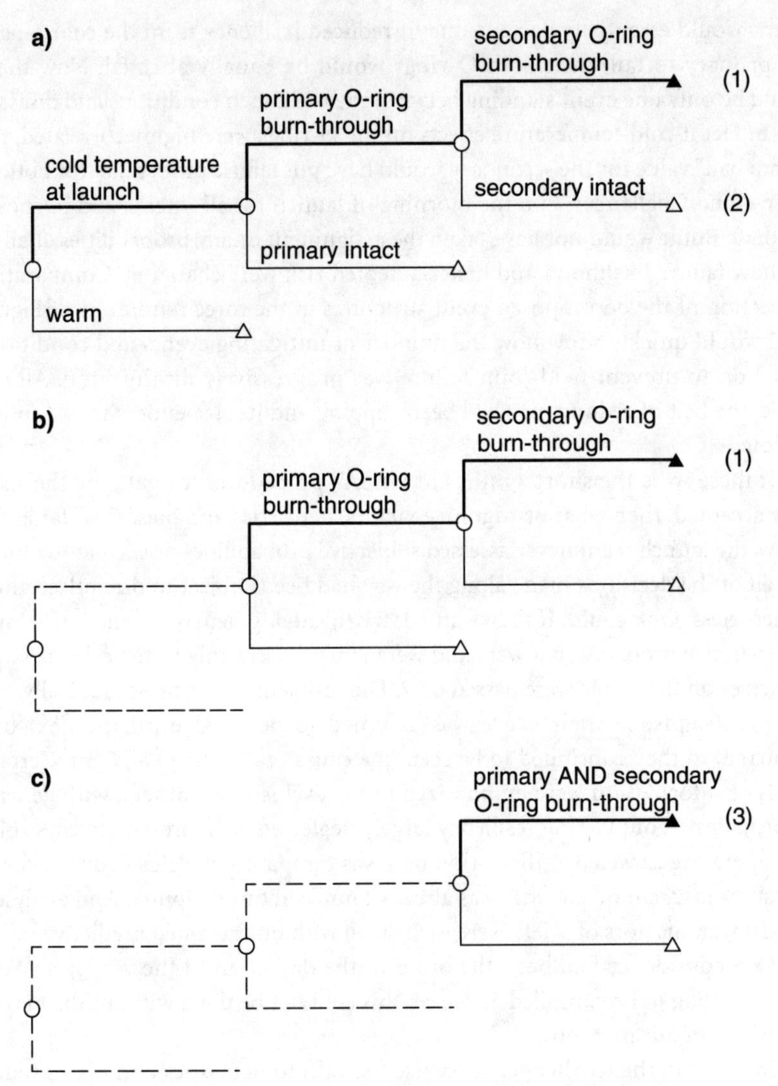

FIGURE 5-12. *Failure sequence representations: a, after flight STS 51-C (January 1985); b, prelaunch teleconference (January 27, 1986); c, common-cause failure (January 28, 1986).*

Note: (1) indicates failure branch pathway; (2) indicates branch pathway experienced on STS 51-C; and (3) indicates branch pathway experienced on STS 51-L (Challenger).

events would essentially become one: If reduced resiliency from the cold caused the primary to fail, then both O-rings would be equally affected. Now there would be only one event standing between the prelaunch conditions and disaster, and in fact if cold-temperature effects on the O-rings were highly correlated, the "jump ball" value for the secondary would have put failure probability for both of them—and Challenger—on the morning of launch for all intents and purposes at 50/50. But it would not have taken the assignment of any probabilities at all to see how failure likelihood and hence accepted risk were changing. Comparative inspection of the decomposed event structures in the three renditions of Figure 5-12 would quickly show how the number of mitigating events and conditions relied on to prevent field-joint failure was progressively diminishing. All the while, the belt of redundancy had been slipping and its suspenders were coming undone.

If these were the shortcomings of NASA's procedures for gauging the risks they accepted, then what of cognitive factors, heuristics and biases? So far as we know, the launch team never assessed subjective probabilities as such at the time, but all of the decisions made along the way had been subject to dire uncertainty nonetheless. One could, if they wanted, try to catalog heuristics and baises and the cognitive processes that were and weren't used. Certainly statistical reasoning strategies on the whole were passed over. The problems came up sporadically, but after each episode their cause was assumed to be fixed until the next one occurred, so they continued to be seen in a causal, not statistical, light. Certain kinds of information were emphasized to the exclusion of others, with general information about O-ring resiliency largely neglected, so representativeness bias was operating as well. Confirmation bias was rampant—the elevated leak check pressures just reinforced what was already known about the putty. And analyses, test data, and factors of safety were each taken with unwarranted predictive validity. Overconfidence had been the order of the day up until the very last. Were such a catalog to be compiled, however, this couldn't be done without the bias of hindsight on our part too.

In the end, the Challenger story itself speaks to heuristics and biases much more eloquently than any attempt to catalog them here ever could. Still, there is one that especially stands out. That is the motivational bias which pervaded NASA's organization going all the way back to the Apollo program. This was uncovered so well during the Rogers Commission hearings in questioning by Richard Feynman that his own account would be hard to pass up. Jud Lovingood was Deputy Manager, Shuttle Projects Office, at Marshall and had been present for the teleconference of January 27, 1986. Feynman (1987) recounted his subsequent exchange with Lovingood and three of the engineers:

> Suddenly I got an idea. I said "All right, I'll tell you what. In order to save time, the main question I want to know is this: Is there the same misunderstanding,

or difference of understanding, between the engineers and the management associated with the engines, as we have discovered associated with the solid rocket boosters?"

Mr. Lovingood says, "No, of course not. Although I'm now a manager, I was trained as an engineer."

I gave each person a piece of paper. I said, "Now, each of you please write down what you think the probability of failure for a flight is, due to a failure of the engines."

I got four answers—three from the engineers and one from Mr. Lovingood, the manager. The answers from the engineers all said ... almost exactly the same thing: 1 in 200. Mr. Lovingood's answer said, "Cannot quantify. Reliability is determined by studies of this, checks on that, experience here—blah, blah, blah, blah, blah."

"Well," I said, "I've got four answers. One of them weaseled." I turned to Mr. Lovingood and said, "I think you weaseled." He says, "I don't think I weaseled." "Well look," I said, "you didn't tell me what your confidence was; you told me how you determined it. What I want to know is: After you determined it, what was it?"

He says, "100 percent." The engineers' jaws drop. My jaw drops. I look at him, everybody looks at him—and he says, "uh...uh, minus epsilon!"

"OK, now the only problem left is, what is epsilon?"

He says, "1 in 100,000." So I showed Mr. Lovingood the other answers and said, "I see there is a difference between engineers and management in their information and knowledge here...."

What Feynman had asked for was a subjective failure probability, the measure of confidence that his line of questioning so persistently extracted. What he got promptly from the engineers, and eventually from the manager, were their subjective probabilities, and he attributed the difference to their varying states of knowledge. But Lovingood too had engineering knowledge by training. And as we know, he had been a participant in the teleconference and thus privy to not just everything discussed there, but presumably from his position at Marshall to the entire preceding chronology. So it was more than a difference in states of knowledge. Mr. Feynman, perhaps unbeknownst to him, had just discovered motivational bias. Feynman does not say whether he realized where the 1-in-100,000 probability came from, but it may well not have been a number that Lovingood devised on the spot. This value, as we will see, had been written into NASA's institutional script for some time. But first, we take up the final factor that was so strong an undercurrent throughout the Challenger launch decision and everything that led up to it: the question of judgment. We have seen that when NASA officially jettisoned probability long before, it was misleading. When they abandoned judgment, it was a disaster.

A Failure in Judgment

At its uppermost level, NASA gave lip service to judgment and may even have thought they were applying it. Or at least so maintained Dr. James Fletcher, NASA's top administrator in Washington D.C. at the time, who provided the following explanation in the Congressional Report of how it was judgment that governed the kinds of decisions that preceded the Challenger launch (Garrick, 1989):

> ...When you get down to the flight team, the launch crew in those last several hours or couple of days, risk management is an entirely different thing. They have to look at the factors that come up just before the launch and assess whether this is a risk we want to take. This is a judgment question; you can't make calculations at this point.

A question of judgment it assuredly was. But NASA's Marshall managers, in their organizational tradition of deductive science and hard data, consistently chose to dismiss it. We can turn to the Thiokol engineers' account of how and why, during the prelaunch teleconference, they interpreted the temperature information in Table 5-4 the way they did.

Thiokol had based its temperature inferences on two data points, both showing O-ring blow-by, with that for the cold conditions of STS 51-C having been much worse. And how they knew it was worse was that they had seen it themselves. It was reasonable to them that the cold had been responsible for this difference in severity—qualitatively at the least, its effects would be hard to deny. But during the teleconference, hard data were available for temperature only from the two flights considered. Moreover, the severity of damage was based on personal, hence subjective, observations from their own postflight inspections, without any quantified measurement. So all things considered, they knew they were in for rough sailing:

> I was asked, yes, at this point in time I was asked to quantify my concerns, and I said I couldn't. I couldn't quantify it. I had no data to quantify it, but I did say I knew it was away from goodness in the current database.

"Away from goodness in the database." What a curiously strange way of expressing one's sense of what was important enough to be of concern. But basing these concerns on data and data alone was what Marshall required, even if it meant such acrobatic contortion of linguistics. And when Thiokol engineers acknowledged that the data included blow-by under not just cold but warm conditions as well, some recalled having felt sheepish, almost apologetic, in having to admit to their uncertainty:

> I recognized it was not a strong technical position, but yes, I basically supported that position ... and I felt that that really weakened our position

> because our whole position was based on the performance of [STS 51-C], and it had been successful and yet it had blow-by and erosion. But there had been other motors that had blow-by and erosion that had been warm shots.... And so it began, to my way of thinking, to really weaken our conclusions and recommendations. And I was already wishy-washy.

Wishy-washy, uncertain, however they might have expressed their predicament, judgment nevertheless did prevail in the face of uncertainty among Thiokol's engineers as they pressed for a launch delay. After all was said and done, inductive reasoning, their sense of what was important, told them that temperature was important, data or not. They had developed, in their words, an unquantified "engineering feel" for the problem, and in the end they were compelled to act on it, knowing full well that making a subjective case like this would not go down easily with Marshall:

> Most of the concerns we had presented were qualitative in nature. At that particular time we had a very difficult time having enough engineering data to quantify the effects that we had been talking about. A lot of it was based on 'engineering feel'.... Engineering, I think, was generally aware that a great deal of our recommendation was subjective and kind of qualitative, but nonetheless it was there, and the engineering judgment was there.

With their feel for the problem and the subjective aspects of their launch recommendation, truly engineering judgment was there on the part of the Thiokol engineers. But notwithstanding Fletcher's expectations that it would be judgment, and not calculations, on which the launch decision would ultimately rest, NASA's Marshall had never seen judgment or any subjective element of interpretation in this light. Referring back to the Flight Readiness Reviews at Marshall and how Lucas had conducted them, one engineer recalled:

> It was our feeling at the time that nothing gets presented to Dr. Lucas unless the people that are doing the presenting are absolutely sure ... and this was a case where clearly there were no answers available because it was just a question of observation as to what we were presenting. I have been personally chastised in FRR at Marshall for using the words 'I feel' or 'I think,' and I have been crucified to the effect that that is not a proper presentation because 'I feel' and 'I suspect' are not engineering supported statements, but they are just judgmental. And so when people go in front of Dr. Lucas, they know full well that if they use words like that or if they use engineering judgment to try to explain a position, that they will be shot down in flames.

Ultimately, of course, it was far more than these engineers and their judgment that went down in flames and not only the Marshall managers who suffered from ignoring it. "Just" observation, "just" judgment indeed. If answers could not be obtained from hard data but just observations, then there were no answers. The absolute certainty of deductive proof was required, and nothing less was admissible: The feelings of doubt that constitute the essence of uncertainty judgments were the object of ridicule. The inductive elements of judgment, the establishment of probable cause, found no place in engineering at all. So if a problem could not be objectively proven to exist, it didn't.

And the matter of judgment returns us to why NASA had never used probability in determining acceptable risk in the first place. Certainly the embarrassingly high failure probabilities much earlier for the lunar Apollo landing must have remained a large factor, but it was not simply the values themselves. No such probability could be based entirely on data but would have to rely in some measure on judgment. It would have to be subjective, not objectively derived from data alone, just as Thiokol's engineers went beyond data to their subjective beliefs about temperature effects. And this was fundamentally not something that NASA, in its organizational culture from the time of Von Braun, had ever been able to abide. At least not openly.

As an epilogue to the Challenger story, we go back to 1983, three years before the fateful launch as the shuttle's operational missions were reaching full swing. As we learn from Cooke (1991), apparently in NASA's darker (and classified) recesses some vestige of probability had managed to survive, although this came to light in a most roundabout way. From the start, a major block of shuttle payload had been reserved for military purposes. The Air Force was then considering sending up a type of radioisotope generator that would pose special ground-based risks in the event of mission failure, so it had asked consultants E.W. Colglazier and R.K. Weatherwax to review an earlier NASA-sponsored study of shuttle failure modes.

In their 1983 Air Force review report, which remained unavailable to the public except for a brief abstract, Colglazier and Weatherwax incorporated operational experience from a variety of solid rocket boosters in deriving their own subjective probability estimate of roughly 1 in 35 for SRB-caused failure of any given shuttle launch (Cooke, 1991). But NASA management is reported to have rejected this in favor of their own value of 1 in 100,000, thus reducing it by more than three orders of magnitude. After all, if subjective probability is whatever you want it to be, then one value could be conveniently substituted for any other, in the process easily avoiding any repeat prospect of the Apollo probability's would-be political debacle. NASA had probably never heard of motivational bias even as they raised it to an art form. The problem was not that NASA had used subjective probability, quite to the contrary. They had abused it with little regard to the bias that dominated it—or to the consequences that would ultimately follow.

Challenger had been the twenty-fifth flight of the space shuttle. If we were to take the 1983 subjective Colglazier and Weatherwax estimate and the assumption of independent Bernoulli trials for each launch, the probability of experiencing at least one SRB-induced shuttle failure by the time of Challenger's launch would have been almost exactly 50 percent from the binomial theorem. In all of the material so exhaustively presented by Vaughan (1996), there is no indication that the Thiokol engineers had knowledge of any of this when they alluded to "jump-ball" failure chances for the crucial secondary O-ring in their 1985 memo. Once more we are left to wonder whether it would have made any difference if they had.

Neither can we know for sure in later providing his 1 in 100,000 failure probability—his "epsilon" that Feynman had so painfully extracted in the Rogers Commission investigation in 1986—whether the NASA manager was simply parroting the very same value NASA had substituted for the Colglazier and Weatherwax estimate three years before. We would have no hard data to prove any such inference, nothing we might point to as factual. Our evidence would only be circumstantial: It might, after all, have been just a coincidence. Nevertheless, the space shuttle continues to fly today with the same basic design. The SRB field joints are still there, and so are the O-rings. In the end, after everything that transpired, all it took was to retrofit them with heaters. The solution had always been as simple as the problem.

Against the backdrop of the Challenger story, this chapter's compendium of heuristics and biases and its collection of cognitive snares might at first appear to present a dismal picture indeed, and lost in this morass might seem any way to overcome or correct them. This is not the intent, nor is it the case, for without recognizing the pitfalls in application of cognitive processes we could scarcely go on to address them. Heuristics are just the everyday tools of cognition, and uncontrolled biases need not be their inevitable result. People need not be condemned for using these heuristics or think that they doom subjective probability from the start. The trick is to adapt them to the way people think. We are now prepared to see how to do so when subjective probabilities are assessed.

6

Assessment of Subjective Probabilities

While the assessment of subjective probabilities ends up with a number, it begins with the way people think, and the uncertainty judgment beneath the number is where the real work is done. This chapter provides the tools for doing this work, and these are of two types: internal and external.

The internal tools are the cognitive processes one uses in formulating the uncertainty judgment and expressing it numerically. These, in effect, are the machinery of subjective probability assessment, and we have already been introduced to how they operate. While so far this has focused on how cognitive mechanisms can get gummed up with heuristics and biases, the emphasis here is on how to oil them and make them work better. So rather than taking heuristics and biases for granted, instead we consider how their effects can be reduced or controlled through various techniques collectively called *debiasing*. It would be a mistake to think that people could be entirely cleansed of these effects, as the word itself might suggest. Yet there is good reason to believe that, when understood and addressed, the effects of heuristics and biases need not dominate subjective probabilities if people are encouraged to use their cognitive skills to best advantage, and exploiting peoples' abilities to compensate for their weaknesses is a major theme throughout. To be successfully applied, the strategies and techniques for subjective probability assessment must be adapted to how people think and to assisting them in the process. This is the primary goal of the chapter.

External tools are procedural. As examined in Chapter 4, subjective probability is most often used in a risk analysis framework. This provides the external structure to the internally derived uncertainty judgments which allows them to be assembled and put to some end. Various protocols have been developed for extracting these quantified judgments in different kinds of structured formats. Most rely on the external assistance of an *analyst* or *facilitator* to guide the process, and they are reserved for discussion later in the chapter. While such formats have their advantages, there is nothing to say that procedural tools are mandatory. Some of the greatest benefit to be derived from subjective probability assessment comes from the understanding of uncertainty it promotes on a personal level, and the internal process of formulating a probability statement—with or without structured assistance—can be of immense value, both for gaining this insight and for communicating it to others. For unassisted assessment of subjective probabilities, the techniques described throughout this chapter would be applied in much the same way.

All of this presupposes, of course, that one wants to assess subjective probabilities in the first place. Several questions can arise prior to undertaking this activity, and they deserve answers. We start with one of the most important.

Why Bother?

Most of us manage to get along quite well without probability in our daily activities despite the uncertainty that accompanies them. It is basic human nature for people to seek certainty and, consciously or not, we use various devices to obtain it. Engineering has developed its own through customary practices that recognize uncertainty, though indirectly. But subjective probability requires that uncertainty be confronted explicitly, personally, and in no small detail. This takes work, and where uncertainty has been dealt with in some way through existing procedures, there must be good reason to characterize it probabilistically. If customary practices for dealing with uncertainty work well enough, then why bother?

Acknowledging Uncertainty

Accommodating uncertainty, no matter how this may be accomplished, cannot proceed far unless its existence is first acknowledged. This is not something easy for people to do. Recall from Chapter 3 that Herbert Einstein (1991) said it took judgment to recognize limitations and uncertainties. And even beyond judgment, Kitchener and Brenner (1990) view acceptance of the limitations and uncertainties in one's knowledge as prerequisite for wisdom, noting that wisdom supersedes logic as a more fundamental requirement for dealing with unstructured problems. If so, those among us with great wisdom will readily acknowledge uncertainty, recognizing that full and complete understanding will always be elusive. For the rest of us though, this can be the first obstacle that the assessment of subjective probabilities must confront. In this, geologists and engineers provide a most instructive contrast.

As with any of the sciences, geology is fundamentally explanatory. Its reason for being is to explain how earth's features got there and why they came to be the way they are. But what sets geology apart from the experimental sciences is that its hypotheses can never really be transformed into theories by experimental confirmation. Because geologic processes do not provide the opportunity for replicable experiments, its hypotheses always remain uncertain. So geologists, by and large, take uncertainty as a matter of course and are quick to admit and embrace it. Uncertainty is the most prominent feature of any geologic terrane, and for geologists it comes with the territory plain and simple. As Princeton geology professor Kenneth Deffeyes described his field's state of affairs:

> At any given moment, no two geologists are going to have in their heads exactly the same levels of acceptance of all hypotheses and theories that are floating around.... All science involves speculation, and few sciences include as much speculation as geology.... You have to deal with partial information. Do physicists do that? Hell, no. They want to have it to seven decimal places on their HPs. The geologist has to choose the course of action with the best statistical chance. As a result, the style of geology is full of inferences, and they change. No one has ever seen a geosyncline. No one has ever seen the welding of a tuff. No one has ever seen a granite batholith intrude. (McPhee, 1980)

While Deffeyes can be forgiven for his notion of chance, the point here is that geology is interpretive, not predictive like engineering, so there are there are different expectations of geologists' cognitive skills. Causal reasoning and causal narratives are used to put observations into context. So the subduction processes of plate tectonics could be used to explain why a volcano might appear or an earthquake occur but not to predict exactly when or where. Geologic mapping is an exercise in constructing patterns in time and space, fitting things together using all manner of varied hypotheses. These skills are second nature to geologists, and they are inductive rather than deductive with uncertainty at their core. It is not ordinarily expected of geologists that they predict the operation of geologic processes with great assurance or detail, to say what will happen in millions of years at locations where place shifts in meaning. So the world of geology exists in concert with uncertainty, not conflict.

Engineering is another matter, and in some ways being asked to provide a subjective probability statement is one of the most intimidating requests an engineer can face. By training and custom, engineers live in a deterministic world where answers are either right or wrong, calculations are correct or they're not, and uncertainty in either one is a sign that something is surely amiss. By comparison to the speculative carnival of geology, designing a reinforced concrete structure is an exercise in precision to seven decimal places, where an answer having uncertainty of plus or minus 20% simply won't do. Characterizing uncertainty can seem entirely beside the point when eliminating it, not evaluating it, is the chief object. So requesting that uncertainty be described with a subjective probability can seem somehow nonsensical, and "How can I when I don't know?" sums up the instinctive response. Altogether, this can be quite an unnatural act.

But a few moments' sober reflection, enhanced by a sense of humility, is usually sufficient to illuminate the host of uncertainties that affect most everything engineers do. Some day, with better information or better technology, the effects of the unknowns and assumptions that influence our predictions may be reduced, but we can seldom await someday's arrival when actions must be taken and decisions made. Life and engineering both must go on, whether we're sure about things or not. So characterizing uncertainty with subjective probability requires frank and candid admission of uncertainty as we see it, not how we would prefer it, given the state of information and knowledge at hand whatever that may be. Admitting to uncertainty is not a somehow shameful sign of incompetence, even if engineers are supposed to know all the answers. Admitting to uncertainty is admitting to the realities that engineers must confront in making predictions and ultimately in doing their jobs.

It is said on Wall Street that there are only two things that motivate people: fear and greed, and fear often has the upper hand. Some engineers have been persuaded to feel the same about probability, for fear of not getting the "right answer." But in any such calculus of punishment and reward, the thing to fear most about uncertainty is ignoring it, and for probability not taking advantage of what it can do. The rewards can be many for those who wish to better come to grips with their uncertainty and thus become better engineers.

Safety, Uncertainty, and Probability

We ordinarily adapt to uncertainty through codes and standards, factors of safety, design criteria, and other established procedures that serve to reduce its effects to what seems a sufficient degree. If these things work, as they usually do, "then why should I care about probability?" is another commonly-heard question that challenges the relevance of characterizing uncertainty at all. But the use of probability, subjective or otherwise, need not be taken as an assault on the deterministic edifice of conventional practice or seen as irrelevant to how we conduct it. Probability is something that extends established procedures to circumstances that are less than routine, and the two can coexist comfortably side by side.

Codes, standards, factors of safety, and various other customary design criteria are, of course, the embodiment of the collective judgment of the profession gained through induction by enumeration—by experience with what works more often than not—and they are how uncertainty is handled on an everyday basis. As simple rules and strategies, these are the heuristics of design that tell us what to do most of the time and in most of the situations we encounter. They shortcut the need to explicitly address uncertainty by encoding it in procedural rules. But some circumstances are unique, some uncertainties are not routine, and relevant codes and standards may not exist. Other cases, particularly those involving existing structures, require going beyond codes and standards to evaluating safety when the original design criteria are no longer met, whether from changes in the structure's condition, departures from design assumptions, or changes in the criteria themselves. Here it is not enough just to say that the design criteria are no longer satisfied. We need to know the safety implications of these possible deficiencies—how important they are and what can be done about them. The criteria themselves cannot tell us these things.

Safety itself is an internal construct, a concept and not a measurable quantity or any objective attribute of a structure. No one has yet come up with a device that displays safety on a digital readout, nor are they likely to anytime soon. We associate certain characteristics with structures that are safe and others with those that are not. Some of these are physical characteristics, or indicators, used to inductively establish safety status, and others are analytical findings that we agree to by convention. But safety is not a property of the structure, it is a property of the evaluator and it therefore exists in the mind. Safety is inevitably a judgment that cannot be proven true by any method of deductive logic. Safety resides in belief, and when we say that a structure is safe, this means we hold some sufficient degree of belief that it is.

The evaluation of safety requires interpretation of these indicators, but they always fall short of perfection. So the very concept of safety inherently invokes not only belief but also uncertainty. And if this uncertainty is to be accounted for and its effects accommodated, its sources must first be understood. This is where probability comes in, and why we should care about probability is why we should care about judgment: It provides an interpretive framework that helps guide and direct how uncertainty is comprehended and subsequently managed. Using subjective probability, or for that matter any probabilistic approach, does not replace the deterministic procedures that serve well in so many cases. Probability supplements and extends them to areas they could not otherwise address.

So how do we assign a subjective probability when we don't know the answer? The response is contained in the question. Not knowing is the very essence of uncertainty, and subjective probability as its descriptor does not require knowing the unknowable or predicting the unpredictable with some act of clairvoyance. It asks us to say honestly and fairly how much we don't know and to what extent we're unsure, and to express these things as a number. It asks that a measure of our uncertainty be provided, not that we find a way to eliminate it. The complaint that "we don't know enough to estimate probability" goes away when probability is recognized as the expression of the limitations in our information and state of knowledge at the moment.

Numerical Quantification

Presuming one acknowledges uncertainty and accepts the need to characterize it, then why should this characterization be in numerical form? There can be many ways to express a likelihood judgment, and numerical probability requires the most effort.

Uncertainty is usually expressed qualitatively, which is to say without numbers as such. We don't ordinarily find it necessary to use numbers in conveying likelihood judgments when words like "possible" or "probable," "likely" or "unlikely" serve just as well. These descriptors have a rich and multidimensional quality in expressing uncertainty, as we will go on to see, and can be easily rank-ordered on a scale of relative likelihood. But this scaling need not be limited to verbal descriptors and could be constructed in any fashion we might choose. It could be simply a set of likelihood categories such as {high, medium, low} or {I, II, III} where membership of event likelihood in any such category expresses its relationship to some other but not its absolute value from 0 to 1.

A variety of qualitative techniques for risk analysis adopt just such a rank-ordered scaling or indexing approach. One we have already encountered is Failure Modes and Effects Analysis, or FMEA. Its basic purpose is to produce some rank-ordered scaling of the various failure modes a system might experience according to their comparative risk, and it was shown through NASA's example that this can be done only if likelihoods are evaluated along with consequences. Although these qualitative techniques are of great value for broad screening of complex systems, they are also sometimes seen as convenient avenues for side-stepping all the effort that quantified probability assessment entails.

In avoiding the probability metric, qualitative approaches allow for greater ambiguity in expressing likelihood judgments according to membership in some broad category. This indeed makes expressing the likelihood judgment simpler. But recall that the numerical expression of this judgment is only the final stage in the cognitive process and all of the preceding steps required for formulating it remain. In some ways, this makes the others more difficult. First of all, there are usually many events which together comprise the failure sequence, but qualitative approaches evaluate their likelihoods together on a lumped basis. These approaches do not provide for decomposing failure modes or sequences, nor does rank-ordering allow individual event likelihoods to be combined. The necessary additive and multiplicative operations for doing so cannot be performed—for example, within a set containing {high, high, low, medium, medium}—in any meaningful way.

Moreover, without decomposition it is more difficult to control the effects of bias, particularly overconfidence. We have seen how overconfidence varies greatly at different likelihood levels, which applies just as much to qualitatively expressed likelihood judgments as numerically quantified ones. And in addition to all this, a qualitative rank-ordering may allow us to say that one thing is more likely or less likely than some other, but it cannot say by how much. Thus, an important dimension of the likelihood judgment is lost.

So the value of quantifying the likelihood judgment as numerical probability lies with decomposition, and this is the payoff that justifies the effort. Decomposition helps overcome our information-processing limitations by evaluating it piece by piece. Likelihood judgments are made more easily for simple events and situations than complex ones, but this then requires that these individual pieces be reaggregated in some way to arrive at the combined result. Producing these aggregations conveniently is what the probability calculus is for, and using it requires that individual likelihood judgments be quantified in numerical form. Numerical probability is what allows these individual judgments to be combined in a richer variety of ways more useful and revealing than any of them separately.

Still, this should not obscure why these likelihood judgments are made in the first place. If there could be some automated process for manufacturing subjective probability numbers, little insight would be produced, and without it the value of the numbers would be nil. It is not enough for subjective probability to express what we believe about uncertainty; we must also discover why we believe it. These insights are derived from the cognitive process of formulating the likelihood judgment and not from the numbers that express it. In arriving at a subjective probability value, it is the cognitive journey in getting there and not the numerical destination that matters most. So having taken up some key questions about subjective probability, we can now proceed to the internal tools for its assessment.

Preparation

Preparing assessors is the first step in probability assessment. They need to know what they are being asked to do, why they are being asked to do it, and the reasons why both of these things are important. Aside from basic fairness, this encourages the assessor to devote the necessary mental effort and concentration to probability assessment. Like anything else, people tend to perform better in probability assessment tasks when they simply try harder.

We have seen that people have a rich repertoire of cognitive strategies they can bring to bear on probability estimating tasks. Nevertheless, the most effective strategy may not be the easiest or the most convenient, and people simply may not go to the trouble of refining their strategy unless they see some good reason to do so (Beach, et al., 1986). But engineers as a rule are not known for taking their tasks lightly, and they will take most things seriously if they know why they are serious things. So it doesn't usually take much prompting to bring forth diligent efforts when the connection between their uncertainty judgments and the significance of these judgments is established. Explaining such matters is the purpose of preparation.

Encouragement and Support

Assessors need to be encouraged to apply their judgment and to know that it is valued as an essential part of the uncertainty evaluation process. The assessment is more comfortable and ultimately more successful if it is conducted in a supportive atmosphere. This may not be as obvious or simple as it seems. Engineers are trained and instructed in deductive thinking, and the inductive processes of judgment can be perceived as an inferior substitute, if indeed they are appreciated at all. Assessors need to be made aware that subjective probability assessment is an inductive exercise as important and essential as any deductive one and that this is why their personal judgment is being sought.

Subjective probability is considered to be based on expert judgment, insofar as the knowledge of engineers is expected to exceed that of laypersons in their areas of technical competence. This is not to say that the assessment of subjective probabilities is reserved for the technical elite. To the contrary, a subject-matter expert is not necessarily the person at the pinnacle of technical knowledge but one who combines technical understanding with a firm grasp of the specific conditions of interest. Thus, an engineer with in-depth knowledge of construction practices, subsurface conditions, or a structure's behavior can be every bit as much of an expert on the problem at hand as the most technically advanced specialist. Experts in this sense come in all shapes and sizes, and probability assessors need to be encouraged to bring their expertise to bear on the problem, wherever it may lie. Knowing that their personal expertise and their judgment are valued and being relied on helps assessors take their task with the gravity it deserves. While there will be much more to say about experts and expertise later on, it is sufficient to note here that a person performs most like an expert when acknowledged as being one.

It is no secret that some more rigidly structured organizations which rely heavily on procedural process may not particularly encourage the exercise of personal judgment ("That's just not how we do things here"), the previous example of NASA being a salient case in point. If the organization asks for judgment to be applied in subjective probability assessment, then the assessor must be reassured that this is not some kind of a trap. Since there can be no "wrong answers" associated with either subjective probability or judgment, no sanctions can be applied for providing one. Being made aware of this also signals to the assessor that no preconceived outcome is sought, which helps discourage motivational bias. Indeed, a common fear of assessors is that their probability values will be misused to some unknown and ulterior motive, and they need to be assured that this is not the case. This can be done by explaining how the values provided will be used to reveal the sources from which uncertainty derives in order to better manage it and that there is no one in a better position to do so than they are.

A related issue concerns organizational control over the process, especially in top-down organizational structures. In light of possible ramifications of risk analysis findings that cannot be anticipated in advance, some organizations may be reluctant to cede control over probability assessment to the individuals involved ("We can't have these people running around doing these things"). But it destroys any incentive for assessors to exert the efforts required if it is perceived that the organization doesn't value and support these efforts.

Control over probability assessment must be clearly distinguished from control over any decision that results—the former is the responsibility of the assessor, while the later is the duty of the organization. The probability assessor is an information provider, while the decisionmaker determines a course of action based on this and other inputs. This separation comes clear by recognizing that probabilities and risk analyses do not themselves make decisions—only organizational decisionmakers do—and probabilistic uncertainty judgments provide a resource that can inform a decision but never mandate its outcome. At a more basic level, an organization may simply not trust the judgment of those it assigns to the task. If so, it has problems that go far deeper than probability assessment, and this may be something it is not cut out for until these matters are resolved. If the assessor is to be committed to the task, the organization must also be committed to the assessor and with this the assessor's judgment.

Dealing with these matters requires personal interaction on the part of the assessor, the facilitator, and the organizational sponsor, and this does not lend itself readily to encapsulation in written instruction. Even so, some simple guidance for probability assessors like that in Table 6-1 has been found to provide useful background.

Probability Concepts

For purposes of probability assessment, we usually consider that to be a subject-matter expert requires substantial experience. Yet someone with 20 years' experience in their technical specialty may have had 20 minutes' exposure to probability concepts. This can hardly do justice to the expert or to the probability. Neither would it be fair to castigate assessors for the deviations of cognitive bias from probability rules without some explanation of what these rules constitute. Preparation for subjective probability assessment also involves providing some background in probability concepts.

Explaining the basic probability axioms is a good place to start. Assessors who have had some exposure to probability and statistics may fear being required to perform some of its more acrobatic manipulations. Others without this exposure need a place to begin. The axioms themselves, notions of independent and conditional relationships, the binomial theorem and Bernoulli trials, and Bayes' theorem are ordinarily all that are really needed to provide assessors with the necessary degree of familiarity, and the Appendix has been constructed with this in mind. Such an acquaintance does not require any comprehensive instruction in advanced probabilistic methods and if nothing else helps assessors feel more securely grounded in the underlying probability calculus.

But even more fundamentally, if assessors are being asked to provide a subjective probability value, they need to know what subjective probability is. Most of all,

TABLE 6-1. *Preparatory Notes for Probability Assessors*

The art of geotechnical engineering has been described as the ability to make sound decisions in the face of imperfect knowledge. These decisions and the predictions used in making them always incorporate uncertainty to one degree or another, so the engineer must apply judgment—the very real interpretive process that derives from the sum total of one's experience, insight, and intuition.

Subjective probability requires applying this judgment to express one's personal degree of belief or confidence in the state of nature, engineering properties, or outcome of a process. You have been asked to assess your subjective probabilities for certain aspects of the case at hand where it is recognized that uncertainties exist. You may feel that there is insufficient information or that your knowledge does not qualify you as an expert in the field. However, these probabilities are based on your personal understanding of the site conditions according to the information currently available to you as well as on your understanding of the engineering, soil behavior, and geologic processes involved. Use all relevant information, including your personal experience and case histories, and do not neglect simple observations or general knowledge. The object of this exercise is to obtain your professional opinion about how much uncertainty these factors together entail.

Do not be intimidated that there may be others who know more in this technical specialty. Remember that right now there is nobody who knows more than you do about both the site and the physical processes involved, and it is for this reason that your assessment of uncertainty is being sought. There is no right or wrong answer, only a numerical probability statement that quantifies your uncertainty judgments as fairly and honestly as possible.

The assessment of subjective probability relies on self-questioning, and it is important that you interrogate yourself carefully about all possible outcomes without discarding any prematurely. Ask why you believe that a particular outcome will occur or state of nature exists, what supports this belief, and how strongly you believe it. At the same time, account for the quality and the quantity of the information, and search for evidence that might counterindicate the most likely outcome. In the end, if you remain highly uncertain do not hesitate to assign probability values that reflect this.

this involves explaining the two probability interpretations, the difference between them, and that the probability calculus does not mandate one or the other. It is usually safe to assume that few will be conversant with subjective probability, so it needs to be placed on an equal footing with frequency interpretations. Assessors cannot be expected to take their task very seriously if they harbor doubts that what they are being asked to assess is something other than "real" probability. We have seen how the early probabilists moved easily back and forth between belief and frequency interpretations and also how these interpretations are mirrored internally by the causal and statistical reasoning strategies people adopt. Assessors will be required to use and synthesize both causal and frequency information in their assessments as well, and this kind of background prepares them for doing so.

What all of these aspects of preparation seek to do is to provide assessors with a certain level of comfort in approaching the probability task ahead and some assurance that they can indeed perform it capably. Even if judgment and uncertainty underlie most if not all engineering activities, they are seldom brought to the fore in the way that probability assessment requires. Knowing that there exists an underlying conceptual basis for doing so helps remove any misperception that it is nothing more than arbitrarily picking a number.

Training

Training of probability assessors goes beyond preparation directly to the matter of heuristics and biases. Training of probability assessors has been the first line of defense against bias in behavioral research, where considerable effort has been directed toward determining what kinds of techniques are most effective.

The first aspect of training is to make assessors aware of the heuristics people use in the cognitive process of forming likelihood judgments and what their effects can be. These are not repeated here, but Table 5-2 has provided a useful summary. In general, however, it will not be apparent from these descriptions alone how heuristics and biases operate, and illustrations like those in Chapter 5 will be necessary to clarify them.

Because overconfidence is the most pervasive cognitive bias, it has also received the most attention in training techniques. From several studies, it appears that merely making people aware of overconfidence bias is not enough to counter its effects. Some studies have gone beyond informing subjects about it to more explicit warnings such as the following instructions in one experiment involving two-choice questions designed to be exceptionally difficult (Fischhoff and Slovic, 1980):

> Remember, it may well be impossible to make this sort of discrimination. Try to do the best you can. But if, in the extreme, you feel totally uncertain … do not hesitate to respond with 0.5 for every one of them.

While improvement in calibration from cautions like these has been observed in some studies, others (Alpert and Raiffa, 1982) have found subjects to be nearly impervious to them. When simply informed that most people are overconfident in assigning excessively high or low probability values, a fairly predictable response goes something along the lines of: "Not engineers, we know how to handle numbers." A more direct approach seems to be more effective, whereby probability assessors are allowed to experience overconfidence bias themselves rather than just being instructed or warned about it. This often takes the form of asking assessors to provide confidence limits on numerical quantities for almanac-type questions of the kind shown in Table 6-2.

Merkhofer (1987) has generalized the results of questionnaires like these to the "2/50" rule: When people are asked to provide estimates of quantities lying between their 1% and 99% confidence limits (that is, with only a 2% chance of

TABLE 6-2. *Questionnaire for Subjective Probability Training*

Estimate your 90% confidence range on the following quantities, such that the actual value should have only a 5% chance of being either higher or lower than the corresponding limits of your range.

a) What is the radius of the earth at the equator in kilometers?
b) At what age did Bill Gates become a billionaire?
c) What was the population of Des Moines in 1992?
d) What year did Amundsen sail the Northwest Passage?
e) How long is the main span of the Brooklyn Bridge in meters?
f) James Buchanan's term as president began in what year?
g) What volume of concrete does Hoover Dam contain in cubic yards?
h) What is the ratio of the area of Rhode Island to that of Georgia?
i) What was the closing Dow Jones industrial average on April 18, 1997?
j) What is the melting point of iron in °C?

a) 6378 b) 31 c) 195,752 d) 1905 e) 486 f) 1857 g) 4,400,000 h) 0.00075 i) 6703.55 j) 1538

being wrong), usually about 50% of the answers fall outside this range. Similarly, Keeney and von Winterfeldt (1991) found that even absolute upper and lower bounds set by subjects with supposedly no chance of being exceeded were actually exceeded about 30% of the time. This, of course, is overconfidence manifested as subjectively assigned probability distributions that are too narrow about the mean or "best estimate" value, the same effect that was illustrated in Figure 5-6. But demonstrating overconfidence on a personal level seems to be key to reducing it.

The difference between demonstrating and instructing in overconfidence bias lies in the feedback that participants receive when told the outcome of their assessments. Lichtenstein and Fischhoff (1980) found that after only a few sequential repetitions of exercises like that in Table 6-2, assessors can learn to reduce their overconfidence bias in these demonstrations fairly quickly and effectively. In this case, well-calibrated 90% confidence ranges would encompass the answer for 9 of the 10 quantities and feedback would take the form of providing the actual number that did. While questions remain about the extent to which this kind of training generalizes to other probability-assessment tasks, there is little doubt it is a step in the right direction.

Feedback

The importance of outcome feedback is clear from its effect in training exercises. However, feedback in the real probability assessment tasks assessors face is a longer-term proposition that can have a pronounced effect on bias if incorporated more routinely in professional activities. Weather forecasters are the prime example. This is not to say they are always accurate, as we are all well aware, but that they become quite well calibrated over the long run in their predictions. What is accurate is the forecasters' sense of how much they know and how reliable their subjective probabilities turn out to be. In the long run over many predictions, their uncertainty judgments prove to be pretty much right on the money. But if forecasters are the shining stars of calibration, physicians are the runners-up by quite a wide margin. The difference, it appears, is all a matter of practice in the kinds of likelihood judgments their professional activities involve, and this requires not just repetition but also the report card of outcome feedback.

Chapter 5 described some of the causal reasoning strategies used by physicians in clinical diagnosis and showed how these strategies sometimes resulted in representativeness bias and insensitivity to the predictive validity of diagnostic indicators. Physicians also exhibit overconfidence bias in repetitive tasks performed routinely for common and familiar problems. Poses, et al. (1985) report a study in which 10 physicians recorded their treatment decisions and estimated subjective probabilities of streptococcal infection for patients with sore throats, which is one of the possible causes. Their diagnoses were based on various signs and symptoms such as fever, cough, swollen glands, and recent streptococcal outbreaks in the area. But of 308 patients diagnosed with the infection, throat cultures showed the presence of the bacterium in only 15, and the probability of streptococcal infection was overestimated for 81% of the patients. Similar overconfidence has also been demonstrated in diagnoses of pneumonia and skull fractures.

On the other hand, we might take some comfort in knowing that physicians were quite well calibrated in estimating the probability of patient survival in intensive care units (Winkler and Poses, 1993). Interestingly, simply averaging the probabilities provided by two or more of the least-experienced interns was almost as good as the values provided by the patients' attending physicians with most intimate knowledge of their condition. This confirms the wisdom of second opinions that would seem to apply not just to physicians. Winkler and Poses (1993) speculated that performance might have been improved still further if the physicians had collaborated in face-to-face discussions to provide a group probability, in the process enhancing their medical judgment more generally.

Notwithstanding these successes, physicians can be among the most persistently overconfident subjective probability assessors to be found. Among the reasons is thought to be the lack of prompt and unambiguous feedback they receive (Lichtenstein, et al., 1982; Ayton, 1992). It can be difficult or costly to confirm a diagnosis. Moreover, patients who get better, whatever the reason, do not know or report the accuracy of the diagnosis unless their condition fails to improve, so there is limited opportunity for followup. Or they may be referred to other physicians who do not report back on the original diagnosis. And if outcome feedback is received, the circumstances and conditions favoring the original diagnosis may be long forgotten. The lack of outcome feedback is one of the things that make it

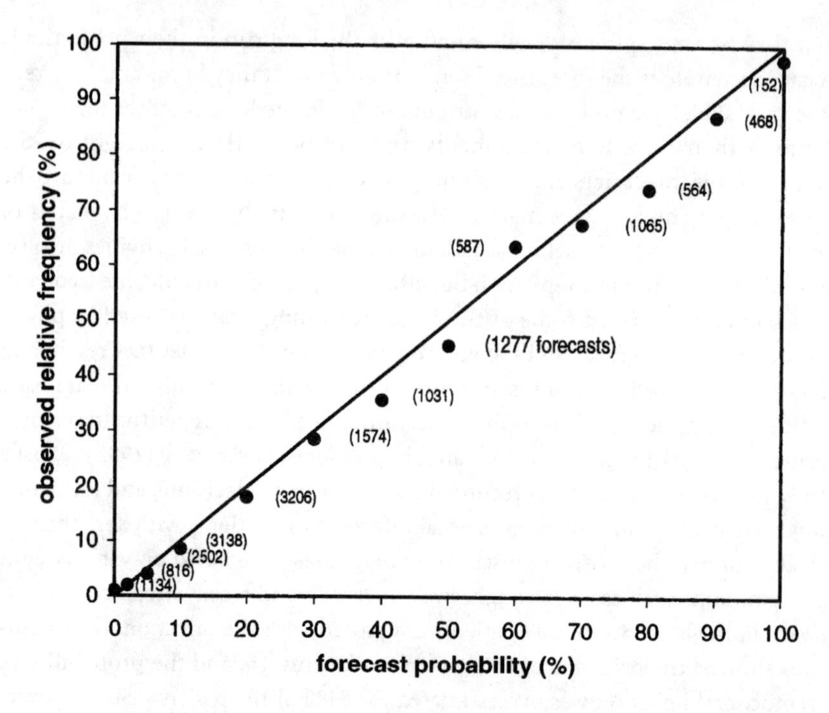

FIGURE 6-1. *Calibration data for precipitation forecasts.*

difficult for subjective probability performance to improve because the effects of the various heuristics at work never become evident. Indeed, perhaps it is feedback that explains the good calibration of the intensive-care physicians in their estimates of patient survival, patient mortality being prompt and unambiguous feedback of a kind particularly hard to overlook.

Contrast the decidedly mixed performance of physicians with National Weather Service forecasters, who have been expressing forecasts of precipitation in probabilistic terms since 1965 (Lichtenstein, et al., 1982). The result is a massive database of probabilities and forecast outcomes, with Figure 6-1 showing four years of these data for over 17,000 precipitation forecasts for Chicago illustrating just how well calibrated they have become (Murphy and Winkler, 1977). Reviewing a number of studies on meteorologists, Wallsten and Budescu (1983) note that they are also remarkably well calibrated in forecasting daily temperature and even tornadoes. Here, the differences in outcome feedback between clinical diagnosis and meteorological prediction are evident. Rain gauges and thermometers at the reporting station provide a clear measure of forecast outcome, and this feedback comes regularly every 24 hours. So practice with subjective probability assessment in real on-the-job settings makes all the difference, and this performance is made all the more impressive by noting that these forecasts require integrating causal reasoning from meteorological knowledge with statistical reasoning from climatological data.

In considering the kind of outcome feedback that geotechnical engineers receive, it would appear that they share some of the same handicaps as physicians. Particularly for predictions involving stability, factor of safety is an ambiguous indicator. Certainly it is easy to determine whether the structure was stable. But this tells only that the mobilized FS was greater than 1.0, and its operative value in the field can never be known. On the other hand, readily measurable quantities like settlement or deformation can be verified, but these measurements are neglected more often than not. When they are made, they may come after long delay for such lengthy processes as consolidation, and by that time the original predictions may be buried in the files. But outcome feedback is not always so difficult to obtain, and when it occurs some remarkable improvements in calibration can come about. Here Focht (1994) describes an interesting and instructive counterpoint.

In 1989 a geotechnical prediction symposium was held at Northwestern University along much the same lines as the I-95 test embankment exercise described in Chapter 5. The object was to predict pile load capacities for two driven piles, one a pipe pile and the other an H-pile, both of which would be loaded to failure. Fifteen participants, and therefore 30 predictions, specified a lower-bound estimate of pile capacity with 90% confidence that the actual value would exceed this minimum. The participants were provided with the subsurface data shown in Figure 6-2a, and as in the embankment symposium they were free to use any predictive method they chose.

Their predictions are shown in Figure 6-2b normalized to observed capacity. Only four of the lower-bound estimates exceeded the measured value, and fully 26 of the 30 or 87% were (correctly) less. This correspondence between the 90% judged probability and the 87% frequency shows an impressive degree of group calibration ordinarily achieved individually only by weather forecasters, and this is readily contrasted with the poor calibration of the embankment symposium predictors shown previously in Figure 5-5. It is hard to imagine that techniques for predicting pile capacity could be so much more informative than those for embankment stability, and clearly something else must be at work.

Prediction of both pile capacity and embankment stability are repetitive tasks carried out in ordinary geotechnical practice, and this certainly would have been so for the participants in both symposia. The difference is that, unlike the difficulties in validating calculated embankment stability, pile load tests are performed routinely to verify predicted capacities in everyday foundation design. Outcome feedback provided by pile load tests is fairly clear to interpret, and it comes with minimal delay since construction is usually waiting in the wings. The pile capacity predictors appear to have become much better calibrated in their probability estimates as a result.

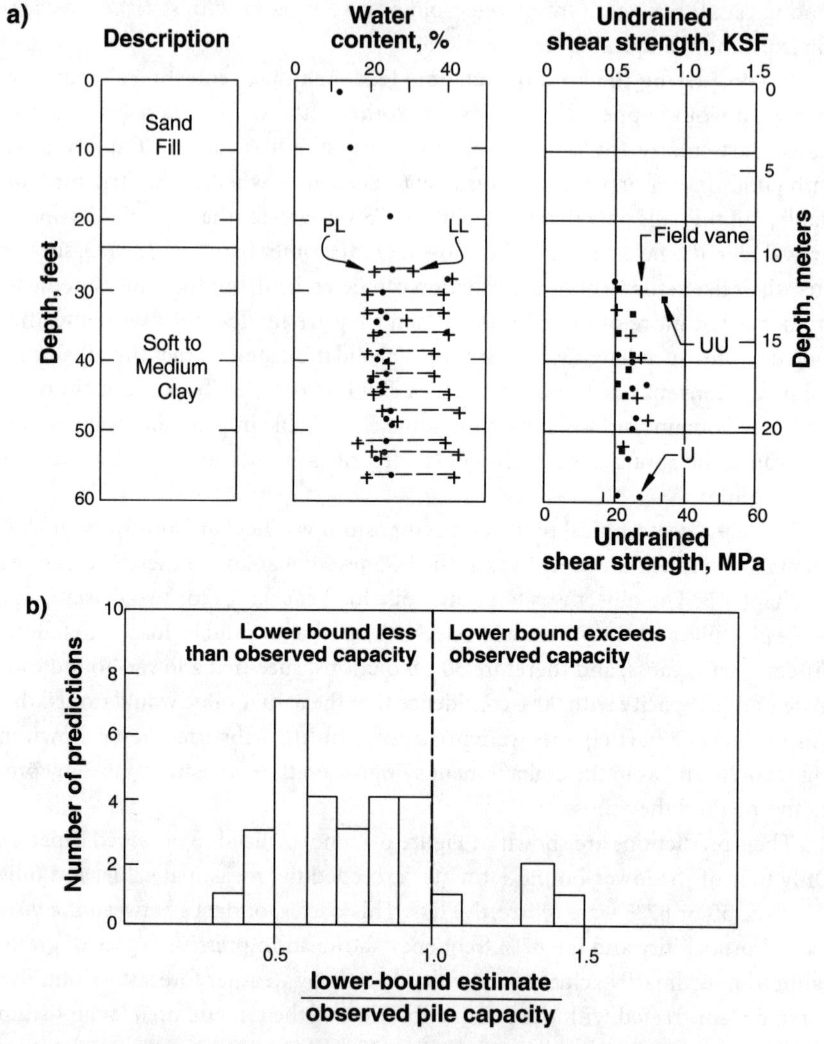

FIGURE 6-2. *Pile capacity predictions: a, subsurface data for test pile site; b, accuracy of lower-bound predictions at 90% confidence of exceedance.*

Such improvements in calibration, with feedback as their cause, offer promise for control of bias in more kinds of geotechnical activities than might be immediately apparent. When making a prediction of some physically measurable quantity, it is easy enough to assign a 50% probability to the mean predicted value. Such a "best estimate" is what we already produce—without the factor of safety—and it is not a huge leap to make similar estimates for the 90% and 10% exceedance values much like the pile predictors did. Taking every opportunity to verify these predictions, keeping a tally of outcomes over time, not only improves calibration, it also improves judgment and in this it is just geotechnical good sense. With experience and practice in subjective probability assessment, along with diligent comparison to measured results, geotechnical engineers can seek to emulate forecasters in at least some of the routine predictions they undertake and can do so with some success. This is a long-run activity that occurs over time and only when opportunities are actively pursued, but it casts the long-recognized value of instrumentation in an interesting new light. Not only do instrumentation programs provide value for the structure and for the profession itself, they provide a fundamental opportunity for outcome feedback and thus for improving uncertainty judgments of individual engineers who carefully digest their results.

Decomposition

We have already encountered decomposition of a failure sequence into its component events in Chapter 4 where it provided a central organizing framework for risk and decision analysis, and in Chapter 5 where it mirrored the causal narrative. Decomposition applies equally as a tool for assessment of subjective probabilities where some particular uncertain event or condition is disaggregated into its more primitive constituents. Here decomposition is essentially an internal mental conceptualization of how an event sequence might proceed which structures the uncertainty problem and identifies its important elements. In the cognitive framework of Figure 5-1, this allows application of the most appropriate causal or statistical reasoning strategy for each element, letting simple arithmetic do the job of recombining the associated uncertainty judgments.

Decomposition is the "divide and conquer" approach common to almost any problemsolving activity. We break a large problem down into smaller parts that are ordinarily more tractable, and engineers are especially adept at applying this technique as almost second nature. It follows that engineers, more so than others, have an intuitive ability to break down a highly aggregated uncertainty into its contributing components, allowing subjective probabilities to be assessed for each of them separately. As Raiffa (1968) encapsulated this process:

> Decompose a complex problem into simpler problems, get one's thinking straight on these simpler problems, paste these analyses together with a logical glue, and come out with a program of action for the complex problem.

From a cognitive point of view, all of the information that must be retrieved and synthesized in formulating a complex uncertainty judgment places a great burden on people's ability to process it. Decomposition aids in overcoming this limitation by reducing the cognitive load, spreading it out over a greater number of simpler situations (Edmunson, 1990). Inasmuch as Murphy's Law constitutes the prime directive of engineering, conceptualizing how things can go wrong is merely an exercise in decomposition, and subjective probability just extends it to the likelihood that they will. This native ability for decomposition is one of the most powerful tools for subjective probability assessment that engineers can exploit, and it works to advantage in a variety of ways. It is especially useful, and all but required, for estimating extreme probability values for single-event occurrences where frequency information is not available (MacGregor and Armstrong, 1994).

Decomposition Strategies

Decomposition of an uncertain event may begin by identifying the prerequisites for its occurrence. These could entail the presence of particular conditions or states of nature that are themselves uncertain and without which the larger event cannot take place. Some external initiating event that triggers a process may be similarly prerequisite, or decomposition may take the form of alternative hypotheses or processes by which the larger event might occur. Either statistical or causal reasoning strategies can be used in any of these circumstances. For example, suppose one is interested in the probability that a friend is wearing a red baseball cap today. This *target probability* might be estimated by starting statistically with the proportion of the U.S. population wearing baseball caps on any given day and the fraction that are red, continuing causally by adjusting for the friend's affinity for caps, whether they are a baseball fan, their favorite color, and so forth. In all such cases, decomposition proceeds by identifying some basic condition then asking what other conditions must be necessary, what other events must also occur, or what must happen next for the event in question to be realized.

There can be many ways to decompose an uncertainty, with none uniquely correct. Baseball caps might be less likely to be worn in cold weather, by women, indoors, or by someone outside the United States, all factors that might be accommodated in alternative decomposition structures for the same target event. The best decomposition structure to use is the one that seems to best fit the information that describes the circumstances, is easiest to think about, and gives the clearest view of uncertainty (Kleinmuntz, et al., 1996). When conceptualizing the progression of a process, it is useful to query not only what has to happen next but also what could happen to terminate the process. This signals that such an occurrence may be a key element of the decomposition structure, and there is an abundance of case histories for "near-misses" where the absence of some condition or occurrence truncated what would otherwise have been a failure sequence.

In these respects, decomposition for purposes of subjective probability assessment mimics the mental conceptualization that constructing a fault or event tree requires. Although it may be internalized, the process is identical, and nowhere is it written that subjective probability assessment must be a closed-book exercise. A simple sketch on the back of an envelope showing the relationship of the constituent events or conditions can easily capture the these mental operations and help in performing the arithmetic that aggregation of their separate probabilities requires.

At its core, decomposition involves visualizing occurrences and conditions. As we have already seen, visualization is a key component of diagnostic judgment, and once more the integral link between judgment and subjective probability becomes apparent. From a cognitive standpoint, we have seen that causal reasoning strategies are centered around the construction of a causal narrative, and the decomposition structure is what describes and gives form to its "story." Indeed visualization, the causal narrative, and decomposition are all much the same thing. Recall the stage model of the process of arriving at a subjective probability value. The first stage involved framing the problem and structuring it to come up with a plausible or favored outcome in a quasi-deterministic way. By refining the mental representation of this first-stage scenario more explicitly and in greater detail, decomposition lifts some of the burden from the subsequent stage of the uncertainty judgment, and consequently from the third stage of numerical assignment as well. In this respect, the decomposition structure becomes an integral part of the numerical probability value it produces. But more importantly, it exploits engineers' cognitive abilities by shifting the emphasis of the process to the things they do best.

As a simple mental strategy for subjective probability assessment, decomposition can be viewed as just another heuristic. But it is among the most helpful ones that engineers can use, and rather than producing bias of its own it can go far toward reducing those from other sources, as we go on to see.

Accuracy of Decomposed Probability Estimates

One measure of the effect of decomposition is to compare probability estimates derived from decomposed events with those obtained directly for the target event they describe. The question then becomes which of these procedures, decomposition or direct estimation, is better.

Ravinder, et al. (1988) considered this problem from the standpoint of error variance by idealizing a probability value as incorporating some fixed measure of a person's uncertainty plus a variable component of random error resulting from various inaccuracies. For independent or conditional probabilities that are multiplied to find the target value, it was shown mathematically that decomposition could reduce random error, provided that the individual decomposed probabilities contain less random error than the target value determined directly. It is reasonable to expect that this condition should be satisfied in most cases, since individual uncertainties can usually be understood more easily than highly aggregated ones. Moreover, there exists an optimum level of decomposition and number of component events. For example, if decomposing a target event can reduce error in each component event by 50%, then just three or four such events are sufficient to produce a similar error reduction in their product. While considerable benefit can be obtained from such modest levels of decomposition, Ravinder, et al. (1988) restricted this demonstration to errors that were equally likely to be either higher or lower than the assessor's actual belief. On the other hand, sys-

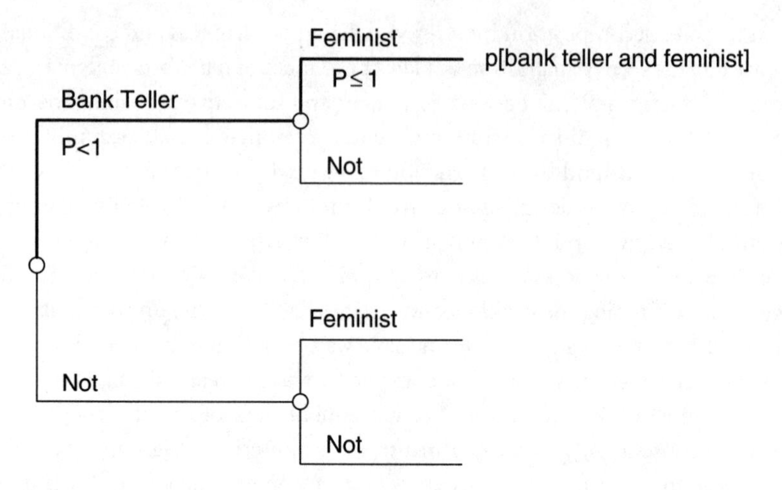

FIGURE 6-3. *Decomposition of the Linda problem.*

tematic errors with the propensity to distort probability estimates preferentially in one direction are another matter, a matter of cognitive bias.

Bias reduction by Decomposition

Decomposition can control the effects of several of the more important cognitive biases. There is no silver bullet for eliminating bias altogether, but decomposition could be the next best thing.

Consider the bias of conjunctive distortion, where the joint probability for some target event is overestimated relative to the combined values for individual components, as in the Linda problem presented in Chapter 5. Decomposition attacks this bias directly by allowing the target value to be calculated from the constituent probability components. Furthermore, recall that the representativeness heuristic encouraged overemphasis on Linda's description at the expense of other information, in this case a statistical observation about the preponderance of feminists within the larger population of bank tellers.

A simple decomposition of the Linda problem is shown in Figure 6-3. It contains two sets of events: that Linda is a bank teller or not, and that she is a feminist or not. These events can be taken as independent since we have no reason to believe that employment in the financial industry should be preferentially attractive to feminists or vice versa. Here we were asked whether Linda is more likely to be a bank teller, or alternatively that she is a bank teller and a feminist. The former is shown as a separate event in Figure 6-3, with the latter represented by the two events on the highlighted branch pathway whose joint probability is obtained by multiplication.

This decomposition structure answers the question even without assigning probabilities as such. We can observe that bank tellers constitute a subset of all occupations, so the probability that Linda is a bank teller must be something less than 1.0. On the other hand, Linda's description might convince us that the probability of her being a feminist approaches 1.0. But no matter what values might be assigned, the probability that Linda is a feminist bank teller can never be greater than the probability that she is simply a bank teller since the product of two numbers less than or equal to unity cannot exceed either one. And so we have the answer. This illustrates how simple decomposition, in this case entirely possible to perform on the back of an envelope or visualize mentally, is an effective antidote for conjunctive distortion. It also shows how considering the component events individually provides a new problem cue. It prompts for the applicable statistical reasoning, helping defeat the representativeness heuristic's neglect of this relevant information.

Useful as it is for these kinds of bias, decomposition is also an effective strategy for combating the most common and intransigent one—overconfidence. Recall here that peoples' subjective probabilities tend to be well calibrated to long-run outcome frequencies only within some limited probability range. As more extreme probabilities extend beyond this range, they become progressively more dominated by overconfidence bias, with less resolution in distinguishing among different degrees of likelihood as cognitive discrimination is reduced.

A number of experimental studies have examined the effects of decomposition on subjective probability estimates, including those by Henrion, et al. (1993), MacGregor, et al. (1988), MacGregor and Armstrong (1994), Edmunson (1990), Wright, et al. (1994), Kleinmuntz, et al. (1996), and Hora, et al. (1993). Virtually all of them document some reduction in overconfidence bias from decomposition, though the amount varies. One reason is that the level of decomposition in some studies was not great, splitting the target event into only two or three components. But both the degree of decomposition and the absolute value of the target probability influence the magnitude of bias reduction achieved. This becomes particularly important when decomposition is used to obtain extreme target probability values, as in many engineering applications concerned with low failure probabilities.

Decomposition of a target event into constituent components tends to make the component probabilities less extreme. As they correspondingly become closer to the well-calibrated range, the contribution of overconfidence bias in each of them individually is reduced, resulting in an aggregated target probability that is less affected. Another way to look at this is in terms of the hard–easy effect described in Chapter 5 whereby overconfidence is often reduced for probability assessment tasks that are easier to perform (Lichtenstein, et al., 1982). Breaking the target probability down into component parts makes each easier to evaluate and hence less susceptible to overconfidence.

This produces effects similar to those demonstrated by Ravinder, et al. (1988) but extends them to systematic errors produced by overconfidence which consistently results in probabilities more extreme than they should be at both ends of the

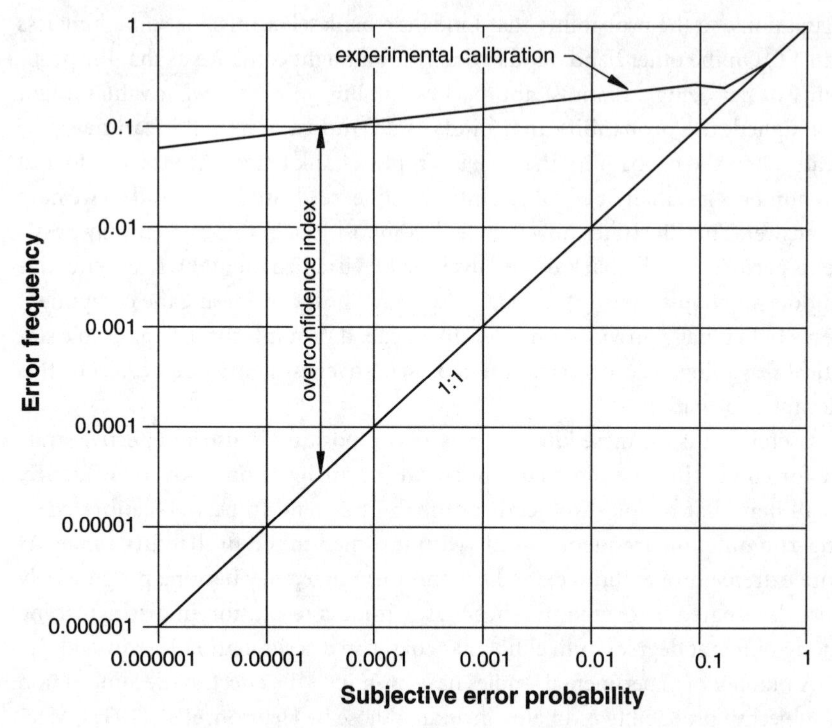

FIGURE 6-4. *Overconfidence index.*

probability scale. It is possible that high and low probability values may be more or less equally represented in a set of such values, with bias tending to cancel out in the collective result. But the most problematic condition will occur for target probabilities that are themselves extreme and either very high or very low in value. The component probabilities will therefore tend to be either high or low as well, so overconfidence bias will affect all or most of them in the same direction. Take, for instance, a target probability of 10^{-6} derived from three component events each having a probability of 0.01. Here, the effects of overconfidence bias will be cumulative in the target value, not canceling. This is illustrated using the experimental results of Figure 5-3, which are replotted in simplified form as Figure 6-4.

If overconfidence is the departure of an estimated subjective probability from well-calibrated conditions according to long-run outcome frequencies, then one way to evaluate it is by the ratio of the actual frequency to the perfectly-calibrated subjective value. By this measure, Figure 6-4 shows that overconfidence varies widely over the subjective probability scale. For a subjective value of 10^{-6}, the corresponding *overconfidence index* would be about 0.06/0.000001 or 60,000, showing it to be dominated by overconfidence effects. But suppose now that the target event is decomposed into components whose subjectively assigned probabilities are multiplied to obtain the target value of 10^{-6}, and assume these are equal in

TABLE 6-3. *Effects of Decomposition on Overconfidence Bias for 10^{-6} Target Probability*

Number of decomposed events	*Component event probability*	*Component overconfidence index*	*Target overconfidence index*
2	1×10^{-3}	0.15/0.001 = 150	$(150)^2 = 22{,}500$
3	1×10^{-2}	0.15/0.01 = 15.0	$(15)^3 = 3375$
4	3.16×10^{-2}	0.21/0.32 = 6.56	$(6.56)^4 = 1852$
5	6.31×10^{-2}	0.26/0.063 = 4.13	$(4.13)^5 = 1201$
6	0.1	0.30/0.10 = 3.00	$(3)^6 = 729$

value. Each will have its own overconfidence index, which together cumulate to that of the target probability in similar fashion. Table 6-3 illustrates the effects of various levels of decomposition as the number of component events increases.

Table 6-3 shows how greater levels of decomposition exponentially reduce overconfidence bias, again because as the number of component events grows, their probabilities each lie closer to the well-calibrated range and are therefore less affected. Figure 6-5 provides the same comparison for other target values, showing that overconfidence bias will be minimized for target probabilities that themselves are least extreme and closest to the well-calibrated range. Here too, a modest number of component events, say three or four, produces a large part of the bias reduction achieved.

The number of component events that can be defined depends on one's state of knowledge about the process involved, and it can sometimes be difficult to decompose mechanisms or processes whose antecedents or stepwise progression may be poorly understood. Nevertheless, making every attempt to decompose uncertainties to a level such that component probabilities lie as close as possible to the well-calibrated range is a primary means for controlling the effects of overconfidence bias. In practical terms, the experimental data of Figure 6-4 suggest that if component subjective probabilities are much less than about 0.01, greater levels of decomposition should be actively pursued unless these values are supported by base-rate frequencies. Interestingly, some studies (MacGregor, et al., 1988) have found that overconfidence bias was reduced more effectively when experimenters provided the decomposition structure than when subjects developed their own or when an algorithm was provided that enforced a certain level of decomposition (Edmunson, 1990). It may be that the experimenters' greater familiarity with the problem allowed them to decompose it more effectively, or perhaps they were simply more cognizant of the effects.

The benefits of decomposition extend to assessed probability distributions as well, where overconfidence is manifested by distributions that are too narrow with insufficient dispersion about the mean or central value. Hora, et al. (1993) show that target distributions obtained by reaggregating individual distributions

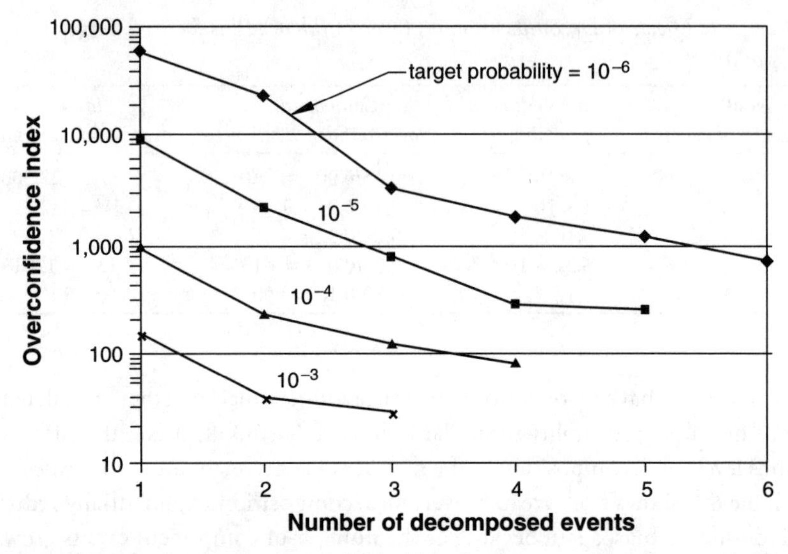

FIGURE 6-5. *Reduction in overconfidence bias by decomposition for experimental data of Figure 6-4.*

on decomposed parameters were much better calibrated than target distributions assessed directly. The target distribution was also decomposed into three to five constituents, a greater level than typically adopted in such studies.

Conformity of Decomposed Probability Estimates

It is often asked whether different individuals or groups, given the same information, would come up with the same subjective probability estimate. Because people may adopt different reasoning strategies and their personal judgment may differ, there is no necessary reason why they should and nothing intrinsic to subjective probability requires it. Agreement is not necessarily a measure of goodness, since greater diversity in perspective often leads to greater accuracy overall, and combining diverse individual assessments by simple averaging can be a superior predictor, as was seen in the geotechnical embankment-stability symposium of Chapter 5 (Einhorn, et al., 1977; Winkler and Poses, 1993). In fact, when individuals work together their group performance can be enhanced by heterogeneity in individual judgments (Snizek and Henry, 1989).

Nonetheless, there are those who might find it reassuring if two or more assessments performed independently were to yield similar values, an effect termed here *conformity.* It has been found that first establishing a common decomposition structure promotes conformity in target probability estimates (Keeney and von Winterfeldt, 1991). This comes about because a shared representation of how the uncertain event is visualized then reduces variation among individuals in this first stage of the cognitive process.

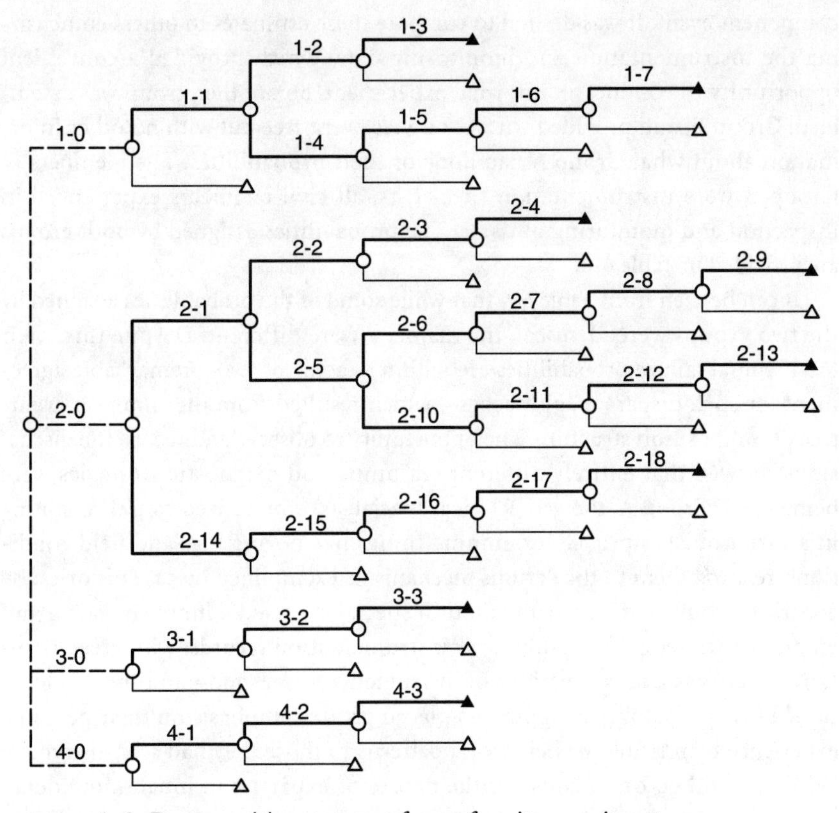

FIGURE 6-6. *Decomposition structure for conformity experiment.*

Conformity in probability estimates obtained by decomposition is illustrated by a case involving a small zoned earthfill dam. Over time it had experienced multiple signs of distress, including cracking, concentrated seepage in the embankment and foundation, evidence of particle transport, deterioration of its outlet conduit, and sinkholes. Of the repairs undertaken over the years, none had been fully effective, but the embankment had been heavily instrumented and good records were available. Then, when inspection and monitoring showing marked changes, it soon became clear that interim measures to stabilize its condition were needed until a more permanent solution could be devised. To this end, a risk analysis for seepage and internal erosion was performed (Vick and Stewart, 1996).

Four potential failure modes were identified, with the decomposition structure shown in Figure 6-6 highlighting the failure branch pathways for each. This event tree was constructed by a group of geotechnical specialists, Group A, after reviewing the relevant information including historical records and subsurface data as well as pertinent field case histories of dams with related conditions and symptoms. They then provided their subjective probability estimates for each component event. It was desired to compare these estimates to others emphasizing the instrumentation and monitoring data, which provided a convenient opportunity to conduct an informal experiment. So another group was assembled, Group B, and provided with Group A's event tree but with no other information about what Group A had done or their probabilities. The members of Group B were instrumentation specialists, all civil engineers experienced in inspection and monitoring of dams. The probabilities assigned by both groups are provided in Table 6-4.

It can be seen from Table 6-4 that while some of the probabilities assigned by the two groups were identical, the majority were different. Despite this, their total annual failure probabilities are within a factor of two—remarkable agreement for such disparate approaches—which resulted from the common event-tree decomposition structure. The opportunity to observe and record the discussions showed that entirely different reasoning and diagnostic strategies were being used. Group A, the geotechnical specialists, emphasized causal reasoning in a variationist approach, examining things like boring logs and field conditions, relating them to the various mechanisms exemplified by case histories. By contrast Group B, the instrumentation specialists, used a more statistical and empiricist strategy. They pored over instrumentation records, with great attention to patterns and correlations of instrumentation response in time and location. Their probability assignments placed greater emphasis on their personal experience by relating the behavioral patterns to those they had seen elsewhere. While we will go on to consider the nature of expertise in much more detail later, it is worth noting here that the two groups were each using separately two of the cornerstones of expertise: The causal reasoning of Group A adopted mental simulation with Group B's statistical strategy relying on pattern recognition. Still, these two different expert skills yielded surprisingly similar results. The failure probabilities estimated by both groups were alarmingly high, but this was consistent with their beliefs after reviewing the information available. Group A, for example, expressed the view that initial mechanisms of failure were already in progress, so failure itself would be only a matter of time—and a short time at that.

The improvement in conformity was no isolated occurrence, with similar effects from common decomposition structures also documented in other cases. Cooke (1991) describes a "benchmark" study in which 10 teams of specialists were asked to independently estimate failure probabilities for a nuclear reactor subsystem. Each team performed its assessments in phases to separate out the effects of common decomposition and common data. In the first phase, the estimates of each team were based on its own data without formally decomposing the problem. In the second phase, each team developed its own fault trees. The third phase used a common fault tree structure, and in the last phase common

TABLE 6-4. *Event Probabilities Assigned by Separate Groups*

Event	*Description*	*Group A estimate*	*Group B estimate*
1-0	**Internal erosion of embankment**	**0.49***	**0.23***
1-1	Loss of core integrity	0.75	0.5
1-2	Pipe/tunnel enlargement	0.5	0.5
1-3	Crest subsidence/cracking/breach	0.9	0.5
1-4	No pipe/tunnel enlargement	0.5	0.5
1-5	Seepage on downstream face	0.9	0.9
1-6	Slumping of downstream face	0.9	0.5
1-7	Progressive instability/breach	0.5	0.5
2-0	**Internal erosion of foundation**	**0.13***	**0.13***
2-1	Foundation cavity forms and communicates with reservoir	0.1	0.9
2-2	Loss of core integrity	0.6	0.3
2-3	Pipe/tunnel enlargement	0.9	0.25
2-4	Crest subsidence/cracking/breach	0.9	0.5
2-5	No loss of core integrity	0.4	0.7
2-6	Silt layer pipes into fill	0.8	0.1
2-7	Phreatic surface rises	0.9	0.5
2-8	Slumping at downstream toe	0.9	0.9
2-9	Progressive instability/breach	0.5	0.5
2-10	No silt layer piping	0.2	0.9
2-11	Artesian conditions at downstream toe	0.1	0.9
2-12	Slumping at downstream toe	0.9	0.3
2-13	Progressive instability/breach	0.5	0.5
2-14	No foundation cavity forms	0.9	0.1
2-15	Buckling/rupture of outlet conduit	0.1	0.1
2-16	Concentrated seepage around outlet with particle transport	0.9	0.9
2-17	Tunnel development and enlargement	0.9	0.5
2-18	Crest subsidence/cracking/breach	0.9	0.5
3-0	**Piping around outlet conduit**	**0.08***	**0.03***
3-1	Concentrated seepage along conduit with particle transport	0.1	0.1
3-2	Tunnel development and enlargement	0.9	0.6
3-3	Crest subsidence/cracking/breach	0.9	0.5
4-0	**Piping into drains**	**0.002***	**0.01***
4-1	High inflow gradient and particle entry	0.1	0.5
4-2	Progressive sinkhole development at downstream toe	0.2	0.2
4-3	Crest subsidence/cracking/breach	0.1	0.1
Total annual probability of seepage/erosion failure		**0.70**	**0.40**

* calculated from respective failure mode event probabilities

TABLE 6-5. *Effects of Common Decomposition and Common Data on Conformity of Probability Estimates*

Group values	*Range of estimates*	*Ratio (max/min)*
Independent data, no decomposition	8×10^{-4} to 2×10^{-2}	25
Independent data, independent decomposition	7×10^{-4} to 2.5×10^{-2}	36
Independent data, common decomposition	1.4×10^{-3} to 1.3×10^{-2}	9
Common data, common decomposition	$\sim 1.4 \times 10^{-3}$	~1

data were also provided to each team. The target probability, in this case the "correct" answer, was 1.4×10^{-3}, and the results are shown in Table 6-5.

Comparing the ratios of the estimates within each group shows again that using the same decomposition structure—here the same fault tree—reduced their variation, to an extent second only to supplementing this with common data.

Defining Events

Regardless of how or how much they may be decomposed, all uncertain events or conditions must be clearly and unambiguously defined before probabilities can be assigned to them. For example, the magnitude of pore pressure can affect any number of processes, and a related event might be defined as "pore pressure increases." But to avoid ambiguity, it will also be necessary to specify where (in what stratum or location) the increase occurs, and at least approximately by how much this "increase" is intended to denote.

In many cases the context within which the event is considered makes its meaning clear. But this context can change as complex event sequences are being developed. An event originally involving pore pressure increase in the context of seepage might later turn out to be relevant to liquefaction as well. Subtle changes in what events are taken to mean often occur during the course of conceptualizing various processes and mechanisms as insights into their progression emerge. An event intended to mean one thing at the beginning may end up meaning something quite different later on as the factors affecting it become more apparent.

Writing down a complete description of each event captures these conceptualizations so they are not lost, and this can be just as important as having decomposed the events in the first place. Subjective probabilities will be affected if ambiguity in event definitions allows these meanings to shift, and it is always worthwhile to review them at the completion of the decomposition process to verify that they reflect the meanings intended in each case. As we will see, there can be vagueness in uncertainty judgments and the probabilities that express them. This can convey important information but only if vagueness results from the uncertainty judgment and not from ambiguity in event definition.

The Internal Process

It is worthwhile at this point to step back for a moment to see where we stand in the probability assessment process. Preparation and training have helped ready the assessor for the task ahead. The larger uncertainty problem has been structured and reduced to its basic elements by decomposition, and these elements have been defined at the necessary level of detail. So far, these activities have been mostly preludes to the main event—the determination of one's degree of uncertainty. This is where the internal activities of formulating then expressing the uncertainty judgment really begin and where the cognitive process reaches full swing. It would stand to reason that those more aware of how this process operates should be able to use it to better effect, so we take the time to walk through it here once more.

Going back to the model of Figure 5-1, the cognitive process starts by perceiving the nature of the uncertainty and establishing from the problem cues it presents whether it is best conceptualized in a causal or statistical light. It then proceeds in the following stages:

1. Memory is searched for relevant information, which is used to determine some plausible scenario or favored outcome;
2. A likelihood judgment (or equivalently, an uncertainty judgment) is formed by assessing one's feelings of doubt about this scenario or outcome by weighing the evidence for and against it or by considering how often it tends to occur; and
3. These feelings of doubt are then evaluated as to degree, which ultimately results in scaling them to the probability metric from 0 to 1.

Recall too that this process may switch back and forth between its causal and statistical forms. It is also applied iteratively, revisiting any previous stage and working through it again if need be.

This all contrasts with the view that a subjective probability exists in one's head just waiting to be tapped (Smith, et al., 1991; Laskey, 1996). Instead, the internal process takes the number as the culmination of a variety of complex cognitive activities. In following this general framework, the intent here is to make probability assessment less intimidating than it might otherwise be, recognizing that there is a structured process involved and thus showing that the numerical assessment stage is not the only or even the most important part. As we go through it, we will revisit the cognitive biases that can enter these stages but at the same time attempt to show how they can be reduced and controlled.

The subsequent discussions parallel the stages of the cognitive process, treating the following elements of subjective probability assessment in turn:

- Assembling information and evidence,
- Synthesizing information and evidence,

- Numerical assignment, and
- Confirmation.

Assembling Information and Evidence

The first stage in formulating the uncertainty judgment is retrieval of relevant information from memory. The catch is in identifying what is relevant, and we have seen that people can be selective in this determination. The cure is to consciously gather all of the information of all kinds and from all sources that can be brought to bear on the uncertainty at hand. Careful reflection often reveals that we may have more information than we think. Given the tendency to neglect disconfirming information, it can be especially important to search out factors that might support each possible interpretation whether plausible or not at first glance.

Information Gathering

Memory is not the only or necessarily the most reliable repository for information. Few engineering activities are conducted from memory alone, and this applies equally to uncertainty judgments. Retrieval of information is as much a physical activity as a mental one, and it requires that data of all types be gathered and compiled. It can be tempting to say "we've done that already" or "we know what's there," but diligent searches of original sources turn up overlooked information of the most influential kind with surprising regularity. This is because those who compile summaries are seldom attuned to the kinds of information that provide the necessary cognitive cues for probability assessment. There can be no substitute for personally examining original sources—field boring logs, photographs, and raw laboratory data to name a few—and summaries or compilations by others cannot replace firsthand information gathering. Similarly, seeing the site firsthand is how any probability assessor develops a sense of what its important features are and how they interact. This, of course, is how one develops a "feel" for the data, its patterns and relationships. The uncertainty judgment sought depends on the assessor's own identification, determination of relevance, and interpretation of information. It cannot rely on someone else's sense of what is important.

We have seen that causal reasoning elaborates the mechanism or process that underlies an uncertain event, while statistical strategies emphasize observed occurrences. Case histories are the raw material for causal strategies, and the information they provide is at least as important as any other kind. We saw in the Challenger story how causal factors were emphasized throughout in attempts to explain and fix the problem, but in the end there turned out to be many other instances of field performance from a wider variety of solid rocket motors. So thorough search of case histories, even if not strictly identical to the case at hand, is central to statistically based reasoning as well.

A causal strategy, statistical strategy, or both are prompted by cues contained in the problem presentation. Unfortunately, reality does not offer up these cues on command, so they must be actively sought out during the information gathering stage. This involves determining what kind of information is relevant to the uncertainty judgment; and if there is anything that the heuristics and bias research consistently shows, it is that frequency information is most often neglected even, or especially, in its simplest forms. Neither statistics as such nor quantified data, however, are required for statistical reasoning. Even if we don't know the exact frequency we may still know something about it, and an observation that "it doesn't happen very often" or "we see that quite a bit" can be among the most significant kinds of information that statistical reasoning can bring to light. Information that is not quantified is information nonetheless. It always tells something, and that cannot be ignored.

As much as anything, subjective probability assessment requires that all information be considered and none discarded prematurely. It seeks to squeeze out every last drop from what constitutes one's state of knowledge. So the information-gathering stage is hard to overemphasize. This is where problem cues are recognized and the feel of the uncertainty problem develops. In this sense, "information" extends far beyond its usual meaning of factual data to general observations and everyday knowledge that are so simple as to ordinarily be thought of as intuitive. Not all of it will subsequently turn out to have the same strength and weight, but it can never be considered if it is not first placed on the table.

The Importance of Lists

If engineering activities are not performed exclusively from memory, neither are they conducted without pencil and paper. As noted throughout, peoples' information processing capacity is limited, and among the most common ways of overcoming this is writing things down. Making a list of evidence, both supporting and contradicting, can greatly enhance the gathering of information and its subsequent evaluation. Words have meaning, and because they do this meaning must be found before they are used. The act of writing things down and seeing what is written forces a critical examination of thoughts as they are articulated, and this prompts for yet other thoughts that uncover things not originally perceived. So the very process of listing evidence can identify more of it in addition to documenting it more completely.

Beyond this, making lists of evidence is important in bias reduction. Listing evidence is a much better way to keep track of it than trying to maintain it all in memory and this provides an essential format for organizing it. While this need not consist of much more than a listing of "pros" and "cons" with regard to a particular outcome, for cognitive reasons it turns out that the cons are of particular significance. They represent disconfirming evidence, that which would counterindicate some initially favored proposition, and it can be recalled from Chapter 5 that confirmation bias results when this evidence is discounted or ignored. This, in turn, leads to overconfidence, and it is only to be expected that one would be more confident than the evidence warrants if disconfirming evidence has not been discovered. But the contents of the lists, as well as when they are compiled in the judgment-forming sequence, seem to make quite a difference.

Koriat, et al. (1980) performed a series of particularly enlightening experiments. In the first experiment, subjects were presented with questions having two possible answers and asked for the probability that their chosen answer was correct. Before responding, however, they were required to list all the reasons they could think of for and against both answers. Along with each entry was a rating of how strongly each reason argued for or against each answer according to a seven-point scale with associated verbal descriptors from weakest to strongest possible. They were asked to do this according to the following instructions, which not incidentally provide an excellent catalog of the various kinds of information relevant to an uncertainty judgment (Koriat, et al., 1980):

> Spell out all the possible reasons you can find favoring and opposing each of the answers. Such reasons may include facts that you know, things that you vaguely remember, assumptions that make you believe one answer is likely to be correct or incorrect, gut feelings, associations, and the like.

The result of having listed all such confirming and disconfirming evidence prior to formulating the uncertainty judgment was to produce a marked improvement in calibration and reduction in overconfidence.

The second experiment was performed along the same lines but with several key differences. This time the sequence of evidence consideration was reversed. Subjects were asked to initially choose their preferred answer, then list the evidence, and finally to provide their probability of this answer being correct. This allowed the effectiveness of listing evidence to be evaluated when performed after instead of before making the uncertainty judgment. In addition, to evaluate the influence of the amount and the nature of the evidence used, the subjects were broken out into three groups. One was asked to provide the single best piece of supporting evidence for its answer, another for both the best supporting and the best disconfirming reason, and the third for only the best disconfirming reason. This time, the subgroup using the disconfirming factor showed improved performance but the other two did not.

In comparing the two experiments, we see that presentation of both confirming and disconfirming evidence—the listing of both pros and cons—is an effective method for reducing overconfidence bias if it is carried out before the uncertainty judgment has been formulated. Afterward, the anchoring and adjustment heuristic makes this more difficult. With the assessed likelihood already established, it takes a stronger and more exclusive emphasis on disconfirming evidence to defeat the effect of insufficient adjustment in accounting for the newly introduced information. The counteracting effect of disconfirming evidence is still achieved without cataloging it in full, and presenting only a few of the most important such factors appears to have the desired result.

The practical significance is twofold. First, the simple act of making pro/con lists of supporting and disconfirming evidence can be an effective aid for improving calibration at an early stage of the probability assessment process. MacGregor, et al. (1988) also found that listing evidence improved the performance of assessors, attributing this to its prompting effect on recall. Fischhoff and MacGregor (1982) essentially duplicated the results of Koriat, et al. (1980), finding that the introduction of disconfirming evidence by listing it reduced overconfidence by discouraging the use of extreme probability values in a substantial proportion of assessors. The second implication is that it is still possible to correct overconfidence bias even after it has been introduced by emphasizing disconfirming evidence. While this will be considered further in more detail, we turn now to how the uncertainty judgment itself is made after the pertinent information has been assembled.

Synthesizing Information and Evidence

Figure 5-1 identifies that part of the probability assessment process where a judgment as to likelihood or uncertainty is formed. A favored or plausible outcome seems to be apparent from the evidence gathered, but the outcome would not be uncertain if this evidence were not to some extent inconclusive or conflicting in some way. The formulation of an uncertainty or likelihood judgment is to assess one's "feelings of doubt" that constitute a degree of belief in the truth of the proposition relating to the outcome. Doubt occurs because questions arise, so the process of belief formation and the formulation of the likelihood judgment is very much one of questioning—of questioning the evidence, and questioning one's assessment of it.

A Stance of Detachment

We often think of evidence in its courtroom context as objects and facts or in its engineering sense as information and data. But evidence in a subjective probability context goes beyond these things to the kinds of inferences that Koriat, et al. (1980) encouraged from their subjects in the instructions cited above. Evidence is not restricted to facts or hard data but may also include even vaguely recalled circumstances or associations, gut feelings, intuition, and the like. All of these things have information content that subjective probability seeks to extract. And when information is synthesized in producing the likelihood judgment, this takes the form of reasons, or arguments, why an outcome might or might not occur.

Smith, et al. (1991) show that evidence, in the way people use it in reaching conclusions, is not strictly factual. In order to become evidence, information must be integrated, and this is not done using computation or rules of formal logic. Instead, evidence packages information in the form of arguments to arrive at conclusions. These conclusions are beliefs, so the formation of belief, and one's degree of belief along with it, is to balance these arguments one against another. Smith, et al. (1991) describe four common types of cognitive arguments. One is *causal argument*, in prediction an argument that a cause produces some effect, or in diagnosis that an effect results from some cause. Causal arguments, of course, invoke causal reasoning, whose operation is clear by now. Another type of argument is *observational argument*, which invokes observed signs and symptoms and relies on similarity with observed occurrences. *Analogical arguments* also use similarity between a particular case and its analog, with similarity at some point becoming sameness according to patterns that infer this sameness by induction. We can see that both observational and analogical arguments conform to statistical as opposed to causal reasoning strategies. The final type is *argument from authority*, the argument of others adopted by conformity or convention, which Smith, et al. (1991) call "parasitic" on the previous three. Belief here comes from the received knowledge or doctrines of others, and this goes back to the primitive notion of probability as the approbation of authority before the concept of evidence was even recognized. So while some might say that what people should do is to evaluate evidence strictly according to rules of deductive logic, we are more concerned with what they actually do. And this is to construct arguments using the causal and statistical strategies of cognition.

If evidence is presented as arguments, then these arguments must be synthesized. So we return to the courtroom and to the verdict of belief rendered in the mind. Assembling information and evidence has produced some position as to a favored or plausible outcome. These are the arguments of opposing advocates, and few who assess probabilities will doubt that these mental debates can be as vigorous as any attorney's. But while advocates plead the case, they do not decide it—the judge or the jury ultimately does. So at that point in the process where the uncertainty judgment comes about, the probability assessor must wear a different hat. Whereas efforts were formerly directed to constructing a plausible argument for an outcome in an advocacy role, the evidence and arguments must now be balanced and judged in a more impartial way to produce the uncertainty verdict. In this, the viewpoint required changes distinctly, now adopting a different perspective. Having already been assembled and presented, the evidence must now be interrogated evenhandedly. It is this stance of detachment perhaps more than anything else, and the self-questioning attitude it promotes, that are central to forming a likelihood judgment as free as it can be from cognitive bias.

It is interesting to note that this change in viewpoint has been incorporated into some formal processes for probability assessment in a most literal way. One such technique is structured so that assessors are first asked in group discussions to be proponents for the personal positions they espouse on uncertain factors, some of a professionally controversial nature. After airing all such views, each assessor is then asked to switch from proponent to evaluator, presumably now detached from the positions they previously advocated. Such a structured distinction between these two perspectives has been reported to directly assist the probability assessment process (Senior Seismic Hazard Analysis Committee, 1997). But formally distinguished or not, at some point the construction of arguments must turn to interrogating these arguments with a detachment that considers both those for and against. This is to weigh the evidence. It is an inductive process that hinges on two attributes of evidence: its strength and its weight.

The Balance of Strength and Weight

The process of weighing evidence and arguments is the essence of inductive reasoning where its departure from deductive determinism becomes most apparent. A deterministic argument is cut and dried, open and shut, Q.E.D. By definition, it does not allow for uncertainty or feelings of doubt. So engineers steeped in deductive tradition may not be well accustomed to a weighing-of-evidence approach, preferring proof instead. This is why only after acknowledging uncertainty to begin with can this approach be successfully employed.

From a cognitive perspective, evidence has two properties—strength and weight. So weighing evidence or arguments requires balancing these attributes. But recall from Griffin and Tversky (1992) that people do this sequentially. The arguments are first evaluated according to how strong they are—how convincingly or persuasively they support one position over another. Only then comes weight—how good this evidence is in quality, quantity, and overall predictive value. The representativeness heuristic, with its emphasis on similarities, seems to pretty much ensure that strength takes care of itself when the favored outcome or scenario is first identified. But when this assessment is adjusted for weight of the evidence, the anchoring and adjustment heuristic makes its appearance with its tendency for insufficient adjustment. This means that weight receives inadequate emphasis. The overall effect is to focus more on how strong the arguments are than on how good they are, so that people are overly confident in strong arguments that may not be very good ones and don't give enough credit to good ones that might not argue very strongly. This shows too that the representative heuristic is what most puts a thumb on the balance between strength and weight. If so, then strength may need a little toning down and weight some extra assistance. More specifically, we can look to one of the outgrowths of the representativeness heuristic from Table 5-2 involving predictive validity.

The results of analytically or empirically derived predictive methods are often taken as intrinsically strong arguments. Indeed, where only one such method is used, its evidence becomes strong by default. But accounting for its predictive validity—its weight—is also necessary. The level-ground liquefaction correlation previously shown in Figure 4-9a can be used to illustrate this. It provides a fairly clear demarcation between the regions of liquefaction occurrence and non-occurrence from reasonably comprehensive data, making it a persuasive and well-accepted predictive method. For example, an $(N_1)_{60}$ of 15 and a cyclic stress ratio of 0.2 would plot comfortably within the region of observed liquefaction and thus provide a strong argument for liquefaction occurrence. But the validity of this prediction is shown in Figure 4-9b, where liquefaction probability would be only about 0.5. While the argument in favor of liquefaction may be strong, in this particular instance the predictive validity of the method and therefore its weight is low, such that one might be overly confident in applying it. This underscores more generally for any such method the importance of questioning how good or how accurate it is in determining the condition it predicts. It also shows that incorporating results from a variety of methods helps moderate strength from just one.

In other circumstances there may be no strong argument. There may be no analytical procedure for predicting the phenomenon, though there may still be several weaker indicators pointing in the same direction. Here it is the number of these indicators that tends to be discounted, and therefore their collective weight. The evidence is not taken in its entirety because our information-processing limitations encourage us to consider it individually, discarding each piece in turn as lacking adequate strength. So in accounting for how much evidence there is, statistical reasoning never has a chance. This can also be considered a form of insensitivity to sample size that operates to the detriment of weight—evaluating the evidence one piece at a time masks its amount. When strength is low, this produces underconfidence in giving less credit to the argument than it deserves.

We can see how this works using the Challenger incident as a case study in balancing the strength and weight properties of datasets. Recall that NASA had never given much credence to the argument that cold temperature causes O-ring problems. The Thiokol engineers thought it was strong—they had seen the cold-launch damage themselves. And besides, the effect of temperature on resiliency has strong intuitive appeal. But this was not enough to prevail against the strength of the putty argument, the pressure testing, and so forth. And when NASA considered the weight of the evidence—its quantity—they meant hard data, experimental data, for temperature and O-ring resiliency. And it wasn't there. So the cold-temperature argument was doomed. Even if were strong, it just didn't carry any weight.

But there was another kind of data, observational data from flight experience, although causal reasoning kept it from being evaluated in its entirety saying in effect, "Sure we occasionally get O-ring problems, but we know the cause so we don't need to look any deeper." Had NASA evaluated the flight data collectively they would have found ample strength, and weight too. This is why we find Figure 5-11 so compelling. In contrast to Figure 5-10, it contains all the flight data—those with problems and those without—separating the evidence for and against the cold-temperature argument at a boundary of 65°F. NASA's opposing argument was that warm temperatures had caused problems too and thus temperature makes no difference. In balancing these arguments, it is the patterns of the data that jump out at us. These patterns contain their strength and their weight.

Inspecting Figure 5-11 more closely, first take the right-hand side for launches warmer than 65°F. We see here that most of the data show that problems tend to occur much less often at warm temperatures, in contrast to NASA's position. Here it is the preponderance of the data—the relative proportion—that gives them their strength. We see too that there are a lot of data—their quantity gives them weight. Turning to the left side of the figure for launches colder than 65°F there aren't enough data to have enormous weight, but their consistency argues strongly. And seeing how much colder Challenger was argues even more strongly that it would be affected. Here it is the degree to which the evidence supports the argument that makes it strong.

In balancing these competing arguments, then, there is a strong case that problems occur disproportionately at cold temperatures and a strong argument of considerable weight that problems are less apt to occur at warm temperatures. So temperature does indeed matter to O-rings, the cold Challenger temperature in particular. We see once again that weighing the evidence could not have taken place without looking to all of its types—flight data, not just experimental data. And in balancing strength and weight, patterns are what make the difference, and these patterns are revealed only by taking all of the evidence collectively.

What we gain from these discussions are several ways to compensate for some of our potential weaknesses in balancing the strength and weight of evidence. It is well to temper the strength of arguments, particular analytical ones, by considering multiple methods and the predictive validity of each in order to bring out uncertainties that go to their weight. Weight can also require more direct attention using a variety of techniques. Listing evidence can better assure that its weight is considered collectively, as can plotting it comprehensively. Either way, the aim is to coax out its patterns and bring balance to the uncertainty judgment.

Some Practical Guidance

A likelihood judgment brings to bear information of different kinds and from different sources. There might be very general or even intuitive factors, other events or conditions similar in some respect to the circumstances at hand, or predictive analyses conducted at various levels. In the end, all of these things must be brought together and synthesized to produce the internal likelihood judgment. In distilling some of the previous concepts into practical guidance, the following are well to keep in mind:

- Actively seek out and list all kinds and sources of information both for and against some outcome—all of the pros and cons—taking care not to overlook simpler observations and knowledge;

- Search diligently for previous occurrences and case histories that, even if not identical, may be related;
- Consider the predictive validity of data with respect to both quality and quantity—how good and how much;
- Account for the predictive validity of analytical methods, closely interrogating their assumptions with respect to what is known about actual field behavior;
- Use analysis in sensitivity studies to reveal the full array of operative influences and to help visualize all the ways a process might progress;
- Evaluate information, data, and analyses collectively to detect patterns of consistency and anomaly; and
- Weigh evidence and arguments with a considered and balanced view, recognizing that neither proof nor conservatism is the object.

USING ALL OF THE INFORMATION. Weighing evidence requires that all of it be accounted for and none of it overlooked or discarded. This is another way of saying again that the representativeness heuristic—the overemphasis on one particular kind or source of information—needs to be consciously avoided, and having written down the evidence in the process of assembling it is a good first step. While every kind of information, every source of evidence, and every argument tells something, it can mislead if used to the exclusion of others. To balance the evidence is to give due consideration to all of it, rather than whittling it down to some subset from the start.

Even if not formally applied, Bayes' theorem still provides a sound perspective. Almost every engineering assessment is conducted in stages, starting with simple approaches and moving to progressively greater levels of refinement. As it progresses, there is an almost unavoidable tendency to discard what came previously: The final analysis is the best analysis that supersedes everything else. This is the representativeness heuristic at work. But reviewing the physicians' diagnosis of cancer in Chapter 5 shows how discarding prior information can distort the likelihood judgment. Bayes' theorem demonstrates clearly that no source of information can be entirely discounted, even if new information represents an improvement in some respect. The application of Bayes' theorem is called Bayesian "updating" for good reason. The previous information is combined with the new rather than being dismissed, an outlook that can go far as an antidote to representativeness bias.

USING PREVIOUS OCCURRENCES. Recall that among the important variants of the representativeness heuristic was base-rate neglect. We have already noted some of the problems in selective retrieval of information, but information can also be processed selectively once it is accessed. This can work to the particular detriment of statistical reasoning strategies, where our preference and facility for causal reasoning can encourage the frequency of related occurrences to be discounted even when their presence is known. Statistical reasoning holds that what has occurred in the past has some relevance to future occurrences, even if certain aspects may be different. Johnson (1988) considers difficulties in combining base-rate frequency with case-specific information one of the main factors limiting the performance of probability assessors because causally derived but unusual exceptions to general or mundane trends all too easily overshadow the trends themselves. These exceptions have been called "broken-leg cues" after a well-known example involving the likelihood that a colleague will go to the movies tonight. Things like his frequency of dating or the number of times he has described movie scenes could be used in developing a statistical strategy. But this information is all thrown out when we learn that the colleague has a broken leg—never mind that people can attend movies on crutches just the same. While exceptions and specific information may modify frequency-based trends, they do not make them irrelevant to single-event occurrences.

While engineers are not specifically trained in diagnostic procedures, physicians are, and the base-rate frequency of diseases is recognized to be an important diagnostic indicator. And for all their heuristic shortcomings, some of the simple adages imparted to physicians in medical training are to encourage them to consider the overall prevalence of a disease and how often it occurs, alongside specific symptoms the patient presents. Two such examples are relevant here in speaking directly to statistical reasoning and base-rate effects (Eddy, 1982):

> When you hear hoofbeats think horses, not zebras.
>
> Common things occur most commonly.

PREDICTIVE VALIDITY OF DATA. The opposite side of the coin from neglecting statistical information is using it without adequately accounting for its quality or quantity. These are the biases of insensitivity to sample size and predictive validity that stem from underemphasis on the weight the evidence displays. Some well-worn geotechnical cautions take on new meaning in the context of the questions to be asked of the data. How much data are there, and to what extent do they provide a meaningful representation of the conditions in question? How many borings have been drilled and how far apart, and how do the unsampled intervals compare with the variability this geologic setting might imply? Equally important is the quality of the data: things that could affect their reliability like sample disturbance or testing procedures, and how conditions imposed in the laboratory compare to those in the field. These are all routine considerations in any geotechnical investigation, but they assume special importance from what we know about weighing of evidence and its effects on uncertainty judgments.

PREDICTIVE VALIDITY OF ANALYSES. It has already been demonstrated how the strength of arguments provided by various kinds of analyses and geomechanical models tends to overshadow their weight, especially predictive validity. This effect has often been observed to increase with analytical sophistication, whereby the strongest evidence is attributed to methods of greatest computational complexity. This is a natural outgrowth of the emphasis on causal reasoning strategies, where more highly refined causal narratives are perceived as more plausible. This, however, has no necessary relationship to their predictive validity, and to counteract this tendency requires questioning their assumptions.

Predictive validity can again be viewed from a Bayesian perspective. Any analysis or geomechanical model can be viewed as a predictive indicator of some potential outcome. Such indicators have two properties: their reliability in predicting the outcome they are supposed to predict and their false-positive rate in predicting that same outcome when they shouldn't. Casting this in terms of a failure-related analysis, one should ask two things. First, how often would the analysis predict failure when failure actually occurs? And second, how often would the method predict failure anyway when it doesn't occur? In this, Bayes' theorem can provide some sense of predictive validity even without numbers.

Even more fundamentally, predictive validity goes directly to the assumptions, approximations, and simplifications any analytical method incorporates both by necessity and mathematical convenience. In everyday practice, we state these as the things the analysis is predicated on, then move on to its predictive conclusion given that they apply. But in weighing evidence in likelihood judgments, none of these things can be left behind. There are no givens for sure, and accounting for the uncertainties assumptions mask is what the likelihood judgment is all about. And as noted here throughout, the assumptions having the greatest effect on predictive validity are not about computational aspects but about soil behavior. A key aspect in evaluating the weight of analytical evidence is to uncover hidden assumptions and the uncertainties they disguise.

Assumptions and approximations come so naturally that they often become buried, embedded in the predictive methods we use. Many are intentionally conservative, especially in procedures developed for use in design. Indeed, conservative interpretations are how we counteract the effects of uncertainty in most routine activities, but the point of probability assessment is to evaluate uncertainty not to avoid it. Inherent conservatism can be equally detrimental to predictive validity as gross inaccuracy, and if an analysis used to guide a likelihood judgment contains conservative interpretations of soil properties or behavior, it will be less useful in informing about performance under field conditions. Interpreting data and performing analyses according to best-estimate or most-likely conditions can seem a discomfiting departure from design practices, but recalling that probabilistic methods serve a different purpose can serve as a useful reminder. Reviewing, and again listing, the assumptions and approximations incorporated in any analysis technique encourages their uncertainties to be considered more carefully. And just as we questioned the data, here we query the analysis. If the analysis is thought to be reliable, what do we know about this and why do we think so? Complex analyses usually incorporate more explicit assumptions, and compared to simplified or lumped-parameter methods this may provide a better opportunity to examine their effects individually. But it does not make these assumptions any more likely to hold true or reduce their net effect. To the contrary, greater complexity often reveals just how many assumptions are really required.

Particularly when sophisticated analytical methods are relied upon in forming the uncertainty judgment, it is important to compare them with other procedures, especially those of a simpler nature. These will incorporate their own assumptions and approximations, but different ones that provide broader perspective. And often the simplest but most effective arguments come from common-sense observations and general knowledge. They might not have the same solid appearance of strength as an analysis of great complexity, and arguments from observations can pale by comparison as Terzaghi deplored long ago. But simple techniques and observations help maintain the balance between strength and weight, and can never be ignored.

USING ANALYSIS EFFECTIVELY. Some predictive analysis that incorporates "best-estimate" parameter values and assumptions will provide a single "most likely" result. Such a prediction constitutes one, but not the exclusive, source of information that can guide the uncertainty judgment. Several cognitive biases can be alleviated by using analysis to go beyond such a single best answer. For the most part, these are familiar techniques applicable to any other analytical approach.

Sensitivity studies can aid greatly in evaluating the predictive validity of assumptions and data. Investigating a variety of what-if scenarios predicated on different computational algorithms, soil behavior responses, boundary conditions, or parameter values can reduce the effects of the representativeness heuristic by avoiding sole focus on any one such condition. At the same time, the effects of these potential variations may produce disconfirming evidence which helps address overconfidence. By revealing the relative significance of the conditions varied, sensitivity studies go directly to developing a sense of which are important and which aren't. Had the embankment stability predictors in Figure 5-5 been specifically requested to carry out parametric assessments along with their best-estimate predictions, their 50% confidence ranges might well have materially improved.

The manner of presentation of analytical results also influences their use in the uncertainty judgment. Graphical formats are almost always superior to numerical tabulations because they reveal patterns and trends in results more effectively through visualization, for example by plotting the predicted factor of safety or calculated deformation against some variable. This also helps detect errors or anomalies of various kinds as deviations in visual trends. But visualization comes into play in another respect that can greatly enrich the causal narrative by better illuminating the operation of processes.

Computational efficiency has developed to the point that postprocessing can allow pictorial presentation of results. This may take the form of an element mesh geometry that progressively charts the development of some process over the course of solution iterations. It might involve increase in pore pressure, changes in stress or mobilized strength, incremental deformation, or progression of a wetting or contaminant migration front. By flipping through these pictures at iterative time steps, one can literally "see" through animation how and where the process develops as it proceeds. While case histories provide two snapshots in time—before and after the process—analytical animation helps fill in the intermediate conceptual blanks. The ultimate value of computational methods depends on how their results are assimilated, and this relies on cognitive factors. For all of our color plotters and three-dimensional bar charts, we have yet to scratch the surface of this fundamental aspect of engineering analysis with computational tools that are equally powerful cognitive tools.

USING JUDGMENT. So we see that in all of the above respects, weighing and synthesizing evidence is best approached with an open-minded but skeptical perspective, taking a devil's advocate point of view. The process requires questioning both data and analytical tools but above all questioning one's self. What do we think we know, why do we think we know it, and how convinced are we in thinking so? This sort of introspection broadens the perspective on uncertainty and better assures that each source of information, each kind of argument, and each reasoning strategy receives its due.

Suppose we have done all these things. The evidence has been assembled and it's fairly complete. We've been careful not to overlook or overemphasize any one kind, we've considered both strength and weight, and we're painfully aware of what we don't know. Some of the evidence argues one way and some another. What happens next to arrive at a sense of how likely the event in question is felt to be?

This is a matter of judgment, and it is where inductive processes carry the load. Weighing evidence means assigning importance to it, and a sense of what is important is how judgment has been defined. So a sense of the likelihood of event occurrence boils down to a sense of which evidence and which arguments are important—and by how much. Recall from its attributes described in Chapter 3 that judgment organizes information and establishes frameworks from theory and observation to form hypotheses about how things happen. It seeks patterns of consistency and anomaly in the information that describe how it fits together. And it interprets information with an instinctive feel assisted by intuition for providing the context in which it makes sense. These tools of judgment are the tools by which evidence is weighed.

Against this backdrop, it is clear that judgment has been at work throughout every stage of the probability assessment process so far. Diagnostic judgment has guided what information to look for and helped determine what is relevant. Its attributes of visualization and mental simulation have operated to decompose events and to form the causal narrative. And its inductive character is what we rely on now to integrate evidence, information, experience, and underlying knowledge, generalizing from other situations and reasoning by analogy. Judgment is not rule-based, and it provides no formula for doing these things. But it supplies many tools for gauging what is important, and engineers perhaps more than others are in a position to exploit these tools to best advantage, adapting them to the case at hand.

Numerical Assignment

The final stage in subjective probability assessment is to convert the likelihood judgment into a numerical probability value. We have already weighed and synthesized the evidence and have formed a likelihood judgment, and we are now required to express this in the probability metric on a scale from 0 to 1. This can be assisted by several heuristics, but as always they can cut both ways. In best exploiting how people ordinarily think and deal with uncertainty they are helpful, if used with awareness of their potential bias-inducing effects. We go on here to examine some of these heuristics and how they can be adapted to assist numerical assignment.

Expressing an uncertainty judgment numerically seems to rather easily become second nature for many. With a little experience and practice in subjective probability assessment, numerical assignment becomes integrated into the synthesis of evidence as if the two were taking place together. People often find themselves carrying the emergent probability value along with the evidence as the uncertainty judgment is being formulated, progressively adjusting it more or less consciously while evidence is sequentially considered. This is often revealed when individuals discuss their numerical assignments out loud, in the process describing how some particular piece of evidence causes them to modify the numerical value.

These skills that come with practice are valuable, provided they are applied with awareness of the anchoring and adjustment heuristic and its propensity for insufficient adjustment. They operate similarly in moving from one probability assessment task to another. Assessors are usually asked to estimate subjective probabilities for more than one event. As the collection grows, their likelihood judgments and corresponding probability values are gauged against each other in a relative or comparative way, where one event is judged more or less probable than some other previously considered. While this is an effective technique for enhancing information-processing capacity, it is also a form of anchoring and adjustment and the same cautions regarding insufficient adjustment apply.

Others may find numerical assignment less straightforward, at least to begin with, but useful heuristics can come to their aid in the form of visual and verbal

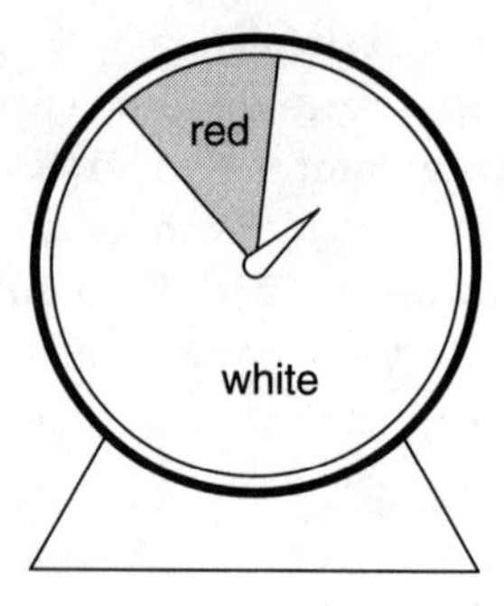

FIGURE 6-7. *Probability wheel.*

devices. These serve as intermediary vehicles for translating the uncertainty judgment into a numerical value.

Visual Devices and Reference Gambles

A *probability wheel* like that illustrated in Figure 6-7 allows the assessor to visually portray the proportional likelihood of an event as adjustable colored sectors of a circle (Spetzler and Staël von Holstein, 1975). Sector sizes are varied until the likelihood that the spinner would land on either one is judged to be equivalent to the likelihood of occurrence and non-occurrence of the event in question, with the area scaling read on the back as the probability value. Many people take quickly to this visual form of representation, with Kinnicutt and Einstein (1996) reporting good results for probability assessment tasks involving geologic site characterization. But as a practical matter, it may be less effective for low probability values, where discriminating between very small sector sizes on the probability wheel can become problematic (Merkhofer, 1987).

The probability wheel and techniques like it have their roots in expected utility concepts long favored by decision theorists with reference gambles in which the assessor receives some imaginary reward (von Winterfeldt and Edwards, 1986). For example, the assessor might be given a choice between a bet and a sure thing. The bet for the event in question, say that it would rain tomorrow, might have a $10 reward if it rains and nothing if it does not. Alternatively, the assessor could choose to avoid the gamble and take $5 instead. The assessor's probability of rain can then be calculated by varying the sure amount until the assessor has no preference between it and the bet, presuming that the expected utilities of the two options are equal. This goes back to Savage (1954), who maintained that people need to thoroughly and conscientiously examine their beliefs in rendering probability statements that express these beliefs, and imaginary rewards for gambles are invoked to this end. While the precept is straightforward, it is less clear that an imaginary reward best accomplishes this purpose. Some people may have attitudes toward risk that affect their preferences for monetary rewards, and most assessors quickly catch on anyway that it is the probabilities and not the bets that are really at issue. In the end, an imaginary $10 bill cannot replace the cognitive processes that underlie the uncertainty judgment.

Verbal–Numerical Transformations

Another technique for numerical assignment involves expressing the likelihood judgment first using verbal descriptors, then translating them into numerical probability values using verbal–numerical mappings. As descriptors of uncertainty, words have a richness of content that makes numbers seem sterile by comparison. Left to their own devices, people usually prefer words rather than numbers in conveying uncertainty, and for good reason. Verbal expressions contain two separate dimensions of uncertainty. They first convey an assessment of likelihood, but along with it a sense of this assessment's precision or vagueness. Recall from the Challenger incident the engineers' expression of "jump-ball" chances of O-ring failure in the memo they produced. There could be no mistaking what this expression was intended to convey, or the precision associated with such a graphic metaphor. Contrast this to "possible" or "probable," whose ambiguity is great by comparison. Here, however, vagueness does not represent inaccuracy in expressing the uncertainty judgment but a way of indicating its range (Hamm, 1991).

People can find words easier and more natural to use than numbers because words conform better to the internal process of weighing arguments, as opposed to computation (Zimmer, 1984). So there is a widespread presumption that people find verbal expressions of likelihood more congenial for communicating uncertainty, but this seems to depend at least partly on which end they are on. Wallsten, et al. (1993) note that those doing the sending—the assessors of uncertainty—prefer to state their uncertainty judgment verbally for the extra information this conveys, while those on the receiving end prefer its numerical form perhaps because they see ambiguity as making their job more difficult. This has come to be known as the *Ellsberg paradox* after its discoverer of Vietnam War notoriety, with its supposed departure from rational decision behavior.

This has led to a whole series of studies on the extent to which decisions involving uncertainty are influenced by its verbal or numerical expression, with a variety of conclusions (Budescu, et al., 1988; Erev, et al., 1993). But what interests us here is how the senders of the uncertainty judgment, with their preference for verbal descriptors, can express it numerically as probability requires. The many studies of verbal–numerical transformations provide fertile ground for using them as a helpful and adaptive heuristic. They show that within reasonable limits, people use mappings of words to numbers in remarkably consistent ways considering the variability in most aspects of human behavior. The numerical equiva-

TABLE 6-6. *Numerical Responses and Ranges for Various Probability Expressions*

Expression	*Single-number probability equivalent, % (median of responses)*	*Specified range, % (median upper and lower bounds)*
Almost certain	90	90–99.5
Very high chance	90	85–99
Very likely	85	75–90
High chance	80	80–92
Very probable	80	75–92
Very possible	80	70–87.5
Likely	70	65–85
Probable	70	60–75
Even chance	50	45–55
Medium chance	50	40–60
Possible	40	40–70
Low chance	20	10–20
Unlikely	15	10–25
Improbable	15	5–20
Very low chance	10	5–15
Very unlikely	10	2–15
Very improbable	5	1–15
Almost impossible	2	0–5

Source: experimental data from Reagan, et al., 1989.

lents of verbal expressions for a given individual appear to be stable over time, though some have broader numerical ranges than others both within individuals and across them (Hamm, 1991). Reagan, et al. (1989) review a number of these studies, adding data of their own. Some typical verbal expressions of likelihood along with median values and ranges for their numerical probability equivalents are provided on Table 6-6, with two types of numerical responses. One is the median single-number probability equivalent subjects provided for the corresponding word, and the other the median upper and lower bounds they gave separately.

Table 6-6 contains several points of interest. One is that the single-number responses were confined within a limited range from 0.02 to 0.90, even for words as definitive as "almost impossible" and "almost certain." This again shows the effect of cognitive discrimination previously illustrated in Figure 5-3. Peoples' ability to verbally express degrees of likelihood does not extend very far out toward the ends of the probability scale because their ability to distinguish these extreme likelihoods is limited. We do not have descriptive words that clearly associate with extreme probabilities because the corresponding likelihoods are not readily conceptualized—there aren't many names for snow to those who live in the tropics.

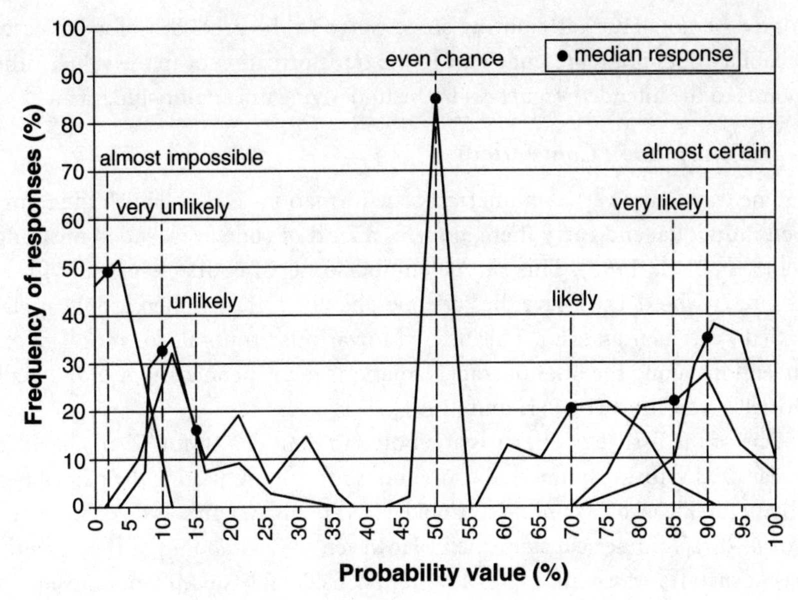

FIGURE 6-8. *Verbal expressions and probability equivalents.*

Source: experimental data from Reagan, et al., 1989.

Reagan, et al. (1989) also found that these mappings have a commutative property: People translate words to numbers and numbers to words in ways that closely match. This means that it is possible to associate numbers with words and vice versa. Thus, Table 6-6 could be used to verbally express some derived probability value according to how most people would interpret it. Closer inspection of several transformations from this perspective also provides some insight into how precise a subjective probability value might reasonably be taken to be.

Figure 6-8 provides median responses and response frequencies for numerical probability equivalents assigned to several of the more common verbal descriptors in Table 6-6. It shows the full range of numerical probability equivalents that can be associated with each likelihood descriptor, where these ranges differ from one descriptor to another. This says that any subjective probability value will have some inherent variability in describing the underlying likelihood judgment, and inferring a great deal of precision may not be warranted. For example, a probability from about 0.05 to 0.3 could be associated with an event whose occurrence is judged to be "unlikely," and the median response of 0.15 viewed in this light would represent roughly 0.15 ± 0.1. Thus, as Larson and Reenan (1979) proposed, a subjective probability can be interpreted as representing some range of values for event likelihood, or *equivalence interval*, and should not be imbued with a degree of precision it may not contain. This applies in general to any subjective probability value whether verbal transformations are adopted or not. Thus, attributing some range to the precision of a subjective probability does not represent a flaw in the transformation or in the value, rather it points to the intended vagueness in the underlying uncertainty judgment.

Transformation Conventions

One way to use verbal–numerical transformations is to establish them for a given individual and carry them along as a kind of code to intended meanings (Wallsten, et al., 1993). This can be cumbersome, of course, especially for an interpreter who doesn't have the codebook handy, and even then it only applies for that particular assessor. This has led to various proposals for more general conventions along the lines of transformations many people adopt individually (Mosteller and Youtz, 1990; Hamm, 1991).

One of the first such efforts is attributed to Kent (1949) for "Kent charts" of standardized verbal-to-numerical conventions for communicating the reliability of military intelligence. According to Cooke (1991), this system was discontinued in favor of direct numerical assessment. However, understanding of the cognitive basis of such transformations has advanced considerably since then. Conventions have provided the basis for the verbal terms used in precipitation forecasts since the National Weather Service started communicating them probabilistically in 1965. These conventions associate a probability of 0.10 or 0.20 with a "slight chance" of precipitation, 0.30 to 0.50 with "chance," and 0.60 to 0.70 with "likely" (Wallsten and Budescu, 1990), although forecasters may supplant them with their own personal meanings depending on context. Indeed, some linguists and others point to the importance of context, where the "likely" occurrence of a tornado carries different intended meaning from its application to rain since the two don't occur with similar frequencies. At the same time, however, this is usually clear to both sender and recipient—we don't use the term "likely" without reference to what.

The design of these conventions would seek to address several criteria. They would not be greatly different from the personal assignments most people would adopt on their own. They would be limited to a few terms of least numerical overlap that could be used to bracket numerical responses. And if they could provide some restriction on bias, then so much the better. Table 6-7 provides such a convention that has been used in a number of geotechnical probability assessments.

Conventions like these adapted from behavioral findings can be particularly useful in group probability assessments where they better assure that individual differences in word/number correspondence are not confused with differences in the likelihood judgments themselves. A select few verbal descriptors are used to first establish upper and lower likelihood limits, then the probability value is interpolated between the corresponding numerical equivalents. This is assisted by using terms in the middle and at the ends of the probability scale with the most well-defined numerical equivalents, avoiding intermediate descriptors that tend

TABLE 6-7. *Verbal to Numerical Transformation Conventions*

Verbal descriptor	Probability Equivalent: Defined convention	Probability Equivalent: Related expressions, values, and ranges from Table 6-6
Virtually impossible, due to known physical conditions or processes that can be described and specified with almost complete confidence	0.01	*Almost impossible*—0.02 (0–0.05)
Very unlikely, although the possibility cannot be ruled out	0.1	*Very unlikely*—0.10 (0.02–0.15)
Equally likely, with no reason to believe that one outcome is more or less likely than the other (two possible outcomes)	0.5	*Even chance*—0.50 (0.45–0.55)
Very likely, but not completely certain	0.9	*Very likely*—0.85 (0.75–0.90)
Virtually certain, due to known physical processes and conditions that can be described and specified with almost complete confidence	0.99	*Almost certain*—0.90 (0.90–0.995)

to contain more vagueness (Hamm, 1991). Such conventions allow discussing likelihood judgments and assigning numerical probabilities to proceed in the way people think and converse. In expressing the likelihood of an event, most assessors would be comfortable making a statement such as "it's very likely but not virtually certain." Arriving at a best-estimate numerical value often occurs quickly after querying the applicability of various verbal descriptors in such a multiple-choice fashion to bracket the uncertainty range. Alternatively, sequentially asking if the event likelihood corresponds to each of the verbal descriptors in Table 6-7, with a yes/no response, can be an adaptive information-processing heuristic in considering them one by one.

Used in conjunction with thorough event decomposition, transformation conventions may purposefully establish numerical limits that discourage assessors from routinely operating beyond the limits of cognitive discrimination where overconfidence bias has its greatest effects. This is reflected in Table 6-7 by transformations bounded within the range of 0.01 to 0.99 which provides a structural control on overconfidence bias. While values outside this range may still be admissible, it flags them and prompts careful examination for base-rate frequencies or other factors supporting the departure. Alternatively, this may encourage review of the decomposition structure to identify ways for further decomposing events that better maintain their probabilities within the well-calibrated range.

TABLE 6-8. *Comparison of NASA Transformations to Common Use of Expressions*

Verbal expression	*NASA transformation convention*	*Average experimental value (Mosteller and Youtz, 1990)*	*Ratio*
Frequent	0.01	0.55 ("frequent")	55
Reasonably probable	0.001	0.70 ("probable")	700
Occasional	0.0001	0.22 ("occasional")	2200
Remote	0.00001	0.09 ("rarely")	9000

As Figure 6-8 shows, correlations between words and numbers can vary among individuals, but probably no more so than directly assigned numerical values would. In any case, the variation would be well within the imputed precision of most any sensibly interpreted subjective probability where the difference between, say, a value of 0.6 and 0.8 is not likely to reflect a meaningful difference in the underlying likelihood judgment. In this respect, Hamm (1991) contends that verbal–numerical transformation conventions sacrifice relatively little in accuracy if they are generally similar to the ones that individuals would define for themselves. But people do not necessarily leave their personal transformations at the door, and grave distortions can result if conventions stray far from ordinary meanings (Wallsten and Budescu, 1990). Once more we return to the Challenger incident to illustrate the importance of this finding. Recall that NASA had come up with a 1-in-100,000 chance of shuttle failure per mission, the number that Feynman extracted from the NASA manager but whose origins remained shrouded. Hamm (1991) reports that NASA assigned this value to words its engineers used in describing component failure likelihoods, but these engineers were never involved in establishing the conventions nor did they even know what they were. The conventions applied retroactively by NASA are shown in Table 6-8 along with the average responses by subjects for the same or similar terms from Mosteller and Youtz (1990) and the ratio of the two numbers. The ratio is similar in nature to the overconfidence index described earlier in expressing the magnitude of distortion. Although the intent here is not to attribute this effect to overconfidence, the end result can be much the same. To this extent, NASA's conventions—quite the opposite from providing a structural control on bias—virtually guaranteed it by building it in.

Convergence Techniques

We have seen that verbal descriptors can contain varying amounts of ambiguity or vagueness arising from the underlying uncertainty judgment. Vagueness pertains equally to a numerical value assigned directly without intermediary transformation, this being the underlying rationale for equivalence interval as a proxy for a particular value (Larson and Reenan, 1979). *Convergence* is a numerical assignment technique that helps cope with vagueness. It starts by first positing plausibly high and plausibly low numerical probability values that bracket what various possible interpretations of the evidence could allow. Examining why these interpretations produce bounding values helps to converge on a best-estimate probability value somewhere between them. This technique promotes the kind of self-questioning that enhances the uncertainty judgment, and it may involve iterating back and forth between the numerical assignment and judgment-forming stages of the process.

Convergence techniques help transform vagueness from a source of discomfort into a useful numerical-assignment device. At least as significant is that they also combat the anchoring and adjustment heuristic where the emerging probability value is insufficiently adjusted for the incremental effects of newly introduced factors or alternative interpretations. Convergence reverses the process, beginning with the "best-case" and "worst-case" probabilities and arriving at a best estimate by anchoring at both ends simultaneously to offset and therefore preclude insufficient adjustment.

Second-Order Probabilities

Second-order probability is used to describe and quantify vagueness. It builds on the equivalence interval's range of values by explicitly defining this range and assigning probabilities to its members. So if vagueness produces ambiguity in what the probability value should be, a second-order probability expresses the likelihood that the assigned probability is that which is most representative of the assessor's actual belief. In this, second-order probability is literally the probability of the probability, corresponding directly to statistically quantified confidence limits. In practice, it is often applied by assigning a high, low, and best-estimate probability to each event, where "high" and "low" are associated with some arbitrarily selected confidence levels (say 10% and 90% exceedance) and the "best-estimate" with 50% likelihood of exceedance. These values are used to define a continuous distribution for each event probability. Propagating the distributions for all such events by any of the solution techniques mentioned in Chapter 4 then produces a measure of confidence in the aggregated probability result.

The entire concept of second-order probability causes discomfort to some theorists because it goes to the philosophical core of subjective probability. If such a probability is a measure of one's actual beliefs, then in principle it must singularly reflect these beliefs (Lehner, et al., 1996). But more pragmatic subjectivists respond that the cognitive processes of belief formation remain open to perturbations in one's state of knowledge and to the context in which these processes are operating. To them it would be equally unrealistic to seek exact measurements of belief as to suppose that a singular frequency value could capture all facets of uncertainty. In describing how second-order probability provides a measure of vagueness, Chávez (1996) attributes ambiguity to missing information—information unavailable at the time of the assessment but knowable in principle. Mosleh and Bier (1996) extend second-order probability to two aspects of the uncertainty judgment: where underlying but unspecified conditions affect this judgment and where cognitive limitations make the assessor unable to draw precise distinctions between the validity of competing arguments. The pragmatic view is adopted here in noting that assigning two probability values—the first to the occurrence of the event and the second to the probability that the first obtains—simply assists some people in expressing their uncertainty by allowing them to quantify their ambiguity separately. For them, two probabilities are better than one. This, in fact, is the underlying rationale for having incorporated the "confidence" categories in association with the qualitative likelihood and consequence judgments in the FMEA format of Table 4-7.

The use of second-order probability has become routine in probabilistic seismic hazard analysis, where it portrays the combined effects of individual uncertainties affecting the seismic hazard curve. Figure 6-9 shows two such curves from data presented by Frankel and Safak (1998), one for Portland, Oregon and the other for New York City. The median, or best-estimate, value in both cases is accompanied by 85th and 15th percentile curves derived largely from subjectively assigned probabilities or distributions on important parameters. In effect, this means that those who generated the curves would have 70% confidence that peak ground acceleration (PGA) would lie between the 85th and 15th percentile curves for any given annual exceedance probability, or *hazard level* as it is customarily termed.

It is quickly seen from the median curves that New York has considerably lower PGA at all hazard levels. However, the confidence limits enhance the content of this information. Both cases show that confidence is greatest at higher exceedance probabilities, reducing for lower probabilities as reflected by the widening confidence limits at higher PGA's. This is an outgrowth of the statistical quality of the database of historic earthquakes and limited ground motion data in any such record of comparatively short duration. But confidence limits at any given hazard level are proportionally much wider for New York due to absence of well-defined seismogenic features and greater uncertainty in eastern ground motion attenuation. Figure 6-9 implies, for example, that designing a structure for 10^{-4} exceedance probability ground motions can mean very different things depending on the confidence associated with such an estimate. In this respect, quantified confidence limits can be necessary to avoid unwarranted and misleading perceptions of precision in median or single-valued estimates (Panel on Seismic Hazard Analysis, 1988).

Some insight can be obtained more generally into confidence associated with subjectively derived probabilities by looking to various risk analysis applications. For seismic hazard and internal erosion of dams, where component probabilities are substantially subjective, the final aggregated value has been found to vary within 15% and 85% confidence limits by roughly one-half to one order of mag-

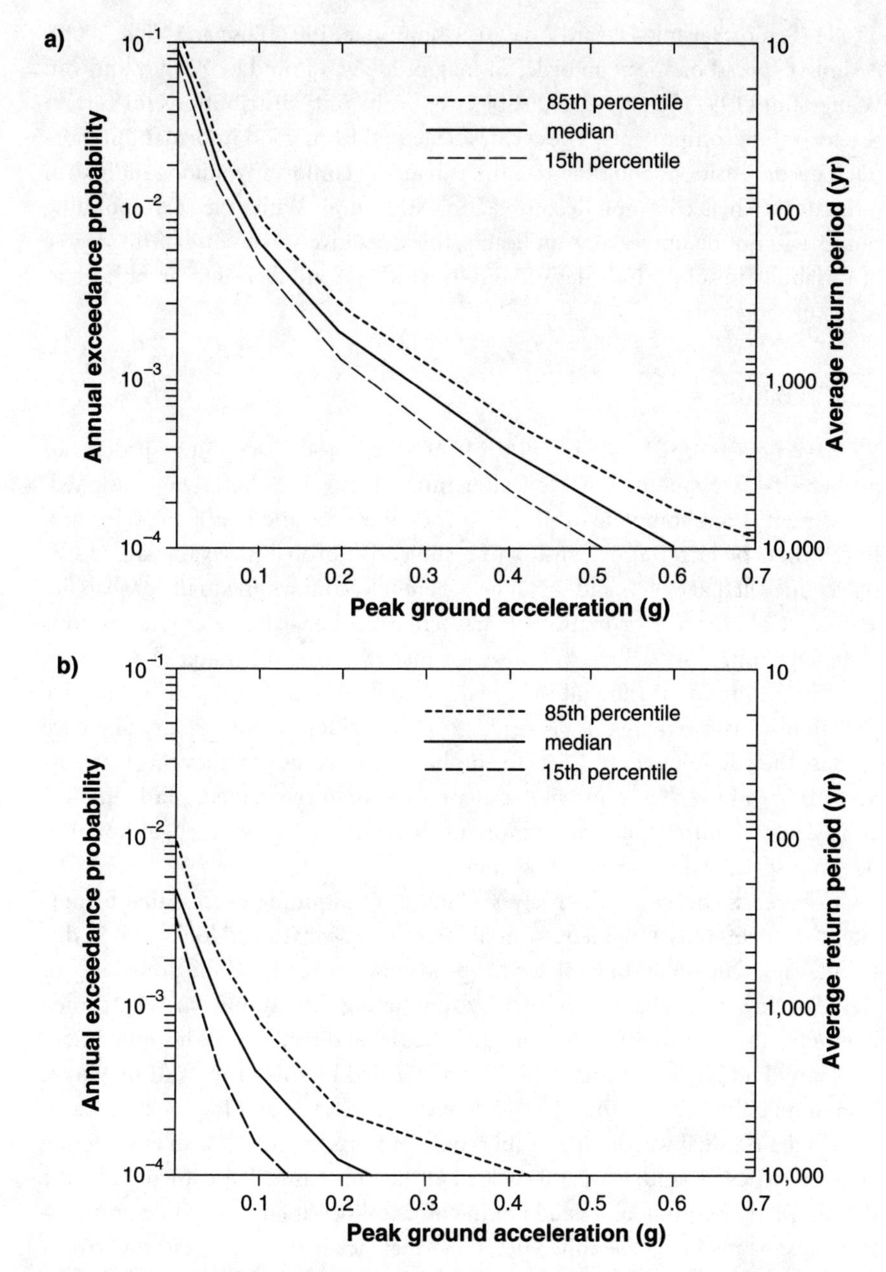

FIGURE 6-9. *Confidence limits on seismic hazard curves: a, Portland, Oregon; b, New York, New York.*

nitude (Senior Seismic Hazard Analysis Committee, 1997; Dise and Vick, 2000). A similar spread of about an order of magnitude was found by Keeney and von Winterfeldt (1991) in aggregated subjective probability distributions for certain reactor-safety components. These cases benefited from good information, thorough decomposition, and other factors enhancing conformity among individual estimates such as common decomposition structures. While the corresponding ranges will not be universally applicable, this does give some sense of the degree of variability in subjectively derived results which can aid their interpretation and use in proper context.

Confirmation

We have now reached the endpoint of the stage-model's cognitive process of probability assessment. But even after numerical values have been assigned, assessment is not complete until checks for coherence and bias are carried out. Recall that the criteria for validity of a subjective probability value are that it reflect one's actual beliefs and that it be coherent in conforming to the probability axioms. Coherence is a matter of mathematical consistency among various assigned probability values. Suppose, for instance, that the material at a given point in the ground is thought to be clay, or sand, or rock and that the respective probabilities assessed are 0.1, 0.4, and 0.3. These values do not display coherence because they do not sum to 1.0 as the probability calculus says they must, at least according to how the permissible outcomes have been defined. Mathematical consistency requires that one or more of these values be modified, although it does not specify which ones or how much.

Coherence checks are themselves a kind of questioning process that formulates the uncertainty judgment from different perspectives and looks at it in different ways. One might first ask for the probability of sand and the probability of clay, deriving the probability of rock by subtracting their sum from 1.0. At some later time, the probability of rock might be assessed directly to see how the values compare. Hidden assumptions are often revealed by this process. It may have been implicitly assumed that the soil deposit was alluvial and the rock unweathered, but a residual weathering profile could produce saprolites with engineering characteristics of sand, clay, and rock all at the same time. Here it would be the definition of event outcomes and not the probabilities at fault. So while probabilities must be made to have coherence, those that lack it are not necessarily wrong. Coherence depends as much on how the uncertainty is conceptualized and the restrictions applied in defining events as it does on the values themselves.

Another example relates to stochastic processes that occur over time, and one particular such case yields several useful insights. It pertains to a risk analysis for a dam, where assessors were asked to estimate the probability that internal erosion might initiate in its abutment in any given year (Dise and Vick, 2000). This event was defined most conveniently as the appearance of at least one bucketful of sand at any of the unprotected seepage discharge faces the abutment contained. The dam was some 30 years old and had displayed significant abutment seepage throughout its life. This seepage had been particularly high on first filling, so much so that a drainage tunnel had been constructed to reduce it, but it now showed signs of increasing again. Seepage, however, is not the only factor that influences particle transport, and a great deal of other subsurface, geologic, and construction information had been assembled and thoroughly reviewed. After carefully considering and discussing how these conditions might be conducive to internal erosion, the assessors as a group provided a probability of 0.1 that the initiating event would occur.

It can be difficult to relate single-event occurrences to a temporal framework because different reasoning strategies pertain. Causal reasoning considers a single-event occurrence without partitioning it into intervals of time, whereas a statistical strategy invokes repeated trials each conducted within some defined period. In this case, the probability desired was not for event occurrence per se but for its occurrence in any given year. The coherence check applied the binomial theorem to the probability provided, taking each of the succeeding 30 years as independent Bernoulli trials as described in the Appendix. If the assessed probability were truly an annualized value, the corresponding probability of particle transport at some time during the next 30 years would be 0.96, implying from Table 6-6 that it would be "almost certain" to occur.

When presented with this implication, the assessors immediately recognized that it was not consistent with their uncertainty judgment and did not reflect their actual belief. This prompted reexamination from a different perspective, using the previous historical period to benchmark the cumulative probability over the next such future period. No particle transport had ever been observed during the past 30 years even under seepage conditions that had at times been worse, and it was clear that conditions overall had not deteriorated to such an extent as to justify a near-certainty of particle transport over the ensuing 30 years. It became evident to the assessors that the value they assigned had not sufficiently accounted for time. Using what they knew about past conditions, and reasoning how and by how much the current conditions might be different, they arrived at a cumulative probability over the next 30 years. This reduced the annualized value, which could now be back-calculated from the binomial theorem by a roughly factor of 10 from what they originally had provided. The coherence check had simply considered the probability statement from a different point of view by framing the question in a different way. But it also revealed something more, which from the preceding discussions we might by now come to anticipate.

Because of its considerable seepage, the available information about the dam was exceptionally detailed and comprehensive. It had been closely monitored over the years, with data from dozens of borings and piezometers, not to mention the many thorough geologic investigations performed along the way. All of this had promoted heavy emphasis on causal reasoning about internal erosion processes, and the representativeness heuristic had been at work. Simpler kinds of information, in this case the dam's performance history with its absence of any previous particle transport, provided base-rate frequency information that had been noted (and listed) from the start. But this had fallen through the cracks. Its significance became apparent only when the coherence check had the effect of prompting for it. In modifying their probability, the assessors were now supplementing causal reasoning with a statistical strategy, combining the two together.

Checks for other biases operate in a similar way by prompting for the kinds of information or reasoning strategies that tend to be overlooked and questioning those typically overemphasized. This is especially so for overconfidence bias, and it involves prompting for disconfirming evidence—that which would be contrary to an outcome judged most (or least) likely. People tend to ignore information inconsistent with their views, or they fail to retrieve it from memory at all. If we're more sure about an event outcome than we should be, a primary reason may be that we haven't fully accounted for the information that would counterindicate our preference for it.

It has already been noted that thoroughly retrieving and assembling evidence can better assure that all of it is considered, and one way to do this is by listing both supporting and disconfirming evidence—all of the pros and all of the cons—for each of the possible outcomes. Recall from the experiments of Koriat, et al. (1980) and Fischhoff and MacGregor (1982), however, that after the uncertainty judgment has already been formulated, disconfirming evidence must be particularly emphasized in order to avoid insufficient adjustment. This can take the form of imagining that a contrary outcome has actually occurred and listing, in hypothetical hindsight, the possible reasons why it could have happened. The greater awareness of disconfirming evidence this brings about can result in modification of the original probability judgment. An effective remedy for overconfidence bias then is to actively search out the reasons one might be wrong (Lichtenstein, et al., 1982).

In a curious way, this technique pits one bias against another. While overconfidence bias is familiar by now, *hindsight bias* is one we encountered only briefly in Chapter 5. Fischhoff (1982) describes hindsight bias as the tendency to exaggerate in hindsight what we knew before the outcome was known, in effect, to say after the fact that we knew it all along. After the home team has won their latest televised game we are convinced we knew they would, even while our recliner may still be damp with perspiration. So if asked for odds that they will win their next game, our response might well be overconfident. But imagining that they have just lost it, describing and listing reasons why, turns the tables with hindsight bias that tends to counteract this overconfidence.

These techniques for evaluating coherence and bias seek to confirm that the probabilities assigned are as valid as we can make them. As much as anything, this requires a final review to confirm that they still make sense. New information and new interpretations can come to light as the assessment proceeds, and how events are defined can undergo subtle changes. After all is said and done, each of the probabilities assigned must fit proportionately within the collective whole, viewing the forest as well as the trees. The final check is a reality check, and judgment provides the ultimate test in confirming that everything makes sense. This often means going back to re-examine base rates for other situations similar to the single-event occurrence being evaluated. Now that we understand the nature and sources of uncertainty, the reality check questions one last time the probability value assigned to it. Now we ask whether and how this understanding can explain why the case at hand is really so much more likely, less likely, or comparable to others. Here as elsewhere, the probability value will satisfy our intuition if it conforms to our judgmental expectations from experience.

Some of the techniques and strategies useful for internally formulating and expressing the uncertainty judgment are compiled in Table 6-9. They have been described here at some length both to show that subjective probability assessment is more than a black box and its product much more than guesswork and equally because all the various heuristics, biases, and their effects might otherwise seem overwhelming. But the same cognitive factors that produce these problems can also help solve them if techniques for actively assisting probability assessors are put to good use by exploiting people's natural abilities. Though framed in a cognitive context, many of the tools summarized in Table 6-9 are not such a radical departure from what engineers do every day. In this, they embody simply good judgment and engineering good sense.

Formal Elicitation

The previous procedures and techniques have been presented in a cognitive setting organized around the ways people think in formulating and expressing uncertainty judgments. Many of these techniques have been packaged in a procedural format called *expert elicitation* or *formal elicitation*, widely used for decision analysis in business decisions (McNamee and Celona, 1990; Ferrell, 1994) and for risk analysis in the nuclear industry (Keeney and von Winterfeldt, 1991; Otway and von Winterfeldt, 1992). Procedural details have been described by Spetzler and Staël von Holstein (1975) and Merkhofer (1987).

Expert elicitation is directed and guided by a decision analyst, or simply the *analyst*, with training and experience in decision analysis techniques, probabilistic methods, and cognitive elements of the assessment process. The analyst elicits the required subjective probability values from *experts*, assessors with subject-matter

TABLE 6-9. *Stages and Techniques in Formulating Uncertainty Judgments*

Process stage	*Applicable techniques*
Assembling information and evidence	Search memory Search original information and data sources Identify and review "type case" histories Identify frequency information List information and evidence
Synthesizing information and evidence	Adopt a "weight-of-evidence" perspective Be introspective and self-questioning Use all information of all types Account for simple methods and observations Incorporate base-rate frequencies Question quantity and quality of data Question analysis assumptions Avoid conservative interpretations Use judgment attributes
Numerical assignment	Converge on the value from both ends Avoid insufficient adjustment Use visual devices Use verbal transformations Limit extreme values by further decomposition
Confirmation	Check for mathematical coherence Adopt different perspectives Prompt for disconfirming evidence Review for changes Confirm that values make sense collectively Do a reality check

knowledge but not necessarily any background in probability, the assessment of subjective values, or the cognitive processes involved. A major element involves questioning—of assumptions, of information, and of reasoning strategies. But here the burden is shifted from self-questioning by the expert to probing interrogation by the analyst. Expert elicitation is conducted in a structured, one-on-one interview between the analyst and the expert. Probability values or distributions are usually obtained in this way from multiple experts whose elicitations are performed separately. The underlying rationale is to generate independent likelihood judgments that are not correlated or otherwise influenced by interactions among the experts, with their individual values later aggregated in some fashion.

Elicitation Procedures

The elicitation interview is conducted in steps that sequentially address different aspects. Although there are variants on these steps and how they are conducted, the breakdown and descriptions in Table 6-10 are typical (Merkhofer, 1987).

TABLE 6-10. *Expert Elicitation Procedures*

Component/step	*Purpose*
1) Motivating	Establish rapport Explain the task and its importance Identify motivational bias
2) Structuring	Discuss the elements of uncertainty Define the uncertain event(s) Establish expert's decomposition structure
3) Conditioning	Draw out relevant knowledge and information Explain cognitive biases Conduct bias demonstration
4) Encoding	Assign numerical values
5) Verifying	Apply coherence checks Calculate target values Review for reasonableness

The first step, *motivating*, presumes that the expert will have had no previous familiarity with risk analysis, decision analysis, or probability principles in general. The initial discussions establish rapport between the analyst and the expert to set the stage for the tasks ahead. They may consider the elements of uncertainty affecting the problem to establish the need and importance of subsequent efforts. The elicitation process is also explained so that the expert can know what is expected and what the task entails. All of this is to motivate the expert to devote the attention and effort the task will require, incorporating the aspects of encouragement and support previously described. While its nomenclature has no intended connotation with motivational bias, it nevertheless is here that the analyst probes to determine whether the expert might have reason to be affected by it.

Structuring advances the interview to the next stage by establishing clear and unambiguous definitions for the events to be considered. Experts may be allowed to develop their own decomposition structures, and this is where these structures are developed by discussing in more detail the uncertainties affecting occurrence of the target event or its progression—in effect outlining a causal narrative or applicable statistical strategy. At the same time, the expert's assumptions about event occurrence are discussed, with attempts to draw out and more closely examine hidden assumptions.

Conditioning comes next, and this is where knowledge and information relevant to the uncertainty judgment are solicited. This is also where the various heuristics and biases are introduced and explained to the expert. This may be accompanied by some simple training exercises or demonstrations like that in Table 6-2.

Encoding is where assignment of the numerical probability value or distribution takes place. The analyst tries to reformulate the problem in different ways by varying the nature of the questions asked without leading or influencing the expert. The expert's rationale behind the numerical assignment is noted to identify any new information, reasoning, or decomposition being applied and to query how this might affect any previous evaluations.

Following the encoding stage, *verification* gives the analyst the opportunity to apply coherence checks to the assigned values and gives the expert the chance to review the results in their entirety. This includes calculating target values from component probabilities. If the expert expresses surprise at the result, the reasons are sought, and this may warrant revisiting any of the previous steps.

Aggregation of Elicited Values

The end result of these procedures will be probability values or distributions elicited from each of the experts, and these must now be aggregated in some way that expresses their judgments collectively. A variety of methods have been proposed, as reviewed by Ferrell (1992), Rowe (1992), Cooke (1991), and von Winterfeldt and Edwards (1986). One such procedure would assign a weighting factor to each expert's value according to how much they know and therefore how influential their subjective assessments should be. The problem lies in identifying some "super-expert" who should do the weighting and then determining whether this weighting needs to be weighted itself and so on. Other methods lumped under the heading of the *Delphi* approach allow some limited feedback of other experts' values (i.e., a peek at the other guy's number) in an attempt to achieve better agreement among experts during subsequent iterations of elicitation. While this may enhance conformity of the probability numbers, it says nothing about the underlying uncertainty judgments. Yet other approaches would apply some correction factor or formula for such things as overconfidence bias. But it is difficult to determine which experts are overconfident or by how much, and these factors can change from case to case so that a better-calibrated expert might be overcorrected and vice versa. In the end, it is hard to show that any of these aggregation techniques is demonstrably superior or of clear-cut benefit, and in practice the experts' individually elicited values are often combined by simple arithmetic averaging that assigns equal weight to them all (Keeney and von Winterfeldt, 1991).

Review of the steps in expert elicitation shows that they provide a systematic procedural framework for applying many of the techniques already described. Some have maintained that these procedures translate to subjective probability values that are more defensible to external challenge and therefore better, hence their preference for calling it "formal" elicitation with the rigor this is taken to imply. However, it is important to distinguish formality from validity. The validity of a subjective probability value derives from its coherence and freedom from bias. But bias reduction occurs within the internal process of formulating the uncertainty judgment, not necessarily from the procedures used to extract it. Others maintain that conformity among the elicited values produces external defensibility, but the effect is often the opposite. Keeney and von Winterfeldt (1991), for example, showed that formal elicitation increased the range of variation in elicited probabilities by 10 to 100 times, as debiasing techniques had their intended effect in reducing overconfidence and more broadly incorporating all information.

There can be any number of procedural formats for eliciting subjective probability values. But they will always be secondary to the cognitive processes that produce the uncertainty judgment, and these internal processes can never be rigorous in conforming to prescriptive rules. Subjective probability assessment is an inductive not deductive exercise, and whether a structured format is adopted or not, the important thing is that cognitive strategies—and the heuristics and biases that can affect them—are recognized and accommodated by probability assessors.

Group Elicitation

Group elicitation shares many similarities with elicitation of individual experts. Here a *facilitator* replaces the analyst, with similar credentials and much the same job in guiding the process and directing it to completion. The process is less structured, however, and no interviewing is involved, the group itself being largely relied on to achieve the same effect. The facilitator still probes in a questioning way and retains primary responsibility for coherence and bias checks. But the intent here is more to encourage group members to question themselves individually, and to question each other.

A basic construct of subjective probability is that it is a property of a particular person; strictly speaking, it cannot be a property of a group. At the same time, subjective probability expresses belief, and groups with shared beliefs are all around us—political parties being one example. Moreover, even if subjective probabilities reside with individuals, expertise may reside in groups (Schmitt, 1997). Group elicitation, therefore, seeks to tap shared beliefs and group expertise as they relate to uncertainty, characterizing this as some subjective probability value on which group members may concur or as a range of values that capture individual opinions.

Group elicitation endorses interaction among experts without attempting to maintain independence in their uncertainty judgments. The presumption is that most experts have similar sorts of professional training and background anyway, so correlation among their opinions is not truly avoidable. Instead the intent is to encourage interactions and their beneficial effects. Because many uncertainty problems cut across the substantive knowledge of any single person, the richest mutually held understanding can come about when the broadest range of perspectives and most varied knowledge are brought to bear. Thus a diverse group of experts might include geotechnical subject-matter specialists along with field geologists, instrumentation specialists, old hands in construction, facility operators, and perhaps a generalist with an overview of these areas who helps the others communicate.

Each of these kinds of experts will bring a different outlook on what could happen, why, and the relevant kinds of information, relying on synergy to best reveal and synthesize the evidence when they freely interact. Underlying this presumption is that different group members will apply different cognitive strategies and that interchange will provide a medium in which a sort of mutual debiasing can take place. If several people are involved in retrieving and assembling information, for example, it is less likely that the representativeness heuristic will allow some kinds to be overemphasized or others overlooked. Similarly, overconfidence bias will be reduced if some group members bring forward evidence disconfirming the interpretations of others. In these ways, the cognitive abilities of individual members can be improved by group interaction, enhancing the group probability estimate (Rowe, 1992). It has also been suggested that the involvement of individuals within a group can have a beneficial effect in motivating them to do well and committing them to provide the best performance they can.

Snizek and Henry (1989, 1990) describe how groups operate in transforming the uncertainty judgments of individuals into a group judgment, which they characterize as *revision and weighting*. This begins with each person forming their own likelihood judgment by the processes previously described. As reciprocal interactions take place, they voluntarily revise their judgments as information and arguments presented by others come to light. At some point, the group ceases to have new ideas or simply runs out of time and as individual judgments begin to stabilize, the usefulness of further information sharing diminishes. The group's focus then shifts from exchanging information to combining individual judgments in a consensus judgment. This is done by some form of weighting, whether by negotiation, selection of the best-qualified members, or some form of aggregation.

This process has been found to increase the accuracy of group judgments compared to those of individuals (Snizek and Henry, 1990). One reason is that first forming then discussing individual likelihood judgments makes areas of disagreement more explicit and better defined. Low disagreement can be less effective because without full presentation of evidence, particularly disconfirming varieties, it does little to counteract bias. In terms of the uncertainty judgment, group elicitation therefore seeks to enhance especially the first two stages in Table 6-9: assembling information and evidence and then synthesizing it. In these activities, group elicitation nods to the old saw that several heads are better than one—when they all work together. To take this one step further, groups can sometimes outperform even their best individual, otherwise called the *assembly effect bonus* (Eils and John, 1980).

Structured Procedures

Procedural formats for group elicitation are not standardized, but one such framework developed by the U.S. Bureau of Reclamation (1999) serves to illustrate how they work. With some 350 dams under its jurisdiction, the agency has developed an active program of risk analysis to better understand and manage dam safety risks to the public, and subjective probability assessment plays an important part in these efforts.

To begin with, the collection of information and data is strongly emphasized, including carrying out relevant supporting analyses and examining instrumentation and surveillance records. From this, potential failure modes are identified in connection with specific features and conditions of the dam along with the types and levels of loading conditions it could experience.

A diverse group is assembled that reviews this information, including specialists in various technical disciplines and those knowledgeable about local conditions and operation of the dam. The group is facilitated by someone familiar with the process and with technical grounding in dam safety practices. In contrast to the analyst in expert elicitation, who has no necessary training in the experts' disciplines, here the facilitator's technical background allows for more active participation in problem structuring and questioning using knowledge and experience corresponding to the experts' own. By comprehending the technical content of the discussions, the facilitator here is better able to guide the group and anticipate its direction, a role that becomes of considerable importance when a number of individuals are involved.

The group begins by together decomposing each identified failure mode into its sequence of component events. This common decomposition structure helps establish a shared causal narrative or mental model of the failure process or alternatively why different and perhaps equally valid conceptualizations may exist. This has been shown to be an essential aspect of effective group performance by fostering communication (Salas, et al., 1997).

Elicitation of the group probability judgment for each such event starts by defining it clearly and then listing pertinent information. This again is the first stage in Table 6-9, and it takes the form of a two-column format with one side showing those factors that would favor event occurrence and those tending to make it less likely on the other. This is, in effect, a listing of supporting and disconfirming evidence, and Table 6-11 provides a typical example for an event involving initiation of internal erosion in an embankment dam abutment. This, in fact, is the same event for which coherence and bias checks were discussed previously, where the observation pertaining to past seepage performance has been highlighted in the "less likely" column.

TABLE 6-11. *Supporting and Disconfirming Evidence for Internal Erosion Initiating Event*

Factors that make the event more likely	*Factors that make the event less likely*
Uncontrolled seepage exits exist	Depth of seepage exits unchanging
Less seepage reduction above D-shale	Historic seeps carry no sand
Lower portion of sandstone more friable	No internal erosion on first filling
Caving occurred in vertical drillholes	**Seepage observed for 30 years without piping incident**
Evidence of some gypsum solutioning above D-shale	No piping observed in gypsum solutioned areas
Horizontal drillholes made sand	Changed seepage pathway required to cause significant observable particle movement
Some tunnel drainholes made sand	No completely uncemented discharge faces
Poor pre-tunnel instrumentation	Friable sandstone tends to be medium and coarse-grained
Some recent changes in seepage areas with decrease in tunnel flows	Measured seepage not increasing
	Most fractures not open
	Currently flowing tunnel drains did not make sand and are not deteriorating

As might be anticipated, this listing of "pros and cons" prompts a good deal of discussion about the evidence and observations, what they mean, and how reliable they are. It provides a focal point for interchange within the group about the strength and weight of the evidence, and in the process other kinds and sources are frequently discovered and added to the list. A sense of relative importance of the factors begins to emerge, although this obviously may vary among individuals. As it does, they also begin to formulate and discuss their likelihood judgments, often in terms of verbal descriptors that the column headings prompt.

In group encoding, the facilitator queries the group for its collective probability judgment, drawing out those with the most divergent views and exploring the reasons. The associated probability values sometimes reflect different hypotheses and may become bounding limits. Convergence techniques are then applied to identify some best-estimate value by informal polling that seeks to determine whether it should lie closer to one limit or the other or whether the group as a whole expresses no clear preference. Throughout this process, probability values are benchmarked to verbal descriptors using conventions along the lines of those in Table 6-7 to reduce miscommunication and make it easier to compare probabilities elicited from different groups for different dams. The facilitator carries out confirmation checks for coherence and bias along the way. With this, the group is afforded the opportunity to review the values in their entirety to establish whether they still make sense as a whole and accurately convey what the group intends.

Group elicitation is demanding in terms of the resources it requires and the effort it entails. Sessions may extend over several days and can be difficult to schedule uninterrupted with many participants. As a result there is a natural tendency to make the most of the time available, with extended hours and concentrated effort. However, if people have limited capacity for processing information, then there is also limited duration over which they are able to maintain these efforts. The mental concentration required is intense and sustained, and people's reservoir of it is finite regardless of how long elicitation sessions may last. Signs of deteriorating group performance can be positive in showing that the necessary efforts are being expended, but they can also be signs it is time to quit.

Group Interactions

Group elicitation does not require any mathematical weighting scheme for aggregating individual values. The group does this internally through the discussions and interactions among its members in what is called *behavioral aggregation.* When no clear preference for any particular value between the bounding ranges is evident, the result is similar to averaging. Otherwise the group, in effect, assigns its own weighting to those it considers to be best informed about the circumstances at hand.

As we have seen, behavioral studies support the potential for improvement in judgments brought about by group interaction (Einhorn, et al., 1977; Snizek and Henry, 1989). It is possible for the group to perform to the level of its best member or even exceed it—the assembly-effect bonus—with parallels to the geotechnical experts of Figure 5-5 whose collective average would have outperformed any one predictor. It appears, however, that group performance frequently falls short of this goal for some of the following reasons (Rowe, 1992):

- Social or cultural pressures may force conformity to majority opinion and reduce the influence of more knowledgeable minorities, sometimes known as "groupthink";
- Dominant, forceful, or vocal members, who may or may not be best informed, may have disproportionate influence—the "squeaky wheel" effect;
- The changing motives of the group may allow reaching agreement to become a goal in itself. This leads to premature closure of discussions and a suboptimal consensus that really satisfies nobody but doesn't overly offend anyone either, an effect often observed late in the afternoon as mental exhaustion sets in;
- The motives of more competitive group members who need to "win" the argument, or at least not lose face, can turn the session into a form of mortal combat; and
- Mutual professional or cognitive biases may be reinforced, especially among those in the same technical discipline or employing the same cognitive strategies ("Not to worry, these are accepted methods that everyone agrees on").

These can all be viewed as various kinds of communication problems, and some may offset others—one person's need to win might cancel out another's forcefulness. But it would be unwise to rely on this effect and it may not enhance group performance when it occurs. Groups are often left to their own devices in communicating among themselves, but there are several strategies for effective group communication that can be used to advantage (Eils and John, 1980):

- Avoid arguing,
- Avoid "win–lose" statements,
- View initial agreement as suspect,
- Avoid changing opinions simply to reduce conflict and reach agreement,
- Avoid conflict-reducing techniques like majority vote or splitting the difference, and
- View differences in opinion as helpful and natural.

The performance of groups specifically encouraged to adopt these strategies can be significantly enhanced. Reviewing several studies, Eils and John (1980) found that groups so instructed roughly doubled their achievement of the assembly-effect bonus compared to controls, and this goes to an important principle of group elicitation. Consensus reached quickly may be less a reflection of agreement than an indication that everything has not been fully considered. Without disagreement, it is not possible to establish why consensus occurs. While each member must honestly and candidly examine the limitations of their own knowledge and temper their contributions accordingly, greater initial disagreement can lead to improvement of group judgment after its origins have been determined (Snizek and Henry, 1989). This may call for re-examination of important sources of information, more explicit elaboration of how different interpretations might lead to different outcomes, or identification of key factors not previously considered. It may reveal entirely different hypotheses for process causation that might not otherwise have been apparent.

Group communication strategies seek to exploit disagreement as positive and helpful by converting it from a stumbling block to a vehicle for understanding why differences exist. In this respect, consensus is best understood with reference to the belief space portrayed previously in Figure 5-7b. While each person will have their own individual belief set, consensus does not require that these be coincident. Rather, to achieve group consensus is to define the belief space that encompasses all such belief sets, which itself requires identifying areas of disagreement. While a probability assignment is sought that reflects the collective judgment of the group, consensus in thinking is not the ultimate aim, and forced agreement among individuals is not required.

Group interactions are where understanding of uncertainty comes about, and if participants know more about uncertainty when they leave the session than when they entered, the goal will have been achieved. This is the true measure of the success of these interactions equal in every respect to the validity of their numerical product, and it establishes their value far more than any procedural formality ever could. While the probability values are important, they are ultimately just a means to this end, and if assessors provided the very best numbers but learned nothing the entire exercise would be a waste of time. Any process that helps better understand where uncertainty comes from is always well worth the trouble.

Summary

The opening paragraphs of this chapter remarked on how subjective probability assessment could seem so unnatural to engineers. To a large extent, this is only to be expected because engineers are hardly ever instructed in its use or the principles that underlie it. Subjective probability requires going beyond deductive reasoning to inductive methods and expressing judgment in a most explicit way. This is no small thing to ask when judgment is so often hidden and implicit.

Yet engineers have intrinsic skills in these areas that they seldom consider because rarely are these abilities called upon in the way subjective probability requires. It is possible to obtain the benefits of feedback for some geotechnical activities that materially enhance uncertainty judgments. Decomposition is a powerful technique that employs problem-solving tools engineers are adept at using. And identifying, retrieving, and assembling information—but most of all integrating, synthesizing, and interpreting it—are things no engineering activity ever goes without.

The cognitive process links these skills and stitches them together. It goes to the heart of subjective probability assessment, but few are aware it exists even as it runs in the background of almost every task we undertake. It has to do with how we think and how we reason in the face of the uncertainty that surrounds us, and there would be little hope of coping with this uncertainty without it. For sure, parts of the cognitive process can sometimes mislead if not properly understood and applied, but the same is true of any engineering calculation or technique. We need not be prisoners of these heuristics and their biases, however, if we learn to handle and use them wisely in ways that counteract their less desirable effects. And if using the tools of the trade is something at which engineers excel, then our cognitive tools must certainly rank with the most formidable among them.

Achieving a better understanding of uncertainty is the point of any probability assessment exercise, and it will be clear by now just how much more this involves than simply picking a number. It should also be clear that much of what probability assessment entails is applying the same techniques central to engineering judgment and common-sense engineering practices. In this, we can come to see that subjective probability assessment is indeed the most natural of acts.

7

Experts and Expertise

As subjective probability has been treated throughout these chapters, it has always been with the understanding that one's subjective probability is predicated, and thus conditional, on one's state of knowledge. We have discussed how information is assembled and compiled and the different cognitive strategies for integrating and synthesizing it. This is normative expertise, and we have seen in particular how it pertains to subjective probability tasks. But what we have not yet considered is that knowledge is more than information alone. Information is something that can be gathered, manipulated, and analyzed in any number of ways but knowledge goes beyond this. Knowledge has to do with the deeper comprehension of what this information means that extends to its interpretation and hence, of course, to judgment. Knowledge sifts information through an interpretive sieve and adds to it with tacit understanding. Substantive expertise is that which pertains to one's knowledge domain, and it is what we now take up.

With regard to the state of knowledge it incorporates, subjective probability has an essentially egalitarian character. It is accessible to anyone without prequalification. Still, it is not hard to appreciate that some assessors will possess greater substantive knowledge than others, so that all subjective probabilities will not necessarily be created equal. The way this is ordinarily acknowledged is by designating probability assessors as "experts," as in "expert elicitation," and it goes without saying that they will possess expertise of one kind or another beyond that of laypersons. But what does this really mean? If we say that subjective probabilities are based on expert judgment, then who is an expert and how might we know? Just what is expertise and how does it work?

We might start by considering who we would take to be a subject-matter expert. Other than identifying them by name—something most anyone could do in their field—an expert might be said to be someone who's smart. Yet the brightest graduate student could not be considered an expert, though they might show promise of becoming one some day. Then perhaps an expert is someone who's experienced. Yet if all experienced engineers were so designated, the ranks of engineering experts would surely be overflowing. Or maybe an expert is the world's foremost specialist, but the restricted focus of specialists is why they are called that to begin with. While an expert may have some or all of these qualities, expertise is plainly something more. Perhaps the best we can do for the moment is to say simply that an expert is someone who knows a lot, which brings us back to one's state of knowledge on which subjective probabilities are based.

There will always be individual differences of both type and amount in domain knowledge, presuming this could be measured; that some know more than others is apparent intuitively. One way to examine this difference is to compare experts with novices and how they perform in various settings. In the same way as for heuristics and biases, these settings can be important, the question as always being how closely expert and novice behavior conform in research to the real-life situations where people actually operate. What might be called the "experimentalists" show that experts like chess masters, prizewinning physicists, or prodigies in the arts use many of the same cognitive processes as novices but use them differently and to greater effect. The "observers" take a different tack in carefully watching on-the-job experts under the pressure of time and stress—people like emergency room nurses, pilots, and battlefield commanders—to see what they actually do. Both approaches are revealing, and from them we can come to learn more about experts and expertise in engineering. And one thing we will see throughout is how many of the attributes of expertise disclosed by experts' cognitive processes, actions, and decisions correspond so closely to what we would otherwise call judgment. So that in knowing a lot, expertise, judgment, and subjective probability all become inseparable.

But note carefully too that a subjective probability is, strictly speaking, based not on one's accumulated mass of knowledge per se, but one's particular "state " or condition of knowledge at the time it is assessed. This implies that knowledge, like information, can change with time and would, we presume, change in the increasing direction. If so, then what does it take to make someone an expert in a knowledge domain, or more to the point, how is one made? And are experts made in the first place, or are they born to the job? While we cannot attempt to fully answer these questions or join the many who've tried, we can examine the path expertise has taken in some who have achieved it to learn more of where it came from and how it was developed. In this, we are not restricted to chess masters, pilots, or generals. We will look as well to some prominent engineers.

In all of these things, it would be too much to suppose that one could become an expert by simply reading about these topics. But it will help to know what to look for when we elicit the probabilities of those we call "experts" that are based on their "expert judgment." And who is to say that in examining these elements of expertise, we might not develop some of them ourselves.

How Expertise Works

The formal study of expertise began in the 1960s as an outgrowth of artificial intelligence, sharing in this some of the origins of contemporary interest in subjective probability. The goal to which artificial intelligence has always aspired has been to capture human expertise then encode its procedures (assuming that it is procedural) in rule-based emulations of human expert performance. Some of the most visible early efforts centered around chess. This focus continues still, with epic chess matches between man and computer evoking tunneler John Henry and the steam-powered drill of an earlier age, though here the machine's representation of expert knowledge and not the machine itself is the true protagonist in today's contest. One reason why chess has always been such an attractive venue for the study of expertise is its ranking system, which allows novices and experts to be easily distinguished. But the ultimate payoff of artificial intelligence has been, and remains, encoding expert knowledge for such complex and difficult tasks as medical diagnosis with greater pragmatic import. At the root of all this is trying to understand how expertise works.

Substantive expertise can be divided into two components. One is how much a person knows—the size of one's domain-knowledge database. The other is how this knowledge is accessed—the search algorithm used to summon it for the problem at hand. Experts plainly possess more domain knowledge than novices, but they access it differently too.

Superficially, since knowledge is stored in memory, then experts should have more memory capacity to hold it. Experts should correspondingly be those with better memory. And if problem-solving knowledge derives from having solved similar problems before, then experts' mental search subroutine should simply screen the database of previous problem solutions to find the right one. But there is much more to it than this.

Chess masters are renowned for their raw intellectual prowess, but they actually rely greatly on their superior memory performance, this being where their vast experience resides. Ericsson and Smith (1991) summarize how what they do with this memory was discovered. It was obvious from the start that Grand Masters made superior moves to novice players, better relating their previous experience to these moves and retrieving from memory those most appropriate for the innumerable possibilities the chessboard might suggest. But beyond this, it also became apparent how rapidly they could memorize, then recall just as quickly, the positions of the pieces on the board. Briefly shown a typical match configuration for as little as several seconds, chess masters were able to reconstruct the positions of all 20 to 30 pieces with almost perfect accuracy, while lesser players could recall only 50% to 70% of them.

The reason emerged that chess masters store information in memory according to meaningful patterns or clusters called *chunks*. When chess pieces were arranged randomly on the board, the chess masters' recall of positions was no better than novices'. Thus, it was the relationship of the pieces to each other—their patterns—that allowed them to be recalled so readily. And as is well known in chess circles, masters can play blindfolded, verbally informed of the moves on the board with little reduction in performance (Ericsson, 1996). They see the chessboard, they say, in their "mind's eye," and they visualize, in the most literal sense of the word, what it looks like as the game progresses.

By grouping elementary fragments in these larger, integrated patterns rather than storing them piecemeal, chess experts could encode and retrieve information much more efficiently. Their memory capacity as such was no greater than anyone else's—it was organized differently. So instead of storing then searching each position on the chessboard separately, they could store and retrieve entire patterns, thus freeing up memory resources. Their mental hard drive, as it were, contained more and better-organized directories, allowing them to search through these directories rather than the contents of each file. It has since been shown that like chess masters, experts in other domains can hold in memory some 50,000 to 100,000 chunks each containing a meaningful pattern or cluster of information about the problem. And if this is extended to possible solutions, the number grows to hundreds of thousands or even a few million (Richman, et al., 1996).

This goes far toward explaining how experts can retrieve so readily the tremendous domain knowledge they possess: They encode this knowledge in the form of patterns. Indeed, in so closely associating expertise with experience, what we refer to is really the patterns that experience provides. And the relationships they contain are not represented internally by correlations or matrices of some sort but rather are visualized as mental images. It will not go unrecognized here that both pattern recognition and visualization are also fundamental attributes of judgment. So it is clear from the outset how the processes used in memory encoding establish the connection between expertise and judgment. That experts possess not only superior knowledge but also superior judgment is a precept of all that follows.

As the elemental unit of how knowledge is structured, patterns or chunks go beyond raw data to include templates for different problem representations—different ways of structuring a problem or looking at it. Visualizing the problem and its representation is what guides the expert's retrieval of knowledge. This is how experts in all fields know what to look for so readily, while novices seem to get lost (Johnson, 1988; Voss and Post, 1988). Experts need retrieve only that information relevant to their problem representation, whereas novices must search through it all. The expert is able to match the problem to a stored problem class it resembles then proceed quickly to a solution by analogy, while the novice conducts a more laborious stepwise search. This fundamental characteristic of experts translates to a number of more specific attributes.

1. Experts are quicker and more accurate. Recall for a moment the strategy physicians use in making diagnoses. First a pivot finding was extracted from the aggregated symptoms, which then suggested a candidate disease. This is called *forward reasoning* where the symptoms are used to infer the disease. Then after a candidate disease was selected, it was compared back to the symptoms to see how they matched. This is *backward reasoning.* Both types of reasoning have an "if–then" character, but they work in opposite ways: If the symptoms, then the disease in the former; and if the disease, then the symptoms in the latter. This is directly analogous to the two different processes used in event-tree and fault-tree construction from Chapter 4. Forward reasoning involves sequences or progressions, while backward reasoning involves checking or screening by a process of matching (Patel, et al., 1996). Although both are useful and necessary, forward reasoning is more efficient.

Experts tend to emphasize forward reasoning, while novices rely more on backward reasoning because they have neither the knowledge base nor its accessibility which forward reasoning requires. This allows experts to arrive at a solution more rapidly without sorting through an extensive list of candidate diseases one by one, comparing the symptoms to each. Experts use forward reasoning to jump more quickly to a solution that appears obvious to them, filtering out irrelevant information. They employ backward reasoning mainly to confirm the diagnosis, to "tie up loose ends" as they say (Patel and Groen, 1991).

Resident physicians, for example, were found to be less able to discriminate between various candidate diagnoses because their initial hypotheses were poorly focused to begin with. Consequently, their list of alternatives proliferated, none of which could be readily ruled out by backward reasoning. Their collection of loose ends overwhelmed them in the end. Experts, on the other hand, arrived at more accurate diagnoses through forward reasoning because their initial hypotheses were generally better targeted from the outset (Patel and Groen, 1991). But while forward reasoning is quicker it can also be more dangerous, relying less on the solution constraints that backward "checking" against the information provides. So only through superior domain knowledge are experts able to use forward reasoning reasonably accurately. Both experts and novices alike make mistakes, of course, but experts proceed with the task more quickly and accurately than novices, who must rely more exclusively on tedious step-by-step search procedures.

We also find an example of forward and backward reasoning in the dialogue between Peck and Terzaghi related in Chapter 3. This concerned the diagnosis of chemical plant subsidence that Peck had attributed to clay compression and Terzaghi to bedrock subsidence. Terzaghi was using forward reasoning from his hypothesis, while Peck was adopting backward reasoning in matching his calculated and observed settlement magnitudes. Terzaghi's hypothesis was more accurate in the first place because his knowledge base was greater—he knew of other such cases, it turned out. But Terzaghi had also done some backward reasoning of his own in confirming his hypothesis by relating it to the abrupt and erratic settlement pattern, thus tying up this "loose end" quite nicely. Here we see something else. Peck and Terzaghi's basic problem representations, their causal narratives, were different. Displaying another characteristic of experts, it was Terzaghi's problem representation of bedrock subsidence that allowed him to make fine discriminations—here the more subtle settlement anomalies—that novices are more apt to overlook or discount.

We can see also how engineering analysis fits into this picture. With no other solution means at their disposal, novice engineers use analysis as a kind of mechanical substitute for forward reasoning. This, in fact, was the very word Terzaghi used in repudiating those who would use his theories as a "substitute" for common sense and experience. Without the knowledge base—the experience—necessary to identify a key pivotal finding and then reason forward from it, they are forced to plod through one or even several analytical procedures. The expert uses analysis in a different way. The expert engineer has already arrived at a solution by forward reasoning, then uses analysis to confirm it—to fill in blanks that remain in the causal narrative, to tie up loose ends. This is exactly what we saw in Chapter 3 where theory was used to aid judgment instead of producing the solution directly.

It would be tempting to take this comparison of forward and backward reasoning one step further in relating them to heuristics and biases. It has been observed that experts work forward on problems that are easy for them, just as the chemical plant subsidence problem was easier for Terzaghi who had seen it before. But they revert to backward reasoning for problems that are hard or novel, where their knowledge base is inadequate and forward reasoning breaks down (Anzai, 1991; Patel and Groen, 1991). In view of the "hard–easy" effect where overconfidence tends to be less for problems that are easy, one might expect experts to be less susceptible to overconfidence when forward reasoning is employed and more so when they use backward reasoning for problems of greater difficulty. But by the same token, using forward reasoning to jump quickly to an answer might just as plausibly cause experts to overlook relevant information and thus be more prone to representativeness bias. There has been little collaboration between the study of expertise as such and the heuristics and biases school, so just how substantive and normative expertise might interact in these matters must remain an interesting topic for speculation until a more unified treatment is developed.

2. Experts have better self-knowledge. Because experts can arrive at a solution more quickly they have more time left over to check their results—as any student who has ever finished an exam early can attest. This is another reason their accuracy is better. With more time to apply it, experts can use backward reasoning to greater effect in checking solutions where it does the most good, again in looking for the "loose ends." Thus, they are more aware of when they may have made a mistake, so are said to be better at *self-monitoring* (Glaser and Chi, 1988). Experts also ask more questions for much the same reason—their speed in arriving at a tentative solution leaves more time for deliberation, while novices have less time for questioning whether their solution is right. Expert physicians, for example, may announce a tentative diagnosis almost immediately on initial presentation of symptoms, then call for additional information before reaching a final conclusion. Likewise, chess experts may recognize a possible move within seconds, then take another 15 minutes to verify or revise it while their novice opponent is still conducting the search (Richman, et al., 1996).

Self-awareness, self-knowledge, self-monitoring, self-questioning—a fancier name for all this is *metacognition*, and as we will see shortly Terzaghi had it in spades. Experts overall have greater conscious awareness of their cognitive processes going on behind the scenes. They can think about their own thinking, to know when it's right and when they're getting off track. Experts can critique themselves without need for coaching, this being one of the ways they learn as much as they do. They know what they're good at and what they're not. This leads to perhaps the single most important implication of self-knowledge, which is that experts better recognize their own limitations, while novices have no such compass. It follows then that a key aspect of expertise lies in recognizing uncertainty, and we need only return to Chapter 3 to see how Herbert Einstein related one's awareness of limitations and uncertainties to judgment.

3. Experts anticipate. The same speed at which experts arrive at a solution allows them to further extend it to its implications. They project it forward in time, anticipating other effects that novices do not foresee. While we often think of an expert as knowing what to do, Salthouse (1991) shows that in fields ranging from chess to management, from music to sports, experts also know what to expect and posits this as a defining characteristic by which they circumvent their information processing limitations. Experts think ahead and they plan ahead to the next move, so that a medical expert would not stop with a diagnosis but would at the same time anticipate a treatment. An engineering expert would go on from solving a problem to devising a fix and still further to anticipating other problems the fix itself might raise. Like top NASCAR drivers and the speed at which they work, experts are always looking car lengths ahead to problems and solutions that blindside the novice. So by the time the next problem arises, experts are already halfway to solving it. Writing of then Princeton basketball superstar Bill Bradley, author John McPhee (1965) notes that what seemed to be his uncanny peripheral vision was really a "sense of where you are"—not just his position on the floor in relation to the pattern of other players but how that pattern would evolve with the anticipation of what they would do. No one having witnessed an alley-oop slam dunk from a halfcourt pass can doubt that these experts know what to expect as situations develop around them. As Bradley himself explained:

> When I was halfway down the court, I saw a man out of the corner of my eye who had on the same color shirt as I did. A little later, when I threw the pass, I threw it to the spot where that man should have been if he had kept going and done his job. He was there. Two points.

While it may have been the corner of Bradley's eye that was doing the perception, it was his "mind's eye" that was doing the work in seeing and visualizing what should happen. But Bradley didn't just watch the situation develop as he visualized it would, he acted on it. In the same way, engineers are above all doers. So not only will visualization provide an anticipatory sense of what to do, the expert engineer will go on to do it, and do it without being told. As we will see, this is a defining characteristic of engineering experts known as initiative. The expert's ability to know what to expect is what makes initiative possible. And this comes from something else too—how the expert represents the problem.

4. Experts see the problem at deeper levels. If the patterns that experts store in memory constitute such a graphically visual image in the mind's eye of basketball and chess players, then this image is less a snapshot than a videotape. This mental image or *problem representation* is the same causal narrative we have seen so many times before. But experts have not just richer and better-developed problem representations than novices, theirs are often different in a more fundamental respect. While novices see the problem according to its outward manifestations, experts see the underlying concept. In physics, this involves abstractions of the physical circumstances to a more general problem class. A student sees a block on an inclined plane, but the expert physicist sees the problem in the context of its class—a force–equilibrium problem—using a free-body diagram as the problem representation (Anzai, 1991). Experts sort, characterize, and recognize the patterns in problems more by their underlying nature than their outward features, storing them in memory accordingly. Expert software developers represent a program to be written not in terms of the literal objects stated in the program description or the particular program application but in terms of the underlying algorithms and protocols, visualizing and anticipating how the program functions might appear to the user (Adelson and Soloway, 1988). Similarly, a geotechnical novice might use a literal problem representation for an open-cut or braced-excavation problem, viewing it in terms of the structure or soil profile alone, while an expert might see it more as a fissured clay (strength), or soft clay (deformation), or critical gradient (seepage) problem using basic soil behavior concepts to categorize it more generally. Used as a vehicle for problem representation, these underlying concepts are what enable the expert to develop a solution more effectively and to further anticipate the additional problems that might be encountered in implementing it (Voss and Post, 1988). So experts see not only the problem but also the nature of the problem, and this allows them to better relate it by inductive analogy to others—they can see the conceptual pattern it fits.

5. Experts have insight. Since experts have access to a greater repertoire of problem representations from their superior knowledge base and more time to apply them, they can "try out" different ones. In fact, experts often devote considerable effort and study to understanding the nature of the problem and how to best represent it before a solution is even attempted, whereas novices dive in without hesitation (Glaser and Chi, 1988). Sometimes, though, the initial problem representation is not successful, and a period ensues when no solution presents itself. No steps to move toward it are evident, and the would-be solver is "stuck." The expert is better able to look at the problem in a new way, to develop a different representation, which often comes in a flash of insight at its moment of occurrence—the wave of crystallization that Poincaré knew and described so well. Richman, et al. (1996) relate this closely to intuition, yet another property of judgment, distinguished by a rapidity of onset fast enough to sometimes be punctuated by a figurative or a literal, "aha!" While we cannot know if this or something else was Newton's utterance when the apple hit his head, others too (including Edmund Halley, who we briefly encountered in Chapter 2 as an early developer of annuities) had conjectured a gravitational force that varied as the inverse of distance squared. But it took the insight that elliptical orbits would have to be represented by infinitely small approximations—the new problem representation provided by his calculus—for Newton to account for the planetary motions (Jardine, 1999). So even if experts have deeper problem representations, they do not necessarily become wedded to them. This is the essence of creative innovation. The ability to change one's problem representation that comes from insight is what makes innovation possible in the interest of deriving a solution. So on top of everything else, experts are creative, as we will go on to see.

6. Expertise is domain-specific. The pattern-based recall that serves expert memory so well does not appear to generalize to knowledge in other than the expert's specialized domain, and experts perform much like novices outside it (Glaser and Chi, 1988). So an engineering expert might use analysis in much the same way as a novice when unable to apply forward reasoning to a new and unfamiliar problem—they have to drop back and "go by the book." Sadly perhaps, or thankfully as the case may be, an expert in one thing is seldom an expert in all things, leading them to sometimes be compared to those with the classical idiot-savant disorder. Thus, the stereotypically absent-minded professor will deliver a brilliant lecture, then failing to have registered the pattern of the parking garage, will find his car to have utterly vanished. It also follows directly from their recognition of limitations that experts acknowledge these restrictions on their knowledge. Even for such a denizen of the cosmos as Albert Einstein, the income tax code remained forever baffling. Understanding it was for him, he said, "the hardest thing in the world." So the folklore about experts does have some merit, and it cautions about reliance on experts beyond the boundaries of their domain knowledge.

How Knowledge Is Acquired

Making use of expert knowledge in these ways is only half the battle in becoming one. The aspiring expert must somehow obtain knowledge to begin with, then gain it in quantity sufficient to qualify. To understand how expertise comes about, we first must consider how and to what extent knowledge can be acquired. One thing seems to stand out, and not surprisingly it is practice. And looking beyond practice to the kinds of things that engineers do, we find another deceptively simple one—reading and writing.

The Necessity of Practice

That it takes some time to become an expert is something few would dispute, despite the occasional prodigy. But prodigy or not, practice is the hallmark of experts and what makes them so good at what they do. In athletics and the arts, the kind of practice we're talking about is called *deliberate practice* as distinct from the more casual sort, basically just playing around. Deliberate practice focuses on a well-defined task at an appropriate level of difficulty, with informative feedback and opportunities for correction of errors (Ericsson, 1996). It is this kind of deliberate practice that produces improvement in performance required to reach expert levels for violinists or pianists, wrestlers or figure skaters (Starkes, et al., 1996). In these or any other activities, the amount and duration of practice needed to reach international levels of competition require personal attributes of persistence, dedication, sacrifice, and of course plain hard work.

Along with practice goes preparation, and across a wide variety of fields the period of experts' preparation is so regular as to constitute what has been called the "10-year rule" (Ericsson, 1996). This is the bare minimum of intense and focused preparation necessary to reach the threshold for world-class expertise, though by no means does it guarantee it. This seems to hold even for notable prodigies. Richman, et al. (1996) observe that although Mozart was composing at age 4 or 5, his first works of true greatness were written late in his teens. An adolescent Albert Einstein was studying physics and had written an unpublished manuscript on electromagnetism 10 years before publishing his first paper on special relativity at age 26.

As prerequisites for expertise, practice and preparation pertain in similar ways to engineering. Indeed, in designating the performance of our activities as the "practice" of engineering, we draw no distinction between practicing and doing what we always do. This implies that practice is a never-ending activity that continues as long as the engineer continues to do engineering. And by any measure, our practice of engineering is nothing if not deliberate—there is most certainly no fooling around about it. So too does the 10-year rule seem to apply. Engineering preparation starts with at least four years of intensely focused university work followed by a statutory eight years of supervised apprenticeship for engineering registration, with the total conforming surprisingly well to the 10-year minimum in other fields. And also as elsewhere, practice for engineers becomes progressively more self-directed, with teachers, coaches, and trainers eventually left behind as sufficient self-knowledge is gained to recognize one's own mistakes and correct them.

Still more, practice in engineering of the kind that promotes expertise is performed at the boundaries of one's capabilities—not with repetition of routine activities but by seeking out and pursuing the most difficult. To become an engineering expert is to become an innovator and the kind of generator of new approaches or ways of thinking that only working at the limits of one's capabilities can inspire. Experts do not spend their time solving textbook problems, at least if they can help it. Working on the most difficult problems is how they advance the field and at the same time extend their own knowledge and capabilities in it. Going beyond the familiar routines of everyday practice is how experts attain expert knowledge in the first place and how they remain experts thereafter. Nor does this come easy. At least as much as in any competitive arena this takes dedication, persistence, and self-sacrifice, virtues that went out with Horatio Alger but all of them qualities required to achieve expert knowledge and performance, as a look back at one widely proclaimed engineering genius will show.

Experts: Born or Raised?

No amount of basketball practice could ever transform any of us into a Bill Bradley. Expertise is often thought to be associated with a certain amount of innate talent that cannot be manufactured no matter how much we try. Is the same true of engineering? And if so, how much of engineering expertise is attributable to raw talent and how much to the practice and hard work it requires? At the pinnacle of expertise lies genius, and Howe (1996) looks to engineering genius, if there is such a thing, for an answer.

There was at least one engineering expert hailed as a genius in his time. His name was George Stephenson, best known for having designed and built the steam locomotive Rocket, or "Stephenson's Rocket" as it became, with his fame indelibly inscribed on the machine he created. With it, George Stephenson did as much as any other single person to open the door to the Industrial Age. Its very name symbolizes the unimagined speed of the era, but the Rocket also stands as a surrogate for Stephenson's achievements far beyond the workings of his famous engine. And as it happened, much of this achievement—and much of his genius—would have to do with what would eventually come to be known as geotechnical engineering.

Stephenson did not invent the steam engine, but was the first to successfully put it on wheels to practical effect. This made mechanized transportation possible for the first time, precipitating a reality no less defining to his age than spaceflight to ours. But if his locomotive were ever to go anywhere or do anything, he would have to build the first real railway of national importance to run it on. This he did in 1830 with the opening of the Liverpool & Manchester connecting two of England's principal cities, in the process confronting some difficult natural obstacles and classic geotechnical problems.

Much is known about Stephenson's early years, making it clear that what would later be proclaimed as genius had its start in some decidedly unpromising beginnings. Born in 1781 in the coal mining region of Newcastle, George Stephenson's first employment was herding cows. The family was dirt poor, none had ever attended school, and his father supported them as an engine stoker at a nearby colliery. Sometimes accompanying his father there, by age 11 young Stephenson was making clay models of steam engines with working hoists and pulleys, and twigs for steam pipes. At 14 he joined his father as an engineman, running, maintaining, and repairing a pump engine that became an adolescent's pet as he applied himself to taking it apart, cleaning it, and understanding its parts.

Through all of this Stephenson showed keen interest in the workings of machines, but as an engineer it was indisputable even to him that he had a long way to go. He was entirely illiterate, unable even to write his own name. Though already engaging in what could authentically be called deliberate practice, his preparation was markedly stunted. So at 18 he began classes in the local village three nights a week, learning to read, write, and perform simple arithmetic. In a scene reminiscent of Abe Lincoln's log cabin, Stephenson is said to have studied at work by the light of his engine's open firebox door.

In demonstrating such a combination of initiative and competence, Stephenson rose through the ranks, becoming recognized as an innovator through things like developing a miner's safety lamp less apt to trigger underground explosions—and this in his spare time. In 1814, however, there was not much reason to expect that the steam engine would ever be anything more than an awkward, unwieldy, and inefficient machine suited for pumping and hoisting perhaps but probably little else. Stephenson had other ideas, and by the time he attained the position of engineer he predicted, accurately it turns out, "I will do something in coming time that will astonish all England." It was just a year later at age 33 that his employer allowed him to try to build his own steam-powered locomotive. He did, and it could push 30 tons up a hill at 4 miles per hour. Named the Blutcher, it was the first successful flanged-wheel traction locomotive.

Stephenson continued his innovations and improvements to his locomotives, establishing with others the world's first company to build them. But as he did, his attention turned to what they would run on, first by patenting his own version of cast iron rails. In 1821 he became chief engineer of the Stockton & Darlington railway, a largely experimental project whose chief outcome was to demonstrate that steam power could move more freight at less cost than the proven horse-drawn rail technology of the time. But it had also demonstrated something else of much greater import. Despite the Blutcher and his other creations, Stephenson had come to see that railway transport would always be limited, both in capacity and destination, by the track gradients his locomotives could sustain no matter how much they might be improved. While working on the Stockton & Darlington, he discovered that a grade of 0.5% would reduce the locomotive's hauling capacity by 50%. The only way to overcome this restriction would be, as Stephenson concluded, to design railways with the object of minimizing this gradient. Obvious to us today, it was no such thing at a time when physical obstacles were even more formidable than mechanical ones, making the wisdom of such a tradeoff far from clear. But in proposing it, what Stephenson had seen as a mechanical engineering problem of optimizing the efficiency and power of his machines now was converted into a civil engineering problem of bridges, tunnels, and embankments. With this, he had transformed his problem representation. He would now see things in a new and different way.

By 1826 Stephenson was put in charge as chief engineer of the proposed Liverpool & Manchester railway. It was here that he first became known for his new Rocket, which won a competition among two other locomotives to win prize money of £500—a handsome amount at the time—and more importantly the right to supply the line's motive power. While this would be what gained him notoriety, Stephenson's real achievement would be constructing the line itself. Here he would put his concept of grade limitation to use, but there were some serious engineering problems. The line would require crossing an unstable peat bog, erecting a nine-arched stone viaduct, excavating a two-mile rock cut, and constructing a major embankment, any one of which would have been no mean feat.

The Olive Mount cut through sandstone was some 70 feet deep and required 480,000 cubic yards of rock excavation, a more than respectable quantity even today. But with Alfred Nobel's adaptation of nitroglycerine and John Henry's steam-powered drilling opponent both still more than a quarter-century away, this would have taken thousands of kegs of black powder and hundreds of men drilling by hand, with the excavated spoil transported and placed as the Roby embankment of corresponding volume.

The Chat Moss bog would if anything be even worse. With its widest part lying directly across the route, no one thought it could be crossed and all previous surveys had skirted it, in the process considerably extending the four-mile length of a direct crossing. So the owners were not alone in their surprise when the chief engineer proposed to try, with one skeptic noting of the plan:

> Everybody knows that the iron sinks immediately upon its being put on the surface. I have heard of culverts, which have been put upon the Moss, which after [being] surveyed the day before, have the next morning disappeared. As fast as one is added, the lower one sinks! There is nothing, it appears, except long sedgy grass, and a little soil to prevent it sinking into the shades of eternal night.

If eternal night were to be the destiny of Stephenson's Chat Moss crossing, then other well-wishers were little more complimentary of Stephenson himself. Another critic went beyond the plan to its originator and what by then was becoming his adoptive geotechnical role:

> It is ignorance almost inconceivable. It is perfect madness. Every part of the scheme shows that this man has applied himself to a subject of which he has no knowledge, and to which he has no science to apply.

True enough. There was then no such thing as geotechnical engineering, nor would it ever be a science. True too that Stephenson lacked the knowledge that would come anywhere close to expertise in this area. But he was willing, as before, to work on such a difficult problem to extend both his knowledge and expertise. Undeterred, Stephenson's first idea evoked the concept of a floating foundation. As he described it to his biographer Samuel Smiles (1857):

> George Stephenson's idea was, that a railroad might be made to float upon the bog. As a ship, or raft, capable of sustaining heavy loads floated in water, so in his opinion, might a light road be floated upon a bog.

Stephenson had certainly developed a unique problem representation maritime in its analogy. He saw beneath the outward appearance of the bog to the underlying concept—Archimedes' principal. Smiles went on to describe how the plan took shape:

> The first thing done was to form a footpath of heather along the proposed road, on which a man might walk without risk of sinking. A single line of temporary railway was then laid down.... Along this way ran the waggons in which were conveyed the materials requisite to form the permanent road. These waggons carried about a ton each, and were propelled by boys running behind them along the narrow iron rails. The boys became so expert that they would run the 4 miles at a rate of 7 or 8 miles an hour without missing a step.

As this narrow "footpath" was laid then widened to form a working pad, some 200 men were laying drains on each side. While drainage enhanced surficial pad stability, it had no effect whatsoever on the deeper peat areas. Better drainage was needed, perhaps something with more capacity. Here Stephenson had to modify his plan, innovating once again and this time using barrels and casks (quite possibly empty powder kegs from the Olive Mount cut), covering them with clay in a pipe-like configuration. This still didn't work—the barrels rose to the surface. This part of the bog was called Blackpool Hole. Its name said it all.

It was with the kind of candor perhaps only possible for a recognized genius that Stephenson would look back on this discouraging but crucial time a few years later. In a public speech in 1837, he openly admitted:

> After working for weeks and weeks we went on filling in without the slightest apparent effect. Even my assistants began to feel uneasy and to doubt the success of the scheme. The directors, too, spoke of it as a hopeless task, and at length they became seriously alarmed, so much so, indeed, that a board meeting was held on Chat Moss to decide whether I should proceed any further. They had previously taken account the opinion of other engineers, who reported unfavourably. We had to go on. An immense outlay had been incurred and a great loss would have been occasioned had the scheme been then abandoned.

Only another engineer having stood alone and outnumbered in Stephenson's shoes can truly imagine that conversation on the bog. The critics, it seemed, had been right. Stephenson, now in serious trouble, still had to continue. But how? He was stuck deep in the bog in every respect.

Together with one of his men, Robert Stannard, Stephenson hit upon a new idea, a wave of insight that accompanied yet another change in problem representation. They would lay out timber in a herringbone pattern on the surface of the peat, together with bundles of moss, heather, and brushwood. This was nothing less than subgrade reinforcement, presaging the use of modern geotextiles. Stephenson had formed a completely new representation of the foundation problem now conceived around tensile strength rather than buoyancy, and he had found one that worked. As recorded in Figure 7-1, the roadbed over Chat Moss was completed successfully, the Rocket at the head of the first train across.

The Liverpool & Manchester was formally christened on September 15, 1830, in attendance the prime minister, the Duke of Wellington, Prince Esterhazy, and a minor kingdom of the titled and privileged. The opening ceremony was choreographed with no less than eight locomotives shuttling back and forth, the occasion marred only by the Rocket having impaled one insufficiently attentive Mr. Huskisson, whose misfortune still could not tarnish the brilliance of Stephenson's achievement.

a)

b)

FIGURE 7-1. *a, Stephenson; b, his Rocket crossing Chat Moss.*

Clearly, from his illiterate beginnings Stephenson was a self-made man, as the technology and culture of his era demanded. But he was also, and above all, self-motivated in achieving his knowledge and accomplishment. In inquiring what it is that inspires such hard work, Winner (1996) calls it the "rage to master." The desire to work hard is something that comes from within, for Stephenson as for others. The successes from the achievements it produces serve to strengthen it even more, with each propelling the other in turn.

It would be easy to say that for those of Stephenson's ilk, hard work is its own reward. But self-motivation goes deeper than that. Underneath the hard work and the rage to master are pleasure, excitement, and a host of human emotions of the most satisfying kind. The motivation for the work of genius, or indeed for any expert, is the simple thrill of discovery that comes during that moment of insight from knowing what was not known before and doing what hasn't been done. This was Poincaré's "true esthetic feel," the pleasure he found in the beauty and elegance of facts when the subliminal self managed to arrange them in the "harmonious whole" that he sought. This pleasure, this satisfaction, this thrill of discovery becomes an opiate to those who have known it, and they continue to seek it in whatever they do. Nor need these discoveries be of the grandiose kind that win Nobel prizes. Young Stephenson found them even in his pet steam engine, and they served him well throughout the rest of his life. But the thrill of discovery does not come from repetition of everyday tasks. It exists only at the frontiers of personal knowledge in doing things that are new. And this is why experts are not just motivated but virtually compelled to work at and beyond their limits in continuing to enjoy this unique form of personal satisfaction.

That pleasure motivates expert performance is not something that comes any more naturally to most engineers than to Poincaré's "profane" colleagues who smirkingly dismissed the idea. But in their more private moments some, like contemporary geotechnical engineer Richard Meehan, have reflected on this pleasure. On how it can come not just from the discovery of new knowledge but from any creative process—like design:

> Engineers are thought to be practical people, and yet the core of engineering contains elements of idealism that are absent in other professions, medicine, and law. I mean idealism in the original platonic sense, the belief that somewhere is the perfect design, the perfect dam. We'll never see it, and we'll certainly not come close to building it, but we know enough about it to be forever seeking it.... I learned the pleasure in it, in this design. For in the end it does not differ from any other art, the satisfaction in making a clay bowl or a painting or writing a sentence or a symphony. First the concept, the trial efforts, the crude shape of a good solution, then refinements, balance and polish until the final arrangement sings with deceiving simplicity and stuns with accuracy of effect. (Meehan, 1982)

To sing with deceiving simplicity and stun with accuracy of effect. There could hardly be a more thrilling description of the expert's reward, with Meehan hard-put to disguise finding it as much in his writing as in the design he describes.

Returning then to Howe's (1996) original question, was it innate talent or something else that produced Stephenson's genius? He maintains that what we see in Stephenson are the same qualities and abilities displayed by others with exceptional levels of expertise. Stephenson always saw the concept underlying the surface appearance, so that beneath the problem of locomotive efficiency lay the road grade itself, and beneath the bog problem was in concept a boat. He got good and stuck at times with back against the wall just as other mortals do, but he was able to change these problem representations when the situation demanded—the floating foundation would become a reinforced one. In these and other things, Howe argues, Stephenson's genius was no miracle of birth and no mystery. His great abilities were built, not born, of such unschooled origins, starting with his childhood toys and progressing from there with unshakable persistence and determination in gaining the knowledge to do all that he did.

Reading and Writing

Though literacy became one of the keystones of Stephenson's genius, reading and writing are not things that spring immediately to mind when we think of expertise. Of all the three R's, it is the last that engineers look to most, with the first two taken largely for granted. Yet it is through reading that most all engineering information is assimilated and report writing, truth be known, that occupies much more of engineers' time than calculation, with hardly a job ad that appears not calling for more than passing skills in written communication. So of course engineers read and they write but often more for utility than anything else in accomplishing daily tasks. Everyone has a memory too. But experts use reading and writing differently, just as they do memory, in ways that enhance their knowledge.

This difference lies in using reading and writing as knowledge tools rather than strictly utilitarian appliances. When we read it is most often for securing information from a text to some specific purpose, and this is so for experts no less than novices. Johnson (1988) shows that in knowing what they are looking for, experts read with economy, extracting that which is most relevant and sometimes skipping over parts a novice would have to search through. But even if more selective, experts recycle what they read back through and into their knowledge base, in the process interpreting it, revising it, and extracting it all over again from this revised format. In the end, not only is richer content derived from the text being read, but the knowledge domain itself is reorganized and restructured, preserving the modified patterns for the next time around. In a similar way, as experts proceed through a text they iterate back through it more often, questioning and revising their previous interpretations in light of subsequent statements. But novices read straight-ahead simply to absorb, with each statement interpreted literally, immediately, and without reconsideration, unmolested by anomalies inconsistent with prior knowledge or inferences that would enrich it. While this may be efficient in a task-oriented sense, there is little to increase knowledge without feedback to the knowledge domain.

Scardamalia and Bereiter (1991) call this the "dialectic" process in both reading and writing. They represent it symbolically as shown in Figure 7-2 and argue that this is an essential part of what it means to be an expert in any field. Literally, this process is an internal conversation between oneself and the text being written or read, between the general case of one's domain knowledge and the particular case of knowledge in the text being composed or comprehended. But more, it is a two-way conversation that goes back and forth between extraction and embedment of knowledge. Take reading, for example. In Figure 7-2a, the immediate

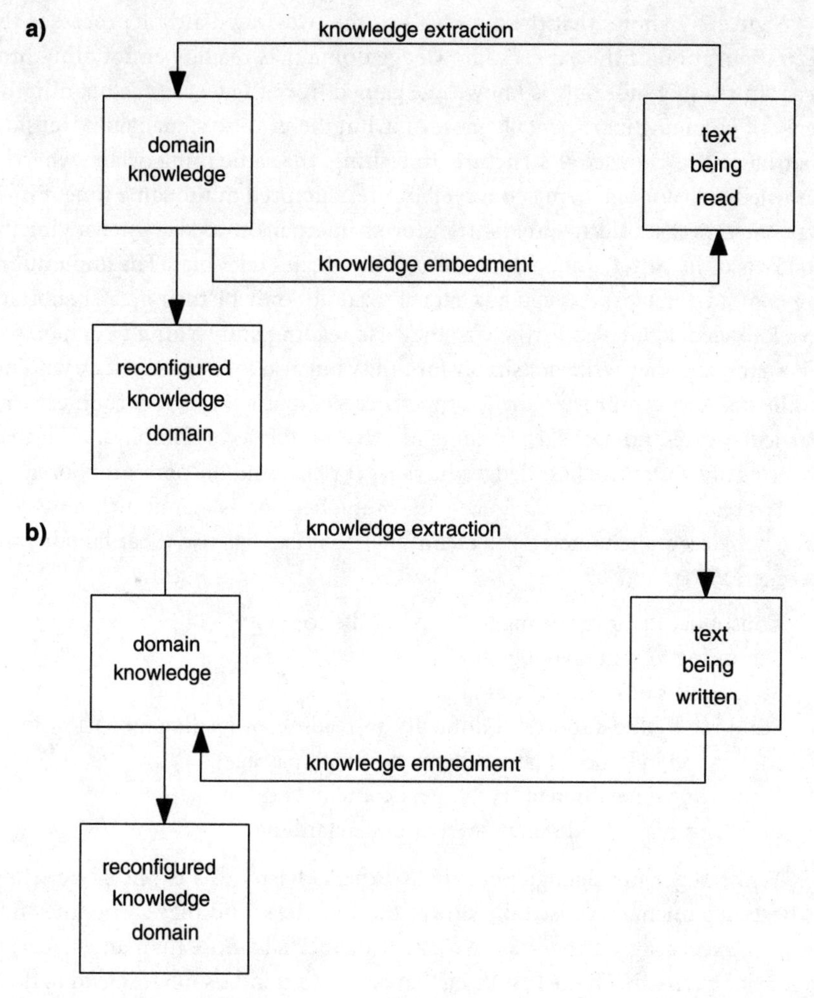

FIGURE 7-2. *The dialectic process in reading and writing: a, expert comprehension; b, expert composition.*

object for expert and novice alike is to extract knowledge. For the novice this is a one-way street, going from right to left only. For the expert, however, the extraction of knowledge from the text is accompanied by incorporating domain knowledge back into it, so that the text becomes interpreted in a broader context that enriches its content. Similarly for writing in Figure 7-2b, the dialectic process is exactly the same but the directions of extraction and embedment are reversed. While the novice extracts domain knowledge to put it down in the text in left-to-right fashion, the expert at the same time feeds back to domain knowledge that which is derived from the activity of composition.

Figure 7-2 shows that the net result of the two-way dialectic process is the reconfiguration of the expert's knowledge domain as reading and writing proceed. In reading, not only is knowledge gained from the text in the quantitative sense of obtaining incrementally more of it, but there is also a qualitative transformation of the knowledge structure. In writing, the same thing occurs whereby knowledge is not just being conveyed but restructured at the same time. Either way, the expert's dialectic process transforms internal knowledge by changing the contents of its patterns and chunks to provide greater relevance. The implications are twofold. First, experts can be characterized not only by their greater substantive knowledge but also by the way they use reading and writing to enhance it. They read and they write not simply for utility but also to increase understanding. While this keeps improving their competence as experts through encounters with particular cases, novices keep running in place. So the second implication is that, in terms of substantive knowledge, the rich get richer and the poor get poorer.

For reading, the dialectic process in comprehension is manifested more concretely by several characteristic techniques that experts use (Scardamalia and Bereiter, 1991):

- Consolidating or reviewing to "firm up" the content;
- Skimming to "put it all together";
- Making lists or notes to "cram it all in";
- Scanning to find a source of difficulty, re-reading, or reading more slowly;
- Moving on in hopes of later resolving difficult passages;
- Refocusing on a different level of text content; and
- Resetting goals at a different level of understanding.

While the same dialogue occurs as experts write, this is one place where experts are uncharacteristically slower than novices, and they sometimes find writing exceedingly strenuous. Writing for experts is more than an exercise of recording pretty much the first thing that comes to mind, as novices tend to do. It becomes not just a matter of extracting domain knowledge but again as on Figure 7-2, recycling it back as the text is constructed, in the process revising and rethinking until this restructuring of knowledge finally makes it come clear. Writing for experts is less the act of telling what one knows than of discovering it. So scientists' notebooks often read like someone thinking aloud rather than recording thoughts already formed. This can be seen in some of the corollary things that happen as experts write (Scardamalia and Bereiter, 1991):

- Rhetorically defending a position shows the relevance of overlooked information;
- Shortening a text leads to critically assessing the importance of related items;
- Constructing transitional statements reveals unrecognized relationships;
- Making a text more interesting leads to new insights by identifying such items; and

- Examining word choice clarifies distinctions and enriches conceptualizations.

Writing was of immense importance for Terzaghi, and he undertook it in ways typical of experts. Peck (1965) remembered having been struck by the sheer volume and detail of Terzaghi's correspondence. He also recalled how Terzaghi looked forward to the end of a consulting assignment when he could fully digest the data and synthesize knowledge gained from the project. This always culminated in the final report which, Peck says, was written at least as much for Terzaghi's benefit as his client's. Neither were these reports written quickly, with Terzaghi spending days or even weeks preparing them. Nevertheless, the bursts of intense concentration and review that Terzaghi's writing required were, in retrospect, "the periods of greatest advancement of the art of applied soil mechanics and engineering geology" (Peck, 1965).

Terzaghi himself recognized how writing caused him to reconfigure his knowledge domain. With the expert's self-awareness of underlying thought processes, he described this much more elegantly in this postscript to a letter to Peck:

> For me, the process of writing a book never meant more or less than a sort of mental housecleaning. It compels me to examine every little piece of my inventory. Much of it is found to be worthless and only fit for being discarded. The rest is cleaned and polished and put into its proper place. If one does not make such a housecleaning from time to time one drifts into chaos. (Peck, 1973a)

Certainly these remarks deserved more than a footnote, echoing as they do the importance of writing, and reading too, expressed by so many others. For in the way that experts undertake them, both have been recognized to lie at the core of thinking itself. It has been said that reading is "thinking with a book in one's hand" (Wagner and Stanovich, 1996). And in speaking of the close ties between writing and thinking, author and former Librarian of Congress Daniel Boorstin once explained, "I write to discover what I think." In making order out of chaos as Terzaghi put it, writing is a process of discovery, of coming to understand what one means, what one believes, and why. Some could not imagine thinking without it. Near the end of his years as blindness overtook him, Jean Paul Sartre went so far as to say that when he could no longer write he could no longer think either. So writing is much more than it appears. It goes without saying that thinking is necessary to write, but the reverse is just as true. The act of writing, it seems, embodies the very essence of thinking. In an interview, Pulitzer prizewinning author David McCullough described it this way:

> Writing forces you to think, to bear down on the subject, makes you think as nothing else does.... It's a way of working out problems, working out your thoughts, and arriving at insights, conclusions, revelations that you never could have obtained otherwise. That's really the reward of it.

Because in the end, he said:

> Writing is thinking. That's what it is. And that's why it's so damn hard.

Oddly enough, what one chooses to write with appears to matter too. Stephen King has always been a prolific writer of fiction, with his audience testifying to his literary expertise. But in 1999 his leg was shattered and he was almost killed by a passing van while walking on the side of the road. His recovery produced a long hiatus in his work, and some even feared he would abandon the craft. But return he did, and in his first novel afterward King described how the experience had affected his work. It had caused him to take writing in a more visceral way, accompanied by a change in technique. He now used a different word processor—a cartridge fountain pen—which he called the finest such machine in the world. And he said that having written the first draft in longhand reconnected him to the language in a way that he hadn't been for years. So it appears that there may be some kind of hard-wired connection between one's thoughts and movement of hand on paper that no keyboard yet devised is fully able to duplicate. This tie may well have to do with the time that experts require for their writing, which not just authors but engineers too have long cherished.

In this, we turn for a final anecdote to noted geologist and mining engineer Ira Joralemon. He and Terzaghi were contemporaries, born just two years apart, and though they never knew each other their careers took parallel paths. At the time Joralemon graduated from one of the early classes in mining engineering at Harvard, like soil mechanics it was also a new field, with his career taking him as well to remote corners of the globe. And too like Terzaghi, he wrote consulting reports, his describing for clients the orebodies and mining properties he inspected. In his memoirs, Joralemon (1976) related how he always relished the long steamship voyages home and train trips of several days that gave him time for his observations to "settle down," to form a "coherent picture" that frequently would modify or change them. This was a time of reflection and thinking when the outline of the report and its contents would gel in preparation for writing.

He contrasted this to his later years of jet travel, when on one occasion a private plane was put at his disposal. Within 10 days of leaving his office, he said, his final report was in the hands of his client. The time for reflective writing and thinking was something he dearly missed. Little could Joralemon have known that today his client would demand that he deliver his report on a laptop computer before being allowed to leave. In this the era of fasterbettercheaper, thinking is mostly an afterthought supposed to occur at one's maximum typing speed, or else on one's own time. But it is not hard to see what happens to the dialectic process when thinking and writing are catapulted to the same terminal velocity.

How thinking and all of the things that underlie it—preparation, practice, reading, writing—come together in a professional sense over time is what we call a career to distinguish it from employment. Yet when we consider career achievement, we envision it as a linear progression on the timeline of professional life. The study of expertise provides insights on how careers develop. In experts, this is not linear but varies according to characteristic trends called "trajectories."

The Trajectory of Expertise

In the career of George Stephenson, we saw how his expertise—indeed his genius—was not something he possessed from the start, nor did it spring up overnight. Instead it accrued over years of preparation, practice, and hard work. Simonton (1996) extends the consideration of how expertise develops to show that it takes on regular and predictable patterns throughout the course of experts' lives. In particular, Simonton examines what he calls *creative expertise*, as distinct from the more garden variety. Experts who display it are those who transform a field's theories or constructs to modify their discipline in some fundamental way from what it was before. These can be disruptive people, slightly off-kilter or somewhat out of step, as a look back at Paracelsus, for one, will show. Their knowledge base is constructed so that they see things differently, and thus they become the "innovators" as opposed to the "adaptors," who do in very much a Kuhnian sense what has not been done before. So those with creative expertise are not necessarily the foremost analysts, but those with more of Poincaré's intuitive insight that precedes analytical thought.

Creative Productivity and Career Landmarks

Simonton (1996) presents a fascinating representation of how creative expertise develops and varies over time. The theory posits that each individual starts their career with a certain fixed quantity of what is called "initial creative potential," designated as the variable m. Much like potential energy becomes converted into work, this store of creative potential is transformed into creative achievement as the career progresses. Creative potential may well involve a certain amount of innate talent, but at least as much the kind of motivation that produces the persistence, dedication, and desire to work hard that Stephenson displayed. It would not be unreasonable to think of one's creative potential, their personal value of m, as the intrinsic reference by which some are judged to be underachievers or overachievers. Clearly then, the value of m will vary from person to person—not everyone who herded cows in Stephenson's time, after all, went on to build railroads. The abilities and qualities that contribute to the development of this initial creative potential are developed and honed during the period of preparation and deliberate practice antecedent to one's full-blown career. So one's initial creative potential is a kind of reservoir of knowledge and motivation accumulated during the developmental period then tapped during the productive years that ensue.

Creative potential is transformed into productive achievement through two processes that Simonton (1996) calls *ideation* and *elaboration* of creative ideas. First an idea is generated then its merit is assessed, in a way directly analogous to the dual scientific processes of discovery and justification. But ideation and elaboration occur at different rates, both across different individuals and in different professions, represented by the variables a and b. One person might turn out new ideas prodigiously but delay following up on them, while another might race to develop ideas occuring less frequently.

The initial fixed quantity of creative potential, m, is converted to achievement at transformation rates a and b to yield a time-varying quantity of creative output or production $p(t)$, where t is the "career time" measured from when the expert's career effectively begins. These variables are then related to $p(t)$ as:

$$p(t) = c(e^{-at} - e^{-bt}) \qquad (7\text{-}1)$$

where the constant c for any individual is given by:

$$c = \frac{abm}{b-a} \qquad (7\text{-}2)$$

Simonton (1996) has shown a correspondence between predicted results from this relationship and observed data, where creative output per unit time has been determined by such readily measurable quantities as papers published, patents held, or musical scores composed. The functional form of the relationship shown in Figure 7-3 can be used to illustrate the sensitivity to its coefficients.

The first thing to note in Figure 7-3 are the two scales for career age t and chronological age T. The correspondence will vary for "early achievers" and "late bloomers" who start their careers at different ages, such that t is shifted from T by the period of preparation and practice, t_0. However, the functional form for $p(t)$ always holds in producing curves of the same general shape, though varying in slope as the time–rate of productive output. These productivity curves are called *career trajectories.* Comparing those in Figure 7-3, trajectories I and II are for high m, a greater value of initial creative potential, while III represents those with some lesser endowment. Their creative productivity never reaches the same level because they simply have lesser potential to begin with. For curves I and II where m is the same, the overall lifetime production will also be the same, just distributed differently over time. Trajectory I reaches an early peak, then productivity drops off markedly in later years. Trajectory II has slightly lower transformation rates a and b, and its shape is very sensitive to small changes in these values. It reaches a somewhat lower peak but with more sustained achievement over life's course.

While career trajectories vary among creative individuals, they also differ among professions. A type I curve with rapid ideation and elaboration rates is characteristic of fields dealing with a finite and bounded array of concepts, as in

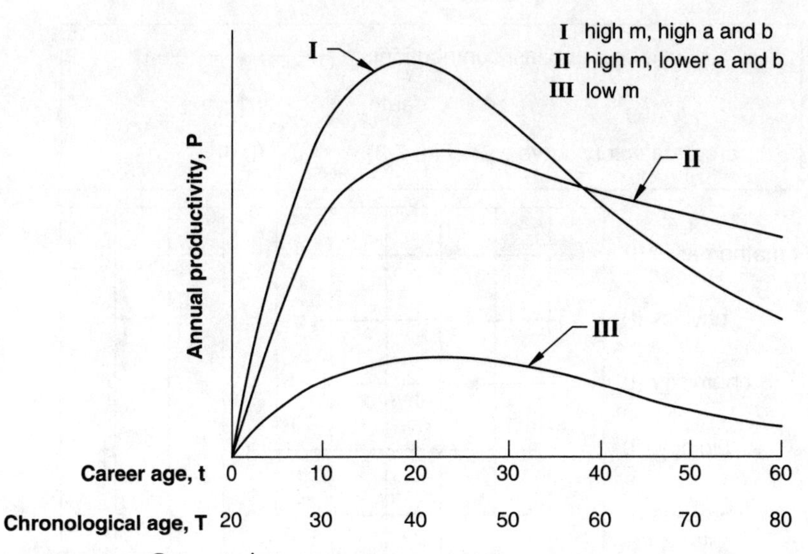

FIGURE 7-3. *Career trajectory curves.*

mathematics for physicists or vocabulary for poets. Type II trajectories are more typical of fields such as history (where the career peak for historians comes 18 years later than poets) and also—notably—geology. These tend to be disciplines that deal in less refined but less restrictive conceptual problem spaces, in imprecise concepts open to varied interpretation.

But behind Figure 7-3 is another intriguing observation. While *p(t)* expresses the quantity of creative output, it does not directly speak to its quality. Each person has their share of successes and failures, "hits" and "misses," experts no less so than others. Only a few scientific journal articles published will ever be widely cited in the literature with the rest for the most part ignored, and only a small proportion of Edison's patents ever achieved commercial success. But as Simonton (1996) shows, the relative proportion of these winners and duds remains fixed throughout the expert's career even though the quantity of output varies; hence the period producing the greatest masterpieces also turns out the most forgettable failures. So if the ratio of hits and misses stays constant, then the relationships in Figure 7-3 hold for quality of creative products as well as their quantity, with only a reassignment in vertical scale. Thus a surprising reason that experts can run up such an impressive score at their peak is not that they target their efforts more accurately but that they simply take more shots at the goal. The payoff of experts' persistence and hard work, then, is realized as their scoring attempts become more prolific. And as their hits diminish in number, this is not from any reduction in knowledge or ability but from reduction in output with age.

Simonton (1996) goes on to show that in various fields of science, the hits of greatest professional impact take on their own form over time, with three that are

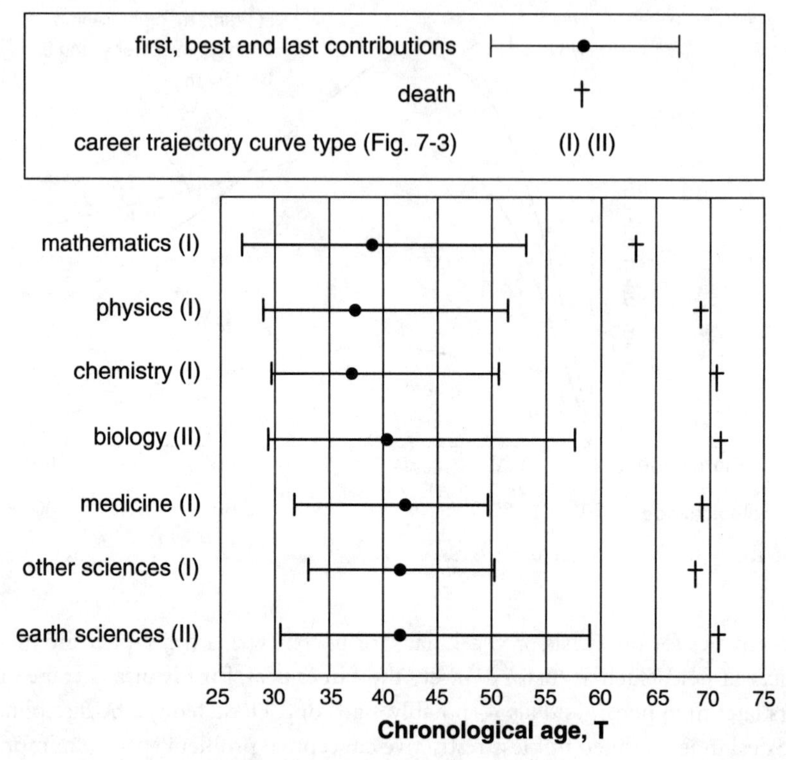

FIGURE 7-4. *Career landmarks in selected technical fields.*

characteristic over a wide variety of activities and individuals. There is typically a first, a best, and a last professional contribution that together mark an expert's career, and these *career landmarks* can be used to benchmark the career trajectory (Simonton, 1991). The first landmark hit occurs on the rising limb of the Figure 7-3 curves. It is the first creative achievement of true importance, where the expert makes their mark and establishes their reputation. The "greatest hit" is the opus magnum that occurs at the peak of productivity and most enhances this reputation. And then there is a final masterpiece produced on the waning slope of productivity, the crowning achievement that caps their career.

Mean career landmarks from over 2000 contributors are summarized for selected scientific fields in Figure 7-4 and can be roughly separated into two groups. The first containing mathematics, physics, and chemistry, is characterized by comparatively more youthful landmark contributions overall, while these are generally more deferred for the second group including biological, medical, and other sciences. At the bottom of Figure 7-4, and contained in this second category, are creative experts in the earth sciences. Their first contributions are deferred until the early thirties, like physicians' with their lengthy preceding preparation. In contrast to the well-bounded sciences of mathematics, physics, and the like, the best contributions of earth science experts come significantly later, not until their early forties. But along with biologists, they have the most extended (and therefore least diminishing) type II career trajectories and the longest productive career durations. Their final landmark contributions are the latest of all. Still, these are only averages. We can examine more closely the careers of three experts in particular who produced some notably creative contributions, all having to do in different ways with some remarkable geotechnical achievements. We can look to see whether the three landmark accomplishments can be identified, whether all are late bloomers as the earth science averages might suggest, and just how their productive career durations compare. And for the first of these experts, we return to someone who by now is a most familiar character.

Karl Terzaghi

Perhaps it would not be overly prescient to think that an engineer with a dueling scar on his cheek might be somewhat out of the ordinary. In Karl Terzaghi we find not only the landmark achievements that charted his career but also clear evidence of preparation and deliberate practice; of writing; of ideation and elaboration; insight; problem representation; and most all of the things that typify creative expertise. Relying extensively on Goodman's (1999) biography, we look here not just to the things that Terzaghi did but also how he did them. It seems strange to think of Terzaghi as a late bloomer, but his period of preparation was an extended one. Nevertheless, once he effectively started his career, its trajectory was little short of ballistic and it continued to the day he died.

Terzaghi was born in 1883 to a comfortably well-off family of minor aristocracy. His father was a career officer in the army of Austria's empire and his grandfather a successful mechanical engineer. By the time young Terzaghi graduated from the Technical University at Graz, he had studied mechanical engineering and switched majors to civil engineering, taking geology courses throughout and publishing an adapted translation of a British field geology manual in 1906.

Specializing in reinforced concrete structures, during the next six years he took a number of brief positions in rapid fire that introduced him to hydroelectric dams and tunnels in Austria, then in Croatia, then to foundation engineering and excavations for buildings in St. Petersburg and then present-day Latvia, then back to Austria—all the while taking careful notice of the geologic conditions that influenced these structures. Somewhere along the way he hatched a plan to travel to America to see and learn about the dams and reservoirs then coming to fruition as a result of the Reclamation Act of 1902 and with it the creation of the Reclamation Service, John Wesley Powell's belated offspring. Thinking that an advanced degree might open doors when he got there, he found time during his six months in Russia to work up an innovative graphical solution for reinforced concrete design and submitted it to the Technical University back in Graz. With this as his dissertation, inside of a month he was awarded a doctoral degree in 1912, embarking for New York a month after that.

Though technically speaking he was employed throughout most of this breathless activity, Terzaghi's creative career had not yet really begun. Visiting various dams, flood control, and irrigation projects in Louisiana, Arizona, Idaho, California, and Wyoming partly on funds wangled from his supportive grandfather, Goodman (1999) called this a "study trip." And that it was. Terzaghi was doing all this in a very deliberate way, seeking to learn everything and anything that could further his understanding of engineering geology. He admitted as much, saying that if everything panned out, "I will be able to become a specialist in dams and one of the best professional engineering geologists in the German community." Nor were these studies restricted to purely intellectual pursuits. Ending up at one point as a blasthole driller for the Celilo Locks high above the Columbia River, he contemplated the jointing patterns of basalt cliffs even while working suspended from ropes—a practice that may have contributed to his nearly being blown to bits from a premature detonation (Goodman, 1999).

He returned to Austria in 1913 to start a consulting firm but found a draft notice instead. Archduke Ferdinand had just been assassinated, the opening volleys of the First World War had just been fired, and Terzaghi had just been ordered to the Serbian front. As a lieutenant in charge of an engineering battalion, he occupied himself constructing roads, trenches, and earthwork fortifications. But meanwhile, the Austrian government had been conducting along with the war a sort of diplomatic mission to win the hearts and minds of the Turks, sending Professor Phillip Forcheimer to the Royal Ottoman College in Constantinople to better cement Austrian influence there. Forcheimer was acquainted with Terzaghi, having been at Graz during Terzaghi's doctoral blitzkrieg. Knowing that the Turks were planning various irrigation schemes of their own and would be interested in developments in America, Forcheimer invited Terzaghi to Constantinople to aid in Austria's cause. For his part, Terzaghi knew that Forcheimer had pioneered the adaptation of heat-flow analogies to porous media flow—what we would now call flow nets—and was only too glad to accept. Terzaghi arrived there in 1916. He was then nearly 33. His gestational period of preparation and deliberate practice had consumed some 16 years.

Terzaghi's creative career can be said to have begun in earnest at the Royal Ottoman College. Working with little more than kitchen utensils, cigar boxes, and an antique balance right out of a Turkish bazaar, he fashioned a series of experiments of considerable import to retaining wall design. It was not just friction that controlled earth pressures but also deformations and therefore the whole stress–strain character of the soil mass. The work was first published in 1919. Reprinted in English a year later, *Engineering News Record* editorialized that in enabling soil to be characterized as an engineering material, it spoke to the "outstanding research problem in civil engineering," with Terzaghi's article heralding "the opening of an avenue of progress" (Goodman, 1999). While allowing for a grain of press hyperbole and perhaps a modest achievement in light of things to come, this was nevertheless Terzaghi's first significant creative accomplishment and the first to gain him widespread recognition. Just three years after starting his career in earnest, at age 36 this was Terzaghi's "first hit," his first landmark career achievement. He had made his mark.

As war came to a close the Turkish government surrendered to the British, and Terzaghi and Forcheimer found themselves out on the street. But help appeared from an unexpected source—Roberts College, a small American university just outside Constantinople that was starting up an engineering curriculum. They both signed on as instructors, allowing Terzaghi to continue his work now in the area of quicksand and seepage phenomena. This should have been right up Forcheimer's alley and Terzaghi solicited his collaboration, but Forcheimer expressed no interest in practical applications of his flow theory, saying he had no capacity to retain observations of actual soil behavior. Terzaghi incredulously complained to his diary:

> He cannot remember that which is not achieved through logical thought ... had he a better technical memory it would not have been necessary for him to derive a theoretical approach. (Goodman, 1999)

These remarks say as much about Terzaghi as they do Forcheimer. With true metacognition, a conscious awareness of how one's thinking works, Terzaghi was distinguishing between deduction from theory and knowledge contained in memory. And clearly in calling it Forcheimer's "technical" memory as opposed to some other kind, the fault Terzaghi found in it was its lack of structure for retrieving relevant information, a structure that Terzaghi's memory already incorporated as he seems to have recognized. It would be Mark Olsen, Terzaghi's field assistant on Mission Dam (which we will come to momentarily), who would later describe Terzaghi's keen observational and writing skills but also his remarkable memory and how "organized" it was (Leonoff, 1994). This points to the structured patterns of Terzaghi's memory, and from what we have learned about experts it would be hard to expect anything less.

Despite Forcheimer's rebuff, Terzaghi continued on his own. Using Forcheimer's theory to give order and meaning to his experimental observations from model dams, he eventually arrived at the concept of a weighted filter for protection against high seepage exit gradients, with a persistence also to be expected of experts. This problem now behind him, he realized that in order to characterize the properties and behavior of soils in any quantitative way he could no longer avoid the problem of clays. His previous efforts on the topic with the Swedish agricultural soil scientist Atterberg had not come to much; he was "stuck" and unsure how to proceed. Goodman (1999) describes how one day Terzaghi had a "sudden burst of inspiration," attributing the following account of this epiphany to Laurits Bjerrum in an image that conjures up Rodin's famous sculpture:

> Sitting in a mood of depression on a piece of rock outside Roberts College, looking out on the Golden Horn, he suddenly visualized what was needed to obtain a rational approach to the problems … [was] the development of testing equipment and methods which could give a quantitative measure of the mechanical properties of the soils involved.

Again from the characteristics of experts, we would come to expect such a moment of insight, and with it the visualization of a new problem representation. So laboratory testing devices would be the ticket to understanding: As the "test" in the drill–test–analyze strategy, this would become central to our entire predictive approach. He did not yet have a solution, but Terzaghi was starting where experts typically start: He was asking the right questions. Still, as he confided, "My plans were ambitious...and I doubted that I would be able to answer all the questions I had posed" (Goodman, 1999).

Terzaghi's new experiments would be conducted in a device we would recognize as an early form of oedometer fitted with a standpipe to measure the hydraulic gradient across his specimens of slurried clay. But compression of the sample began only after some initial delay. Why this was happening must have something to do with the kind of intergranular forces and water pressures whose importance he had established in the quicksand experiments. Terzaghi was by now exhausted in every respect—out of time, energy, and money too—but through 1920 he persisted in moving toward the conclusion of what had become his "detective novel," as he now called the clay problem.

It is not altogether clear exactly when his comprehension of the physical process going on in the test reached its zenith, but Goodman's account suggests it may well have been as Terzaghi was in the process of writing. In 1923 he was preparing the manuscript of *Erdbaumechanik auf bodenphsikalisher Grundlage* (Earthwork Mechanics Based on Physics of Soils). In it, Terzaghi explained that the external stress applied to the clay at first produced an equivalent increase in pressure of the water within the interparticle voids. As this pressure began to dissipate during the outflow of water being squeezed from the sample, more of the externally applied load was progressively transferred to the clay particles themselves. It was this interparticle stress that governed the consolidation behavior of the soil. With this, Terzaghi had struck on the principle of effective stress upon which all of soil mechanics is ultimately based.

In terms of Simonton's explanation, this was the ideation of the concept of consolidation. Its elaboration would require expressing the process mathematically and finding the solution. Terzaghi made little headway until brushing up on some of the mathematics of heat transfer Forcheimer had earlier introduced. Another new problem representation was now at hand, this time driven by thermodynamic analogy, and Terzaghi pounced on it. The problem of pressure dissipation in clays could be represented by the analogous process of time-dependent heat diffusion in solids using much the same derivation of the governing differential equations. This mathematical formulation was published in 1924 at almost the same time *Erdbaumechanik* appeared. Terzaghi was then 41 years old. Coming only five years after his first career milestone, the theory of consolidation would be his greatest career landmark and its highest achievement. Its elaboration, however, was almost—but not quite—complete. This would have to wait for an astonishing climax.

Some years passed, and as Terzaghi worked on an increasing number of projects he realized that *Erdbaumechanik* needed to be updated, which he proposed to do with a second edition in 1936. The same was true of his theory of consolidation, and also in 1936 he and long-time colleague Otto Frölich published *Theorie der Setzung van Tonschichten: eine Einführung in die Analytische Tonmechanik* (Theory of the Settlement of Clay Layers: An Introduction to the Analytical Mechanics of Clay), intended to be a practical guide. De Boer, et al. (1996) recount what happened next.

Born in the same year and receiving his diploma in mechanical engineering at the Technische Hochschule in Vienna the same year Terzaghi got his at Graz, Paul Fillunger's interests had long paralleled Terzaghi's. He too was interested in the behavior of concrete and especially how internal water pressures operated in producing uplift in concrete gravity dams. A consummate theoretician, Fillunger was already a professor at the Technische Hochschule when Terzaghi arrived there in 1929 to accept a similar post, having returned to Austria after an unsatisfying stint at the Massachusetts Institute of Technology. It did not take long for the two to lock horns.

Terzaghi, it might be said, unwittingly fired the first shot. He had already addressed the topic of uplift pressures and effective stresses in dam foundations in *Erdbaumechanik* in 1925. By 1933 he had extended this work in Vienna to the internal water pressures within the concrete itself, which varied somewhat from Fillunger's earlier views. Showing Fillunger his manuscript of a paper, Terzaghi proposed that they publish jointly, but Fillunger was unmoved. Terzaghi went to print anyway, and a series of vitriolic exchanges ensued in the literature.

If we go back to Thomas Kuhn and his explanation of how changes in science come about, we see an almost perfect description of what Terzaghi had done. In a revolutionary upheaval of previous thinking, Terzaghi had created a new paradigm. With perhaps one not-so-small exception, and that was the principle of effective stress. Although he had not applied it to soils, Fillunger himself had originated an all but identical concept as early as 1913 as it pertained to the uplift problem in concrete dams. If there had been a previous paradigm it was Fillunger's, and neither were prepared to go quietly into the night just as Kuhn would predict. So by the time Terzaghi and Frölich came out with their new consolidation volume in 1936, they were walking directly into a classic academic ambush (de Boer, et al., 1996).

This took the form of an immediate rejoinder by Fillunger in a 47-page pamphlet he published himself, entitling it "*Erdbaumechanik?.*" Not only was it sarcastic (its description of "Happy-Terzaghi mathematics" pretty much gives the idea), Fillunger also presented his own formulation of the governing differential equations for consolidation, claiming they were impossible to solve. Terzaghi was incandescent at this challenge to his work, not to mention the smear to his personal honor and integrity. It was not for nothing that he wore the dueling scar still from his college days, and Terzaghi demanded that the matter be taken up by the university's Disciplinary Board. It was, and a technical evaluation by a committee of experts completely vindicated Terzaghi's theory. Its elaboration was now complete. As for Fillunger, he was by then out of the picture. Getting wind of the committee's forthcoming decision, he and his wife left their apartment one afternoon in a disturbed and agitated state, returned, and opened the gas jets (de Boer, et al., 1996).

Terzaghi returned to America in 1938 scarcely one step ahead of World War II, this time to Harvard where he headquartered his practice throughout the remainder of his career. We could, if we wanted, go on to catalog Terzaghi's creative achievements as his career progressed, but Goodman (1999) has done this so aptly that nothing we might do here could add much more. Terzaghi's productive output was truly prodigious, as a visit to the Terzaghi Library at the Norwegian Geotechnical Institute in Oslo will attest, and were we to go through his voluminous files, papers, and consulting reports housed there, we could expect to find the "misses" as well as the "hits" that Simonton (1996) would predict. But instead, we fast forward to the 1960s to what would be Terzaghi's last career milestone and his final creative achievement.

By any measure, the published account of the design and construction of Mission Dam (Terzaghi and Lacroix, 1964) contains an engineering saga of epic proportion. Terzaghi began work on the dam in 1955 at the age of 72, and one of his early reports was characteristically candid: It was one of the worst damsites he had ever seen (Goodman, 1999). This said a lot, for he had surely seen hundreds by then. According to Charles Ripley who worked frequently with Terzaghi on his consulting projects in British Columbia, Mission Dam was Terzaghi's "crowning achievement," doubting that anyone else then or now could have built such a dam at its site (Leonoff, 1994).

Of the three basic engineering properties of soils—strength, compressibility, and permeability—the foundation had dire problems with every one and then some. There were two highly pervious strata in the foundation separated by an exceptionally soft and weak layer of normally consolidated clay. The dam's upstream slope would have to be exceedingly flat to maintain stability, but this meant that the clay core it supported would be subject to severe settlement and cracking. Not only that, but the dam's alignment was constrained to that of a small existing diversion structure which the new dam would have to incorporate, with some serious problems of its own. Its cutoff was inadequate and its filters equally so, with a tendency for sinkholes to form in its deteriorated and damaged core with disturbing regularity.

But if the problems were legion, Terzaghi's design solutions were equally heroic and innovative. The upper pervious stratum would require a sheetpile cutoff and the lower one a grout curtain some 500 feet deep. What Terzaghi surmised to be the self-healing character of the zoned fill materials would have to be relied on to reduce the effects of internal erosion. Protection from the inevitable settlement-induced cracking of the clay core would be provided by one of the first applications of geomembranes in a dam of this scale. The sheetpiles would have to be installed within a slurry-filled braced excavation. In one of the first applications of the slurry trench method, hard-hat divers whose names have gone unrecorded were—incredibly—sent down to torch off the bracing struts in complete and total darkness, working blind in the thick viscous slurry entirely by feel.

The diversion tunnel was closed in 1960 and the reservoir began to fill. Terzaghi was then 77. It had been 44 years since he began his career at the Royal Ottoman College in 1916, and he would die just three years hence. This was the final and the crowning landmark achievement of his career, and Mission Dam was renamed Terzaghi Dam shortly thereafter. No other dam save one with the same name in Brazil comes to mind that celebrates its engineer–designer in preference to a politician. The act spoke for itself.

Terzaghi's three career milestones shine through as beacons to his creative achievements. At age 36 he first made his mark with his earth pressure experiments which helped put soils on equal footing with other engineering materials. He was 41 when, with his effective stress principle and consolidation theory, his *Erdbaumechanik* manuscript for all intents and purposes invented soil mechanics. Then at 77 he created the dam that stands in every way as the monument to his career. And for all of Terzaghi's intuition and insight into matters of engineering, he seems also to have possessed an intrinsic notion of how expertise and the creative process work, long before these were elaborated in any formalized way. With deep introspection and self-awareness, he appears to have been very much cognizant of how his own creative expertise was developing and its importance in his life. To mark the year 1923 in which he turned 40, standing on the threshold of *Erdbaumechanik* and all the rest, he reflected that it had not been "a decisive year as far as events are concerned, but a year of decisive preparations" (Terzaghi, 1964). Even then, so close to his pinnacle of achievement, he considered himself to still be cultivating his knowledge in anticipation of things to come. More than that, in this same year he had discovered the meaning of happiness. It was for him, he said, nothing less than "continuous creative activity," the kind of activity that produces the thrill of discovery, and for Terzaghi to equate this with happiness itself has as much to say about what motivated him as anything else he wrote or did. Indeed, his entire career trajectory reflected this creative activity in a way remarkably consistent with what Simonton's model would come to describe. Many are aware of the things Terzaghi did. What we have only begun to appreciate is how he did them.

Terzaghi's career lasted long and it began late too, much the same as the earth scientists' in Figure 7-4. Before concluding, however, that this is typical of geotechnical experts, we might look to another engineer whose career started much earlier, then was cut short by one of the very landmark achievements it produced.

Washington Roebling

The Brooklyn Bridge was not the creation of any one engineer. John Roebling was its designer, Washington Roebling its builder, and Emily Roebling its savior. Before it was finished it would destroy two of the three.

John Augustus Roebling was considered a genius of American bridgebuilding as much as Stephenson was of English railways. A staid German immigrant, to say he was stiff as a board would do injustice to boards. John A. Roebling was the first to manufacture wire rope in America, which led him naturally to a chief market for his product, suspension bridges, and though his were by no means the first in America, it would have to be said they were different. Roebling bridges tended to stay up. Among his better known was the Niagara Bridge, completed in 1855, which survived a number of less fortunate predecessors elsewhere. John Roebling's designs, no less than his personality, emphasized the importance of rigidity, with stiffening deck trusses and diagonal stays his trademarks. But while we could surely track the landmarks of his career, he is not the Roebling we focus on here.

Washington Roebling was by all accounts a modest and unassuming man, more genial and lacking the dark side his father was known to display (McCullough, 1992). Perhaps the son's more approachable persona is why the two were, and still are, so often confused, with Washington saying years after his father's death, "Long ago I ceased my endeavor to clear up the respective identities of myself and my father. Many people think I died in 1869." But John and Washington were always closely entwined in this and other ways. The son became more than an assistant to the father, emerging as engineering confidant and collaborator. This may explain why Washington's career started so early as it built to its greatest achievement—the construction of the most enduringly magnificent suspension bridge in the world, with none since, it is said, standing entirely clear of its shadow. It is historian David McCullough (1972) who brings to us the remarkable account of this bridge and the Roeblings who created it.

Washington Augustus Roebling was born in 1837 in Pennsylvania and was 12 when the family moved to Trenton, New Jersey. At age 15, Washington accompanied his father from there on business to New York in the winter of 1852 and to

Brooklyn on a side trip. In the process the East River ferry became icebound, and with that abhorrence for inefficiency known only to the engineer it was there that John Roebling first envisioned a great bridge in his mind's eye. Washington had been present from the moment of the bridge's conception and so would remain until its delivery.

Within little more than a year, Washington Roebling, now 17, went off to Rensselaer Polytechnic Institute in Troy, then one of the few places an aspiring bridgebuilder could study civil engineering. We cannot claim that he ever became a geotechnical engineer or even studied the subject—in 1854 there was no such thing, let alone in Rensselaer's curriculum. But Washington was always smitten by geology and he excelled in it there, having started what would become one of the world's finest mineral collections (McCullough, 1992). This erstwhile hobby would later serve him deep below the East River when it came time to make the most crucial decision of his career. On the whole, however, Roebling's experience in Troy seems to have been as much a trial as an education, and when he emerged from it just three years later at age 20 he was among only 12 of the 65 in his class who did. His father by then had started work on the Cincinnati Bridge, which Washington would later take over. But in 1858 he was off to the Allegheny Bridge in Pittsburgh to work with his father there. So by the age of 21, his engineering career had already begun in earnest.

Washington Roebling's career was at least as much a product of circumstance as preparation and practice. In his model of career trajectories, Simonton (1996) acknowledges the perturbations that things like war or illness can cause. Terzaghi seems to have tolerated his war experience well enough, though it didn't do all that much for his career. But war seems to have worked in the opposite way for Roebling, and the Civil War produced the first creative achievements for which he became recognized.

Washington enlisted in 1861, quickly becoming an engineer officer on the staff of General Irwin McDowell who headed the stunned Union forces in their initial engagement at First Manassas. Roebling was ordered in 1862 to construct a suspension bride across the Rappahannock at Fredericksburg, which he did with little more than a few reels of cable sent down from the John A. Roebling & Sons mill in Trenton. It must have taken no little ingenuity to build such a structure in the field without proper tools or skilled labor and with Confederate forces at times only five miles away (McCullough, 1972). His Rappahannock bridge had 14 spans, was 1200 feet long (exceeding his father's at Niagara by half again as much), and was completed within only a month. Almost immediately he was ordered to build another across the Shenandoah at Harper's Ferry and yet another at Waterloo. Not long afterward, Union General Burnside, in his tearful retreat from the even more calamitous defeat at Fredericksburg, blew up Roebling's first Rappahannock creation behind him. Confederate Generals Jackson, Lee, and Early would repeat the favor for the others, some more than once after

Roebling had rebuilt them. As the war progressed, portable and quickly erected pontoon bridges would become the method of choice for the Union Army's crossings for reasons that by then had been made evident. But this did not diminish Roebling's achievement in having constructed suspension spans in almost as little time.

Later following the battle at Chancellorsville, Roebling, then a staff officer for McDowell, began going up in a hot-air reconnaissance balloon every morning to observe Confederate positions. He was the first to discover on one such occasion that Lee was beginning to move in the direction of Gettysburg. Roebling, as well as the vanguard of Meade's Army of the Potomac, would be there when Lee arrived.

The battle at Gettysburg was Lee's high water mark and the turning point of the Civil War, with the taking of Little Round Top by Union forces widely regarded to have been the crux of the engagement's outcome. Meade ordered General G.K. Warren, Engineer-in-Chief of the Union Army, together with Roebling, to perform a reconnaissance from the summit. Arriving there, they saw the hill to be occupied by only a sprinkling of Union troops and instantly recognized that without immediate action to secure it, the Union line would be exposed to lethal flanking fire from above that would soon make the entire army's position untenable. Roebling remained there to await the reinforcements Warren went to dispatch, and helped muscle the first crucial piece of artillery to the summit just as the Confederate forces arrived (McCullough, 1972; Ward, 1990).

Roebling was not only a participant in the war but a keen observer of it. His military duties as an aide to several high-ranking generals and his encounters with Lincoln (whom he described with some bemusement) on two occasions produced some strong views on the qualities that counted most in a leader. Among them, he said, was "the intuitive faculty of being at the vital spot at the right time" (McCullough, 1992), a fairly apt description of his own experiences at Little Round Top. Roebling, it seems, had already developed a considered appreciation for the expert's ability to anticipate, along with the intuition this required.

Washington Roebling's bridges and other engineering works during the Civil War were his first career accomplishments achieved entirely independently of his father, and we consider them here together as his first career landmark, beginning with the Rappahannock bridge in 1862 when he was 25. If we look back from this vantage point, we can see that his preparation and practice had indeed started 10 years earlier that winter day on the Brooklyn Ferry. It is hard to say whether the recognition Roebling won during the war was due more to his engineering prowess or his military gallantry, but the two became largely indistinguishable. Roebling himself was not much impressed by the war's handiwork, writing at one point to General Warren's sister Emily, "They must put fresh steam on the man factories up North. The demand down here for killing purposes is far ahead of the supply..." (McCullough, 1972). Even so, Roebling would be awarded three brevet promotions, ending the war with the rank of Colonel and the title by which he would thenceforth be known.

He returned to Trenton in 1864 to marry his correspondent Emily Warren, who would now be surrounded entirely—brother, husband, and father-in-law—by civil engineers. So it was only natural that when Washington and Emily left for the grand tour of Europe in 1867, it would be more than a belated honeymoon trip. Primarily it would be an engineering excursion to gather information for the Roebling project then taking form.

Most of the structural aspects of the Brooklyn Bridge had already been worked out on John Roebling's bridges at Niagara, Cincinnati, and Pittsburgh, and the technique of cable spinning was by now something the Roeblings knew well. Instead, the bridge's success or failure would hinge on the foundations, and these in turn on something not done before on anywhere near this scale. To find out about pneumatic caissons and excavating within them under compressed air was why Washington and Emily had gone to Europe at his father's behest, and by the time they returned Colonel Roebling would know more about the subject than any engineer in America (McCullough, 1972). In this, the Roeblings were unanimous in their sense of what was most important for the bridge's success. It would not be so much the gargantuan gothic towers, or the cables, or even the foundations themselves. Rather, it would be getting these foundations in place. Sinking the caissons to establish the bridge's foundations would be Washington's part of the job.

A French engineer by the name of Triger had been the first to sink an iron cylinder in saturated sand for a mine shaft in 1841, using compressed air to combat groundwater inflow and quick conditions at the base long before Terzaghi would discover the theoretical basis of the problem at Roberts College. Ten years later, British engineers William Cubitt and John Wright sunk a similar caisson for the Rochester Bridge. The renowned bridgebuilder Brunel was using a wrought iron caisson to reach bedrock for the central pier of the Royal Albert Bridge in Wales at the time of the Roeblings' visit. And even as Roebling was becoming familiar with caisson techniques, another authentic engineering genius, James B. Eads, was planning his own for his bridge spanning the Mississippi at St. Louis.

But all of these were essentially just iron or steel boxes meant to penetrate only the softest of sediments, and for Roebling this would not answer. The massive size and weight of the tower it would support after filling with concrete meant the Brooklyn caisson would have to be 168 by 102 feet in plan dimension, with that on the New York side even larger, a size no pneumatic caisson would surpass for more than a century—until 1990 to be exact. Consequently, side friction to be overcome during sinking would be enormous. For the caisson to penetrate to bearing grade would require that it sustain tremendous gravity loads of up to 30,000 tons, the equivalent of a modest battleship sitting atop it. It would therefore be constructed from yellow pine 12 × 12's, its roof 15 feet thick and its walls 9, tapering to six inches at the bottom and fitted with a cast iron cutting shoe. Inside, internal walls two feet thick would divide the chamber into six work compartments housing laborers who would excavate the internal area by hand.

Much like today's offshore oil platforms, the massive structure would be launched on a shipway and floated into place. While men and supplies would pass through an airlock, it also contained an ingenious innovation and the first of its kind. Excavated muck would be lifted through a "water shaft" using a column of water to balance the air pressure inside. Figure 7-5 shows these features of the Brooklyn caisson from a contemporary illustration. Its scale can be appreciated from the masons placing the limestone blocks that would rise to form the base of the tower, with the laborers excavating inside. Still, what cannot be appreciated is the misery of the working conditions or the unseen dangers they would entail.

By 1869 preparations were well under way, and on June 28th John and Washington were completing the final survey for the Brooklyn pier at the Fulton Street ferry slip. In what at first seemed a minor mishap, John Roebling stepped back on a fender as the ferry came in and caught his foot just as the boat landed, crushing the toe of his boot. He went right on shouting directions until he promptly collapsed. Three weeks later John A. Roebling, the master builder of bridges, was dead from tetanus—the first victim of the design he created.

The great bridge was itself suspended in the balance, its backers as distraught as the family. Washington was really the only choice to carry on, though some wondered privately if he was up to the job. Yet as Washington recalled years later, he knew the project as intimately as his father did and was the only man living with the experience to spin the cables. But just as important was his trip to Europe and the knowledge of pneumatic caissons he brought back. So in 1869 Colonel Washington A. Roebling became Chief Engineer of the Brooklyn Bridge at all of 32. There were at that time only some preliminary sketches, no working drawings and "nothing fixed or decided," as he would later acknowledge (McCullough, 1992). The bridge as he visualized it existed almost entirely in his head. But it was less this image than the problem of constructing it that concerned him. And here, his problem representation went to a far deeper level, well below the bottom of the East River in fact.

The Brooklyn caisson would have to be sunk through boulder till. At first, each rock under the cutting shoe was removed by hand and hydraulic jacks. But then they had to be blasted, an operation whose effect on the compressed air containment and the men inside was highly uncertain. It was not just this danger but the swirling powder smoke, the sucking mud, the suffocating heat, and the stench of the riverbed slime that made work in this airtight hell subject to a complete turnover of each 112-man shift on average every three weeks—despite the daily pay of two dollars. For each man who quit there were newly arrived immigrants lined up to replace him, but as the workers—and Roebling—were to learn, there would be problems more serious than any of these.

FIGURE 7-5. *The Brooklyn caisson.*

Source: *Harper's New Monthly Magazine.*

One Sunday morning, the only day with no workmen inside, a blowout of compressed air erupted in a volcano of water, mud, and rock 500 feet into the air, raining its ejecta on ships and warehouses for blocks around. Certainly the workmen were grateful for such Sabbath good fortune, an inducement to piety among them, no doubt. Roebling was too, but for reasons of a different kind. The caisson had survived the impact of its 18,000 tons dropping 10 inches as the supporting pressure was abruptly released, albeit with boltheads sheared off and two inches of compression over the nine solid courses of timber. Thankful for the structure's survival but also for the design information provided, he was able to back-calculate that the caisson had a factor of safety of 4.6 with respect to its ultimate design loading (McCullough, 1972).

But the trials weren't over yet. Roebling had always been anxious about fire in the pressurized atmosphere and had gone to great lengths to control flames from candles and lamps. But during a change of shifts, it appears, a workman left a candle close enough to the oakum roof caulking to set it smoldering. By the time the fire was discovered the next morning it had burned a cavity through 15 feet of timber in the caisson's roof now supporting 28,000 tons of limestone blocks overhead. It was clear that the caisson would not survive the impact of another rapid depressurization in its weakened condition. With a few hand-picked men, Roebling entered his burning caisson. When he emerged some 20 hours later it seemed he had personally saved it, almost a repeat of his performance on Little Round Top. He was in a state of collapse and had to be pulled up through the airlock, with his condition put down to exhaustion and smoke.

A few hours passed and Roebling had to go down again. Carpenters, boring four feet into the overhead timbers, had discovered a glowing mass of live coals. In a last-ditch effort to save it, Roebling on the spot and without a moment's hesitation ordered the caisson to be flooded immediately. This was no easy matter, and even with pumping from a fireboat, two tugs, every fire engine in Brooklyn, and three more brought over by ferry from New York, it would take another five agonizing hours for the water inside to reach the smoldering roof. The Chief Engineer remained there, outwardly calm, saying only that everything would turn out satisfactorily although the work would naturally be somewhat delayed (McCullough, 1972). It would later be found that the fire had meandered through the timbers like ant tunnels, with the charred material mined out by hand and replaced with dental concrete.

The Brooklyn caisson would reach bedrock at 44 feet two weeks later, and filling with mass concrete began. During the pour, with the caisson still pressurized, another blowout occurred, this time with Roebling among those trapped inside. With water rising to their knees in a matter of seconds, he was able to restore the pressure by clearing a blocked supply shaft door. But by now it was almost as if there were something about his Brooklyn creation that had never wanted him there.

There were two caissons, of course, and the New York one would have every appearance of treating him more kindly. The riverbed soils were different here, mostly sand with some gravel and few of the boulders afflicting its twin. This would make for much easier excavation, this time with another innovative system to blow the fine muck out through pipes by compressed air in a better-controlled version of Brooklyn's Sunday spectacle. But the New York caisson would have to descend 110 feet to bedrock, an unprecedented depth under correspondingly unparalleled air pressures. The work proceeded quickly, almost dull by comparison, then at 68 feet progress all but halted. The point of a crowbar could still be hammered into the silty sand, but just barely. The first rock appeared at 75 feet, but soundings revealed its surface to be sloping and highly irregular. With Terzaghi's soil mechanics still far on the horizon, there was no way to tell if the whole New York tower, should it be founded at this depth, would settle differentially and tilt, or worse yet slip off to the side. But to go any deeper would mean many more feet, and months, of unendurable blasting.

Roebling had to use his judgment, and McCullogh (1972) notes that his entire professional career, not to mention the bridge, would rest on its outcome. Relying now on his former geologic diversion, Roebling observed that down to 60 feet they had encountered the occasional remains of domestic animals and shards of brick and pottery. This material was clearly alluvial and recently deposited. The soil underlying it, however, was dense as concrete, geologically much older and glacially overconsolidated, with attempts at a crude penetration test battering an iron rod to pieces. The New York caisson, he declared, would rest where it was.

With the second and last great caisson in position, this part of the bridge was completed in 1872. Its foundations were Washington Roebling's greatest creative achievement, his second career landmark, and it had come at the age of just 35. There can be no doubt that this was an innovation the likes of which had never been seen and whose story astonishes still. The only thing to top it would be the bridge's completion, but this would be hailed by others much more than by Roebling himself.

The New York caisson had done its insidious work on Roebling. He had first been hit by a temporary paralysis after emerging from the Brooklyn caisson that day of the fire, and some thought he had never been quite the same. The malady, or signs of it, had been observed in Triger's first caisson, and by now it was called simply "caisson disease" or "the bends." The workmen were mostly all affected, few being much inclined to linger as they exited the airlock at the end of a shift. It was known to be related to compressed air in some way, but not exactly how. Eads' men working in St. Louis had been similarly afflicted, with symptoms including cramps of such rapid and debilitating onset as to be likened to being struck by a bullet. Indeed, the first death in the New York caisson had occurred at a depth of 71 feet as pressures increased, and the prospect of continuing mortality had figured in Roebling's decision to terminate it where he did.

Yet another attack struck Roebling during the pour of the New York pier, this one much worse, and he collapsed in excruciating pain. He was not expected to live until morning, as he lay near death in the same house where his father had died, with little hope of recovery. But he gradually strengthened somewhat, with the public told nothing of this, though the attacks of paralysis kept recurring. As it happened, it was in the same year of 1872 that French professor Paul Bert discovered that the rate of decompression was what caused nitrogen bubbles to form in the bloodstream, but this would come too late to help Roebling and the others.

It would be 11 more years until the Brooklyn Bridge was officially opened on May 24, 1883 in a pandemonium of celebration not seen since the driving of the last spike was telegraphed cross-country the same year the bridge had been started. Washington Roebling would remain debilitated through much of his remaining years, physically and emotionally. Charges of scandal and corruption, like those accompanying the completion of the transcontinental railroad, were the order of the day, and they sucked him down into a quagmire of Tammany intrigue far deeper than anything ever penetrated by his caissons. Even would-be colleague Eads attacked and then sued him over an airlock chamber detail Eads claimed to have invented, in a series of exchanges scarcely less vitriolic than Terzaghi and Fillunger's.

During much of this time it was Emily who took over day-to-day construction oversight of the bridge and she more than anyone, truth be known, who completed it. She had lived with the bridge, and with the Roeblings, long enough to have become as familiar with it as she was with them, and it was she after all who had accompanied her husband in Europe in the quest for knowledge of caissons. Even though an invalid and bedridden, however, Washington continued to direct the work, as legend has it observing it through the window of the house where he lived, with Emily at his bedside as indispensable field liaison who at that point understood the work as well as he did. And as McCullough (1972) points out, Roebling, as we have come to expect of an expert, had always been possessed of remarkable clarity of observation and recall, with an uncanny ability to effortlessly commit vast quantities of information to memory that went all the way back to the meticulous notes he constructed after each day's European observations. The bridge still existed in his head as much as it did over the East River by then, with Washington tracking details of its progress that Emily reported. This was all regularly documented through his equally prodigious writing skills, that other characteristic of experts, with his annual reports to the bridge trustees still standing today as emblems of clarity and thoroughness (McCullough, 1992).

Through all of this Roebling's mind seems not to have been affected in the slightest, and he continued to do everything mentally even if no longer able to work it out on paper. But by now having established that Roebling was an expert in every sense that matters, we would easily imagine that it was less what he viewed through the window that guided his work than his powers of visualization and what he was seeing in his mind's eye. Washington Roebling would live on until 1926 as the hero of the Brooklyn Bridge indistinguishable and indistinguished from John Roebling. Assuming the mantle of his father or becoming him, either way it was all the same to the public. After the bridge, he would never actively practice engineering again. He returned to John A. Roebling's Sons where he would turn out wire and cable to be used, among a great many other things, on the Panama Canal.

As career trajectories go, Roebling and Terzaghi are a study in contrasts. Roebling began his career much earlier at the age of just 21, as opposed to Terzaghi's 33. The first career landmark followed quickly for both, with Roebling reaching his at 25 and his greatest creative achievement by 35. But Roebling's career was cut short, and he was a casualty of the bridge no less than the 27 who died in its building. No one can know what his crowning achievement might have been.

Roebling was by all accounts bright but not brilliant, without the same creative potential (Simonton's coefficient *m* we might say) as his father, or Terzaghi perhaps. Even though he shared with Terzaghi the expert's memory, powers of observation, and writing abilities, the two engineers could not have been more different in temperament, with Roebling as understated as Terzaghi was brash. Nor do we see in Roebling the same sustained period of preparation. The Civil War interrupted it, but he had already absorbed much from his father by then. Indeed, his career trajectory was propelled always by circumstance: being born to John Roebling, the happenstance of war, his father's death, then his own incapacitation. But what he might have lacked in creative potential, he more than made up for in raw physical courage which caused him to act instinctively and decisively at the crucial moment, and in persistence which allowed him to continue through years of agonizing pain. Washington Roebling set out to build a bridge. It is far from clear that he ever set out to be an expert, but he indisputably became one just the same.

Well over a century later in 2001, the Brooklyn Bridge stood tall, providing sanctuary to Lower Manhattan on that eleventh day of September with a strength surpassing anything measurable in pounds per square inch. The Roeblings, each and every one, would have been enormously proud of it. We can be just as proud of them.

John F. Stevens

We return now to the place where we started with Stephenson, to railroads and the engineers who built them. John F. Stevens was another such expert, also with great initiative, and the building of railroads constituted his first and last career landmarks. Sandwiched between them was his greatest, but this one involved a canal. It was Stevens' highest achievement, and his career's creative masterwork, to see this canal differently from how so many other engineers before him had seen it. It would be a railroad problem to him.

If Terzaghi was a commanding presence and Roebling genial and modest, then Stevens was a construction hand's engineer, by his own description tough and plain-spoken, an important man who had no need to act important and would much rather get his hands dirty. With hard work, he said, being the only "open sesame" he ever knew, Stevens was exactly the kind of man Teddy Roosevelt would have admired. And Roosevelt did, describing Stevens in classic Rough Rider portraiture as a "rough and tumble westerner ... a big fellow, a man of daring and good sense, and burly power" (McCullough, 1977). And though not the sort of thing that a man like Stevens would have easily admitted, he was a creative innovator and a pathfinder in every sense of the term.

John F. Stevens was born in 1853 on a small farm in Maine just 23 years after the maiden run of Stephenson's Rocket. Starting out as a teacher, he figured engineering would suit him better and took up surveying, first as rodman then instrument man, eventually arriving in Minneapolis as assistant city engineer. Along the way he had studied mathematics, physics, and chemistry at night on his own, following in Stephenson's footsteps of self-inspired preparation and training.

It was there in Minnesota in 1875 that John Stevens first found railroads, or they found him, and this is when we can say his career truly started at age 22. Shortly he left there for Texas and the Subine Pass & Northwestern. But as Stevens headed south the railroad did too, with a change of financial direction that left its youthful Engineer in Chief back working as a trackhand. From there his résumé would read like a train of passing boxcars: the Denver & Rio Grande; the Chicago, Milwaukee & St. Paul; Canadian Pacific; Duluth, South Shore & Atlantic; Spokane Falls and Northern; and from Assistant to Division to Principal Assistant Engineer.

In what has been described as the turning point of his career and the most important decision of his life, Stevens signed on with James Hill in 1889 to build the Great Northern. At that time every line had a nickname, a kind of license plate slogan emblazoned on its crack passenger express. The Great Northern's was a double-entendre: "The Route of The Empire Builder." The Great Northern's route to the Pacific would be entrepreneur and magnate James J. Hill's path to personal empire. Problem was, in 1889 no one knew how it would get there.

A most unusual hotel stands today off by itself, concealed in a stand of lodgepole pines on the outskirts of tiny Essex, Montana bordering Glacier National Park. It serves mostly railroaders, and from a picture window behind the lobby a guest can watch out back while throbbing diesel-electric helper locomotives are coupled to trains for the push over the Continental Divide—or go couple them as the case may be. In the cracking cold of the winters here a few hardy cross-country skiers stop by, and it was during just such a winter in 1889 not far from here, on foot and alone and at forty below, that John Stevens discovered Marias Pass and with it the Great Northern's summit passage through the Rockies. Stevens said he kept from freezing to death that night by stomping back and forth in the snow until daybreak. This is the stuff of Robert Service. It is how Northwest legends have always been made, and Stevens became one—the "Hero of Marias Pass" he was called. This was his first creative achievement, the one that first brought him recognition, and he produced it at age 36.

This was truly a landmark year, a remarkable one in the career of John Stevens, who then went on to discover another pass through the Cascades later dedicated to him, with his name appearing today on maps of the world. And it would not be long until he would go one better by building the 2.63-mile long Cascade Tunnel beneath it. But more important to Stevens was that this earned him Hill's unwavering respect and gratitude; the two were in many ways peas in a pod, with Stevens always reciprocating in kind. Along with this came a promotion in 1895 to become the Great Northern's Chief Engineer, with Hill saying, "He is always in the right place at the right time and does the right thing without asking about it"—the description of an expert's anticipation to be sure. But Stevens left Hill in 1903 for the Chicago Rock Island and Pacific and a chief engineer position there. In what would become a curiously and uncharacteristically circumspect pattern, he never said much of his reasons, only that Hill was planning a higher post for him that would have required "a diplomacy which I was temperamentally unfit to exercise" (McCullough, 1977). With the expert's self-knowledge, Stevens was clearly a man aware enough of his own capabilities to know what he was good at and what he was not.

A canal across Panama had already ruined all who had come near it, striking down thousands of laborers with yellow fever and governments of nations with its politics. Engineers had fared no better. In again providing us with the story of the Panama Canal, David McCullough (1977) calls it a path much like Stevens had already charted across mountains, a "path between the seas."

In 1879, Frenchman Ferdinand de Lesseps had come to see Emily Roebling in Brooklyn, ostensibly to inspect construction of the bridge but more to drum up enthusiasm, support, and of course money back home for his own daring project in Panama. In reality a charismatic promoter, de Lesseps was billed as Europe's reigning engineer for having opened the Suez Canal a decade before. With the bridge at the time half completed, he announced that the cut for his sea-level canal would rival the height of its towers. With such inducements, the stock issues his backers floated in Paris were irresistible to flocks of small investors—the dot-com public offerings of their day.

There were many things that would ultimately defeat the French, not all of them technical. There were the incessant rains; the slides in the cut at Culebra that grew increasingly worse; the disorganized efforts of dozens of contractors, each with their own excavation techniques and equipment of such uncoordinated disparity that McCullough (1977) calls it "a mechanical Noah's Ark." Then there was the earthquake that shook the confidence of investors an ocean away and the fearsome toll from disease they were told nothing about. Yet ironically and perhaps most of all was the frenzied obsession with digging. This would seem to be plain enough; excavation was what building a canal was all about. But without ever having any well-conceived plan for the mountains of soil, rock, and mud it produced, spoil would be hauled to some convenient valley in dinky dump-car trains with little regard to stability or blockage of drainages. Dump sites would quickly turn to seas of flowing mud and crevasse fields of moving scarps, bending tracks into a giant roller coaster and suspending the entire operation until both dump and trackage could be reconstructed.

By 1885 things were falling apart and even de Lessep's charismatic charm could no longer keep them together. In 1889—with Stevens still high in the frigid northern passes—the company went under, taking with it the savings of almost a million investors. The scandal brought down the government of France, where the ill-fated enterprise would no longer be called the Panama Canal but the "Panama Affair." Yet if the canal could destroy governments it could create them too, which Roosevelt managed to do with the "insurgency" covered by American gunboats that carved out the Republic of Panama from Colombia and allowed the United States to take control of construction.

If it was his charge up San Juan Hill that had first made Theodore Roosevelt famous, then it would be his image in the maw of that dark, looming steam shovel in Panama, teeth gleaming as white as his linen suit, that as much as anything made him immortal and bigger than life. Still, mindful of his edict to "make the dirt fly" the first American attempt at the canal started off no better than the French, and to much the same effect. With a U.S. government commission now in charge, digging would again be the sole priority, though not just through mountains but through mountains of paper as well. It was said that a carpenter needed a signed permit to saw planks over 10 feet long, and two weeks' payroll required filling out 7500 forms. This all made the proper equipment a nightmare to get from the States: A shipment of 15,000 new doors managed to arrive in Panama—followed by 240,000 pairs of hinges (McCullough, 1977). But despite all these procedures, or perhaps because of them, the construction efforts were undertaken almost randomly and there remained no comprehensive plan for the work. Moreover, the laborers were ill housed, ill fed, and thousands just plain ill from the malaria and yellow fever still rampant.

When Roosevelt could no longer abide this bungling he characteristically took charge himself. Who better then to turn to for advice than that other great builder of empires, James J. Hill. There was no better construction engineer in Hill's estimation than John F. Stevens, and with this Roosevelt appointed Stevens Chief Engineer of the Panama Canal, assigning him complete control of the work and reporting if need be directly to the president himself. Stevens insisted on this authority and the responsibility that went with it, observing later (with George Stephenson smiling approvingly over his shoulder) that "such a policy encourages initiative, which is a most valuable asset to an engineer" (Stevens, 1936). As for Roosevelt, his great vision and the country's prestige were both on the line, doubtless for him in that order.

It did not take much of an expert to appraise the situation in Panama in 1905. Roosevelt put it bluntly "a devil of a mess," with Stevens little more impressed, facing he said on his arrival "about as discouraging a proposition as was ever presented to a construction engineer" (McCullough, 1977). What he did first was simply to observe and do it firsthand. He was out daily, slogging through the mud, swinging up onto a passing switch engine, stooping to inspect machinery, in his bib overalls and high rubber boots distinguishable from the laborers only by the signature cigar he smoked or chewed ("Big Smoke" they called him). He said very little except to ask questions about everything he saw, having quickly arrived at the expert's solution it might be surmised, and now taking time to verify it. But privately he was appalled, once looking out over the Culebra Cut to see seven trains derailed and every steam shovel in view standing idle. He watched as the entire available labor force struggled to get the trains back on the tracks, calling this "an unwise proceeding, for they were of more value where they were" (Bennett, 1915). Informed, however, that there had been no collisions in over a year, without missing a beat this taciturn Maine downeaster laconically remarked that even train wrecks had their good points. "It indicates there is something moving," he said (McCullough, 1977). And, true to form, shown the grand plans for the chief engineer's residence, Stevens wanted nothing to do with it, preferring instead a tin-roofed house at Culebra close to the work and his men.

What Stevens quickly realized within those first days was that the problem was not what the French or the Americans who followed thought it had been. The Canal altogether was not a question of excavation; this was but one part and a small one at that. It was a project management problem, a problem of systematically and efficiently organizing the work. And this work was itself a materials handling problem—a problem of moving food and supplies for thousands of men, of hauling materials to build proper shelter, of bringing in equipment and bringing out dirt on the return. This meant transportation, and that meant a railroad. So where the others had seen trenches Stevens saw tracks, and this would lead to the greatest creative achievement of his career. It came about because he brought with him to Panama exactly what an expert would bring: a problem representation that conceived things on a deeper and more fundamental level. With it, Stevens was the first to diagnose the true problem. "The digging," he pronounced, "is the least thing of all."

The Chief Engineer's first official act was to do the unthinkable—he shut down the project completely. He sent craftsmen and steam shovel operators to wait back in the States, giving Dr. William C. Gorgas just four months to finish ridding the isthmus of mosquitoes, yellow fever, and malaria. Stevens himself would be making his own preparations during this time, both for the work and

FIGURE 7-6. *Stevens' canal—the railroad shops at Paraiso.*

Source: National Archives.

for the barrage of criticism sure to follow, saying "...as long as I am in charge of the work ... I am confident that if this policy is adhered to the future will show its absolute wisdom" (McCullough, 1977). His preparation for digging would be, of course, a railroad.

Stevens set out to build a real railroad and all that went with it: housing, repair shops and warehouses, even a cold-storage plant for perishables and that most precious of jungle commodities—ice. And he did it to Great Northern standards, with double tracking and heavier rail, strengthened bridges, switchyards, and a hundred new freight locomotives. As Stevens' vision of the canal began taking shape, it materialized in the form of trainyards like the one in Figure 7-6, the very image of ordered readiness. With construction superintendents and trainmen brought down from the Great Northern, Stevens in the next five months would have 24,000 men at work on his 50-mile railroad, more than the Union Pacific in its race to Promontory. With all the previous delay, some questioned yet another diversion from digging. But from a construction expert's perspective, Stevens did not find much out of the ordinary, maybe in scale but not concept, so the solution was quite obvious to him. The canal was in many ways simpler than his previous projects, just bigger: "...the problem is one of magnitude, not miracles," he said. If there was anything miraculous, though, it was again the way Stevens visualized the problem. He could see through the canal to the railroad beneath, with a problem representation that had eluded the others.

When excavation resumed there was yet another dramatic change in how Stevens viewed the problem that McCullough (1977) calls a revelation. The canal had long since been premised on a sea-level route, although the alternative merits of a lock-and-dam system had been debated at least since de Lessep's trip to New York. But with his railroad in place and working efficiently, trains loaded by the 102 steam shovels could roll out of the Culebra Cut on their own by gravity according to plan. Stevens could now haul dirt anywhere he wanted, literally cross-continent from coast to coast, and once it was excavated and loaded the distance made virtually no difference in cost. With this in mind, a lock-and-dam system began to make much more sense than gouging away at a sea-level route. We might speculate that for Stevens this was a quite natural analogy: Locks could gain elevation on water much like his switchbacks did on Marias Pass. And why waste the excavated material when it could be used for an enormous dam to create Gatun Lake, reducing the depth of excavation still further and with it the nightmarish Culebra slides. This was Stevens' second great change in problem representation, and after supporting the plan before Roosevelt and Congress, it brought the canal to fruition. The year was 1907, Stevens was 52 years old, and this was his finest hour. The highest creative achievement of his engineering career had been accomplished, the plan and everything needed to achieve it were in place, and the canal could be completed by others.

Stevens always maintained that his agreement with Roosevelt was to stick with the job until its success was assured. The time had come to go, but intense speculation accompanied his resignation to Roosevelt, who appointed U.S. Army Colonel George Goethals to replace him. Stevens would deny all rumors, saying his reasons were purely personal. And just as so oddly before with Hill, he vowed never to say what they were, nor did he. Stevens' last week in Panama was devoted to departure speeches as the conquering hero of Panama, an accolade he could now add to that of Marias Pass. In one of the early histories of the canal, Stevens would write that he had turned over to Goethals a "well-planned and built machine, one that was running fairly smoothly" (Bennett, 1915). And he meant no disparagement, he said, in noting that "a good executive, with ample experience in construction, possessed of a clear head and a strong arm could have turned the crank...," which Goethals himself endorsed. But Roosevelt saw Stevens as the worst kind of failure: a quitter, the commander abandoning his army. McCullough claims that the president harbored no resentment toward Stevens. Yet in the section on the canal in Roosevelt's sweeping autobiography, his name would never appear—an empire builder's petulant payback to his master builder it would seem. Roosevelt, however, would not be the last American president to call on John Stevens.

Stevens' career was far from over, and he returned to railroads as vice-president of the New York New Haven and Hartford before rejoining Hill to build yet another, the Spokane Portland and Seattle. But in 1917 a new opportunity would appear, one that would lead to his crowning achievement and his last career landmark—on the Trans-Siberian Railway. In a little-known episode right out of *Doctor Zhivago*, this plainspoken construction engineer would be caught up in political intrigue that would color America's relations with Russia for years to come. Few today are aware that this country once had an occupying force there. John Stevens, a man by his own reckoning having no truck with diplomacy, would soon find himself up to his neck in it.

The Trans-Siberian Railway was then and remains today the single most important man-made feature in Russia, linking its endless spaces and extending some 4700 miles across five time zones from the Urals of eastern Europe to Vladivostok on the Sea of Japan. In 1917, with World War I not yet over, German U-boats had all but closed the Baltic supply routes to Russian forces supporting the Allied effort. The one backdoor and tenuous link remaining was the Trans-Siberian Railway, where American supplies shipped to Vladivostok could be taken west (Griffin, 1998). That is, until Alexander Karensky established a provisional government after overthrowing the Czar. This was the same Czar Nicholas whose coronation Emily Roebling attended in 1896, and the same coronation where he said of his new position, "It is a job I have feared all my life." And fear it he might. Russia, and the railroad along with it, was now in chaos.

The significance of all this, militarily and strategically, was not lost on President Wilson. His response was to order the formation of the Russian Railway Service Corps (RRSC), purportedly to improve the condition and operation of the railroad at the request of the Karensky government. The choice to head it was obvious by now: John Stevens, whose railroad expertise had accomplished so much in Panama. Stevens for his part responded the same way as before, by calling on the talents of some 200 railroaders of the Great Northern. If these Minnesotans, North Dakotans, and Montanans could do what they had under his direction in steamy Panama, then how much more at home they would be in the frozen mountains and tundra of Siberia. But as the RRSC was preparing to depart St. Paul, Lenin was at that very moment seizing control of Moscow. By the time they arrived, the Bolsheviks would also control the railroad.

When Stevens reached Russia, it must have been complete déjà vu. Just like when he arrived in Panama 12 years before, the railroad operations were a shambles of bureaucracy and ineptitude. Again it was foremost a matter of organization, with one of Stevens' men writing home to compare the Russian way of running a railroad to the comic opera of South American armies "with fifty generals and ten soldiers" (Griffin, 1998). Within weeks, however, Stevens had the RRSC inspecting track, supervising repairs, and working to improve conditions all along the line. Yet as he did so, a presence obscure even by the standards of the Russian Revolution was in the process of dramatically changing events.

Ever since the outbreak of the First World War, a group of Czech and Slovak prisoners of war together with deserters from the Austro-Hungarian Army (Terzaghi's old outfit) had been cut off and stranded in the Ukraine. Calling themselves the Czech Legion, the aim of this unlikely ragtag group was to return to France to fight the Germans on the western front. But by 1918 any direct route for getting there was blocked by the Germans and now by the Bolsheviks too, ever since Lenin's separate peace with the Central Powers in the treaty of Brest-Litovsk had allowed his more pressing activities to proceed undistracted. The only remaining avenue of the Czech Legion's return was eastward through Vladivostok, and as Stevens was beginning his work thousands of them were strung out along the Trans-Siberian line. In one of those minor incidents pivotal in retrospect, a rock hurled by no one knows who precipitated a riot, and open hostilities now broke out between the Czech Legion and the Bolsheviks. The Czechs quickly threw in with the counterrevolutionary anti-Bolshevik forces led by Russian Admiral Aleksandr Kolchak, with the predictable result of total anarchy all along the line Stevens was still trying to restore. Only it was no longer a restoration job. As the young Colonel Roebling had found, one of the things armies do best is to blow up bridges, and railroads run a close second. What had begun as restoration would now be reconstruction of the Trans-Siberian Railway directly through a no-man's-land of combatants.

Meanwhile, President Wilson, hearing of this, found in it an opportunity. Though no great fan of armed conflict, neither was he enamored of Bolshevism. The nascent uprising could perhaps be nipped in the bud. So under the guise of clearing the way for the Czech allies to reach Vladivostok and protecting American supplies stockpiled there, Wilson in 1918 dispatched an American expeditionary force under Major General William S. Graves to Siberia (Griffin, 1998). Almost incidentally, this allowed Stevens and his RRSC to complete their work, albeit under near-constant gunfire and raids. But Stevens had brought to the Trans-Siberian railroad much the same problem representation of organization and infrastructure he had used so effectively in Panama. By the time they left Russia in 1920, the RRSC had repaired bridges, cleared blocked tunnels, reconstructed depots, introduced modern train dispatching systems, and generally returned the Trans-Siberian Railway to a functioning entity. Nor did it go without notice in some quarters that their departure had overstayed the armistice by nearly two years.

By now Stevens was 67 years old, and he had completed the crowning achievement of his career in Siberia. He would receive the Order of the White Lion from the Czechoslovak Republic and decorations from the United States, France, and Japan. He would go on to still more railroad ventures, though none quite as exciting, and would eventually preside over the American Society of Civil Engineers. And as for the Russians, it is perhaps understandable that they declined to shower him with commendations, but they did show him grudging respect even while questioning his motives in a way that may not have been entirely off the mark. Though Stevens' career landmarks are little known today even among most civil engineers, his place in Russian history is secure. As late as 1979, Soviet historians would note that "the American railroad mission headed by John F. Stevens spared no effort in organizing transportation for the Kolchak 'government'.... To Stevens the 'value' of his Russian Railway Service Corps was in the help it gave to Kolchak" (Griffin, 1998).

John Stevens lived to be 90, and his productive career of creative achievement continued almost until the end. It is said that he summoned his son shortly before, telling him, "The next time you come home I shall not be here. On the mantel are pictures of the only two men who ever influenced my life and I wish you to have them." On that mantel stood the photographs of James J. Hill and Teddy Roosevelt whose empires Stevens had both built, and he who had surely influenced their lives as much as they did his. John F. Stevens' engineering expertise did more than change how things were viewed. It changed the paths of nations.

The Attributes of Engineering Experts

We can go back now through this gallery of engineering experts to see how they conform to the various constructs of expertise presented earlier. We start by considering how they display some of the key characteristics of experts more generally, as shown in Table 7-1. Although the biographical sketches compiled here cannot hope to have captured all of their abilities, Table 7-1 shows that taken individually and together, these engineering experts display most or all of the attributes of expertise displayed by experts across the board. Furthermore, their career trajectories conform to what Simonton (1996) has generalized.

Figure 7-7 plots their three career landmarks—the first, best, and last outstanding creative contributions. As previously noted, Terzaghi's first landmark was delayed and his best came quickly thereafter, while those of Roebling both came earlier in terms of chronological age. Stevens' first contribution came at the same age as Terzaghi's, but his greatest achievement was the latest of the three. Yet despite these individual differences, Figure 7-7 shows that on average their first and best contribution ages are remarkably close to those of the earth scientists in Figure 7-4. The engineers' last contributions, however, were substantially deferred, save Roebling whose career was truncated prematurely. Granted, none of these engineering experts was subject to mandatory retirement, and they were blessed with great longevity. Neither did their careers shift more toward administration at the peak of their productive achievement, as many engineers' would today. Still, this alone does not seem enough to explain why they were able to produce their crowning achievements and maintain their creative productivity so

TABLE 7-1. *Qualities of Engineering Experts*

Attribute of expertise	*Terzaghi*	*Roebling*	*Stevens*
Quick response and problem solutions		X	X
Posing questions	X		X
Self-awareness/self-knowledge	X	X	X
Deeper problem representation/visualization	X	X	X
Change in problem representation/innovation	X	X	X
Observation and memory skills	X	X	X
Writing skills	X	X	
Preparation/hard work/persistence	X	X	X
Anticipation/initiative	X	X	X

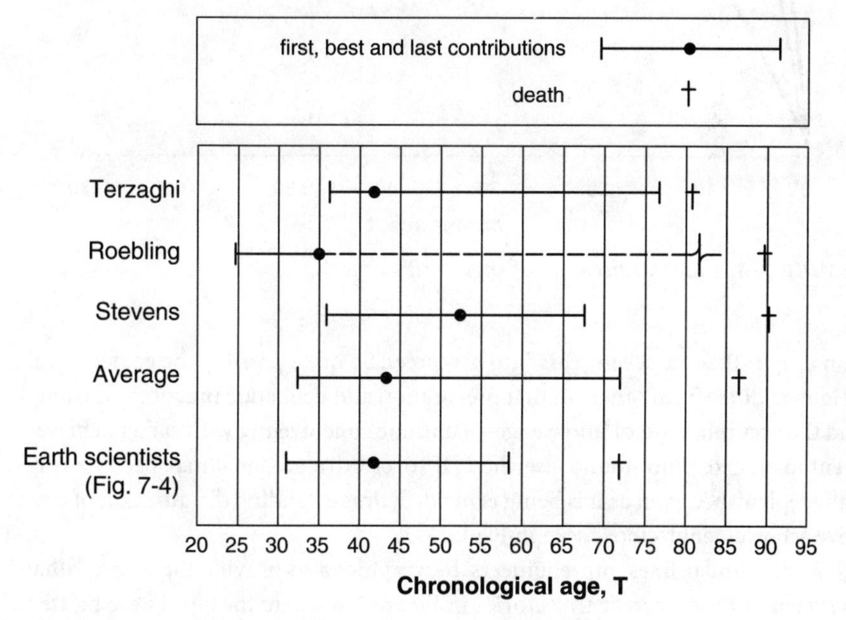

FIGURE 7-7. *Career landmarks of engineering experts.*

late into their careers, almost 15 years after the earth scientists and as much as 20 after some others on Figure 7-4. Terzaghi (1964) has provided us with a clue, and this goes back to Simonton's coefficient *m*.

Recall that Simonton (1996) showed that fields dealing with less-structured concepts like geology tend to produce slower rates of ideation and elaboration that delay and disperse career milestones more widely. But he also posited that each individual's initial store of creative potential, their initial value of *m*, peaked at some maximum value attained at the end of training and preparation, to then be expended as achievement during the ensuing career. Terzaghi himself, however, with his uncanny self-awareness, said that he was still engaged in prepara-

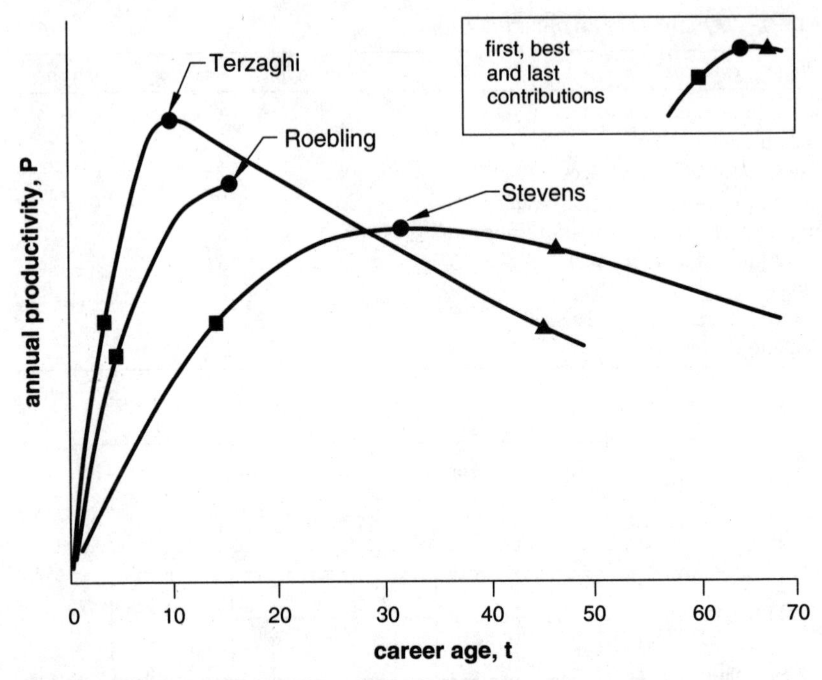

FIGURE 7-8. *Career trajectories of engineering experts.*

tion at age 40, some seven years into his career. So quite possibly the geotechnical field may differ from others in that preparation and deliberate practice—learning and the accumulation of knowledge—continue concurrently with career achievement to an extent not found elsewhere. If so, creative potential may become partially replenished even as it is being expended, thus extending the duration of creative achievement to those seen in Figure 7-7.

Along similar lines, our engineers' biographies also provide the wherewithal to sketch out their career trajectories in Figure 7-8 where the curves are plotted according to career age, t. It is not intended that significance be attached to the relative magnitudes of creative output $p(t)$ shown here—few would be so bold as to compare the Brooklyn Bridge, the Panama Canal, and the Theory of Consolidation on any absolute scale of achievement. But the first, best, and last achievements allow the form of the three curves to be established. Comparing them to the generalized curves in Figure 7-3, we find evidence for both type I (Terzaghi and Roebling) and type II (Stevens) career trajectories. This too suggests that the creative careers of engineering experts conform to Simonton's thesis, although their individual trajectories may differ.

What we learn from all of this has much to do with how we might recognize in the careers of engineering experts the kind of substantive, knowledge-based expertise that underlies the formulation of subjective probability or for that matter any other task that might call for considerable domain knowledge. We ordinarily conceive of expertise as something that grows at a fixed rate over the years as one's career takes its course. But expertise has its hits and its misses, with the greatest accomplishments corresponding to the highest productive output long before the crowning achievement. Productive output, in turn, results from the intense training and preparation that accumulate creative potential in the first place, but just as much from the simple virtues of persistence and hard work that turn this potential into reality, as each of our engineering experts has attested. Simply put then, these experts achieved the most when they produced the most, and they produced the most after they had learned the most. While some part of creative expertise must come from the "rage to master" that resides within, persistent hard work over a lifetime is what produces knowledge and converts it into achievement, what transforms novice into expert. We have found here no secrets and no mysteries, none of Stevens' "open sesames" to release genius from a bottle. The experts we seek reveal themselves in their preparation, their deliberate practice, their hard work. These are the things that enable experts to build and sustain their superior knowledge base, and together we call them experience.

Then too, it is only fair to note that the experts whose portraits are hung here were products of the late 19th century, and as such participants in the most explosive period of technological innovation the world has ever known. From electricity and communications to the miracles of transportation, it was civil engineers who made it possible for ordinary people to make use of these things, turning them from curiosities into truly transforming experiences. Still, McCullough has long cautioned against presuming the inevitability of such historical outcomes. We take them so much for granted it can be hard to remember that nobody knew at the time how any of it was going to turn out. Things could all so very easily have been different—the New York tower might have settled or the canal been abandoned—and they would have, were it not for dedication, determination, and self-sacrifice of those whose sheer force of will made things turn out the way that they did.

So it was these engineers, along with untold thousands of workers, who changed peoples' lives every bit as much as as the inventions themselves. Terzaghi, Roebling, and Stevens stood at the very epicenter of all this as the Bill Gateses of their era. And at the risk of mixing metaphors, such Gullivers of civil engineering were not tied down with conflicting demands from society as they would be today. Yet there is no reason to think that such giants should not still be among us, though perhaps in other disguises. To further seek them out, we turn to an altogether different form of the study of expertise conducted in the most contemporary of settings.

Experts and Decisionmaking

In a certain sense, all expertise involves making decisions of one sort or another. Expert chess players must decide what is the best move, physicians what diagnosis to make, engineers what design measures to adopt. If it weren't for uncertainty, these decisions would follow automatically from prescriptive rules and expertise would hardly be required. Such decisions are ordinarily made with some deliberation after varying degrees of reflection and study, yet just as a timed exam most pointedly puts academic performance to the test, the pressures of time and stress do much the same thing with decisions that experts must make. Examining experts and novices in real-world situations where time is a critical element has become a specialty in the study of experts known as *naturalistic decisionmaking.* Decisions and actions in these settings call for creative solving of unstructured problems under various forms of duress, marked by the following characteristics (Zsambok, 1997):

- Uncertain and dynamic environments with missing or ambiguous information,
- Time pressure that forces actions to be determined and taken quickly,
- High stakes and outcome consequences,
- Ill-structured problems without any single correct solution,
- Shifting or poorly defined outcome objectives, and
- Interrelated decision sequences where the outcome of one defines another.

Even cursory inspection shows that these are not the kinds of situations that easily allow for ponderous cogitation, but nevertheless they characterize the environment of crucial real-world decisions. The experts who make them must act quickly and decisively based on their judgment.

There is one view of decisionmaking which holds that it is, or at least should be, a systematic exercise conducted under rules of procedure applied step by step to determine a course of action. There is also the classic view of systems analysis that provides a systematic structure for decisionmaking, whereby decision objectives are defined along with measures of effectiveness for establishing the degree to which these objectives are met. Alternative courses of action are identified then compared to produce the preferred decision outcome (de Neufville, 1990), and indeed this provides the basic structure for the decision analysis techniques described in Chapter 4. This is also called the *rational choice* model involving selection among various alternatives, though some protest that it would reduce decisionmaking in complex and dynamic environments to a form of grocery shopping on the run (Smith, 1997). It is primarily these kinds of structured and rule-based procedures that artificial intelligence has so diligently sought in emulating the performance of experts. But experts do not always work this way. Take, for instance, Mr. Thomas Kelly.

Vermillion Dam rests on a pervious foundation of fluvio-glacial and morainal deposits so intricate and complex that its geologic maps more than anything resemble works of modern art. With no impervious stratum that a cutoff could reach, the design contained an upstream blanket of compacted silt when Terzaghi was called in during the early 1950s (Terzaghi and Leps, 1958). Despite Terzaghi's misgivings, the blanket did prove to have a beneficial if fortuitous effect on the very high seepage quantities experienced on first filling—the silt eroded into the pervious foundation strata, partially plugging and sealing them (Goodman, 1999). The dam, needless to say, has been carefully observed and closely monitored ever since. And through the years it has developed its own personality and idiosyncrasies as all such structures do. But Vermillion Dam is a problem child, and Tom Kelly is the engineer who had always looked after it.

In time, the dam's owner brought in a team of analysts in artificial intelligence. Mr. Kelly's expertise with the dam could never be duplicated, but perhaps an expert system could be constructed to mimic it somehow. The trick would be to encapsulate his judgment in rule-based form for interpreting the dam's sometimes quirky behavior—an attempt, in effect, to extract his memory chunks and encode them. But the results were less than encouraging and definitive rules were not forthcoming, almost as if Kelly were concealing something. His reasoning was intuitive and hard to explain, even after he showed them the dam and things like the rodent holes and the vegetation that, along with the instrumentation, would color any decision he might make to lower the reservoir should there ever be some emergency that would cause him to question its safety (Rose, 1988).

Clearly, his decision reasoning was not rule based, or at least not on simple if–then rules he could readily articulate. If the time ever came to lower the reservoir he would know it, but he would be using something else. Terzaghi himself not long after he left Vermillion Dam would say that soil mechanics is not entirely ruled based. Noting that the geotechnical field had arrived at the interface between science and art, he used "art" to describe the opposite of such step-by-step reasoning (Terzaghi, 1957). Experts may sometimes disregard ordinary stepwise procedural rules entirely, and to find out more about these mental processes that Terzaghi invoked is the aim of studying what experts like Tom Kelly do in the most difficult terrain of their natural habitat when time and pressure close in.

Rather than conforming to the rational-choice decision model, with its alternatives and systematic stepwise comparisons, experts often engage in what Nobel economics laureate Herbert Simon called *satisficing.* Their decisions and actions may not be optimal, and the course they choose may not be the perfect one in the full universe of such choices. But with limited time and imperfect information, they choose wisely the first option they come across that serves well enough. They find something that works and they do it. The contrasting rational-choice decision process has its place, and experts use it too. A physician might carefully size up various treatment options when helping a patient decide on one. It's just that this is not how the expert surgeon decides what to do while the patient is lying on the operating table.

Neither is it farfetched to consider engineering experts in this light. They may not work in operating rooms or cockpits, but engineers still must confront emergency situations where life is at stake and quick and decisive action is crucial, like the situation Tom Kelly might someday face and as Roebling did more than once in the Brooklyn caisson. And as mining engineer Ira Joralemon observed, the pace of engineering keeps accelerating, with fast-track clients demanding expert decisions on the spot and engineering firms promising to outdo each other in completing an assignment the fastest. Should Tom Kelly ever have to make a decision on Vermillion Dam, it might well be under the circumstances we go on to describe here. And even if engineers today are becoming more caretakers than builders, this makes it all the more important that they know instinctively what to do and when.

Experts in Action

We have already seen how experts think and the attributes they bring to the task. Equally important is what they think about—and what they don't—as revealed in their actions under fire. So what better experts to turn to than those who deal with fire all the time. Fire commanders must make on-the-spot decisions. They must decide where the fire is; how long it has been burning what fuels; whether to try to extinguish it; and if so, which firefighting techniques to use. They must determine whether to fight the fire immediately on arrival or first conduct search and rescue. And all this without endangering their personnel. Klein (1998) shows that they do this not by any rule-based procedure or comparison of alternatives. They spend the bulk of their time sizing up the situation, and once they do there is only one option that enters their mind. The solution to them is obvious. It is arrived at intuitively.

Klein (1998) relates the case of a simple house fire, using the ensuing narrative of the fire commander on subsequent interrogation. Smoke is coming from the kitchen when the firefighters arrive. The lieutenant leads his hose crew through the living room and sprays water on the fire, but it doesn't seem to be having much effect. "Odd," he thinks and tries again, then they drop back to regroup. The lieutenant starts to sense that something is not right. He can't put his finger on it, but it just doesn't feel right being there. So he orders his crew out of the building. Seconds later, the floor where they had been standing collapses: The fire in the basement has just burned completely through.

The lieutenant later attributed this to his "sixth sense" that every expert fire commander develops. He had no inkling that the fire was burning directly beneath them, nor did he know that the house even had a basement. But upon close questioning it was apparent that from the very start he was wondering why the fire did not react as expected. Certain problem cues that should have been there weren't, and some were there that shouldn't have been. The living room was hotter than expected for a fire in the kitchen. And it was too quiet—fires like this usually roar. Though not consciously aware of any of this at the time, he had pulled back to get a better sense of what was going on. There was a pattern of anomalies. And the pattern just didn't fit.

In another case related by Klein (1998), the leader of an emergency response team is called to the scene of a car accident with the driver still inside and unconscious. Other rescuers are already there, and they begin trying to extricate the victim by forcing open the driver's side door with the hydraulic Jaws of Life. As the commander investigates, he walks around the car to find both doors so jammed and crumpled that the operation will be difficult and time-consuming to say the least. But he also notices that the impact has severed three of the posts holding the car's roof. He begins to wonder if it can be removed and tries to imagine doing this: cutting the remaining support, lifting out the driver, how the rescuers would position themselves to support the driver's neck. He has heard of this before but has never done it himself, so he runs through this mental simulation of the sequence again but can't find any problems. He goes through it once more, this time aloud and explaining it to the others. They try it. It works.

In neither case were there any rules or programmed procedures to go by. Nor did the commanders identify alternative courses of action and select one by multiple choice. When the problem became clear (or for the house fire, even before), only one course of action was considered. The solution was arrived at by perceiving the situation and not by studying the options. According to Klein (1998), these are the hallmarks of experts in such circumstances, while novices' actions are governed by procedural rules and deliberative comparisons. The expert can identify a reasonable action as the first and only one considered, using the forward reasoning processes described earlier.

Procedure and Intuition

Dreyfus (1997) illustrates how rule-directed novices move toward expert performance with the stages in learning to drive. There is first the progressive attainment of beginner standing, followed by competence, proficiency, and finally expert status. The novice initially does everything by rote from received knowledge, learning rules like to shift into second when the speedometer reaches 10 miles per hour. They are notoriously slow and uncoordinated while they fumble to remember everything and apply it in proper sequence, as many a grinding clutch will bear witness. The novice progresses to beginner with more experience in recognizing situational rather than just rule-based cues, learning to shift when the engine sounds like it's straining. As a basic level of competence is reached, they begin to pick out which rules apply rather than adopting them all mechanically, and a sense of what is important begins to take hold. When exiting a freeway, the competent driver learns that speed, position, and road conditions are the things to account for in deciding whether to let up on the gas or step on the brake. When proficiency is reached, rule-based procedures are gradually replaced and intuitive behavior takes over: It is instinctively sensed that speed is too fast when approaching a curve on a rainy day. Finally, for the expert, feel and familiarity govern what actions to take without conscious awareness—the foot, not the brain, is what knows and anticipates how far in advance to let off on the accelerator when approaching the off-ramp.

So the acquisition of expert skills is a process of going beyond rule-based prescriptions. It's not that the expert has a better appreciation of the rules than the novice. Experts know what the rules are, of course, and can probably recite them if prompted, but they also know what these rules apply to, what they don't, and when there are no applicable procedures at all. The expert develops intuition that supplements and eventually replaces straight-ahead procedural mechanics. Experts outgrow the training wheels, as it were, though they may well return to step-by-step procedures for problems that are novel to them or for justifying a decision to others (Dreyfus, 1997). So while it is sometimes thought that novices can be trained to be experts by simply posting the same procedural rules in bold letters, what this really requires are different, more intuitive, and knowledge-based skills.

Even when comprehensive procedural rules are in place, experts spend much more time assembling and evaluating information than they do going through the procedures themselves, with air traffic controllers, for example, spending about 90% of their time tracking and understanding the unfolding situation before issuing instructions (Kaempf and Orasanu, 1997). The corresponding actions become simple, even obvious, once the situation is comprehended. Experts also incorporate a greater variety of information, or problem cues, than procedural rules are able to account for. Pilot training places heavy emphasis on procedural decisionmaking: In situation X, take action Y as prescribed by detailed checklists. Many of these decision rules are relatively simple, so much so that to novice trainees they are "no-brainers." For them, the apparent simplicity derives from the rules themselves rather than the situation that prompts their application. But accident investigations have shown that these "simple" decisions can lead to pilot error when the situation gets complex.

The most elementary kind of procedural rules in aviation are those that apply to "go/no go" decisions, as in the so-called Rejected Takeoff decision. As the aircraft accelerates down the runway, the pilot must decide to continue the takeoff or terminate it, depending on whether there is some malfunction prior to reaching a predetermined airspeed. The student pilot is taught to reject the takeoff if an alert occurs before reaching this velocity. More experienced pilots learn that other factors come into play, like aircraft weight, tire pressure, and runway condition. They look beyond airspeed and system status checklists to these other contextual cues, sometimes resulting in actions that run contrary to procedure (Kaempf and Orasanu, 1997). It is for this reason that even expert pilots log many hours in flight simulators designed specifically to mimic the contextual cues that procedures cannot address. In fact, in a National Transportation Safety Board (NTSB) assessment of accidents attributed to "tactical decision errors," two thirds were for the type of "go/no go" decisions where procedural rules are the simplest. This was not because checklists were ignored, but because there were other more ambiguous cues that were not fully taken into account. It was the pertinent aspects of the situation, not the rules, that were disregarded (Orasanu and Fischer, 1997).

The most difficult decision situations call for creative problemsolving. The nature of the problem may be unclear, there may be no applicable guidance, and the situation may have to be stabilized first before it can be corrected. There were no rules or checklists to tell the crew of United flight 232 what to do when their DC-10 experienced a catastrophic failure of the rear engine that severed all hydraulics and caused complete loss of flight controls. But together with a flight instructor who happened to be on board, they were able to improvise a way to crudely steer the plane by throttling the remaining two engines. Though the plane cartwheeled down the runway on landing, their desperately creative actions saved the lives of the majority on board.

Expert pilots consider a broader range of cues and learn when these cues are relevant. This enables them to employ checklists with discretion, while the novice sees them as absolute. The engineering equivalent of checklists are guidelines and "cookbooks" containing stepwise and prescriptive problemsolving procedures simple enough for anyone to use, and with their affinity for rule-based deductive procedures, engineers seem particularly attracted to canned methods in general.

The development of guidelines proceeds in familiar fashion. Some new technology or variant on an old one is introduced by its developers, and as it gains notice others begin to apply it. There comes a point as applications proliferate that the developers start to become nervous. The would-be appliers (for want of a better term we can call them the "amateurs," one step below novices) are producing all kinds of different results. But consistency demands that any procedure seen to be credible yield the same answer no matter who might apply it, so a move toward standardization is required lest it fall into disrepute. For their part, the amateurs want to adopt the new technology but only with assurance that they can do so without untoward liability—an instruction manual to go with the new toy as it were. Guidelines are the eventual result and they would seem to serve both parties, with credibility for the developers and cover for the amateurs.

Guidelines, cookbooks, and prescriptive rule-based procedures are the great levelers. They cut off the abjectly incompetent on one end and the expert on the other without distinction. But more than this, they can prevent real expertise in the technology from ever taking hold. Consider Dreyfus' (1997) progression from novice to expert in driving. The rank amateur will obtain instruction sufficient to reach the novice stage from received procedural rules. But without at some point departing from these rules, the novice can never progress to the stages of competence or proficiency, let alone become expert in the technology's application. The rules do serve as a kind of basic training in turning the amateur into a novice. But the novice is condemned to remain a student driver, trapped there by the same rules of instruction that must inevitably be supplanted but unable to learn how or when. There is nothing in the guidelines to encourage this progression—their purpose in achieving uniformity demands quite the opposite. But the novice is not made aware of this. The guidelines do not, indeed cannot, instruct the novice in the need to go beyond them or inform how to do so. No "student driver" warning appears in big letters atop the car. And guidelines come equipped with no flight simulator to enhance further learning from situational rather than rule-based instruction.

Most sets of guidelines do come complete with the obligatory disclaimers to remind that they be used with discretion and so forth—none of which, of course, the novice has the comprehension necessary to heed. So eventually the day comes when something goes wrong, possibly like the NTSB's "tactical decision errors" from some situation or combination of circumstances the guidelines didn't fully anticipate or something in them that turned out to be inapplicable. Our amateur-turned-novice is hauled into court with an expert witness, perhaps one of the original developers, to aid in the defense. The plaintiff's attorney seizes on the guidelines like a shark on a mackerel. The exchange goes something like this:

> "Mr. Expert, what I'd like you to help me understand please if you would, I'm not too clear on this, is whether you are aware of any, say, rules, or standards of practice, or guidelines that are currently used in your profession in connection with this procedure?"
>
> "Yes, I helped write them myself."
>
> "Good. Then you're familiar with them I take it?"
>
> "Yes I am, very familiar."
>
> "Now Mr. Expert, would you say, with your expert knowledge and your familiarity with these guidelines, that Engineer So-and-so applied them correctly in this case?"
>
> "Well no, but they were never inten..."
>
> "So what you're saying is that Engineer So-and-so didn't follow the rules?"
>
> "No, not exactly."
>
> "So then if I understand, and please correct me on this if I'm wrong, it is your testimony here today that Engineer So-and-so broke the rules, violated the guidelines, whatever you wish to call them, that govern how experts like yourself have determined this procedure should be conducted. [pause]. Thank you Mr. Expert, that will be all."

Despite all good intentions, the guidelines in this entirely fictitious exchange protected neither the technology nor those who applied it. They could never have provided more than basic training, and if anything they did the opposite by inhibiting the development of expertise needed to know when to conform to prescriptive procedures and when to go beyond them. In this, procedural rules seek to transfer judgment by codifying it. This may work some of the time, but it cannot work all of the time because judgment is not procedural. There can be no substitute for judgment, and as well no escaping that the expert has it but the novice does not.

In all of this it becomes apparent that in going beyond rule-based strategies, experts are not just extending them somehow but are doing something altogether different. It's not just that experts are better at reasoning but that they use two entirely different kinds. Moreover, experts have the ability to best match the type of reasoning they use to the characteristics of the task (Shanteau, 1992). Sloman (1996) calls one kind of reasoning *associative*, having to do with associations or loose relationships between characteristics and observations. The other is *rule-based*, using causal factors and relationships in a serial manner. These two kinds of reasoning do different things and apply to different tasks, and experts can more readily distinguish both reasoning and task characteristics, which allows for a better match. Hammond, et al. (1987) call these types of reasoning *intuitive* and *analytical*, respectively. So associative or intuitive reasoning is roughly analogous to the statistical reasoning strategy in forming an uncertainty judgment, while rule-based or analytical reasoning parallels causal strategies. Recall that both appear in the cognitive process of Figure 5-1 and that problem cues are what direct people to one or the other or both. Much like the lieutenant in the house fire, experts recognize these cues from problem content more effectively, so they can better match the kind of reasoning they apply to what the problem calls for. They know which reasoning tools are right for the job.

Hammond, et al. (1987) extend this in proposing that there exists a continuum of reasoning processes between these two end members, with what they call *quasi-rationality* somewhere in the middle and containing elements of each. And to see how good experts are at positioning their reasoning methods in relation to the corollary task characteristics, they called upon an especially interesting group of expert subjects—civil engineers.

A group of 21 experts in highway engineering were chosen for their training and demonstrated ability in intuitive, quasi-rational, and analytical reasoning problems. They were given three tasks. The first involved highway aesthetics and required intuitive cognition. They were asked to view scenes of some 40 actual highways, and their aesthetic judgments were compared to those of 91 members of the general public. Of course, there are no rules for specifying what looks good or even any way to define it. The intuitive nature of this task was illustrated by one engineer's remark: "When you ask me how I did this I will say 'I don't know.'"

The second task involved assessment of highway safety to draw on quasi-rational reasoning. Although there is no established theory or algorithm for determining accident potential, there are some general factors such as lane width, curves, intersections, and traffic levels that are related. In this task, the engineers were provided with the values for 10 such factors and asked to estimate accident rates for each of the same 40 highways. These were compared to actual accident rates averaged over seven years. Here some degree of analytical thinking was involved, but the engineers had to devise their own algorithm inductively and select what parameters to incorporate without any strict rule-based procedures for doing so. As one of them put it, "I will select the most important points that have a tendency to constitute accidents.... Probably the most important thing would be curves per mile.... Shoulder width is the next important thing." So a sense of what was important guided construction of this hierarchy, as we will subsequently see.

Their final task was purely analytical. It was to determine the maximum hourly capacity of each of the 40 stretches of highway under ideal conditions. The engineers were given measured values of key parameters, and their answers were judged according to the results of procedures in the Highway Capacity Manual. This exercise was prescriptive and rule based: "The idea of taking a maximum capacity...and then multiplying it by factors is the Highway Capacity Manual way of doing it," as one of them said, in other words a straightforward and literal application of guidelines.

The results were revealing. Of all people, engineers would be thought to perform best in tasks calling for analytical reasoning, but this was not the case. These experts actually did better on the whole for the intuitive and quasi-rational tasks where judgment is called for, and the ranges of their individual scores widened dramatically in moving from intuitive to analytical tasks, with the worst as well as the best performances produced by analytical cognition. Their errors in analysis were worse than their errors in judgment because when analytical mistakes were made they were big ones.

This contrasts with the argument that intuition produces faulty judgments and should be replaced whenever possible with analytical methods. It all boils down to the task. Through this study, Hammond, et al. (1987) determined that experts achieve expert performance not only because they have a broader array of reasoning tools at their disposal to begin with but also because they know which to use where. In translating this into practical guidance, Hammond, et al. (1987) suggest better awareness of the nature of the task and the kinds of reasoning tools it requires, more consciously selecting those that are right for the job and not opting for analysis in problems that call for intuitive reasoning or vice versa. Also important is balance, with diminishing returns in accuracy when either intuition or analysis is employed exclusively. We have learned this already from Peck, and Focht, and Terzaghi, who advocated and practiced the synthesis of analysis and judgment. Achieving this balance means melding intuition and analysis, using one to complement the other, and this requires a sense of which is important in relation to the problem at hand. So in the end it is judgment that makes all the difference for experts.

Pattern Recognition and Mental Simulation—Seeing the Invisible

In reflecting on experts like Terzaghi and Roebling and Stevens, those who knew them best often comment on their powers of observation. It is curious that this should be called a "power," implying something more than just acuity of visual or mental perception. In a very real sense, experts can look at the problem and see things that others cannot. They can listen to the data and it talks to them. As Klein (1998) puts it, they can see the invisible.

In this brilliantly expressive passage, Robert Pirsig (1974) captures how being "stuck" on a problem of diagnosing a broken motorcycle is overcome by reflecting on what is important, querying judgment with deliberate efforts to watch and listen for what the problem is trying to tell you:

> What you have to do is … to slow down deliberately and go over ground that you've been over before to see if the things you thought were important were really important and to … well … just *stare* at the machine. There is nothing wrong with that. Just live with it for awhile. Watch it the way you watch a line when fishing and before long, as sure as you live, you'll get a little nibble, a little fact asking in a timid, humble way if you're interested in it…. Be interested in it.

Seeing the invisible was the fire commander's intuitive "sixth sense" that told him to get out of the building. A geologist has to "see through the topography" as they say, to the geologic structure beneath. Seeing the invisible is the expert's intuition that engineer Tom Kelly was unable to put down in rules, and Klein (1998) shows that this is no product of metaphysics. Intuition is pattern recognition and pattern matching—the patterns of previous fires, the patterns of stratigraphy and juxtapositions of rocks, the patterns of a dam's behavior. Intuition is the reason that, once the pattern is discerned, the solution becomes "obvious" even though the expert may not be able to immediately explain why. But beyond patterns is another requirement of seeing the invisible, mental simulation—visualizing how the problem has come to be and how it might progress. Through mental simulation, the solution to the problem becomes obvious as well, without going through step-by-step reasoning or examining various options, like the emergency response leader who walked through the process of extricating the accident victim from the car.

With regard to pattern recognition, Klein (1998) identifies the things experts look for in seeing what novices do not:

- Typicality of events and conditions;
- Anomalies and violations of expectations;

- Relative importance of information—what is relevant and what is not;
- Fine discriminations, distinctions, and details too small for novices to detect; and
- Key or pivotal findings that define the problem.

Patterns are the template, the context by which typicality is determined. The expert has experienced, read, or heard about other situations similar to the one that presents itself. Typicality has a frequency connotation: The basic elements of the situation have been repeated enough times for a pattern to have been established and for this pattern to have been encoded as memory chunks. Here again, what we mean when we say that experts have experience is that they recognize and use its patterns. So John Stevens had built enough railroads, goodness knows, to see quickly the pattern that the canal conformed to. Moving materials over mountains and dirt over distances was so routine that the canal was not all that different from the typical pattern of railroads ("just magnitude, not miracles"). All this was obvious—just another day at the office.

The elements and structures of the pattern that define its typicality at the same time specify what its important aspects are. Thus, the pattern tells the expert what to look for and how to quickly pick out what information is relevant and discount what is not, while the novice must sort through it all piece by piece. So patterns, not rules, establish which things are important. And these include anomalies, things that are present but shouldn't be, those that should be but aren't—the dog that didn't bark in the night. Experts notice the anomalies that novices gloss over because their patterns are better developed from their superior domain knowledge and memory structure. This is how experts take note of details and can make fine discriminations. They use the more subtle indicators of conformity or anomaly contained in the richer patterns they have stored.

Largely for this reason, experts are drawn to certain problem cues and not others, the kernel notion or seminal idea that leads directly to the problem solution. The solution is constructed around these key findings, and identifying them becomes one of the most powerful tools experts use. Recall how the physicians in Chapter 5 found these "pivot findings" in groupings of symptoms then used forward reasoning to determine the diagnosis. These findings come from the patterns of experience and are what enable experts to devise innovative ways for attacking the problem no cookbook contains. Such was Terzaghi's epiphany on the rock overlooking the Bosporus—the laboratory devices he envisioned would be key to cracking the consolidation problem. This pivotal finding that ultimately led to its solution was contained in a pattern of previous experiences—he had already tackled the earth-pressure problem in a similar way with his cigar box and shoestring apparatus. And the pivotal finding sometimes comes in that wave of crystallization or flash of insight when its pattern jumps out in discovery. This is why experts ask questions in searching for subtle problem cues that may turn out to be pivotal. As Terzaghi said, he knew he was asking the right questions, even if he didn't yet know the answer.

We now turn to the second aspect of seeing the invisible—mental simulation, the other way patterns are used. Patterns of events stored in memory are encoded as "stories," the prototype narratives that give structure and sequence to how things have taken place and events have unfolded before. We have already seen that causal reasoning constructs a plausible narrative around one or more of these stored prototypes, and Klein (1998) shows that the expert puts this narrative to work by means of mental simulation. This would seem simple enough, yet experts consistently adopt much richer, extended, and more comprehensive mental simulations than novices.

Damage control on board ships requires mental simulation in triage decisions. The threat to the vessel determines what damage has to be addressed now and what can wait, and this threat must be visualized by simulating what could happen. Schraagen (1997) describes an exercise given to two naval damage control officers. A missile has just struck their ship, and they are asked for their responses aloud. The experienced officer relates this narrative:

> There's a missile impact in the officer's mess. OK, let me think. This is close to the Sea Sparrows ammunition storage depot. There will certainly be a fire in the officer's mess, so this fire may spread to the ammunition storage depot, or at least because of the heat it may cause the ammunition to explode. So there is a high priority to contain this fire.

While the less experienced officer comes forth with the following:

> A missile impact in the officer's mess? Then there will be a frying pan in the officer's mess that may fall and catch fire.

The expert officer has stored as a mental pattern the map of the ship and the spatial relationships of its vulnerable components. Superimposed on this spatial pattern is the potentially calamitous sequence of events that could put the ship in jeopardy, and this goes to the underlying nature of the problem. It is not the missile strike per se whose damage is critical but the explosion of nearby ammunition it could set off. The expert sees this, but by contrast the novice sees from the direct hit of a missile only a grease fire in a frying pan.

Geologists may well be the all-time world champions of mental simulation. There are but a handful of geomorphic processes—volcanism, glaciation, fluvial erosion and deposition—that can be observed in a contemporaneous setting in anything remotely resembling their operation in geologic time scales. Yet geology demands that the full range of these processes be mentally simulated in constructing even the simplest of geologic maps. This truly requires seeing the invisible, and what we call geologic observation is not just visually recording geologic features or even perceiving their patterns. For these patterns are not purely spatial—the stratigraphic sequences have to make sense as temporal patterns by mentally simulating how they got there. This simulation is so important that when it doesn't work geologists have a name for it. They call the missing link an "unconformity": It marks something that should be there that's not, and the simulation can't explain why.

In this sense, observation means seeing in a way that goes beyond visual perception. Many geotechnical engineers take a camera on an inspection to document site conditions. But this is recording, not observation, and the activities of photographic composition may even distract from what true observation requires. Observation requires first identifying the things that are important to observe by recognizing and matching their patterns to those of experience. It then requires mentally simulating the effects of the things seen. This is what Terzaghi meant when he said that in their powers of observation, most engineers were "incurably blind." Without any pattern in memory to guide them, they couldn't know what to observe—they couldn't pick out the pivotal findings. And without these findings, they had no way to simulate their effects. Far from seeing the invisible, they weren't seeing anything at all.

The first thing Stevens did on reaching Panama was to inspect and observe the work. It would not be a huge leap to suppose that he was mentally simulating how his railroad would improve things, since he shut down the project to build it. It did not take Roebling long to simulate the effects of another blowout when he flooded the Brooklyn caisson after the fire. And Terzaghi too went one step further with the geologic observations for which he was renowned by simulating their effects on the structure, not just cataloging them. This became formally incorporated into the observational approach, whose central element requires simulating what departures from design premises would produce. Herein then was the power in Terzaghi's powers of observation, and without it the observational approach he developed with Peck on the Chicago subway could never have been successful. Klein (1998) describes mental simulation as key to achieving the following:

- Explaining past events,
- Projecting the situation forward in time to form expectancies of future behavior,
- Explaining how the process works,
- Using analogies and metaphors, and
- Recognizing uncertainty.

So experts have a mental model of a process that captures the fundamental aspects of its mechanisms to explain how it works. This allows them to interpret past events and also to project them forward in anticipating what to expect, the essence of initiative. Mental simulation was Roebling's "intuitive faculty of being at the vital spot at the right time." It is what allowed Stevens to "always be at the right place at the right time and to know what to do without being told." The similarity in these descriptions is no accident—it's how experts do their job.

Mental simulations of a process can be derived from underlying knowledge of causality that theory provides. They can also be derived from inductive analogy where the process is similar to some other, as Terzaghi did in adopting Forcheimer's heat flow analogy in his consolidation equations and George Stephenson in his boat analogy for the crossing of Chat Moss bog. In both of these instances it was simulation from related cases that showed how the process would work. Experts' richer repertoire of mental models or prototype narratives—plus their greater ability to project them backward and forward to explain past and predict future events—also allows them to apply their knowledge to new and different situations. They can improvise when necessary to derive workable solutions. In this respect, experts have been called "opportunists." Analogy allows them to remain flexible rather than rule driven, open to opportunities for devising innovative solutions to unique, unfamiliar, or ill-defined problems.

Information in complex situations seldom presents itself in one package. It changes and evolves along with the unfolding situation. And as it does, the new data can be increasingly incompatible with and hard to incorporate into the mental simulation. This gives rise to uncertainty, and the hardest part of using mental simulation can be distinguishing when certain data can be passed over as irrelevant, from the point at which inconvenient data are being explained away. While experts can fall into this trap as well as novices (representativeness bias being a prime example), their self-awareness and recognition of their own limitations can enhance their ability to acknowledge uncertainty in their mental simulation and to change it when it becomes untenable (Klein, 1998). Mental simulation is not foolproof. But by recognizing disconfirming information and distinguishing it from the merely irrelevant, mental simulation can be used not only to explain how a process works but also how this explanation might be wrong.

Situation Awareness—Seeing the Big Picture

While engineering situations rarely develop as quickly as those engaging fighter pilots or fire chiefs, they still unfold over time. In most any engineering investigation, information is developed in stages; it comes in progressively, and its meaning and implications must be interpreted along the way. Decisions arise in guiding the course of information gathering, in determining when this information is enough, and in using it to shape the design taking form. These activities are dynamic and changing in time. But sometimes the information comes in too rapidly to be fully assimilated or does not comply with expectations. The designer and the field engineer don't communicate enough, or the message gets garbled. There is not enough information to fully support a preliminary conclusion, or it subsequently counterindicates the course already taken. These are the situations where pilots crash, fires rage out of control, and engineering mistakes happen.

The key factor in coping with such dynamic and changing environments is what Klein (1998) calls *situation awareness*, and it refers to three things. First, it requires keeping track of large masses of data, especially critical details, how these details relate to each other, and how these relationships change as events are unfolding. Its second aspect requires drawing inferences from the data in the context of evolving events and decisions. And the third is projecting these inferences to their implications for future events and executing corresponding tasks (Endsley, 1995).

As Endsley (1995) describes it, situation awareness is a state of knowledge and as such goes directly to subjective probability, among other things. Only here, knowledge is not static or wholly stored in memory but includes a comprehension of how surrounding circumstances are developing. As a state of knowledge, situation awareness means knowing what is happening—happening in the present tense. With its roots in aviation, situation awareness was first recognized as a crucial commodity as far back as World War I. The constantly changing interrelationship of factors in space and time caused pilots in close combat, while keeping track of enemy aircraft, to neglect their instruments and all too frequently fly into the ground. Situation awareness is recognized as essential to things like air traffic control and has been used to explain such incidents as the mistaken downing of an Iranian airliner by the U.S.S. Vincennes in 1988 (Klein, 1998).

Situation awareness is the framework in which pattern recognition and mental simulation operate, and experts characteristically display much more of it than novices. For our purposes, we could just as well call it *problem awareness*, and what it means is that not only do experts see the invisible, they see the problem in its totality. They see the Big Picture and it is what they base their actions on, keeping it always in mind as information and decisions develop. The big picture keeps the expert from becoming bogged down in information, dead-ended or sidetracked by detail. It is the difference between knowing what to do and not knowing. Between knowing what to do and what not to. Between knowing when to do nothing, as opposed to knowing nothing to do. And as we saw from the NTSB's aviation accident reports, there are many circumstances when understanding the situation can be more important than the procedural rules for addressing it.

There are several representations of how situation awareness works, but for our purposes the one described by Serafaty, et al. (1997) serves best. Developed in the context of battlefield command decisions, the process is shown in its three parts by the "hourglass model" in Figure 7-9. First comes problem diagnosis or problem recognition, where observations are linked to causal factors to recognize the key issues the situation presents. Starting with some undifferentiated mass of information, it is funneled down using pattern recognition and matching to identify its significant elements and the pivotal findings. What remains is used to construct some workable mental representation of the problem, again from the

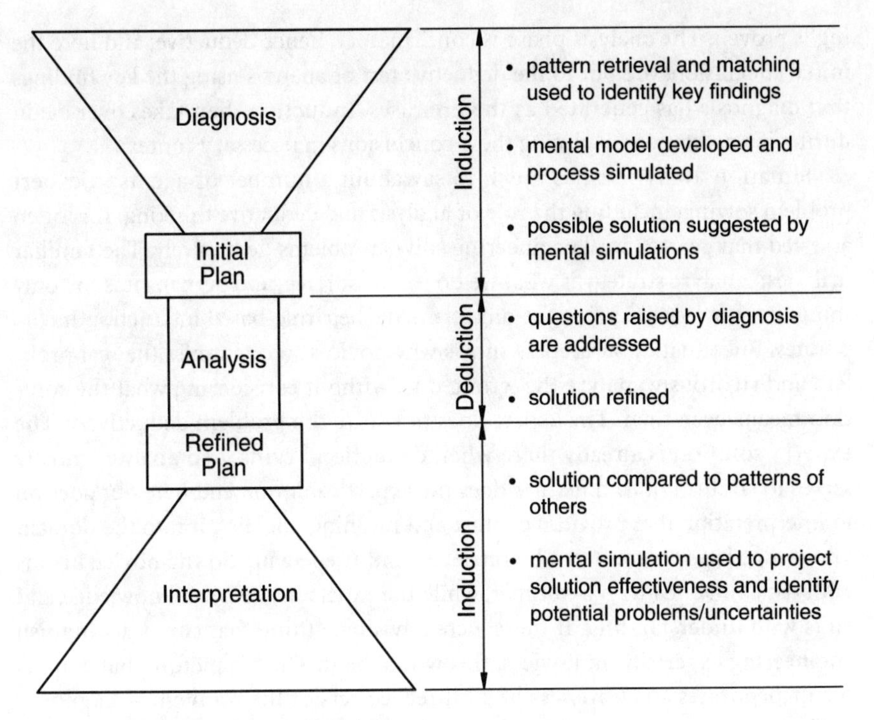

FIGURE 7-9. *Hourglass model of situation awareness.*

prototypes stored in memory, and mental simulations are then conducted to link the key findings to a causal explanation. What results is an initial plan or candidate problem solution.

We now think we know what the problem is and what causes it, and have generated at least the crude outline of its solution. But the diagnostic phase has also identified certain questions, and these are evaluated in the second phase—analysis. The analytical phase produces a better developed and more refined solution. It serves to check that the plan meets certain tests or otherwise to fill critical gaps that remain, here again to fill in the blanks and connect the loose ends.

With this accomplished, the interpretation phase visualizes the solution to see how it will work, again by mental simulation and matching it to the patterns of previous outcomes. Here problem awareness widens in focus to broaden the scope of factors considered. There might be uncertainties that impose conditions on the effectiveness of the solution or outcomes that would call for different strategies. These scenarios are subjected to "what-if" mental simulation to explore various ways the solution might play out, and contingencies are developed to deal with them.

We see too in this model of situation or problem awareness the kinds of reasoning its three stages adopt. The diagnosis stage is inductive: At this point nothing is proven. The analysis phase is confirmatory, hence deductive, and here the initial conclusions are put to the deductive test of analysis using the key findings that diagnosis has generated as the premises. Induction then takes over again during interpretation by placing these conclusions in necessary context.

Situation awareness has much to say about a number of agents in expert problem solving, including the role of analysis and deductive thinking. It is often asserted that geotechnical engineering solves problems deductively. The familiar drill–test–analyze strategy is what we do on the surface, and perhaps it is the only thing that novices do, being the only element their rule-based instruction incorporates. But situation awareness shows why novices so often miss the real problem and rigorously analyze the wrong one, without considering what the solution means even then. The expert does not solve the problem deductively. The expert's solution is already there when deduction begins, and analysis mostly serves to confirm it. And neither does the expert's solution end here but goes on to interpretation that provides context and meaning, melding it into the domain knowledge base where it can be retrieved and used again. So the novice begins with data and ends with a number, while the expert begins with knowledge and ends with understanding. If there were any single thing that could distinguish engineering experts from novices, this would be it. The big picture that experts see incorporates an awareness of all three aspects of the problem—diagnosis, analysis, interpretation—while the novice sees only one. The expert does whatever it takes to keep the ship afloat, while the novice fiddles with the fire in the analytical frying pan.

Pattern recognition, mental simulation, situation awareness, these and more are the tricks of the expert's trade. But it would be a mistake to think that the novice could become an expert simply by learning to jump through these hoops. There is no circumventing the years of hard work and deliberate practice—the experience as we put it more loosely—required to attain the expert's domain knowledge and memory skills in the first place, and to retain them after that. With these things, the expert earns the right to apply certain shortcuts and the wisdom to know when they can be, while the same for the novice would amount to mere sloppiness. In expertise as in other things, there is no free lunch, and what truly distinguishes an expert is what we have seen throughout—in a word, judgment.

Expertise and Judgment

What all of this gives us is a sense of what expertise is and how it works, and we have seen too how it develops in engineers who are unquestionably expert. In identifying those from whom we might wish to solicit subjective probabilities, we cannot wait for a Roebling or Terzaghi, a Stevens or Stephenson to come along, and nothing presented here is intended to imply that we must. Still, the attributes of their expertise and the qualities they display can serve as guideposts to the things we might look for in others and what we would aspire to ourselves, be this for purposes of subjective probability or anything else. And if these aspects of expertise sound familiar, they should. Some if not all of them we have seen before in various forms. Many first appeared in Chapter 3 as the attributes of judgment expressed through the voices of Peck and Terzaghi and Focht, though we were not then prepared to place them in the setting that the study of experts provides. But against the backdrop of these discussions, the elements of judgment that were encapsulated back in Table 3-1 can be seen to correspond closely to the skills of pattern recognition and mental simulation—the art of "seeing the invisible"—in the ways that experts go about using them. It is this that lends a sense of what is important to the things experts perceive, and it is why they perceive them in the first place. When experts act with an awareness of the situation around them, they perceive its important aspects, how they interact, and how they are developing. So in the end, the nature of expertise and the essence of judgment are largely one and the same.

Indeed, if we go back to Focht's (1994) description of how judgment applies to the predictive process, he said that it entered in three places. The first was in using judgment to determine the critical components of the problem and select a solution approach, the last was in interpreting the results, again according to judgment, with judgment working as well throughout the various steps of data analysis in between. What Focht was describing, in effect, were the three components of situation awareness in Figure 7-9. So this "situation awareness," "problem awareness," the "big picture," can simply be considered judgment by another name. And as we have said all along, subjective probability is judgment's quantified expression. We have come a long way to now see by concrete example of flesh-and-blood engineering experts what this means. Subjective probability uses the expert's knowledge, yes, but it also does something much more. Subjective probability captures those qualities most valued in experts to condense and convey them in numbers. Subjective probability is the quantified embodiment of their sense of what is important.

8

Judgment, Probability, and Thinking

With this chapter we arrive at how the subjective, and engineering judgment, and probability all come together. In the simplest terms, this might be viewed as a pyramidal hierarchy. At the base is the subjective, which underpins everything else. The subjective is personal, defined by the qualities of individual engineers that produce expertise and govern how knowledge is structured. At the next level is judgment, the ability to distinguish what is important and to interpret with meaning and context. And at the top sits probability, the quantified expression of this judgment as it relates to uncertainty. Here probability does more than just rest on these layers; it extends back down to reinforce and strengthen them with comprehension.

But thinking is the mortar that cements all this together. Thinking is the most highly personal of all human activities, engaged in by people—not algorithms or machines (at least not yet as far as we know). And every bit as much their accomplishments, it is the thinking of the engineers encountered here that is so truly impressive. From Agricola to Stephenson to Stevens, from Terzaghi to the whole Roebling family, we have seen the thinking contained in their achievements and heard it expressed in their voices. It is people like these who have given us the heritage of civil engineering, and their thought is our legacy. Although some of their methods may have been superseded, their thinking has not, and without helping it flourish we would forfeit much of our professional birthright. More than any structure or analysis we might ever produce, more than any pretensions to science, it is this thinking that makes us a profession and defines our engineering identity.

So if thinking is personal and the personal is necessarily subjective, then thinking itself is subjective. There are no immutable laws that make thinking correct, and it varies from person to person. Still, we recognize good thinking when we see it, so we should be able to look to some of its traits. What we will see here is that good thinking has much to do with how uncertainty is accommodated, which ties it to probability—and to judgment.

The Dualities of Thinking and the Pursuit of Certitude

A persistent theme throughout this book has been that methods and strategies used in thought-related processes contain vastly different elements, so much so that they may seem diametrically opposed. These contrasts were first encountered in the two interpretations of probability that Hacking (1975) called a duality, and they appeared once more in the split personality of geotechnical engineering's two paradigms and views of itself. They extend to the ways for drawing inferences, to expert/novice distinctions, and to the conduct of scientific and engineering activities. Table 8-1 reviews some of these separate dichotomies, but it is apparent that they contain a common thread. In every case they are linked by the overarching duality of the objective and the subjective.

The very notion of duality requires that the objective and subjective views of the world coexist. It is not a matter that one is correct while the other is not, and like Hacking's Janus face of probability they are two sides of the same coin. The objective lies within that part of the world where things can be proven, the subjective occupies the rest, and to restrict one's residence to one or the other would

TABLE 8-1. *Objective–Subjective Dualities*

Methods and strategies	*Objective element ⇔ Subjective element*
Predictive approach	Deterministic/analytical ⇔ Observational/interpretive
Probability interpretation	Relative frequency ⇔ Degree of belief
Probabilistic methods	Reliability analysis ⇔ Risk analysis
Inference method	Deductive ⇔ Inductive
Diagnostic strategy	Variationist ⇔ Empiricist
Expert strategies	Rule-based ⇔ Associative
	Analytical ⇔ Intuitive
	Elaboration ⇔ Ideation
Scientific method	Hypothesis testing ⇔ Hypothesis formulation
	Justification ⇔ Discovery
	Logic ⇔ Intuition
Engineering approach	Theory ⇔ Judgment
	Analysis ⇔ Interpretation
	First principles ⇔ Empirical observations
	Problem solution ⇔ Problem diagnosis
	Problem solution ⇔ Solution interpretation

make for a small world indeed. Yet we have seen throughout that the objective tends to dominate the subjective, thus suppressing many of the dualities in Table 8-1. And the reason for this more than anything else has to do with uncertainty.

In the enclave of the objective, what constitutes knowledge is only that which can be known for sure from what can be proven to be true, or at least what is accepted as proof. That which is not known for sure—which is to say, that which we're uncertain about—has no place in objective truth. In the objectivist's relative-frequency version of probability, there exists a knowable, provable, and singular true value invariant from one person to another. So this objective side of the duality is a comfortable place free from uncertain knowledge, where one can exist content in the security of truth. Because everyone here thinks deductively they think alike, and their conclusions are always identical. This is the ultimate haven of consistency, and while banishing the unknown sharply restricts its boundaries the attraction would be hard to deny. The assurance of predictability in an unpredictable world is such a basic human craving as to rank with food and shelter.

The realm of the subjective is a much less tidy place. There is no pretense of incontrovertible truth in this much larger sphere where proof is not possible. Then too, neither is there any ordinance to command that if we don't know something for sure then we must know nothing at all. Knowledge is qualified to one degree or another, and facts do not constitute knowledge devoid of their subjective interpretation. Knowledge here is not certain knowledge, only probable knowledge, where induction provides some likelihood of its truth or correctness. Here is where judgment resides, and since judgment relies on induction it can never be entirely certain or fully consistent. So anyone inhabiting this world must be adaptive indeed, and much more ambitious as well. Continually coping with uncertainty is no easy thing, and in using adaptive heuristics the unwary can fall prey to their biases.

Much of this book has been, in truth, an engineering epistemology, an inquiry into what can be known, to what degree of certainty, how, and why. The perspective has been unapologetically subjective and purposefully so, since the world of the objective is more than amply represented throughout engineering. Yet engineering is fundamentally not about objective proof but rather what works or what we believe will work, and the certitude of the objective is a scarce commodity as much as we wish it weren't. But at the same time, to have extolled the subjective is by no means to reject the objective, which is instrumental in making quantified predictions and therefore in making engineering possible. And to use both the objective and the subjective together, to do engineering with one foot in both worlds, is what good engineering—and good thinking—are really all about.

Consulting most any dictionary shows that "thinking" has a vast multiplicity of meanings all having to do in various ways with forming ideas, mental constructs of some sort. To think can be to invoke formal rules of logic in reaching a conclusion—to prove that something is so. To think can also be to form a judgment of opinion or belief—to think that something is so. Thinking therefore applies both objective and subjective strategies of various kinds. But the act of thinking requires selecting among these strategies, and since no two people think alike this is ultimately personal and subjective itself. So the essence of thinking requires recognizing and accommodating both the objective and subjective, and this is to reconcile the objective–subjective duality in any of its forms.

Here, reconciling does not mean reducing this duality cleanly to one side or the other. The objective cannot be made into the subjective or vice versa. It means acknowledging the duality and preserving it, synthesizing objective and subjective viewpoints to a common end and adopting them both in ways best suited to each. To think is therefore to accept both the objective and subjective, moving back and forth comfortably between them in much the same way as the Classical probabilists moved so fluidly between the two probability interpretations using opposing concepts as complementary and not conflicting.

So just what can we say is "good thinking"? Good thinking is first of all balanced, using both deduction and induction in their places to diagnose the problem, solve it, and bring meaning and context to this solution. This extends the solution with insight that leads to understanding, and understanding which increases one's store of knowledge in turn. Good thinking uses judgment, one's sense of the important, to distinguish the essential from the irrelevant with awareness of cognitive processes. And good thinking applies theory to assist this judgment, not as the sole arbiter of truth. Good thinking admits to uncertainty, communicating it to others so that they may judge for themselves. And uncertainty brings us back to probability and how it not only expresses good thinking but aids it as well.

Probability and Thinking

Again we can look to the history of probability to find a problem that has long plagued it, a kind of riddle it seems. It too derives from the pursuit of certitude, it has to do with expectations, and it came about because the calculus of probability was conceived prior to establishing its meanings. At first, this didn't bother the Classical probabilists, who remained content to adopt probability as either objective frequency or subjective belief as application might warrant. But probability eventually became entrenched as purely an exercise in mathematics, whose "real opprobrium" was belief and to "exclude all opinion" its first rule. And as we all know, mathematics is above all precise and provides precise answers. So an unspoken expectation arose that continues to this day, which is that probability will characterize the unknown with mathematical precision. And nothing about precision is vague or ambiguous or variable. Precision is certain, so probability must be certain too. In this, such a mathematician of note as Poincaré (1905) was among those to recognize that the very name "calculus of probability" is a paradox:

> Probability as opposed to certainty is what one does not know, and how can we calculate the unknown? ... How can we explain this apparent contradiction?

If this does seem paradoxical or at least somehow counterintuitive, it is. Probability is supposed to describe the uncertain, or that which we do not know for sure. But how can we be sure about what we're not sure about? How can we be certain about uncertainty? If we possessed sufficient knowledge to specify the extent of this uncertainty precisely, we wouldn't be uncertain to begin with. So to characterize uncertainty with certainty—to demand that this be mathematically precise and invariant—is to replace uncertainty with certainty. The Mad Hatter could not have done better. With this, probability becomes accompanied by an expectation of all but clairvoyance in specifying the unknown unambiguously, and doubtless more damage has been done by this expectation than by the most grievous of mathematical transgressions. Probability was never intended to turn uncertainty into certainty or to make it disappear, and herein lies the answer to Poincaré's contradiction. In substituting mathematical precision for the unknown, probability has been looked to for things it cannot provide.

The precise and the proven define the world of the objective, and only here can unique and invariant probability values be found. But the great seething mass of uncertainty lies outside the objective statistics of hard data, in the world of subjective understanding of mechanisms and processes. So there can be great consternation when probability ventures into the arena of expert opinion, where the most voracious uncertainties lay in wait, only to discover with shock and dismay that experts don't always hold the same opinions and their assessed probabilities can vary. And if probability cannot provide precise answers but only vague or unreliable numbers, then what use can it possibly be?

The answer again lies in thinking. The very act of formulating a probability value forces thinking—synthesizing both objective and subjective information using both objective and subjective processes. With this comes richer comprehension of the nature and sources of uncertainty quite apart from any numerical result. And for all of the discussions here about heuristics and biases, it is clear that probability itself is nothing more than a heuristic device. Used to best advantage, it provides a simple tool to conceptualize and structure uncertainty. Seen in this light, probability becomes a tool for thinking and correspondingly an engineering tool of the most indispensable kind. Mathematics plays a key role here but as servant not master—just the means to an end, not the object. It provides the vehicle for the "divide and conquer" approach to uncertainty just like for any engineering problem, whereby constituent uncertainties can be identified and evaluated separately then mathematically reaggregated. And identifying these separate constituents, along with their probabilities, shows how important each one is with respect to the whole and to each other—a relative importance that draws upon judgment while at the same time enhancing it too.

At the start, judgment informs probability. It is judgment that allows us to say that one component of uncertainty is more important than some other. It is then probability that allows us to say by how much. The sense of what is important is refined by this quantification, so that in the end probability informs judgment as well. And if knowledge requires the interpretive context of judgment, then probability is also a tool for gaining knowledge about uncertainty—without having to transmute it into certainty through the alchemy of mathematical preciseness.

The use of probability does not end here. The practical consequence of understanding uncertainty leads as a matter of course to means for reducing it, or means for reducing its effects. If lack of information or data is the source of uncertainty, then this points the way to obtaining more of it—and more of the right kind. If gaps in basic understanding of mechanisms or processes are at issue, then perhaps these uncertainties can be circumvented in some way. This is simply good thinking, of course, not to mention down-to-earth engineering good sense. While these things can be formalized through reliability and risk analysis or any of their constituent probabilistic methods, the techniques themselves are not really the point. As a roadmap for charting out possible courses of action, simply thinking about uncertainty in probabilistic terms can often produce much of the desired result. In all of this, the whole exercise of using probability is fundamentally diagnostic in nature. It doesn't produce proof, and it can't make certainty out of uncertainty. What it can do, used sensibly, is to show where the most serious uncertainties lie. And with better understanding of any problem, along with better judgment, better engineering cannot help but result.

Probability and Informed Consent

While all of this may aid the engineer, what about others? The days when engineers had exclusive authority over engineering decisions are long gone. Others now make them in large part, where foremost among these "others" is the public itself, or those who act on its behalf. Still, others cannot make sensible decisions in matters of engineering without at least obtaining input from engineers, and this input must include the associated uncertainty in order to be truly fair and accurate. In this respect, probability is an information tool—it is used to communicate uncertainty judgments to others. But communicating requires more than dropping information on someone's desk or slipping it under the door. Any communication device requires some kind of protocol in order for sender and recipient to be mutually understood. Though this need not be rigid, communicating uncertainty benefits from some strategy for conveying it.

In a scene played out in hospitals every day, a physician enters a patient's room. The diagnosis is heart disease, and the doctor offers up a prognosis by way of medical opinion. If nothing is done the patient has a one-in-five chance of suffering a fatal heart attack, judging from applicable statistics. But various treatment methods are available. A promising new drug might reduce this chance tenfold, though with no guarantee of success. There is also a surgical procedure that could fix the problem altogether—but with an accompanying 1% mortality rate. The patient asks what the physician would advise. In the doctor's medical judgment, the best thing would be to try the drug first then operate if the patient fails to respond. The patient concurs, and a course of treatment is established.

This is the model of informed consent, and it describes most aptly the responsible use of probability as an information tool. The physician is but an advisor, there to transfer medical knowledge and convey medical judgment in a form that can be used by the patient. And note carefully that while the physician may offer advice, it is the patient who decides on the treatment. The doctor is there to inform this decision, and probability statements are essential to doing so. But nothing in these probabilities mandates the decision outcome, no medical decision criterion to say that probability X prescribes action Y. Note too that informed consent requires not just communicating the prognosis, it also carries a duty to inform of treatment options. Thus, no probability associated with adverse consequences can be communicated responsibly without identifying possible means for reducing it as well. And some of these alternatives carry their own uncertainties requiring a balance or tradeoff in choices.

The informed consent model has other implications. In rendering a medical opinion on drug treatment, it should go without saying that the doctor would not be beholden to the pharmaceutical company—medical ethics would surely preclude even the appearance of any such motivational bias. In addition, ranges of likelihood are provided to convey uncertainty in the estimate, as for the drug treatment's implied probability range from 1 in 50 to the original 1 in 5 depending on its effectiveness. And in providing these likelihood assessments, the patient fully expects—indeed demands—the exercise of medical judgment, since few would knowingly consult a physician without it. So the patient has every right to expect that the assessed outcome likelihoods will reflect not only general statistics but that these statistics will be adjusted for the patient's condition using the doctor's medical judgment. Then too, the patient might seek—with the physician's encouragement—a second opinion, both accepting without reservation that medical opinions can vary and tacitly acknowledging the wisdom to be found in diversity of opinion and judgment. Patients don't seek proof, they seek cures, and they don't expect odds to be guaranteed.

If these principles of informed consent can structure the use of probability in life-and-death situations, then they should apply equally to engineering. Crucial decisions in medicine recognize that it is essential to draw upon the objective principles of medical science, the subjective opinions of medical judgment, and every other knowledge resource that can possibly be brought to bear. Too much is

TABLE 8-2. *The Principles of Informed Consent*

1. The decision rests with the decisionmaker.
2. Probabilities communicate information to inform the decision.
3. Judgment is expected to be incorporated in probability estimates.
4. Probability estimates incorporate ranges where appropriate.
5. Probabilities for adverse consequences are accompanied by means to reduce either or both.
6. The decisionmaker is encouraged to seek other opinions.
7. Motivational bias in probability estimates is not tolerated.

at stake to do otherwise. There is no illusion that there can be a single correct or precise answer for every case or any unique way of obtaining one. Consistency is not expected either in likelihoods or outcomes because every case, every patient, and every physician's medical judgment, are different.

It is fair to say that the engineering use of probability has not yet achieved the maturity embodied in these principles as summarized in Table 8-2. But those who recognize the true value of probability will inevitably gravitate in this direction. For regardless of how probability is applied, it is not in the end about numbers, it is about judgment. Not just about how much uncertainty, but why. Establishing the contributing sources of uncertainty, evaluating their relative significance, raising new questions and seeking new information—all of these things go directly to a sense of what is important about uncertainty. No one has yet put it better than Laplace when he said that probability was good sense reduced to a calculus. What he did not say but might well have is that it works the other way too—that the calculus reduces to good sense. And herein lies the recursive nature of probability: It always leads back to judgment.

The Future of Judgment

It has been some time now since geotechnical engineering was pronounced a "mature" field, with all that this implies. It is said there is little new to be discovered anymore, mostly just refinement of existing technique. Indeed, civil engineering as a whole suffers a certain self-consciousness as it watches itself being left in the silicon dust of today's high technology. It can be uncomfortable in this society in this day and age to admit finding in concrete and steel and earth and rock the same allure microprocessors hold for others. So in an effort to find something to hold on to and pull ourselves up, we look to computational methods as the technological cure for what ails us. Small comfort and slim pickings this, for progeny of the likes of Terzaghi and Stevens and all the rest.

Perhaps it is no accident that it was about the same time as the field's youth was reportedly lost that Peck (1980) asked if its judgment had departed along with it. Judgment has become a vestigial limb bypassed in computational evolution, and to examine it as we have here may seem more a regression to some relic preanalytical state. Still, perhaps in our search for relief from the aches and pains of technological maturity, it is possible we have been looking in the wrong place. In looking to the future of the profession, we must also look to judgment.

We return here to Kuhn (1977c) and his chronicle of the progression from youth to maturity in scientific fields. What characterizes this transition is the reversal, from external to internal, in the profession's view of itself. When a field starts out, it is defined by the needs of society that determine the problems it solves. This is a time of great achievement and, according to Kuhn, a time marked by pragmatism, common sense, or as we might say judgment, as its theories arise hand in hand with its accomplishments. A profession at this stage views itself externally from the outside in, much as society does in looking to it for solutions to problems affecting peoples' basic human aspirations and everyday lives. But as the field matures its perspective turns inward, becoming driven by needs of its own. Its members develop their own subculture immersed in refinements of theory. They become increasingly insulated as they mainly talk to themselves, their former external challenges supplanted by internal incentives to increase theory's precision and scope.

There are those who might find Kuhn's characterization to fit our field today. The great civil engineering achievements of the past profoundly changed the way great numbers of people lived and improved their daily existence in the ways the Panama Canal, the Brooklyn Bridge, and Terzaghi's hydroelectric projects so demonstrably did. But having become, in a way, victims of our own success, this mantle has now largely been passed to biomedicine, genetics, communications, and computer science. The last product of American civil engineering to so directly affect people's lives may well have been the interstate highway system, half a century now since its inception. But a casual perusal of today's technical journals does not evidence any great clamor from society for very much they contain. Society depends on us still to be sure, but the last thing it wants is great changes. Maintaining its standard of living and quality of environment will be quite enough, thank you, and the less we build from now on the better has become society's prime directive. And if external demands from society have diminished, then so too has the profession's external perspective—and the judgment that always came with it.

So it is indeed no accident that the role of judgment has diminished and theory advanced as the profession has changed in historical pattern. But the question remains as to where we go from here and whether the profession's attainment of maturity must inevitably mean loss of vitality. In considering this question, we can look to civil engineering's chief product.

Society's infrastructure that others put in place must now be preserved, with this in some ways harder than first building it. A great deal is invisible, buried below ground or in places seldom seen, with the rest in the background of the everyday landscape. There is little to inspire any great outpouring of society's resources for maintaining its condition and function, and any such efforts require a kind of triage. Exactly what is it that most needs to be maintained, replaced, or rebuilt and how, where, and when? And how to set priorities for doing these things in the face of limited funds? In short, how do we determine what is most important? If this weren't enough, the uncertainty in evaluating anything already built can be magnified from what it was during design. Conditions of both the structure and its foundation once taken as design assumptions can be superseded by reality. While uncertainty may originally have been overcome with conservatism, it can reappear with time and actual behavior. Deterioration is not something design can ever fully anticipate, nor changes brought about by new knowledge.

Tackling these problems of infrastructure will require judgment, every bit as much as its builders possessed and then some. But judgment was always gained as the product of building. If we are to go forward more as caretakers than builders, the question becomes not so much where judgment has gone, but where it will come from tomorrow.

If Kuhn's ideas about revolutions in science are anywhere near right, then incremental advances in theory and computation will only take us so far. Tomorrow's problems of infrastructure will still need analysis but perceptive diagnosis even more, and with this a greater overt reliance on inductive reasoning than ever before. Diagnostic advances will not come about simply from new high-tech devices, though we may continue to seek and develop them, because diagnosis arises from thinking, not hardware. So Kuhn would predict some fairly dramatic change in geotechnical thinking if the field is to escape ending up as much more than a withered appendage of engineering mechanics. And that change, if not a barn-burning revolution, might just be a re-emergence and revitalization of engineering judgment, with deeper appreciation for the kinds of expertise and thinking it requires. If these are the frontiers of aviation, artificial intelligence, battlefield command, and real-life decisionmaking, then so might they be pathways for us to pursue at least as fruitful as refinements in computation or theory.

For theory propagates itself through the generations as its paradigm specifies, while judgment dies with each engineer who has managed to acquire it. Yet judgment is a renewable resource, and there is nothing to stop the paradigm of practice from more actively assisting its continuing generational rebirth. This has to start with a clearer understanding of what judgment is, where it comes from, and how it works. It must then go on to application. For all of their judgment, it is here we possess something the great builders did not—a knowledge of how their ideas and projects turned out. This knowledge of outcomes has accumulated into a critical mass of the whole profession's experience, which expands with each passing year. We are all serious students of theory and analysis, as we must be to become engineers in the first place. If we were to immerse ourselves as deeply in the profession's wealth of case histories, and our own, we could hardly avoid becoming equally serious students of judgment and how to use it.

Certainly judgment can be fallible. So can any other engineering technique. It is axiomatic that good decisions don't guarantee good outcomes, and neither does good judgment in much the same sense. But this is no excuse for discarding it, quite the opposite. And if we sometimes suffer from faulty judgment then what we need is better judgment, which goes to its operation more than its outcomes. Judgment can no longer be dismissed with a wave of the hand as "subjective," so vague as to be impervious to improvement. We have discovered here some of its internal processes and have compared it in experts and novices. With these foundations in place, what remains is to build on them.

So then where will tomorrow's judgment come from? While it may spring from all of these things, it might also come from probability. To use probability sensibly with appreciation of what it can do is to use it to gain understanding of uncertainty. And central to this is which uncertainties are most important in the very same diagnostic and interpretive senses that constitute the essence of judgment. So judgment can indeed be enhanced and even developed by the thoughtful use of probability. Probabilistic and risk-based methods will never replace judgment—they require it to begin with. Instead they structure, foster, and enhance it like nothing else can. And if probability can serve judgment, then so too can it promote the kinds of expertise that rely on judgment so intimately.

Judgment is at the heart of engineering, but our appreciation of its inner workings is still embryonic. So perhaps the reasoning strategies and cognitive processes we use are where the next revolution might start. In the end, this cannot be handed down from research but rather must rise up from grass-roots practitioners in the things we do every day. If so, what we have seen here might just be the first tentative step down a well-worn path toward revitalizing the tradition of engineering judgment. And with it the soul of our profession.

Appendix

Probability Axioms, Theorems, and Bounds

The axioms and essential theorems of probability constitute the probability calculus. They provide the means for combining and manipulating individual probability values, regardless of how these values are determined or which probability interpretation they adopt. For subjective probabilities, the axioms also provide for coherence, one of the tests for their validity. And along with these axioms and theorems go probability bounds, which are useful when relationships among events and probabilities become too numerous or complex to specify completely and unambiguously.

Subjective probability is most often applied in its discrete or single-valued form, and the discussions here are restricted to these cases. They do not require intimate knowledge of more refined aspects of statistical analysis or continuous probability distributions, though this can certainly be helpful. But competence as a probability assessor greatly benefits from some familiarity with a few basic mathematical constructs. The intent here is to provide a cursory background sufficient to acquaint subjective probability assessors with these principles—especially in a risk analysis setting—that will allow them to function at a reasonably informed level. Likewise, this background will be equally beneficial to those interpreting probabilities and probabilistic analyses, whether subjective or frequency in nature. The essential works on these topics in a civil engineering context remain the venerable references by Benjamin and Cornell (1970) along with the two by Ang and Tang (1975, 1984), all rich with instructive examples. Those seeking deeper treatment of the mechanics of probability, risk, reliability, or decision analysis are encouraged to consult them.

Venn Diagrams and Set Nomenclature

The probability axioms propose certain requirements for collections of events based on set theory. As a schematic representation of set membership, Venn diagrams are used to establish and portray set relationships among events, but without incorporating probability values. Here, an *event* refers to the undetermined outcome of some occurrence, condition, state of nature, or parameter value. The set of all possible outcomes is the *sample space,* which contains every possible outcome regardless of likelihood. A set that includes each and every such member of the sample space is said to be *collectively exhaustive.*

Thus at any specified moment and location, an earthquake either occurs or it doesn't, two events which together constitute a collectively exhaustive set. These events are also said to be *mutually exclusive* since one cannot pertain in the presence of the other. Similarly, if earthquake magnitude were grouped according to the intervals $M < 5.0$, $5.0 \leq M \leq 7.0$, and $M > 7.0$, these would be mutually exclusive and collectively exhaustive because any particular magnitude must be contained in one and only one of the three sets. The basic idea of a Venn diagram is to portray sets of events as nondimensional "bubbles," where regions of overlap designate common set membership and no overlap for mutually exclusive sets containing no members in common.

The Venn diagram of Figure A-1 depicts two events in the sample space of material type existing at some point in the ground, where this might be either soil or rock. These two states of nature are taken to be mutually exclusive, so they contain no overlap in their Venn diagram representations. Additionally, the sample space contains all other events not specifically identified but nevertheless pos-

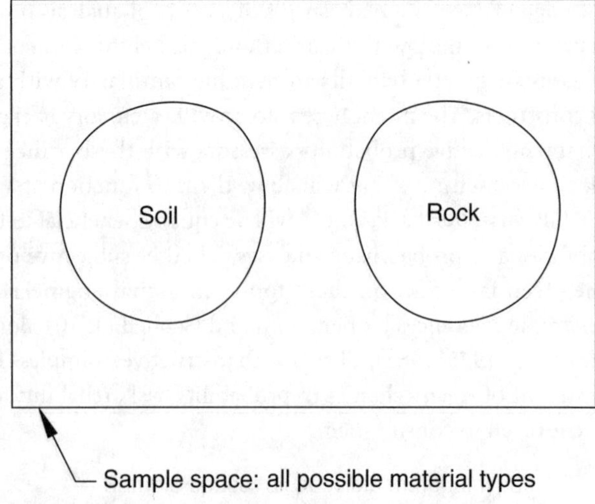

FIGURE A-1. *Venn diagram for mutually exclusive events.*

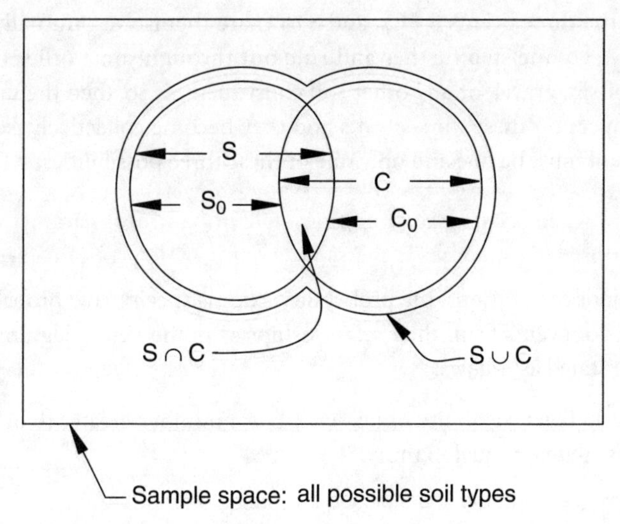

FIGURE A-2. *Venn diagram for intersecting events.*

sible, such as the presence of residual saprolite, karst cavities, tree roots, or old car bodies at that particular location. Note here that Venn diagrams are not drawn to any scale of relative proportion and also that event relationships and sample spaces can involve certain implied states of knowledge, frames of reference, and underlying assumptions. For instance, were we to "zoom in" on this particular subterranean point, other members of the sample space would be interstitial voids and pore water, along with rock joint openings and infillings.

Suppose it were determined for sure, say through some geophysical method, that the material is soil of some kind. Now the existence of soil is no longer uncertain, but its constituents still are. The event that soil exists can now be expanded to become the new sample space shown in Figure A-2, the set whose members include all possible soil types. Of these, two in particular are shown: that the soil consists of sand (event S) and that it consists of clay (event C). It may be entirely one or the other, as represented by the regions S_0 and C_0, but it may also contain both at the same time as sandy clay or clayey sand. So events S and C are not mutually exclusive, and their *intersection* or joint occurrence is indicated where these two events overlap on the Venn diagram. This intersection of events S and C is expressed symbolically as:

$$S \cap C \qquad \text{(with } \cap \text{ read as "AND")}$$

The Venn diagram portrays the event that the soil is composed of either sand or clay materials as the region defined together by S and C. This is the *union* of these two events expressed as:

$$S \cup C \qquad \text{(with } \cup \text{ read as "OR")}$$

Note that the three events S_0, C_0, and $S \cap C$ are themselves mutually exclusive. Suppose we go one step further and rule out through some other method the presence of silt, gravel, or any other soil constituent. If so, then the sample space again collapses on these three events and they become collectively exhaustive as well: The soil must be one and only one of these three possibilities.

Axioms

As convenient conventions, the probability axioms specify how probability operates on sets of events from their relationships as in the Venn diagram. The three axioms are stated as follows:

AXIOM I. The probability of an event E is a number greater than or equal to zero but less than or equal to unity:

$$0 \leq p[E] \leq 1$$

AXIOM II. The probability of a certain event E_c is unity, where E_c is the event associated with the occurrence of all members in the sample space:

$$p[E_c] = 1$$

AXIOM III. The probability of the union of two mutually exclusive events is the sum of their probabilities:

$$p[E_1 \cup E_2] = p[E_1] + p[E_2]$$

This is sometimes called the *addition rule* for the union of mutually exclusive events. Where E_i represents the set of all events in the sample space, Axiom III implies that:

$$\Sigma p[E_i] = 1$$

It also implies that for *complementary* events E and $\bar{E}$, where $\bar{E}$ (read as "NOT E") represents the set of all events in the sample space except E:

$$p[\bar{E}] = 1 - p[E]$$

Returning to the Venn diagram of Figure A-2 after sand and clay have been determined to be the only soil constituents present, the event S (sand) can be taken as the union of the intersection $S \cap C$ (sand and clay) and the non-overlapping region S_0 (sand only). Likewise, C is the union of the intersection $S \cap C$ and the non-overlapping region C_0 (clay only). So from Axiom III:

$$p[S] = p[S \cap C] + p[S_0] \tag{A-1}$$

and:

$$p[C] = p[S \cap C] + p[C_0] \tag{A-2}$$

Also from Axiom III, for the three mutually exclusive events S_0, C_0, and $S \cap C$:

$$p[S \cup C] = p[S_0] + p[S \cap C] + p[C_0] \tag{A-3}$$

Solving Eqs. A-1 and A-2 for $p[S_0]$ and $p[C_0]$ and substituting in Eq. A-3, the probability of either sand or clay at the location in question, or again their union, is:

$$p[S \cup C] = p[S] + p[C] - p[S \cap C] \tag{A-4}$$

The right-hand side of Eq. A-4 can be interpreted as the sum of the probabilities of sand and clay considered individually, with the third term subtracted to prevent "double counting" of the overlapping region in Figure A-2 where both occur together. For small values of p[S] and p[C], the third term is commonly neglected.

It is worthwhile noting that the probability axioms are not the only such conventions for establishing relationships among event likelihoods. Others include the so-called Dempster-Shafer theory of evidence (Gillette, 1990; Stein, 1993) and fuzzy-set theory (Zhang and Tumay, 1999). Their assumptions are generally less restrictive than the axioms of probability, and both have found useful application to problems of geotechnical uncertainty but are not as widely known.

Independent and Conditional Probabilities

Suppose now that we put down a wash boring without obtaining a sample. The cuttings show sand in the drilling return, but we can't tell whether there is any clay—which would be important in terms of, say, liquefaction susceptibility. What we want to evaluate then is the *conditional* probability of the existence of clay given the known presence of sand, expressed as:

$$p[C \mid S] \qquad \text{(with | read as "GIVEN")}$$

This is the proportioned ratio of the probability of the overlapping region $S \cap C$ to the probability of S as depicted schematically in the Venn diagram of Figure A-2. So the conditional probability of clay given sand is defined as:

$$p[C \mid S] = \frac{p[S \cap C]}{p[S]} \tag{A-5}$$

In this illustration, the presence of sand has already been determined with certainty. But a conditional probability of clay given sand is equally applied to the postulated presence of sand as if its existence were already known. In general, then, conditional probability refers to the probability that some event will occur, predicated on either actual or presumed knowledge that some other event has occurred.

To take one example, a deposit of colluvial origin might suggest an association of clay and sand. Here, the presence of sand would say something about clay.

On the other hand, an alluvial origin might say nothing at all, or the depositional history might be unknown. In this case, the probabilities of clay and sand could reasonably be taken as *independent*, and for independent events:

$$p[C \mid S] = p[C] \tag{A-6}$$

Now the probability of clay is the same whether sand is present or not. The test for independence is whether the known outcome of one event would modify in any way the probability assigned to the other. This determination depends on one's state of knowledge about the underlying conditions and mechanisms that influence event relationships.[1] So independence and conditionality are not intrinsic to either events or the probabilities assigned to them. It is customary to assume, for instance, that floods and earthquakes are probabilistically independent from the apparent absence of any causal connection between geology and meteorology.[2] In practice, independence among events and event probabilities is often assumed for the sake of convenience because it makes otherwise intractable problems easier to solve. We will go on to further examine some of the implications.

Basic Theorems

Continuing from the definition of conditional probability, Eq. A-5 provides that for any two events A and B, the probability of their joint occurrence or *joint probability* is:

$$p[A \cap B] = p[A \mid B]\, p[B] = p[B \mid A]\, p[A] \tag{A-7}$$

This is known as the *multiplication rule*, and if there are several such events including, say, B_1 and B_2:

$$p[A \cap B_1 \cap B_2] = p[A \mid B_1 \cap B_2]\, p[B_1 \mid B_2]\, p[B_2] \tag{A-8}$$

and so forth for events $B_1 \ldots B_n$.

If these events are independent, then from Eq. A-6, $p[A \mid B_1 \cap B_2] = p[A]$, and $p[B_1 \mid B_2] = p[B_1]$. So for independent events, the multiplication rule for joint probability accommodating any number of such events becomes:

---

[1]We do not go so far as to say that the probability calculus requires a subjective interpretation. Nevertheless, independent and conditional relationships could not be established without the subjective element of one's state of knowledge, making it otherwise all but impossible to apply the calculus to most situations of practical importance.

[2]This has not always been so, with established thought in the scientific community during the opening of the American West having been firmly persuaded of the influence of soil texture on precipitation ("rain follows the plow"). The early-day geotechnical engineer would therefore have been instrumental in determining the conditional probability of flood occurrence given the soil conditions.

$$p[A \cap B_1 \cap \ldots B_n] = p[A]p[B_1] \ldots p[B_n] \tag{A-9}$$

In an event or fault tree, each separate branch pathway contains those events that together must all be realized for the consequences associated with that particular pathway to be incurred. Hence, the probability of pathway occurrence is the joint probability of the events it contains. Eqs. A-8 and A-9 show that these individual event probabilities are multiplied, regardless of whether they are independent or conditional on some other event in the same pathway.

Total Probability Theorem

Also in an event or fault tree representation, the desired end result is the probability of occurrence of all branch pathways that could produce some specified consequence, or their union. By way of illustration, consider the simplified event tree in Figure A-3 for seismic liquefaction. The peak ground acceleration (PGA) is incremented into three mutually exclusive and collectively exhaustive ranges designated a_1, a_2, and a_3. These are called the *parameter states* that PGA can assume. The conditional probability of liquefaction for each one, $p[L \mid a_i]$, is derived from the branch pathways enclosed in dashed lines using Eqs. A-8 and A-9 to find the joint probability of the events they contain. What is needed is the total probability of liquefaction, $p[L]$, over all three PGA states as the intersection $[L \cap a_i]$. From Axiom III:

$$p[L] = p[L \cap a_1] + p[L \cap a_2] + p[L \cap a_e] \tag{A-10}$$

Also, from Eq. A-7:

$$p[L \cap a_1] = p[L \mid a_i]\ p[a_i] \tag{A-11}$$

Accordingly, the desired result for $p[L]$ is obtained as:

$$p[L] = p[L \mid a_1]\ p[a_1] + p[L \mid a_2]\ p[a_2] + p[L \mid a_3]\ p[a_3] \tag{A-12}$$

This is known as the *total probability theorem*, and it provides the basic arithmetic of event and fault tree calculations for multiplying along branch pathways and summing the products.

Bayes' Theorem

Bayes' theorem provides a formal algorithm for combining various kinds of information pertaining to event occurrence. This information is often in the form of some test, observational finding, or predictive procedure determined to have accompanied the occurrence of similar events in the past. These serve as indicators of event occurrence, albeit imperfect ones where their predictions are not always realized. These indicators are also derived retrospectively from previous tests or observations on some sample population of similar cases, so that if A were the

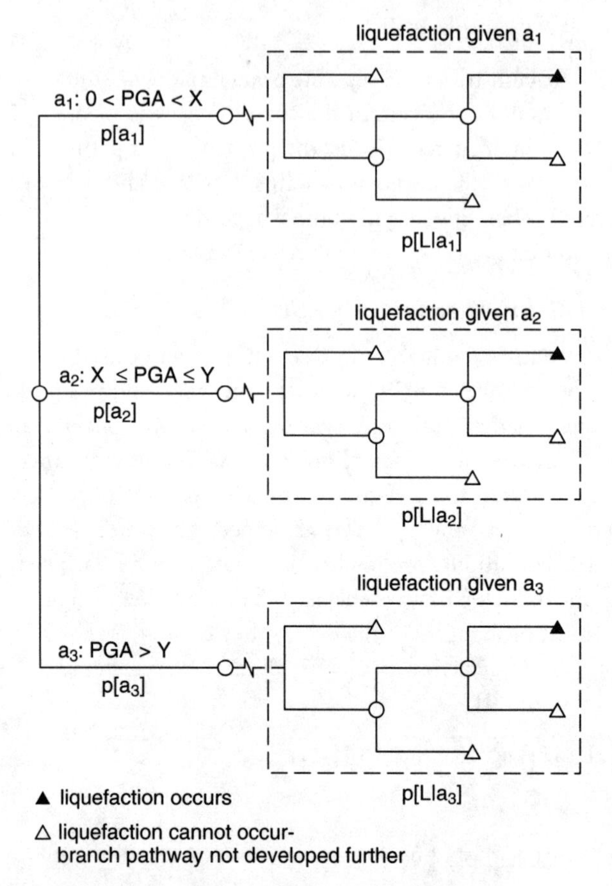

FIGURE A-3. *Simplified event tree for seismic liquefaction.*

indicator and B the event of interest, the conditional probability of the indicator given that the event occurred, or p[A | B], can be determined in advance from the sample. But for the indicator to be useful in making future predictions what is needed is the opposite: the probability of event occurrence given the indicator, or p[B | A]. Bayes' theorem allows the sense of conditionality to be reversed.

The use of Bayes' theorem requires three things. One is the probability of event occurrence assigned using pre-existing information of some kind, before incorporating that provided by the indicator. This is p[B], the *prior probability* of event occurrence, or simply the *prior.* The indicator itself has two properties, both determined from the sample population. The first can be termed its reliability, or p[A | B], which expresses the degree of indicator A's association with event B. In a sample of repeated trials, this would be represented by how many times A occurs when B occurs. The second property of the indicator is its false-positive rate, or $p[A \mid \bar{B}]$, which reflects how often indicator A occurs anyway even when event B does not. After the prior has been updated by this information from the indicator, the result is $p[B \mid A]$, the *posterior probability.*

More formally, let B_i represent one of several outcomes associated with the event of interest. For the joint occurrence of the outcome and the indicator as given by Eq. A-7:

$$p[A \cap B_i] = p[A \mid B_i]\, p[B_i] = p[B_i \mid A]p[A] \qquad \text{(A-13)}$$

So the posterior probability is:

$$p[B_i \mid A] = \frac{p[A \mid B_i]p[B_i]}{p[A]} \qquad \text{(A-14)}$$

Eq. A-14 is the simple form of Bayes' theorem. Where event B has a *binary outcome*, meaning that it has only two possible states, this can be expanded using the total probability theorem of Eq. A-12 to become:

$$p[B \mid A] = \frac{p[A \mid B]p[B]}{p[A \mid B]p[B] + p[A \mid \bar{B}]p[\bar{B}]} \qquad \text{(A-15)}$$

noting here that $p[\bar{B}] = 1 - p[B]$ by Axiom III.

As an example, consider the case of a dam with a broadly graded core having experienced formation of a sinkhole on the crest symptomatic of internal erosion. Such internal erosion incidents, although undesirable, are not entirely uncommon for this kind of core material (Sherard, 1979) and are often arrested by its "self-healing" properties. Accordingly, it is of interest to determine the probability that another internal erosion incident could occur at some time in the future given the occurrence of a previous one. Call a previous incident event A and a future incident event B. From the condition and characteristics of this particular dam, an annual probability $p[B] = 0.05$ provides the subjective prior. What is desired is $p[B \mid A]$, and the performance history of similar dams can be used for estimating the properties of the Bayesian indicator.

Norway has a large number of well-instrumented rockfill dams with cores of broadly graded morainal material that are uncommonly uniform in design and construction. There have been 23 cases of self-healing internal erosion manifested by sinkholes and transient increases in seepage, some repeated in the same dam after the initial occurrence. These cases of internal erosion can therefore be used as a sample population in evaluating the occurrence of an internal erosion incident (event A) as an indicator of some subsequent occurrence (event B).

From this performance experience, $p[A \mid B]$—the probability of an incident given at least one subsequent repetition—has been found to be 0.70. Correspondingly, the probability of a single incident with none thereafter, or $p[A \mid \bar{B}]$, is 0.13. Substituting in Eq. A-15:

$$p[B \mid A] = \frac{(0.7)(0.05)}{(0.7)(0.05)+(0.95)(0.13)} = \mathbf{0.22} \quad \text{(A-16)}$$

So in this case the posterior is significantly increased from the prior of 0.05 after Bayesian updating incorporating observations from the sample population.

Binomial Theorem

Many events have binary outcomes with only two possible states: Either they occur or they don't. Sometimes these events are repeatable over some interval of time or within some spatial area or volume where their probabilities are expressed in terms of these units. The binomial theorem and the associated binomial distribution apply to the special case where these occurrences are independent and their probabilities are the same. Repetitions of such events are known as Bernoulli trials, a convenient shorthand for independent events of constant probability produced by probabilistically stationary or stationary stochastic processes in time or space.

If p_i is the probability of occurrence in any one such trial, then Axiom III specifies that the probability of nonoccurrence is $(1 - p_i)$. From the multiplication rule for independent events, the probability $\bar{P}_n$ of no occurrences in n trials is:

$$\bar{P}_n = (1-p_1)(1-p_2)\ldots(1-p_n) \quad \text{(A-17)}$$

And because p_i for Bernoulli trials is constant, $p_1 = p_2 = \ldots p_n$. So the probability of at least one occurrence (or equivalently, the probability of one or more occurrences) in a series of n Bernoulli trials becomes:

$$P_n = (1-\bar{P}_n) = 1-(1-p_i)^n \quad \text{(A-18)}$$

Also, the mean number of occurrences is $(n)(p_i)$.

Vick and Bromwell (1989) describe the use of Eq. A-18 to represent the spatial occurrence of karst sinkholes as Bernoulli trials, with p_i as occurrence probability per unit area. Similarly, temporal Bernoulli trials are commonly expressed in units of time, often as occurrences per year, where p_i is their annual probability of occurrence. Where the event refers to some quantity such as flood flow or earthquake magnitude, p_i is usually the annual probability that some such value is equaled or exceeded—its annual *exceedance probability*. Here, the value of n in Eq. A-18 is the *exposure period* and $1/p_i$ is said to be the *average recurrence interval* or simply the *return period* of event occurrence. A common mistake is to assume, for instance, that the 100-year return period flood occurs exactly one time in each and every 100-year interval. Although subtle, the misinterpretation is seen by noting that the 100-year flood is that which would be exceeded once on average over many such 100-year periods if enough of them were strung together. In fact, it is easily shown by substituting $1/p_i$ for n in Eq. A-18 that, for large *n*, the probability of event occurrence within its own return period is always 0.63 universally for any return period duration. Thus, the capacity of a structure designed to resist some return-period event will have about a two-out-of-three chance of this event being exceeded at least once during a corresponding period of time, with the mean number of exceedances being $(p_i)(1/p_i) = 1$.

It is well to point out here too that Bernoulli trials have what is called a "memoryless" character. Each successive trial neither knows nor cares about the outcomes of trials that preceded it. While the principle of "regression to the mean" provides that the long-run proportion of actual occurrences will eventually converge to the probability in any one trial, the "gambler's fallacy" asserts the reverse whereby the probability of the next occurrence is supposed to somehow reflect the previous outcomes. But again, from the definition of probabilistic independence, knowledge of the outcome of any one Bernoulli trial cannot modify the probability that pertains to any other. This is not to say that every process in nature conforms to Bernoulli assumptions—far from it—and Bernoulli trials are a special case. But these assumptions can make application of the binomial theorem convenient, provided they are properly understood and accounted for. Some of the instances that can violate them will be reviewed after the following example.

To illustrate both spatial and temporal Bernoulli trials in application of the binomial theorem, consider the two-lane alpine highway shown in Figure A-4. It crosses the runout path of a well-defined avalanche chute extending from the head of a cirque high above. It is desired to determine the probability that at least one vehicle will be struck by an avalanche during the winter.

In formulating the problem, note first that it is equivalent to the probability that at least one vehicle and at least one avalanche will attempt to occupy the same space at the same time. So it is useful to first establish a spatial frame of reference of fixed location and constant dimension where the avalanche runout zone crosses the highway. Note also that drivers typically conform to some minimum intervehicle spacing depending on their speed. For the average speed on the highway in Figure A-4, call this distance L where this includes the length of a typical vehicle (inset B). The width of the runout zone is 3L, so that considering traffic in both directions, there are six fixed spatial increments of length L (inset A).

Any such increment may or may not be occupied by a vehicle at any given time. The probability p_v that an increment is occupied can be found from traffic count, average speed, and distance L, where the duration of vehicle presence is Δt. If the presence of a vehicle in any increment is taken as independent of any other, then vehicle presence in any of the six increments within the runout zone can be

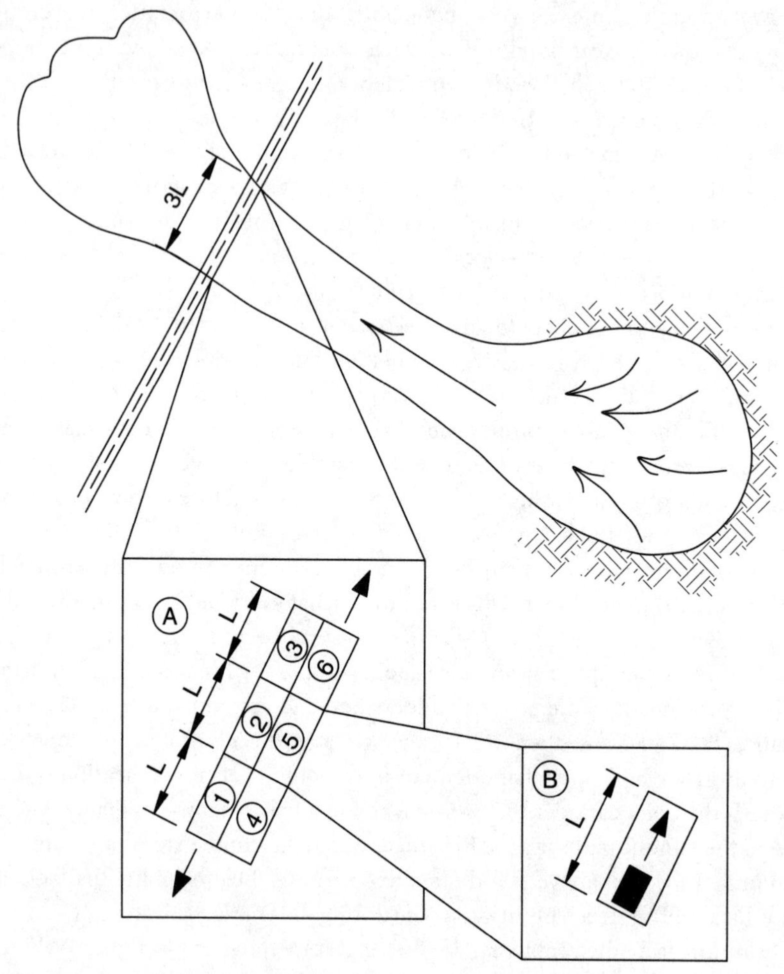

FIGURE A-4. *Avalanche example.*

taken as corresponding Bernoulli trials with constant probability p_v. Denoting the presence of at least one vehicle anywhere in the runout zone over time Δt as event V, and its complement $\overline{V}$ that no vehicles are present:

$$p[\overline{V}] = (1 - p_v)^6 \tag{A-19}$$

and correspondingly,

$$p[V] = 1 - (1 - p_v)^6 \tag{A-20}$$

Now consider the occurrence of an avalanche during time Δt. Call this event A and its probability p_a. Highway maintenance records provide the frequency of avalanche occurrence over the winter, and partitioning this frequency into time increments of Δt provides p_a, where there are n such increments over the winter's duration. The probability of the complementary event $\bar{A}$ that no avalanche occurs is then simply:

$$p[\bar{A}] = 1 - p_a \qquad \text{(A-21)}$$

Let event C represent the contact of an avalanche with at least one vehicle during any given time increment Δt. Its complement $\bar{C}$ is that there is no contact, and this can occur in three ways: if there is an avalanche but no vehicle, a vehicle but no avalanche, or no avalanche and no vehicle. Hence by Axiom III:

$$p[\bar{C}] = p[A \cap \bar{V}] + p[V \cap \bar{A}] + p[\bar{A} \cap \bar{V}] \qquad \text{(A-22)}$$

Assuming that avalanche occurrence and vehicle presence in the runout zone are independent and substituting:

$$p[\bar{C}] = (p_a)(1 - p_v)^6 + (1 - p_a)[1 - (1 - p_v)^6] + (1 - p_a)(1 - p_v)^6 \qquad \text{(A-23)}$$

where $p[C] = 1 - p[\bar{C}]$. However, event C pertains only to duration Δt, so it has n opportunities to occur throughout the winter. Again taking these as independent Bernoulli trials with constant $p[C]$, we arrive at the probability that at least one vehicle is struck during at least one avalanche:

$$p[C_n] = 1 - (1 - p[C])^n \qquad \text{(A-24)}$$

The mean number of avalanches that occur while the runout zone is occupied is $(n)(p[C])$ from Eq. A-24, and the mean number of vehicles present within the runout zone from Eq. A-20 is $6(p_v)$. So the mean number of vehicles struck by avalanches during the winter can be found as $6n(p[C])(p_v)$.

Revisiting the assumption of Bernoulli trials in this example can show some of the various ways in which it might apply only as an approximation. First of all, if the section of highway were on a steep grade, the probability of a spatial unit being occupied at any given time may not be independent of that for its neighbor—cars backing up behind slow-moving trucks could make one unit more likely to be occupied if another were too, thus making these probabilities conditional. Similarly, avalanches might be less likely immediately after some previous occurrence until subsequent accumulation of snow with increasing likelihood thereafter, or they might occur preferentially at the end of the season when spring thaws approach. In short, there can be any number of reasons why some kinds of occurrences may not conform to the probabilistically stationary processes that independent Bernoulli trials and their constant probabilities represent, but they often provide a convenient approximation for making the problem tractable.

Exceptions to Bernoulli assumptions are often encountered for mechanisms that produce some kind of progressive effect or cumulative damage, causing occurrence probabilities to systematically increase over time. Examples might include strain accumulation on a fault that could increase annual earthquake probability, or deterioration of a structure due to repeated episodes of earthquake shaking that would magnify failure probability with each successive event. In these cases the binomial theorem can provide only a snapshot of the physical conditions existing at the moment. While occurrence probability can be cumulated over some future exposure period, the deteriorating physical conditions cannot be reflected or projected forward in time because probabilities for independent Bernoulli trials must remain constant. The binomial theorem can, however, provide a lower bound, so that for situations involving cumulative damage Eq. A-18 becomes:

$$P_n \geq 1 - (1 - p_i)^n \tag{A-25}$$

So if internal erosion were to produce cumulative damage over time, then the 0.22 annual probability of Eq. A-16 makes it possible to state by Eq. A-25 that there would be at least a 92% chance of the dam again experiencing another internal erosion incident at some time within the ensuing 10 years. For comparison, the prior probability of 0.05 before Bayesian updating would have shown this chance to be greater than 40%. Where the influence of previous occurrences on a subsequent trial's probability can be specified, an alternative to the assumption of Bernoulli trials is to adopt a *Markov process* representation (Benjamin and Cornell, 1970).

Probability Bounds

Derived probabilities can also be bounded in other ways that turn out to be quite useful. This applies especially for events that are not truly independent, but whose conditional probabilities would be too difficult or awkward to specify explicitly. Two particular kinds of such cases stand out: spatial variation, and common-cause or common-mode failures, both of which involve correlations among events and their probabilities.

Correlation

The concept of correlation is loosely related to conditionality. While a conditional probability expresses the probability of one event given the known occurrence of some other event, correlation implies a less specific association, hence some relationship between their probabilities. Correlation may be ascribed to an inferred causal dependence that relates the events in some way. But it can also be simply a tendency for their probabilities to be related according to empirical observation without clear-cut knowledge of cause and effect. Correlation can be either positive or negative. Probabilities for positively correlated events (or positively correlated probabilities) tend to move in the same direction: If one probability is higher the other is likely to be higher too, and if one is lower so may be the other. Negative correlation works in the opposite sense, where a higher likelihood for one occurrence tends to imply a lower likelihood for another.

Another property of correlation is the strength or degree of the tendency for association. Here, correlation is anchored to conditional probability at three points. If two events F_1 and F_2 are perfectly positively-correlated, then if one occurs the other surely will too. That is, if $p[F_1] = 1.0$, then $p[F_2 \mid F_1] = 1.0$ as well. And similarly for events that are perfectly negatively correlated, if $p[F_1] = 1.0$, then $p[F_2 \mid F_1] = 0$ and F_2 is impossible. The weakest degree of correlation, of course, is none at all, and here F_1 and F_2 are probabilistically independent. Between these three cases are varying degrees of both positive and negative correlation. This is where probability bounds come in, as illustrated by the following example.

Consider a proposed embankment to be constructed on the foundation soil profile of Figure A-5 containing two clay layers separated by a thin seam of sand. Suppose it has been determined that failure will occur through the foundation if either layer has an undrained shear strength less than 25 kPa, and call these events F_1 and F_2. The probability that the strength falls below this value has been estimated as 0.05 in both cases. Referring back to Eq. A-4, the probability of failure is then represented as:

$$p[F_1 \cup F_2] = p[F_1] + p[F_2] - p[F_1 \cap F_2] \tag{A-26}$$

From Eq. A-5, and substituting the two values:

$$p[F_1 \cup F_2] = 0.05 + 0.05 - p[F_2 \mid F_1]p[F_1] \tag{A-27}$$

which is then:

$$p[F_1 \cup F_2] = 0.10 - 0.05p[F_2 \mid F_1] \tag{A-28}$$

The result of Eq. A-28 now depends on whether the strengths of the two clay layers are independent or correlated in some fashion. Assume first that both were deposited in different lacustrine environments at different times, each with its own parent sediment sources, conditions, and post-depositional stress histories. Here the strengths can be taken as independent, such that:

$$p[F_2 \mid F_1] = p[F_1] = p[F_2] \tag{A-29}$$

So if F_1 and F_2 are independent events, then:

$$p[F_1 \cup F_2] = 0.10 - (0.05)(0.05) = \mathbf{0.0975} \tag{A-30}$$

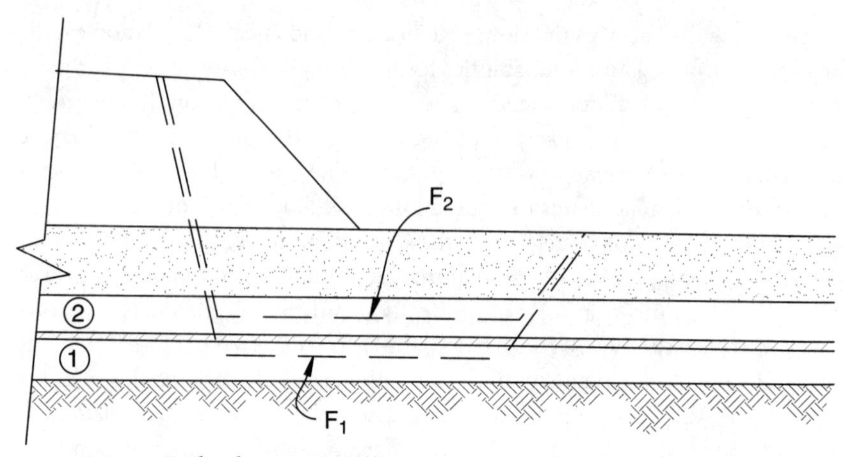

FIGURE A-5. *Embankment example.*

On the other hand, the two layers might be part of a single deposit, thus ensuring identical conditions and stress histories. In this case, the strengths should exhibit perfect positive correlation, hence $p[F_2 \mid F_1] = 1.0$ and:

$$p[F_1 \cup F_2] = 0.10 - (0.05)(1.0) = \mathbf{0.05} \qquad \text{(A-31)}$$

This result is the same, of course, as if the two layers had been considered as one from the start.

Suppose now it is determined that the surface of the deposit was at one time exposed to desiccation to produce a strength profile dependent on depth, such that if the undrained strength of layer 1 were less than 25 kPa then the strength of layer 2 would surely be greater. Now F_1 and F_2 become perfectly negatively correlated where the occurrence of one would preclude the other, or $p[F_1 \mid F_2] = 0$. So for perfect negative correlation:

$$p[F_1 \cup F_2] = 0.10 - (0.05)(0) = \mathbf{0.10} \qquad \text{(A-32)}$$

And if nothing at all were known about geologic origin and no assumptions whatsoever made about independence or correlation, $p[F_1 \cup F_2]$ could still be said to lie somewhere between 0.05 and 0.10. Observe in this case that failure probability is bounded by the assumptions of perfect positive and perfect negative correlation, as might be expected. In the more usual case, failure probability can be bounded within narrower limits given the assumption of independence on one hand, and either positive or negative correlation on the other. Oka and Wu (1990), for example, illustrate probability bounds for independence and positive correlation for failure of an excavation in layered clays. But how these bounds are applied depends on the system and how its components are structured.

Series and Parallel Systems

Any engineered system can be considered as a collection of physical parts or components that must successfully function for it to perform as intended. This applies no less to geotechnical systems where components can include not only structural elements and parts but also geologic features, so that in the previous example the embankment and the two foundation clay layers represent components of such a system. Predicting system performance requires establishing how the components interact, and where this performance is to be predicted probabilistically it also requires establishing the relationships among the component failure probabilities.

Figure A-6(a) shows a simple *series system* with physical components A through N. Here, for the system to perform successfully all components must function, so the failure of any one or more (in other words, A or B or C or ... N) results in system failure. By contrast, Figure A-6(b) shows a simple *parallel system* (or more precisely, a *standby-parallel system*) with redundant components where system failure requires that all of them malfunction. Systems can contain both series and parallel components, as in the *mixed system* of Figure A-6(c). These representations of component interactions are called *reliability block diagrams.* Note that system failure for series systems is the union of the failure modes or component failures, whereas for the parallel system it is the intersection of component failures.

A *failure mode* for a system is defined as some unique combination of events whose joint occurrence would cause the system to fail. In a geotechnical setting these events may include the presence of certain physical or geologic conditions, the failure of a physical feature, or the occurrence of some external initiator such as an earthquake. Moreover, failure modes of the same nature can pertain separately to different spatial regions of their physical setting. Reliability block diagrams can be used to represent failure modes of a system as well as its physical components, and all but the simplest of systems have more than one failure mode. As shown in Figure A-6(a), multiple failure modes always occur in series because by definition any one of them causes system failure. The malfunction of components A through N in parallel would together represent one such failure mode in Figure A-6(b). In the mixed system of Figure A-6(c) there are two failure modes—failure of component A alone, and failure of components B through N inclusive. This is shown as the equivalent series representation in Figure A-6(d), where the second failure mode also represents joint malfunction of the components in parallel.

Geotechnical systems may contain redundant physical components in parallel, this being the "belt and suspenders" principle of design. Adverse effects of seepage for an earth dam, for example, might first require a flaw in the core and then inadequate capacity of a drain. Failure of such redundant components can be reduced to one failure mode in the manner of Figure A-6(d) in cases where there are rela-

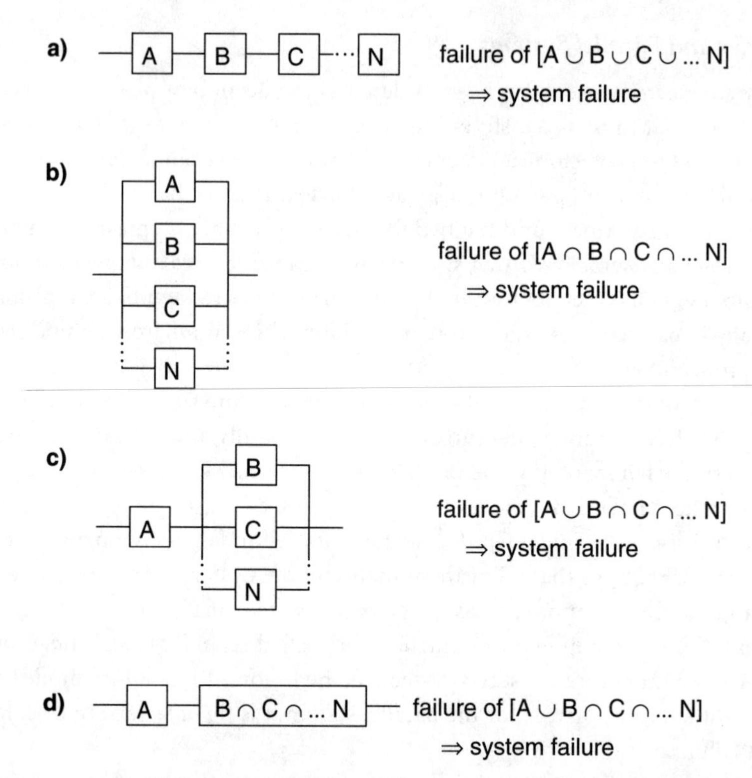

FIGURE A-6. *System types: a, series system; b, parallel system; c, mixed system; d, series representation of mixed system.*

tively few such components and their physical interactions are reasonably well understood. This is the case for most geotechnical systems, which can usually be represented by failure modes in series after accounting for redundancies.

If all of the remaining failure modes are truly independent, or can be assumed so with little argument, then determining the probability of system failure is straightforward. But a problem arises when failure modes may be correlated in some way such that one is related to another, as in soil-structure interaction problems where the response of the structure to the soil influences the response of the soil to the structure. In complex situations, it may not be possible from a practical standpoint to fully specify these interactions or explicitly address them with conditional probabilities, and this is where bounds on system failure probability become useful. The reason for having so carefully distinguished between series and parallel systems is that probability bounds work differently for the two, and it is important to keep this in mind in applying them to a system's failure modes as distinct from its physical components. Because failure modes (though not necessarily failures of individual system components) by definition are always in series, further discussion is restricted to series systems.

Unimodal Bounds for Positive Correlation in Series Systems

For series systems, it can be shown that the greater the degree of positive correlation among failure modes, the lower the probability of failure compared to no correlation—which is to say independence (Cornell, 1967). Probabilistic independence therefore provides an upper bound for positively correlated failure modes, as was demonstrated in the example of Figure A-5. This important observation can be understood by noting that positive correlation implies that if one failure mode is unlikely, the others will be unlikely as well, where series system failure requires that one or more occur. At the other limit, the lower-bound failure probability must be at least as great as for perfect positive correlation, represented by single most likely such event. More formally, Ang and Tang (1984) provide the *first-order* or *unimodal* bounds for positively correlated events in series systems as:

$$\max p_i < p_F < 1 - (1 - p_1)(1 - p_2) \dots (1 - p_n) \quad \text{(A-33)}$$

where $p_i = p_1 \dots p_n$ are component or failure mode probabilities and p_F is the probability of at least one. For the case of small p_i, the right-hand side can be approximated as Σp_i so in this case Eq. A-33 reduces to:

$$\max p_i < p_F < \Sigma p_i \quad \text{(A-34)}$$

Eq. A-34 can be visualized with reference to a chain, whose links might snap with probabilities that would be positively correlated if, say, it were rusty. The left-hand side of Eq. A-34 represents failure of that link most likely to be weakest, while the right-hand side represents failure of any link. It is useful to note for both Eq. A-33 and Eq. A-34 that bounds on p_F will be narrowest for systems with few failure modes or where one dominates, and widest when there are many failure modes all of similar occurrence likelihood.

These unimodal bounds are not restricted to failure modes and can be applied to situations involving spatial variability. For long structures spanning considerable distances, like levees, highways, or pipelines, probabilities of various subsurface conditions or soil properties are often estimated individually for different geologic conditions over discrete segments of their lengths. Here, spatial variability often gives rise to correlation effects among these segments much like those in the example of Figure A-5. Take the dam foundation shown in Figure A-7. The condition of interest is whether and where permeable materials might be present, and various regions of the dam alignment are defined to reflect differences in geologic conditions. The left abutment (S_1) contains moderately jointed rock, on the right side (S_5) are colluvial deposits, and alluvial soils exist in between (S_2, S_3, and S_4). Their distinct geologic origins suggest that the regions of rock, colluvium, and alluvium could be treated as probabilistically independent

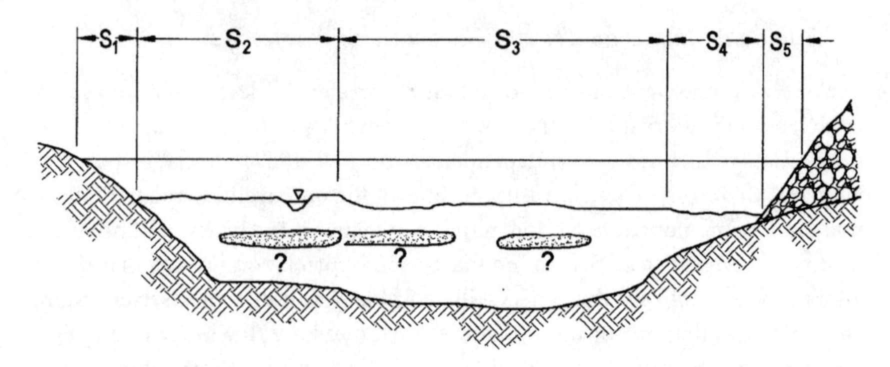

FIGURE A-7. *Dam foundation example.*

because the occurrence of pervious conditions in one would not be relevant to their presence in any of the others.

For the alluvial material, however, the concern is for pervious buried streamchannel deposits, and the location of the existing streamchannel suggests they would be most likely to be found beneath it (segment S_2). But they might also extend from there within the overbank deposits (S_3) or could occur as ancient meanders, with their likelihood decreasing with distance toward S_4. Thus, there are different degrees of correlation among all of the segments, ranging from none (independence) to positive correlations between the three alluvial segments such that if buried streamchannel deposits were found in one they would be more likely to be found in one or more of the others as well.

In principle, these spatial correlations could be represented explicitly as a matrix of conditional probabilities for a Markov process, where conditional probability of occurrence in each segment is related to all the possible conditions in each of the others. Einstein, et al. (1996) describe the use of these Markov *transition probabilities* for tunnel segments where the probability of encountering certain geologic conditions in advance of the face is conditional on their presence in the heading itself. But if the problem concerns whether some condition occurs within at least one spatially-defined segment, the unimodal bounds of Eq. A-33 or A-34 can be conveniently applied. Again, note that the probability of such a condition occurring anywhere is at least as great as its highest probability in any segment but less than its probability in one or more taken independently.

Unimodal bounds provide a useful but highly simplified technique for handling spatially correlated conditions. More sophisticated treatment is described by Vanmarke (1977), who presents the basic elements of statistical analysis of spatial variability where measured data are available. DeGroot and Baecher (1993) describe related statistical properties such as *autocovariance* and *autocor-*

relation and their determination, with Christian, et al. (1992) providing a comprehensive example of their use that also includes isolating the effects of true spatial variation from other sources of data uncertainty.

Aside from spatial variation, unimodal bounds are often useful in accounting for correlations that occur for other reasons, such as interrelated failure mechanisms. Failure modes are often taken as independent even though there may be some underlying mutual condition, physical process, or interaction that results in positive correlation among them. This results in *common-cause* or *common-mode* failures, which come about from some shared condition that affects each failure mode or when one enhances another, circumstances that can be difficult to explicitly identify or specify. This can be illustrated by taking the example of Figure A-7 one step further. The conditional probability of high foundation pore pressures given the presence of a buried channel would be of interest for segments S_2 through S_4. To the extent that these channels are interconnected and have access to the reservoir, they would have some degree of positive correlation calling for the application of Eq. A-33 or A-34.

Failure modes treated as independent can also interact physically, where the occurrence of one could promote another, or by the same token where the nonoccurrence of one failure mode would make some other less likely. Examples in Figure A-8 include sequential rockslope toppling failures, tieback pullout and "zippering" from transfer of load, or instability of one part of a slope destabilizing an adjacent area, all cases of *progressive failure.* In principle, the simple situations of Figure A-8 might be represented explicitly in a fault or event tree where one failure mode is conditional on the other. But there may be too many separate possibilities, or their physical dependencies may be too difficult to describe. Here, unimodal bounds can be applied to estimate the probability of at least one such failure mode occurring.

Finally, separate failure modes from the same failure initiator, like an earthquake, are usually assumed to proceed independently of each other once that initiator occurs. Embankment crest deformations and slope instability would ordinarily be taken as independent failure modes, or lateral spreading as independent from flowsliding. However, this is mostly for analytical convenience, since these are end-members of what is really a continuum of effects from the same underlying mechanism. Similarly, it is hard to specify exactly where excessive foundation settlement leaves off and bearing capacity failure begins when they both have in common the stress/deformation response of the soil. So correlated conditions can come about through spatial variability, physical interactions or shared mechanisms among failure modes. For all these reasons, common-cause or common-mode effects among multiple failure modes may be more the rule than the exception, with Vick and Stewart (1996) providing several examples of the use of Eq. A-33 to address them.

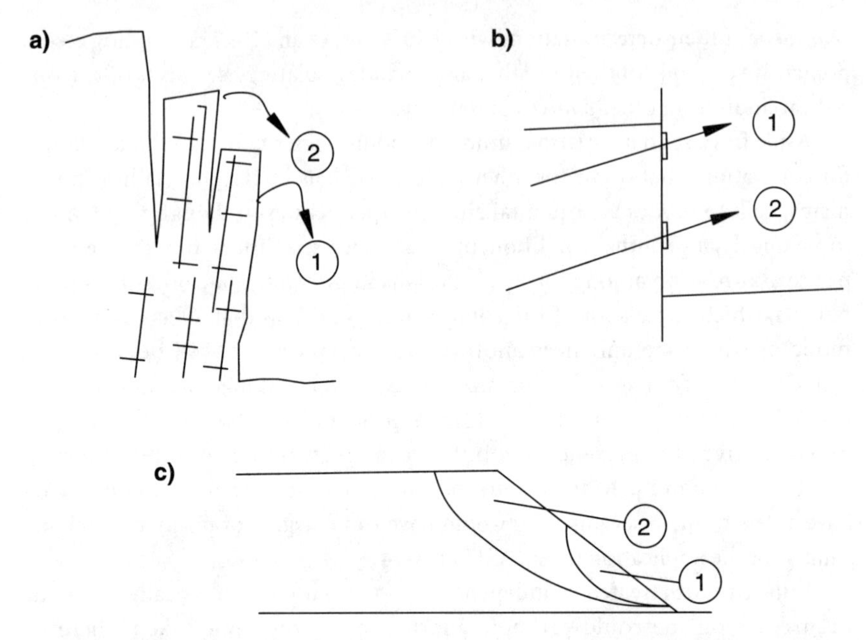

FIGURE A-8. *Progressive failure examples: a, rockslope toppling; b, tieback pullout; c, slope instability.*

Negative Correlation in Series Systems

Correlated events are, of course, not always positively correlated, and negative correlation pertains when an event that is more likely implies that some other event will be less likely. In terms of spatial variability, negative correlation affects some feature, condition, or property that tends to be preferentially concentrated in some spatial region, such that its occurrence in one place would reduce its likelihood elsewhere. For example, if stress-relief joints had formed in a steep valley sidewall, they would be less likely to form farther back. Or if karst sinkholes align preferentially with regional joint lineations, they might be less likely to be found in between. And since overconsolidation effects on clay strength generally decrease with depth, a higher probability of encountering soft material at the bottom of such a deposit will be associated with a lower probability at the top.

In terms of failure conditions, and returning to the example in Figure A-7, suppose one were interested in the conditional probability of a seepage-induced sand boil given the occurrence of high pore pressures in the alluvial foundation segments S_2 through S_4. These events will be negatively correlated if a sand boil in one segment relieves the pressure within a buried channel interconnected with that in some other segment. Similarly, this negative correlation means that if a sand boil does not occur in one segment, then high pressures will persist in them all.

Negatively correlated failure modes are often found in cases where the occurrence of one would relieve stress or produce some kind of stabilizing effect on others. For parallel systems this would have a favorable "fail-safe" influence by enhancing the performance of redundant components. But the effect on systems in series is the opposite because here by the time the first component fails the system has failed already, and so long as the first component has not yet failed the others remain more likely to do so.

Negative correlations are especially important because they produce probabilities for series system failure that are higher than those that result from assumptions of independence. This can be confirmed from the example of Figure A-5. Specifically, for negatively correlated events in series systems, the unimodal bounds on p_F (Ang and Tang, 1984) result in:

$$p_F \geq 1 - (1 - p_1)(1 - p_2) \ldots (1 - p_n) \tag{A-35}$$

The similarity to Eq. A-25 for cumulative damage can be explained by noting that both Eq. A-25 and Eq. A-35 reflect a propensity for component probabilities to increase in the aggregate.

References

Adelson, B., and Soloway, E. (1988). "A model of software design." *The nature of expertise*, M. Chi et al., eds., Erlbaum, Hillsdale, N.J., 185-208.

Agricola, G. (1556). *De re metallica*, translated by H. Hoover and L. Hoover, 1912, The Mining Magazine (publisher), London; reprinted 1950, Dover, New York.

Alpert, M., and Raiffa, H. (1982). "A progress report on training of probability assessors." *Judgment under uncertainty: Heuristics and biases*, D. Kahneman et al., eds., Cambridge Univ. Press, Cambridge, U.K., 294-305.

Ambraseys, N. (1988). "Engineering seismology." *Earthquake Engrg. and Struct. Dynamics*, 17(1), 1-105.

Ang, A., and Tang, W. (1975). *Probability concepts in engineering planning and design - Vol. I: Basic principles*, Wiley, New York.

Ang, A., and Tang, W. (1984). *Probability concepts in engineering planning and design - Vol. II: Decision, risk, and reliability*, Wiley, New York.

Anzai, Y. (1991). "Learning and use of representations for physics exercise." *Toward a general theory of expertise*, K. Ericsson and J. Smith, eds., Cambridge Univ. Press, Cambridge, U.K., 64-92.

Arlin, P. (1990). "Wisdom: The art of problem finding." *Wisdom: Its nature, origins, and development*, R. Sternberg, ed., Cambridge Univ. Press, Cambridge, U.K., 230-243.

Atkinson, G., Finn, W., and Charlwood, R. (1984). "Simple computation of liquefaction probability for seismic hazard applications." *Earthquake Spectra*, 1(1), 107-123.

Ayton, P. (1992). "On the competence and incompetence of experts." *Expertise and decision support*, G. Wright and F. Bolger, eds., Plenum Press, New York, 77-105.

Baecher, G. (1972). "Site exploration: A probabilistic approach." Ph.D dissertation, Massachusetts Institute of Technology, Cambridge, Mass.

Baecher, G. (1983). "Professional judgement and prior probabilities in engineering risk assessment." *Proc. Fourth Int. Conf. on Applications of Statistics and Probability in Soil Struct. Engrg.*, Pitagora Editrice, Bologna, Italy, 635-650.

Baecher, G. (1999). "Discussion of 'Inaccuracies associated with estimating random measurement errors.'" *J. Geotech. and Geoenvir. Engrg.*, 125(1), 79-80.

Baecher, G., Einstein, H., Vanmarke, E., Veneziano, D., and Whitman, R. (1984). "Discussion of 'Conventional and probabilistic embankment design.'" *J. Geotech. Engrg.*, 110(7), 989-991.

Bar-Hillel, M. (1982). "Studies of representativeness." *Judgment under uncertainty: Heuristics and biases,* D. Kahneman et al., eds., Cambridge Univ. Press, Cambridge, U.K., 69-83.

Baron, J., and Frisch, D. (1994). "Ambiguous probabilities and the paradoxes of expected utility." *Subjective probability,* G. Wright and P. Ayton eds., Wiley, Chichester, U.K., 273-294.

Barry, T. (1989). "An overview of health risk analysis in the Environmental Protection Agency." *Risk analysis and management of natural and man-made hazards,* Y. Haimes and E. Stakhiv, eds., ASCE Press, Reston, Va., 50-71.

Beach, L. (1992). "Epistemic strategies: Causal thinking in expert and nonexpert judgment." *Expertise and decision support,* G. Wright and F. Bolger, eds., Plenum Press, New York, 107-127.

Beach, L., and Braun, G. (1994). "Laboratory studies of subjective probability: A status report." *Subjective probability,* G. Wright and P. Ayton, eds., Wiley, Chichester, U.K., 107-127.

Beach, L., Barnes, V., and Christensen-Szalanski, J. (1986). "Beyond heuristics and biases: A contingency model of judgmental forecasting." *J. Forecasting,* 5(3), 143-157.

Benjamin, J., and Cornell, C. (1970). *Probability, statistics, and decision for civil engineers,* McGraw-Hill, New York.

Bennett, I. (1915). *The history of the Panama Canal, its construction and builders,* Historical Publishing Company, Washington, D.C.

Benson, P., Curley, S., and Smith, G. (1995). "Belief assessment: An underdeveloped phase of probability elicitation." *Mgmt. Sci.,* 41(10), 1639-1653.

Bernstein, P. (1996). *Against the gods: The remarkable story of risk,* Wiley, New York.

Boorstin, D. (1983). *The discoverers,* Random House, New York.

Brehmer, B., (1980). "In one word: Not from experience." *Acta Psychologica,* (45), 223-241.

Brinch-Hansen, J. (1962). "Relationships between stability analysis with total and effective stress." *Sols-Soils,* (3), 28-41.

Budescu, D., Weinberg, S., and Wallsten, T. (1988). "Decisions based on numerically and verbally expressed uncertainties." *J. Exp. Psychology: Human Perception and Performance,* 14(2), 281-294.

Burke, J. (1985). *The day the universe changed,* Little Brown, Boston.

Byrne, P. (1991). "A model for predicting liquefaction induced displacement." *Proc. 2nd Int. Conf. on Recent Advances in Geotech. Earthquake Engrg.*, Vol. 2., Univ. of Missouri-Rolla, 1027-1035.

Byrne, P., Imrie, A., and Morgenstern, N. (1993). "Results and implications of seismic performance studies - Duncan Dam." *Proc. 46th Ann. Canadian Geotech. Conf.*, BiTech Publishers, Vancouver, B.C., 271-281.

Cai, Z., and Bathurst, R. (1996). "Deterministic sliding block methods for estimating seismic displacements of earth structures." *Soil Dynamics and Earthquake Engrg.*, 15, 225-268.

Casagrande, A. (1936). "Characteristics of cohesionless soils affecting the stability of slopes and earth fills." *J. Boston Soc. Civil Engrs.*, 23(1), 13-32.

Casagrande, A. (1980). "Discussion of 'Liquefaction potential: Science versus practice.'" *J. Geotech. Engrg. Div.*, 106(GT6), 725-727.

Chávez, T. (1996). "Modeling and measuring the effects of vagueness in decision models." *IEEE Trans. on Systems, Man, and Cybernetics - Part A: Systems and Humans*, 25(3), 311-323.

Christian, J., Ladd, C., and Baecher, G. (1992). "Reliability and probability in stability analysis." *Stability and performance of slopes and embankments - II*, R. Seed and R. Boulanger, eds., Geotech. Spec. Pub. No. 31, Vol. 2, ASCE Press, Reston, Va., 1071-1111.

Clancey, W. (1988). "Acquiring, representing, and evaluating a competence model of diagnostic strategy." *The nature of expertise*, M. Chi et al., eds., Erlbaum, Hillsdale, N.J., 343-418.

Clark, D. (1992). "Human expertise, statistical models, and knowledge-based systems." *Expertise and decision support*, G. Wright and F. Bolger, eds., Plenum Press, New York, 227-249.

Combs, B., and Slovik, P. (1979). "Newspaper coverage of causes of death." *Journalism Quarterly*, 56(3), 837-844.

Cone, J. (1991). *Fire under the sea*, William Morrow, New York.

Connell, E. (1991). *The alchymist's journal*, North Point Press, San Francisco.

Cooke, R. (1991). *Experts in uncertainty: Opinion and subjective probability in science*, Oxford Univ. Press, New York.

Cornell, C. (1967). "Bounds on the reliability of structural systems." *J. Struct. Div.*, 93(ST1), 171-200.

Curley, S., and Benson, P. (1994). "Applying a cognitive perspective to probability construction." *Subjective probability*, G. Wright and P. Ayton, eds., Wiley, Chichester, U.K, 185-209.

de Boer, R., Schiffman, R., and Gibson, R. (1996). "The origins of the theory of consolidation: The Terzaghi-Fillunger dispute." *Geotechnique*, 46(2), 175-186.

de Finetti, B. (1937). "Foresight: Its logical laws, its subjective sources." translated in: *Studies in subjective probability*, H. Kyburg and H. Smokler, eds., 1964, Wiley, New York.

de Mello, V. (1977). "Reflections on design decisions of practical significance to embankment dams." *Geotechnique*, 27(3), 279-355.

de Neufville, R. (1990). *Applied systems analysis: Engineering planning and technology management*, McGraw-Hill, New York.

DeGroot, D. (1996). "Analyzing spatial variability of in situ soil properties." *Uncertainty in the geologic environment: From theory to practice*, C. Shackelford et al., eds., Geotech. Spec. Pub. No. 58, Vol. 1, ASCE Press, Reston, Va., 210-238.

DeGroot, D., and Baecher, G. (1993). "Estimating autocovariance of in-situ soil profiles." *J. Geotech. Engrg.*, 119(1), 147-166.

Dise, K., and Vick, S. (2000). "Dam safety risk analysis for Navajo Dam." *20th Cong. on Large Dams*, Int. Commission on Large Dams, Q.76, R.22, 337-350.

Dorbath, L., Cisternas, A., and Dorbath, C. (1990). "Assessment of the size of large and great historical earthquakes in Peru." *Bull. Seis. Soc. of Am.*, 80(3), 551-570.

Dreyfus, H. (1997). "Intuitive, deliberate, and calculative models of expert performance." *Naturalistic decision making*, C. Zsambok and G. Klein, eds., Erlbaum, Mahwah, N.J., 17-28.

Duncan, J. (1996). "State of the art: Limit equilibrium and finite-element analysis of slopes." *J. Geotech. Engrg.*, 122(7), 557-596.

Dushnisky, K., and Vick, S. (1996). "Evaluating risk to the environment from mining using failure modes and effects analysis." *Uncertainty in the geologic environment: From theory to practice*, C. Shackelford et al., eds., Geotech. Spec. Pub. No. 58, Vol. 2, ASCE Press, Reston, Va., 848-865.

Eckersley, J. (1990). "Instrumented laboratory flowslides." *Geotechnique*, 40(3), 489-502.

Eddy, D. (1982). "Probabilistic reasoning in clinical medicine: Problems and opportunities." *Judgment under uncertainty: Heuristics and biases*, D. Kahneman et al., eds., Cambridge Univ. Press, Cambridge, U.K., 249-267.

Eddy, D., and Clanton, C. (1982). "The art of diagnosis: Solving the clinicopathological exercise." *The New England Journal of Medicine*, 306, 1263-1268.

Edmunson, R. (1990). "Decomposition: A strategy for judgemental forecasting." *J. Forecasting*, 9(4), 305-314.

Edwards, W. (1975). "Comment on 'Cognitive processes and the assessment of probability distributions.'" *J. Am. Statistical Assn.*, 70(350), 291-293.

Eils, L., and John, R. (1980). "A criterion validation of multiattribute utility analysis and of group communication strategy." *Organizational Behavior and Human Performance*, 13(2), 268-288.

Einhorn, H., and Hogarth, R. (1981). "Behavioral decision theory: Processes of judgment and choice." *Ann. Rev. of Psychology*, 32(1), 53-83.

Einhorn, H., and Hogarth, R. (1982). "Prediction, diagnosis, and causal thinking in forecasting." *J. Forecasting*, 1(1), 23-36.

Einhorn, H., Hogarth, R., and Klempner. E. (1977). "Quality of group judgment." *Psychological Bull.*, 84(1), 158-172.

Einstein, H. (1991). "Observation, quantification, and judgment: Terzaghi and engineering geology." *J. Geotech. Engrg.*, 117(11), 1772-1778.

Einstein, H., Halabe, V., Dudt, J., and Descoeudres, F. (1996). "Geologic uncertainties in tunneling." *Uncertainty in the geologic environment: From theory to practice*, C. Shackelford et al., eds., Geotech. Spec. Pub. No. 58, Vol. 1, ASCE Press, Reston, Va., 239-253.

Endsley, M. (1995). "Toward a theory of situation awareness in dynamic systems." *Human Factors*, 37(1), 32-64.

Erev, I., Bornstein, G., and Wallsten, T. (1993). "The negative effect of probabilistic assessments on decision quality." *Organizational Behavior and Human Decision Processes*, 55(1), 78-94.

Ericsson, K. (1996). "The acquisition of expert performance: An introduction to some of the issues." *The road to excellence: The acquisition of expert performance in the arts and sciences, sports and games*, K. Ericsson, ed., Erlbaum, Mahwah, N.J., 1-50.

Ericsson, K., and Smith, J. (1991). "Prospects and limits of the empirical study of expertise: An introduction." *Toward a general theory of expertise*, K. Ericsson and J. Smith, eds., Cambridge Univ. Press, Cambridge, U.K., 1-38.

Fanelli, M. (1997). "The scientific definition of dam safety: Is the Emporer fully clothed?" *Int. J. Hydropower & Dams*, 4(2), 54-58.

Ferguson, E. (1992). *Engineering and the mind's eye*, MIT Press, Cambridge, Mass.

Ferrell, W. (1994). "Discrete subjective probabilities and decision analysis: Elicitation, calibration, and combination." *Subjective probability*, G. Wright and P. Ayton, eds., Wiley, Chichester, U.K., 411-451.

Feynman, R. (1987). "Mr. Feynman goes to Washington." *Engineering & Science*, LI(1), Fall.

Finn, W. (1998). "Seismic safety of embankment dams, developments in research and practice 1988-1998." *Geotechnical earthquake engineering and soil dynamics III*, P. Dakoulas et al., eds., Geotech. Spec. Pub. No. 75, Vol. 2, ASCE Press, Reston, Va., 812-835.

Fischhoff, B. (1982). "Debiasing." *Judgment under uncertainty: Heuristics and biases*, D. Kahneman et al., eds., Cambridge Univ. Press, Cambridge, U.K., 422-444.

Fischhoff, B., and MacGregor, D. (1982). "Subjective confidence in forecasts." *J. Forecasting*, 1(2), 155-172.

Fischhoff, B., and Slovik, P. (1980). "A little learning...: Confidence in multicue judgment." *Attention and performance VIII*, R. Nickerson, ed., Erlbaum, Hillsdale, N.J, 779-800.

Fischhoff, B., Slovik, P., and Lichtenstein, S. (1977). "Knowing with certainty: The appropriateness of extreme confidence." *J. Exp. Psychology: Human Perception and Performance*, 3(4), 552-564.

Fischoff, B., Lichtenstein, S., Slovik, P., Derby, S., and Keeney, R. (1981). *Acceptable risk*, Cambridge Univ. Press, Cambridge, U.K.

Focht, J. (1994). "Lessons learned from missed predictions." *J. Geotech. Engrg.*, 120(10), 1653-1683.

Folayan, J., Höeg, K., and Benjamin, J. (1970). "Decision theory applied to settlement predictions." *J. Soil Mech. and Found. Div.*, 96(SM4), 1127-1141.

Frankel, A., and Safak, E. (1988). "Recent trends and future prospects in seismic hazard analysis." *Geotechnical earthquake engineering and soil dynamics III*, P. Dakoulas et al., eds., Geotech. Spec. Pub. No. 75, Vol. 1, ASCE Press, Reston, Va,, 91-115.

Franklin, J. (2001). *The science of conjecture: Evidence and probability before Pascal*, Johns Hopkins Univ. Press, Baltimore.

Freudenthal, A. (1947). "The safety of structures." *Trans. ASCE*, 112, paper no. 2296, 125-159.

Fullwood, R., and Hall, R. (1988). *Probabilistic risk assessment in the nuclear power industry: Fundamentals and applications*, Pergamon Press, Oxford, U.K.

Gammack, J. (1992). "Knowledge engineering issues for decision support." *Expertise and decision support*, G. Wright and F. Bolger, eds., Plenum Press, New York, 203-226.

Government Accounting Office (1985). "Probabilistic risk assessment: An emerging aid to nuclear power plant safety regulation." *GAO/RCED 85-11*, U.S. Government Accounting Office, Washington, D.C.

Garrick, J. (1989). "Risk assessment practices in the space industry." *Risk Analysis*, 9(1), 1-7.

Gigerenzer, G. (1994). "Why the distinction between single-event probabilities and frequencies is important for psychology (and vice-versa)." *Subjective probability*, G. Wright and P. Ayton, eds., Wiley, Chichester, U.K., 129-161.

Gigerenzer, G., Swijtink, Z., Porter, T., Daston, L., Beatty, J., and Krüger, L. (1989). *The empire of chance: How probability changed science and everyday life*, Cambridge Univ. Press, Cambridge, U.K.

Gillette, D., (1990). "Dempster-Shafer modeling of judgment in geotechnical engineering." Ph.D dissertation, Univ. of Colorado, Boulder.

Glaser, R., and Chi, M. (1988). "Overview." *The nature of expertise*, M. Chi et al., eds., Erlbaum, Hillsdale, N.J., xv-xxviii.

Goodman, R. (1999). *Karl Terzaghi: The engineer as artist*, ASCE Press, Reston, Va.

Gordon, J., and Duguid, D. (1970). "Experiences with cracking at Duncan Dam." *10th Cong. on Large Dams*, Int. Commission on Large Dams, Q.36, R.14, 469-485.

Gould, S. (1991). *Bully for brontosaurus: Further reflections on natural history*, Penguin Books, London.

Grayson, C. (1960). *Decisions under uncertainty: Drilling decisions by oil and gas operators*, Harvard University Press, Cambridge, Mass.

Greenberg, H., and Cramer, J. (1991). *Risk assessment and risk management for the chemical process industry*, Van Nostrand Reinhold, New York.

Greenberg, M., Sachsman, D., Sandman, P., and Salamone, K. (1989). "Network evening news coverage of environmental risk." *Risk Analysis*, 9(1), 119-126.

Griffin, D., and Tversky, A. (1992). "The weighting of evidence and the determinants of confidence." *Cognitive Psychology*, 24(3), 411-435.

Griffin, F. (1998). "An American railroad man east of the Urals." *The Historian*, 60(4), 812-819.

Hacking, I. (1975). *The emergence of probability: A philosophical study of early ideas about probability, intuition, and statistical inference*, Cambridge Univ. Press, Cambridge, U.K.

Halim, I., and Tang, W. (1993). "Site exploration strategy for geologic anomaly characterization." *J. Geotech. Engrg.*, 119(2), 195-213.

Hamada, M., Towhata, I., Yasuda, S., and Isoyama, R. (1987). "Study on permanent ground displacement induced by seismic liquefaction." *Computers and Geotechnics*, 4, 197-220.

Hamm, R. (1991). "Selection of verbal probabilities: A solution for some problems of verbal probability expression." *Organizational Behavior and Human Decision Processes*, 48(2), 193-223.

Hammond, K., Hamm, R., Grassia, J., and Pearson, T. (1987). "Direct comparison of the efficacy of intuitive and analytical cognition in expert judgment." *IEEE Trans. on Systems, Man, and Cybernetics*, SMC-17(5), 753-770.

Hanks, T. (1997). "Imperfect science: Uncertainty, diversity, and the experts." *EOS, Trans. Am. Geophysical Union*, 78(35), 371-377.

Harr, M. (1987). *Reliability-based design in civil engineering*, McGraw-Hill, New York.

Henley, E., and Kumamoto, H. (1992). *Probabilistic risk assessment: Reliability engineering, design, and analysis*, IEEE Press, Piscataway, N.J.

Henrion, M., Fischer, G., and Mullin, T. (1993). "Divide and conquer? Effects of decomposition on the accuracy and calibration of subjective probability distributions." *Organizational Behavior and Human Decision Processes*, 55(2), 207-227.

Hogarth, R. (1975). "Cognitive processes and the assessment of subjective probability distributions." *J. Am. Statistical Assn.*, 70(350), 271-289.

Honjo, Y. and Veneziano, D. (1989). "Improved filter criteria for cohesionless soils." *J. Geotech. Engrg.*, 115(1), 75-94.

Hora, S., Dodd, N., and Hora, J. (1993). "The use of decomposition in probability assessments of continuous variables." *J. Behavioral Decision Making*, 6(2), 133-147.

Horwich, P., ed., (1993). *World changes: Thomas Kuhn and the nature of science*, MIT Press, Cambridge, Mass.

Howe, M. (1996). "The childhoods and early lives of geniuses: Combining psychological and biographical evidence." *The road to excellence: The acquisition of expert performance in the arts and sciences, sports and games*, K. Ericsson, ed., Erlbaum, Mahwah, N.J., 255-270.

Hungr, O. (1988). "Notes on dynamic analysis of flowslides." *Proc. Fifth Int. Symp. on Landslides*, Balkema, Rotterdam, 679-683.

Hungr, O. (1990). "Mobility of rock avalanches." *Nat. Res. Inst. for Earth Science and Disaster Prevention Report No. 46*, Tskuba, Japan.

Hynes, M., and Vanmarke, E. (1976). "Reliability of embankment performance predictions." *Proc. First ASCE Engrg. Mechanics Div. Spec. Conf.*, Univ. of Waterloo Press, Waterloo, Ontario, 367-384.

Idriss, I. (1985). "Evaluating seismic risk in engineering practice." *Proc. Eleventh Int. Conf. Soil Mech.*, Balkema, Rotterdam, 255-320.

Indraratna, B., and Vafai, F. (1997). "Analytical model for particle migration within base soil-filter system." *J. Geotech. and Geoenvir. Engrg.*, 123(2), 100-109.

Jardine, L. (1999). *Ingenious pursuits: Building the scientific revolution*, Anchor Books, New York.

Jeyapalan, J., Duncan, J., and Seed, H. (1983). "Analyses of flow failures of mine tailings dams." *J. Geotech. Engrg.*, 109(2), 150-171.

Johnson, E. (1988). "Expertise and decision under uncertainty: Performance and process." *The nature of expertise*, M. Chi et al., eds., Erlbaum, Hillsdale, N.J., 209-228.

Johnson, S. (1975). "Analysis and design relating to embankments." *Proc. Conf. on Analysis and Design in Geotechnical Engineering*, Vol. II, ASCE Press, Reston, Va., 1-48.

Joralemon, I. (1976). *Adventure beacons*, Society of Mining Engineers, AIME, New York.

Junger, S. (1997). *The perfect storm*, HarperCollins, New York.

Kaempf, G., and Orasanu, J. (1997). "Current and future applications of naturalistic decision making in aviation." *Naturalistic decision making*, C. Zsambok and G. Klein, eds., Erlbaum, Mahwah, N.J., 81-90.

Kahneman, D., and Tversky, A. (1973). "On the psychology of prediction." *Psychological Rev.*, 80(4), 273-251.

Kahneman, D., and Tversky, A. (1982). "Subjective probability: A judgment of representativeness." *Judgment under uncertainty: Heuristics and biases*, D. Kahneman et al., eds., Cambridge Univ. Press, Cambridge, U.K., 32-47.

Keeney, R., and Raiffa, H. (1976). *Decision with multiple objectives: Preferences and value tradeoffs*, Wiley, New York.

Keeney, R., and von Winterfeldt, D. (1991). "Eliciting probabilities from experts in complex technical problems." *IEEE Trans. on Engineering Management*, 38(3), 191-201.

Kempton, J., Locke, W., Atkins, D, and Nicholson, A. (2000). "Probabilistic quantification of uncertainty in predicting mine pit-lake water quality." *Mining Engineering*, 52(10), 59-64.

Kent, S. (1949). *Strategic intelligence*, Princeton Univ. Press, Princeton, N.J.

Kinnicutt, P., and Einstein, H. (1996). "Incorporating uncertainty, objective, and subjective data in geologic site characterization." *Uncertainty in the geologic environment: From theory to practice*, C. Shackelford et al., eds., Geotech. Spec. Pub. No. 58, Vol. 1, ASCE Press, Reston, Va., 104-118.

Kitchener, K., and Brenner, H. (1990). "Wisdom and reflective judgment: Knowing in the face of uncertainty." *Wisdom: Its nature, origins, and development*, R. Sternberg, ed., Cambridge Univ. Press, Cambridge, U.K., 212-229.

Klein, G. (1998). *Sources of power: How people make decisions*, MIT Press, Cambridge, Mass.

Kleinmuntz, D., Fennema, M., and Peecher, M. (1996). "Conditional assessment of subjective probabilities: Identifying the benefits of decomposition." *Organizational Behavior and Human Decision Processes*, 66(1), 1-15.

Koné, D., and Mullet, E. (1994). "Societal risk perception and media coverage." *Risk Analysis*, 14(1), 21-24.

Koriat, A., Lichtenstein, S., and Fischhoff, B. (1980). "Reasons for confidence." *J. Exp. Psychology: Human Learning and Memory*, 6(2), 107-118.

Krause, P., and Clark, D. (1994). "Uncertainty and subjective probability in AI systems." *Subjective probability*, G. Wright and P. Ayton, eds., Wiley, Chichester, U.K., 501-527.

Kuhn, T. (1962). *The structure of scientific revolutions*, Univ. of Chicago Press, Chicago.

Kuhn, T. (1977a). "The historical structure of scientific discovery." *The essential tension*, Univ. of Chicago Press, Chicago, 165-177.

Kuhn, T. (1977b). "Objectivity, value judgment, and theory choice." *The essential tension*, Univ. of Chicago Press, Chicago, 320-339.

Kuhn, T. (1977c). "The history of science." *The essential tension*, Univ. of Chicago Press, Chicago, 105-126.

Kulhawy, F. (1992). "On the evaluation of static soil properties." *Stability and performance of embankments and slopes II*, R. Seed and R. Boulanger, eds., Geotech. Spec. Pub.No. 31, Vol 1, ASCE Press, Reston, Va., 95-115.

Lacasse, S., and Nadim, F. (1996). "Uncertainties in characterizing soil properties." *Uncertainty in the geologic environment: From theory to practice*, C. Shackelford et al., eds., Geotech. Spec. Pub. No. 58, Vol. 1, ASCE Press, Reston, Va., 49-75.

Ladd, C. (1991). "Stability evaluation during staged construction." *J. Geotech. Engrg.*, 117(4), 540-615.

Ladd, C., and Foott, R. (1974). "New design procedure for stability of soft clays." *J. Geotech. Engrg. Div.*, 100(GT7), 763-786.

Lambe, T. (1967). "Stress path method." *J. Soil Mech. and Found. Div.*, 93(SM6), 309-331.

Lambe, T. (1973). "Predictions in soil engineering." *Geotechnique*, 23(2), 149-202.

Larson, J., and Reenan, A. (1979). "The equivalence interval as a measure of uncertainty, *Organizational Behavior and Human Performance*, 23(1), 49-55.

Laskey, K. (1996). "Model uncertainty: Theory and practical implications." *IEEE Trans. on Systems, Man, and Cybernetics - Part A: Systems and Humans*, 26(3), 340-347.

Layton, E. (1986). *The revolt of the engineers: Social responsibility and the American engineering profession*, The Johns Hopkins Univ. Press, Baltimore.

Le Mehaute, B. (1986). "Risk analysis due to wave statistical uncertainties in the design of offshore structures." *Risk-based decision making in water resources*, Y. Haimes and E. Stakhiv, eds., ASCE Press, Reston, Va., 240-257.

Ledbetter, R., and Finn, W. (1993). "Development and evaluation of remedial strategies by deformation analysis." *Geotechnical practice in dam rehabilitation*, L. Anderson, ed., Geotech. Spec. Pub. No. 35, ASCE Press, Reston, Va., 386-401.

Legget, R. (1979), "Geology and geotechnical engineering." *J. Geotech. Engrg. Div.*, 105(GT3), 342-391.

Lehner, P., Laskey, K., and Dubois, D. (1996). "An introduction to issues in higher order uncertainty." *IEEE Trans. on Systems, Man, and Cybernetics - Part A: Systems and Humans*, 25(3), 289-293.

Leonards, G. (1982). "Investigation of failures." *J. Geotech. Engrg. Div.*, 108(GT2), 187-246.

Leonoff, C. (1994). *A dedicated team: Klohn Leonoff Consulting Engineers 1951-1991*, Klohn Leonoff Ltd., Richmond, B.C.

Liao, S., Veneziano, D., and Whitman, R. (1988). "Regression models for evaluating liquefaction probability." *J. Geotech. Engrg.*, 114(4), 389-411.

Lichtenstein, S., and Fischhoff, B. (1980). "Training for calibration." *Organizational Behavior and Human Performance*, 26(2), 149-171.

Lichtenstein, S., Fischhoff, B., and Phillips, L. (1982). "Calibration of probabilities: The state of the art to 1980." *Judgment under uncertainty: Heuristics and biases*, D. Kahneman et al., eds., Cambridge Univ. Press, Cambridge, U.K., 306-334.

Linnerooth-Bayer, J., and Wahlström, B. (1991). "Applications of probabilistic risk assessments: The selection of appropriate tools." *Risk Analysis*, 11(2), 239-248.

MacGregor, D., and Armstrong, J. (1994). "Judgmental decomposition: When does it work?" *Int. J. of Forecasting*, 10(3), 495-506.

MacGregor, D., Lichtenstein, S., and Slovik, P. (1988). "Structuring knowledge retrieval: An analysis of decomposed quantitative judgments." *Organizational Behavior and Human Decision Processes*, 42(3), 303-323.

Mackworth, N. (1965). "Originality." *The American Psychologist*, 20, 51-66.

Marcuson, W. (2000). "Soil mechanics and U.S. national defense: A mutually beneficial relationship." *J. Geotech. and Geoenvir. Engrg.*, 126(9), 767-774.

McClelland, A., and Bolger, F. (1994). "The calibration of subjective probabilities: Theories and models 1980-94." *Subjective probability*, G. Wright and P. Ayton, eds., Wiley, Chichester, U.K., 767-774.

McCormick, N. (1981). *Reliability and risk analysis: Methods and nuclear power applications*, Academic Press, Orlando, Fla.

McCullough, D. (1972). *The great bridge*, Simon and Schuster, New York.

McCullough, D. (1977). *The path between the seas*, Simon and Schuster, New York.

McCullough, D. (1992). *Brave companions: Portraits in history*, Prentice Hall, New York.

McNamee, P., and Celona, J. (1990). *Decision analysis with supertree*, Scientific Press, San Francisco.

McPhee, J. (1965). *A sense of where you are*, Farrar Strauss and Giroux, New York.

McPhee, J. (1980). *Basin and range*, Farrar Strauss and Giroux, New York.

McWhorter, D., and Nelson, J. (1979). "Unsaturated flow beneath tailings impoundments." *J. Geotech. Engrg. Div.*, 105(GT11), 259-289.

Meehan, R. (1982). *Getting sued and other tales of the engineering life*, MIT Press, Cambridge, Mass.

Merkhofer, M. (1987). "Quantifying judgmental uncertainty: Methodology, experiences, and insights." *IEEE Trans. on Systems, Man, and Cybernetics*, SMC-17(5), 741-752.

Mitchell, J. (1986). "Practical problems from surprising soil behavior." *J. Geotech. Engrg.*, 112(3), 259-289.

Morgenstern, N. (1985). "Geotechnical aspects of environmental control." *Proc. Eleventh Int. Conf. Soil Mech.*, Balkema, Rotterdam, 155-185.

Morgenstern, N. (1995). "The role of analysis in the evaluation of slope stability." *Proc. Sixth Int. Symp. on Landslides*, Balkema, Rotterdam, 1615-1629.

Mosleh, A., and Bier V. (1996). "Uncertainty about probability: A reconciliation with the subjectivist viewpoint." *IEEE Trans. on Systems, Man, and Cybernetics - Part A: Systems and Humans*, 25(3), 303-310.

Mosteller, F., and Youtz, C. (1990). "Quantifying probabilistic expressions." *Stat. Sci*, 5(1), 2-34.

Murphy, A., and Winkler, R. (1977). "Can weather forecasters formulate reliable probability forecasts of precipitation and temperature?" *Nat. Weather Digest*, 2, 2-9.

National Research Council (1994). "Science and judgement in risk assessment," Committee on Risk Assessment of Hazardous Air Pollutants, National Academy Press, Washington, D.C.

Newman, J. (1988). *The world of mathematics: A small library of the literature of mathematics from A'h-mosé the Scribe to Albert Einstein*, Tempus Press, Redmond, Wash.

Nuclear Regulatory Commission (1983). "PRA (probabilistic risk assessment) procedures guide: A guide to the performance of probabilistic risk assessments for nuclear power plants." *Rep. No. NUREG/CR-2300*, Office of Nuclear Regulatory Research, U.S. Nuclear Regulatory Commission, Washington, D.C.

Oka, Y., and Wu, T. (1990). "System reliability of slope stability." *J. Geotech. Engrg.*, 116(8), 1185-1189.

Okrent, D. (1989). "Safety goals, uncertainty, and defense in depth." *Risk analysis and management of natural and man-made hazards*, Y. Haimes and E. Stakhiv, eds., ASCE Press, Reston, Va., 268-282.

Orasanu, J., and Fischer, U. (1997). "Finding decisions in natural environments: The view from the cockpit." *Naturalistic decision making*, C. Zsambok and G. Klein, eds., Erlbaum, Mahwah, N.J., 343-357.

Oreskes, N., Shrader-Frechette, K., and Belitz, K. (1994). "Verification, validation, and confirmation of numerical models in the earth sciences." *Science*, 263, 641-646.

Otway, H., and Thomas, K. (1982). "Reflections on risk perception and policy." *Risk Analysis*, 2(2), 69-82.

Otway, H., and von Winterfeldt, D. (1992). "Expert judgment in risk analysis and management: Process, context, and pitfalls." *Risk Analysis*, 12(2), 83-93.

Panel on Seismic Hazard Analysis (1988). *Probabilistic seismic hazard analysis*, National Academy Press, Washington, D.C.

Panel on Seismic Hazard Evaluation (1997). *Review of recommendations for probabilistic seismic hazard analysis: Guidance on uncertainty and use of experts*, National Academy Press, Washington, D.C.

Paté-Cornell, E. (1984). "Fault trees vs. event trees in reliability analysis." *Risk Analysis*, 4(3), 177-186.

Patel, V., and Groen, G. (1991). "The general and specific nature of medical expertise: A critical look." *Toward a general theory of expertise*, K. Ericsson and J. Smith, eds., Cambridge Univ. Press, Cambridge, U.K., 93-125.

Patel, V., Kaufman, D., and Magder, S. (1996). "The acquisition of medical expertise in complex dynamic environments." *The road to excellence: The acquisition of expert performance in the arts and sciences, sports and games*, K. Ericsson, ed., Erlbaum, Mahwah, N.J., 127-165.

Peck, R. (1962). "Discussion of 'Engineering geology on the job and in the classroom,'" *J. Boston Soc. Civil Engrs.*, 49(1), 73-78; reprinted in *Judgment in geotechnical engineering: The professional legacy of Ralph B. Peck*, J. Dunnicliff and D. Deere, eds., 1991, BiTech Publishers, Vancouver, B.C., 77-80.

Peck, R. (1965), "Reflections on Dr. Karl Terzaghi." *Proc. 6th Int. Conf. Soil Mech.*, III, 77-78; reprinted in *Judgment in geotechnical engineering: The professional legacy of Ralph B. Peck*, J. Dunnicliff and D. Deere, eds., 1991, BiTech Publishers, Vancouver, B.C., 43-45.

Peck, R. (1969a). "A man of judgment." *Second R.P. Davis lecture on the practice of engineering.* W. Va. Univ. Bull. Series 70, No. 5-2; reprinted in *Judgment in geotechnical engineering: The professional legacy of Ralph B. Peck*, J. Dunnicliff and D. Deere, eds., 1991, BiTech Publishers, Vancouver, B.C., 191-197.

Peck, R. (1969b). "Advantages and limitations of the observational method in applied soil mechanics." *Geotechnique*, 19(2), 171-187.

Peck, R. (1973a). "Soil mechanics in engineering practice: The story of a manuscript, 1942-1948." *Terzaghi memorial lectures*, S. Tezcan and A. Yalçin, eds., Bo_aziçi Univ., Istanbul, 50-77; reprinted in *Judgment in geotechnical engineering: The professional legacy of Ralph B. Peck*, J. Dunnicliff and D. Deere, eds., 1991, BiTech Publishers, Vancouver, B.C., 46-55.

Peck, R. (1973b). "The direction of our profession: presidential address." *Proc. 8th Int. Conf. Soil Mech.*, Moscow, 4, 156-159; reprinted in *Judgment in geotechnical engineering: The professional legacy of Ralph B. Peck*, J. Dunnicliff and D. Deere, eds., 1991, BiTech Publishers, Vancouver, B.C., 23-28.

Peck, R. (1979). "Liquefaction potential: Science versus practice." *J. Geotech. Engrg. Div.*, 105(GT5), 393-398.

Peck, R. (1980). "Where has all the judgment gone?: The fifth Laurits Bjerrum memorial lecture" *Can. Geotech. J.*, 17, 584-590.

Peck, R. (1982). "Comments on risk analysis for dams." *Proc. Dam Safety Res. Coord. Conf., Interagency Comm. on Dam Safety, Res. Subcomm.*, Denver; reprinted in *Judgment in geotechnical engineering: The professional legacy of Ralph B. Peck*, J. Dunnicliff and D. Deere, (eds.), 1991, BiTech Publishers, Vancouver, B.C., 156-160.

Pennington, N., and Hastie, R. (1986). "Evidence evaluation in complex decision making." *J. Personality and Social Psychology*, 51(2), 242-258.

Peterson, C., and Beach, L. (1967). "Man as an intuitive statistician." *Psychological Bull.*, 68(1), 29-46.

Pirsig, R. (1974). *Zen and the art of motorcycle maintenance*, William Morrow, New York.

Poincaré, H. (1905). *Science and hypothesis*, Walter Scott Publishing Co. Ltd.; republished 1952, Dover, New York.

Poses, R., Cebul, R., Collins, M. and Fager, S. (1985). "The accuracy of physicians' probability estimates for patients with sore throats." *J. Am. Medical Assn.*, 254(7), 925-929.

Raiffa, H. (1968). *Decision analysis: Introductory lectures on choices under uncertainty*, Addison-Wesley, Reading, Mass.

Ramsey, F. (1931). "Truth and probability." *The foundations of mathematics and other logical essays*, R. Braithwaite, ed., Kegan Paul, London, 156-198.

Rasmussen, J. (1993). "Diagnostic reasoning in action." *IEEE Trans. on Systems, Man, and Cybernetics*, 23(4), 981-992.

Ravinder, H., Kleinmuntz, D., and Dyer, S. (1988). "The reliability of subjective probabilities obtained through decomposition." *Mgmt. Sci.*, 34(2), 186-199.

Reagan, R., Mosteller, F., and Youtz, C. (1989). "Quantitative meanings of verbal probability expressions." *J. Applied Psychology*, 74(3), 433-442.

Richman, H., Gobet, F., Staszewski, J., and Simon, H. (1996). "Perceptual and memory processes in the acquisition of expert performance: The EPAM model." *The road to excellence: The acquisition of expert performance in the arts and sciences, sports and games*, K. Ericsson, ed., Erlbaum, Mahwah, N.J., 167-187.

Roberds, W. (1990). "Methods for developing defensible subjective probability assessments." *Trans. Res. Record No. 1288*, Transportation Research Board, National Research Council, Washington, D.C., 183-190.

Ronold, K., and Bjerager, P. (1992). "Model uncertainty representation in geotechnical reliability analyses." *J. Geotech. Engrg.*, 118(3), 363-376.

Rose, F. (1988). "Thinking machine: An 'electric clone' of a skilled engineer is very hard to create." *The Wall Street Journal.*, 119(30), Aug. 12, A1.

Ross, M., and Sicoly, F. (1982). "Egocentric biases in availability and attribution." *Judgment under uncertainty: Heuristics and biases*, D. Kahneman et al., eds., Cambridge Univ. Press, Cambridge, U.K., 179-189.

Rowe, G. (1992). "Perspectives on expertise in the aggregation of judgments." *Expertise and decision support*, G. Wright and F. Bolger, eds., Plenum Press, New York, 155-180.

Sage, A., and White, E. (1980). "Methodologies for risk and hazard assessment: A survey and status report." *IEEE Trans. on Systems, Man, and Cybernetics*, SMC-10(8), 425-446.

Salas, E., Cannon-Bowers, J., and Johnston, J. (1997). "How can you turn a team of experts into an expert team?: Emerging training strategies." *Naturalistic decision making*, C. Zsambok and G. Klein, eds., Erlbaum, Mahwah, N.J., 359-370.

Salgado, F., and Pillai, V. (1993). "Seismic stability and deformation analysis of Duncan Dam." *Proc. 46th Ann. Canadian Geotech. Conf.*, BiTech Publishers, Vancouver, B.C., 259-269.

Salmon, W. (1966). *The foundations of scientific inference*, Univ. of Pittsburgh Press, Pittsburgh.

Salthouse, T. (1991). "Expertise as the circumvention of human processing limitations." *Toward a general theory of expertise*, K. Ericsson and J. Smith, eds., Cambridge Univ. Press, Cambridge, U.K., 286-300.

Savage, L. (1954). *The foundations of statistics*, Wiley; republished 1972, Dover, New York.

Scardamalia, M., and Bereiter, C. (1991). "Literate expertise." *Toward a general theory of expertise*, K. Ericsson and J. Smith, eds., Cambridge Univ. Press, Cambridge, U.K., 172-194.

Schlaiffer, R. (1959). *Probability and statistics for business decisions*, McGraw-Hill, New York.

Schmertmann, J. (1991). "The mechanical aging of soils." *J. Geotech. Engrg.*, 117(9), 1288-1330.

Schmitt, N. (1997). "Naturalistic decision making in business and industrial organizations." *Naturalistic decision making*, C. Zsambok and G. Klein, eds., Erlbaum, Mahwah, N.J., 91-98.

Schraagen, J. (1997). "Discovering requirements for a naval damage control system." *Naturalistic decision making*, C. Zsambok and G. Klein, eds., Erlbaum, Mahwah, N.J., 227-232.

Seed, H. (1968). "Landslides during earthquakes due to soil liquefaction." *J. Soil Mech. and Found. Div.*, 94(SM5), 1055-1122.

Seed, H. (1973), "Stability of earth and rockfill dams during earthquakes." *Embankment-dam engineering*, R. Hirschfeld and S. Poulos, eds., Wiley, New York, 239-269.

Seed, H. (1979). "Considerations in the earthquake-resistant design of earth and rockfill dams." *Geotechnique*, 29(3), 215-263.

Seed, H., and Idriss, I. (1967). "Analysis of soil liquefaction: Niigata earthquake." *J. Soil Mech. and Found. Div.*, 93(SM3), 83-108.

Seed, H., and Lee, K. (1966). "Liquefaction of saturated sands during cyclic loading." *J. Soil Mech. and Found. Div.*, 92(SM6), 105-134.

Seed, H., Seed, R., Harder, L., and Jong, H. (1988). "Re-evaluation of the slide in the Lower San Fernando Dam in the earthquake of Feb. 9, 1971." *Rep. No. UCB/EERC-88/04*, Univ. of Calif., Berkeley.

Seed, H., Idriss, I., Lee, K., and Makdisi, F. (1975). "Dynamic analysis of the slide in the Lower San Fernando Dam during the earthquake of February 9, 1971." *J. Geotech. Engrg. Div.*, 101(GT9), 889-911.

Seed, H., Tokimatsu, F., Harder, L., and Chung, R. (1985). "Influence of SPT procedures in soil liquefaction evaluation." *J. Geotech. Engrg.*, 111(12), 1425-1445.

Seed, R., and Harder, L. (1990). "SPT-based analysis of cyclic pore pressure generation and undrained residual strength." *H. Bolton Seed Memorial Symp.*, Vol. 2, J. Duncan, ed., BiTech Publishers, Vancouver, B.C., 351-376.

Senior Seismic Hazard Analysis Committee (1997). "Recommendations for probabilistic seismic hazard analysis: Guidance on uncertainty and use of experts." *Rep. No. NUREG/CR-6372*, Office of Nuclear Regulatory Research, U.S. Nuclear Regulatory Commission, Washington, D.C.

Serafaty, D., MacMillan, J., Entin, E.E., and Entin, E.B. (1997). "The decision-making expertise of battle commanders." *Naturalistic decision making*, C. Zsambok and G. Klein, eds., Erlbaum, Mahwah, N.J., 233-246.

Shafer, G. (1989). "The unity and diversity of probability." Inaugural lecture, November 20, 1989, University of Kansas.

Shanteau, J. (1992). "Competence in experts: The role of task characteristics." *Organizational Behavior and Human Decision Processes*, 53(2), 252-266.

Sherard, J. (1979). "Sinkholes in dams of coarse, broadly graded soils." *13th Cong. on Large Dams*, Int. Commission on Large Dams, Q.49, R.2, 25-35.

Shieh, L., Johnson, J., Wells, J., Chen, J., and Smith, P. (1985). "Simplified seismic probabilistic risk assessment: Procedures and limitations." *Rep. No. NUREG/CR-4331*, Office of Nuclear Regulatory Research, U.S. Nuclear Regulatory Commission, Washington, D.C.

Simonton, D. (1991). "Career landmarks in science: Individual differences and interdisciplinary contrasts." *Developmental Psychology*, 27(1), 119-130.

Simonton, D. (1996). "Creative expertise: A life-span developmental perspective." *The road to excellence: The acquisition of expert performance in the arts and sciences, sports and games*, K. Ericsson, ed., Erlbaum, Mahwah, N.J., 227-253.

Sloman, S. (1996). "The empirical case for two systems of reasoning." *Psychological Bull.*, 119(1), 3-22.

Slovik, P., and Lichtenstein, S. (1971). "Comparison of Bayesian and regression approaches to the study of information processing in judgment." *Organizational Behavior and Human Performance*, 6(6), 649-794.

Slovik, P., Fischhoff, B., and Lichtenstein, S. (1976). "Cognitive processes and societal risk taking." *Cognition and Social Behavior*, J. Carrol and J. Payne, eds., Erlbaum, Mahwah, N.J., 165-184.

Slovic, P., Fischhoff, B., and Lichtenstein, S. (1982). "Facts versus fears: Understanding perceived risk." *Judgment under uncertainty: Heuristics and biases*, D. Kahneman et al., eds., Cambridge Univ. Press, Cambridge, U.K., 463-489.

Smiles, S. (1857). *Life of George Stephenson*; reprinted 1881, Murray, London.

Smith, G. (1997). "Managerial problem solving: A problem centered approach." *Naturalistic decision making*, C. Zsambok and G. Klein, eds., Erlbaum, Mahwah, N.J., 371-380.

Smith, G., Benson, P., and Curley, S. (1991). "Belief, knowledge, and uncertainty: A cognitive perspective on subjective probability." *Organizational Behavior and Human Decision Processes*, 48(2), 291-321.

Snizek, J., and Henry, R. (1989). "Accuracy and confidence in group judgment." *Organizational Behavior and Human Decision Processes*, 43(1), 1-28.

Snizek, J., and Henry, R. (1990). "Revision, weighting, and commitment in consensus group judgment." *Organizational Behavior and Human Decision Processes*, 45(1), 66-84.

Sobel, D. (1999). *Galileo's daughter*, Walker, New York.

Spetzler, C., and Staël von Holstein, C. (1975). "Probability encoding in decision analysis." *Mgmt. Sci.*, 22(3), 340-358.

Stark, T., and Mesri, G. (1992). "Undrained shear strength of liquefied sands for stability analyses." *J. Geotech. Engrg.*, 118(11), 1727-1747.

Starkes, J., Deakin, J., Allard, F., Hodges, N., and Hayes, A. (1996). "Deliberate practice in sports: What is it anyway?" *The road to excellence: The acquisition of expert performance in the arts and sciences, sports and games*, K. Ericsson, ed., Erlbaum, Mahwah, N.J., 81-106.

Stein, R. (1993). "The Dempster-Shafer theory of evidential reasoning." *AI Expert*, August, 26-31.

Stevens, J. (1936). *An engineer's recollections*, McGraw-Hill, New York.

Terzaghi, C. (1929). "Effect of minor geologic details on the safety of dams." *Trans. Am. Inst. of Mining and Metallurgical Engineers*, 215, 31-44.

Terzaghi, K. (1936). "Relation between soil mechanics and foundation engineering." *Proc. First Int. Conf. Soil Mech.*, Vol. 3, 13-18.

Terzaghi, K. (1939). "Soil mechanics - a new chapter in engineering science." *J. Institution of Civil Engrs.*, 12(7), 106-142.

Terzaghi, K. (1955). "Influence of geological factors on the engineering properties of sediments." *Economic geology* (50th anniversary vol., Part II), 557-618.

Terzaghi, K. (1957). "Opening address to the fourth international conference on soil mechanics and foundation engineering." *Proc. Fourth Int. Conf. Soil Mech.*, Vol. III, 55-58.

Terzaghi, K. (1964). "About life and living." *Geotechnique*, 14(1), 51-56.

Terzaghi, K. and Leps, T. (1958). "Design and performance of Vermillion Dam, California." *J. Soil Mech. and Found. Engrg.*, 84(SM3), paper 1728.

Terzaghi, K., and LaCroix, Y. (1964). "Mission Dam: An earth and rockfill dam on a highly compressible foundation." *Geotechnique*, 14(1), 14-50.

Terzaghi, K., and Peck, R. (1948). *Soil mechanics in engineering practice*, Wiley, New York.

Thenhaus, P. (1983). "Summary of workshops concerning regional seismic source zones of parts of the contiguous United States covered by the U.S. Geological Survey 1979 - 1980, Golden, Colorado," *U.S. Geological Survey Circular 898*, U.S. Geological Survey, Alexandria, Va.

Tokimatsu, K., and Seed, H. (1987). "Earthquake settlements in sands due to earthquake shaking." *J. Geotech. Engrg.*, 113(8), 861-878.

Tversky, A., and Kahneman, D. (1974). "Judgment under uncertainty: Heuristics and biases." *Science*, 185, 1124-1131.

Tversky, A., and Kahneman, D. (1982). "Availability: A heuristic for judging frequency and probability." *Judgment under uncertainty: Heuristics and biases*, D. Kahneman et al., eds., Cambridge Univ. Press, Cambridge, U.K., 163-178.

Tversky, A., and Kahneman, D. (1983). "Extensional vs. intuitive reasoning: The conjunction fallacy in probability judgment." *Psychological Rev.*, 90(4), 293-315.

United States Bureau of Reclamation (1997). *Guidelines for achieving public protection in dam safety decisionmaking*, U.S. Bureau of Reclamation, Denver, Colo.

United States Bureau of Reclamation (1999). *Dam safety risk analysis methodology*, Version 3.3, Technical Service Center, U.S. Bureau of Reclamation, Denver, Colo.

United States Environmental Protection Agency (1992). "Final guidelines for exposure assessment." *Rep. No. FRL-4129-5*, U.S. EPA, Washington, D.C.

Van Zyl, D. (1987). "Health risk assessment and geotechnical perspective." *Geotechnical practice for waste disposal '87*, R. Woods, ed., Geotech. Spec. Pub. No. 13, ASCE Press, Reston, Va., 812-831.

Vanmarke, E. (1977). "Probabilistic modelling of soil profiles." *J. Geotech. Engrg. Div.*, 103(GT11), 1247-1265.

Vaughan, D. (1996). *The Challenger launch decision: Risky technology, culture, and deviance at NASA*, Univ. of Chicago Press, Chicago.

Veneziano, D. (1995). "Uncertainty and expert opinion in geologic hazards." *The earth, engineers, and education*, Dept. of Civil and Envir. Engrg., Massachusetts Institute of Technology, Cambridge, Mass.

Vick, S., and Bromwell, L. (1989). "Risk analysis for dam design in karst." *J. Geotech. Engrg. Div.*, 115(6), 819-835.

Vick, S., and Stewart, R. (1996). "Risk analysis in dam safety practice." *Uncertainty in the geologic environment: From theory to practice*, C. Shackelford et al., eds., Geotech. Spec. Pub. No. 58, Vol. 1, ASCE Press, Reston, Va., 586-603.

Vick, S., and Watts, B. (1994). "Decision analysis for liquefaction ground improvement." *Risk and reliability in ground engineering*, B. Skipp, ed., Thomas Telford, London, 241-253.

von Winterfeldt, D., and Edwards, W. (1986). *Decision analysis and behavioral research*, Cambridge Univ. Press, Cambridge, U.K.

Von Thun, J. (1985). "Application of statistical data from dam failures and accidents to risk-based decision analysis on existing dams." U.S. Bureau of Reclamation, Engineering and Research Center, Denver, Colo.

Voss, J., and Post, T. (1988). "On the solving of ill-structured problems." *The nature of expertise*, M. Chi et al., eds., Erlbaum, Hillsdale, N.J., 261-285.

Wagner, R., and Stanovich, K. (1996). "Expertise in reading." *The road to excellence: The acquisition of expert performance in the arts and sciences, sports and games*, K. Ericsson ed., Erlbaum, Mahwah, N.J., 189-225.

Wallsten, T., and Budescu, D. (1983). "Encoding subjective probabilities: A psychological and psychometric review." *Mgmt. Sci.*, 29(2), 151-173.

Wallsten, T., and Budescu, D. (1990). "Comment on quantifying probabilistic expressions." *Stat. Sci.*, 5(1), 151-173.

Wallsten, T., Budescu, D., and Zwick, R. (1993). "Comparing the calibration and coherence of numerical and verbal probability judgments." *Mgmt. Sci.*, 39(2), 176-190.

Ward, G. (1990). *The Civil War: An illustrated history*, Knopf, New York.

Whitman, R. (1984). "Evaluating calculated risk in geotechnical engineering." *J. Geotech. Engrg.*, 110(2), 145-188.

Winkler, R., and Poses, R. (1993). "Evaluating and combining physicians' probabilities of survival in an intensive care unit." *Mgmt. Sci.*, 39(12), 1526-1543.

Winner, E. (1996). "The rage to master: The decisive role of talent in the visual arts." *The road to excellence: The acquisition of expert performance in the arts and sciences, sports and games*, K. Ericsson, ed., Erlbaum, Mahwah, N.J., 271-301.

Wright, G., Rowe, G., Bolger, F., and Gammack, J. (1994). "Coherence, calibration, and expertise in judgmental probability forecasting." *Organizational Behavior and Human Decision Processes*, 57(1), 1-25.

Youd, T., and Noble, S. (1997). "Liquefaction criteria based on statistical and probabilistic analysis." *Proc. NCEER Workshop on Evaluation of Liquefaction Resistance of Soils*, T. Youd and I. Idriss, eds., *Tech. Report NCEER-97-0022*, Nat. Center for Earthquake Engrg. Research, Buffalo, N.Y., 201-215.

Youngs, R., Swan, F., and Power, M. (1988). "Use of detailed geologic data in regional probabilistic seismic hazard analysis: An example from the Wasatch Front, Utah." *Earthquake engineering and soil dynamics II - Recent advances in ground-motion evaluation*, J. Von Thun, ed., Geotech. Spec. Pub. No. 20, ASCE Press, Reston, Va., 156-172.

Zhang, Z., and Tumay, M. (1999). "Statistical to fuzzy approach to CPT soil classification." *J. Geotech. and Geoenvir. Engrg.*, 125(3), 179-186.

Zimmer, A. (1984). "A model for the interpretation of verbal predictions." *Int. J. of Man-Machine Studies*, 20(2), 121-134.

Zsambok, C. (1997). "Naturalistic decision making: Where are we now?" *Naturalistic decision making*, C. Zsambok and G. Klein, eds., Erlbaum, Mahwah, N.J., 3-16.

Index

acceptable risk, 136, 139
addition rule, 410
affirming the consequent, 58–60, 195
Agricola, Georgius, 26, 28–30, 90, 92
aleatory uncertainty, 31–38, 51
ambiguity: see vagueness
analysis (see also: theory, models), 56–70, 74, 107, 108, 110, 111, 115–117, 120, 287, 288, 290, 291–294, 326, 392
  predictive validity of, 204, 287, 288, 290, 291–293
  use in subjective probability, 293, 294
analytical reasoning, 383, 384
anchoring and adjustment, 201, 202, 224, 287, 295
Annacis case study, 169–175
anomalies, 72, 97, 112, 197, 206, 229, 239, 241, 290, 294, 338, 379, 385, 386
Apollo program, 226, 250
Aquinas, Thomas, 21, 22, 31, 33, 51
Aristotle, 21, 23, 95
artificial intelligence, 48, 49, 323, 376, 377
assembly-effect bonus, 315, 317, 318
associative reasoning, 383
assumptions, 58, 69, 74, 120, 204, 207, 212, 290, 292, 293, 306, 404
autocovariance, 112, 426, 427
availability, 200, 201, 203, 224

backward reasoning, 133, 193, 325–327
Bacon, Francis, 23
base-rate frequency, 13, 127, 128, 201, 204, 205, 209, 210, 291, 301
base-rate neglect, 204, 205, 224, 290, 291
Bayes, Thomas, 36, 37
Bayes' theorem, 13, 37, 38, 48, 128, 183, 194, 201, 261, 290, 292, 413–416
Bayesian indicator, 37, 292, 413–415
Bayesian probability: see subjective probability
belief space, 220, 221, 318
Bernoulli, Jacob, 34–36, 43
Bernoulli theorem, 36
Bernoulli trials, 261, 307, 417–420
Bert, Paul, 362
binary outcomes, 129, 415
binary regression, 129–131
binning, 167, 168
binomial theorem, 145, 251, 261, 307, 416–420
Blutcher (locomotive), 333
Boolean algebra, 134

Boorstin, Daniel, 341
Bradley, Bill, 328
broken-leg cues, 291
Brooklyn Bridge, 354–363, 403
Brooklyn caisson, 357–360
Brunel, Isambard Kingdom, 357
Bush, Vannevar, 78

caissons, pneumatic, 357, 358
calibration, 208, 265–269, 273, 274, 301, 312
capacity and demand, 108
career landmarks, 346, 347, 372, 373
career trajectories, 343–375
Cascade tunnel, 365
case-history information, 99, 101, 188, 282, 290, 294, 405
causal narrative, 186–188, 269, 271, 292, 311, 315, 326, 328, 387
causal reasoning, 184–188, 189, 190, 191, 193, 195, 202, 205, 239, 241, 246, 255, 262, 265, 270, 278, 282, 283, 286, 288, 290, 292, 308, 387
Challenger incident, 224–251, 282, 288, 289, 297, 302
Chat Moss bog, 334, 335
chess, 323, 324
chunks, memory, 324, 377
Classical probabilists, 31–38, 39, 43, 47, 53, 104, 398
cognitive discrimination, 208, 209, 273, 298, 301
cognitive processes, 48, 181–192, 221–224, 253, 281, 283, 313, 314, 319, 322, 405
coherence, 9, 182, 218, 306–308, 313, 407
coin toss, 11, 12, 14, 15, 24
collectively exhaustive events, 408
common-cause failures, 244, 420, 427
common sense, 47, 77, 79, 82, 87, 88, 95, 326, 403
complementary events, 410
conditional probability, 37, 201, 241, 261, 271, 321, 411–413, 419
confirmation bias, 201, 224, 246
conformity, 276–280, 306, 313
conjunctive distortion, 202, 224, 272
consensus, 220, 318
consequences, failure (see also: expected value, risk cost), 123–125, 132, 136–138, 146–148, 151, 159, 160
consistency (see also: subjective probability, variability of), 56, 69, 93, 136, 138, 381, 402
Consolidation, Theory of, 350–352
contingency model, 185, 189, 190
convergence techniques, 302, 303, 316
correlation, 420–429
Cournot, Antoine, 44
creative expertise, 329, 343
creative productivity, 343–346, 372–375
Cubitt, William, 357
Culebra cut, 365, 367, 369
Czar Nicholas, 370
Czech Legion, 371

dam engineers, 207
damage control, naval 387
data scatter, 120
de Lesseps, Ferdinand, 365, 366
de Mello, Victor, 84
de Witt, Johann, 41
debiasing, 253
decision analysis, 48, 139–142, 148, 171–175, 177, 309, 376
decision criteria, 135–139, 140, 148, 149, 179
decision trees, 140, 141, 173, 174
decisionmaking, 135–142, 260, 261, 376–392, 405
  expert, 376–392
  informed consent, 400–402
  naturalistic, 376
  rational choice, 376, 377
  risk-based, 135–142, 260, 261
  rule-based, 376, 377, 379–385, 392
decomposition, 132–135, 153, 154, 187, 244, 258, 259, 269–280, 301, 311, 312, 315
deductive proof, 10, 16, 33, 94, 104, 114, 229, 250, 287, 397, 400

deductive reasoning, 1, 9, 10, 16, 21, 92, 94, 95, 103, 186, 197, 286, 319, 392, 397
  in reliability analysis, 114
  in theory and analysis, 56, 63, 68, 74
degree-of-belief probability: see subjective probability
Dempster-Shafer theory, 411
design, 59, 62, 78, 80, 81, 90, 123–125, 158, 207, 292, 337, 404
deterministic approaches, 3, 21, 108, 110, 119, 138, 177, 179, 189, 255, 257, 287, 396
diagnosis, 69, 82–91, 103, 121, 123, 126, 157, 192–200, 239, 241, 242, 367, 390–392, 400, 404
  in judgment, 83–91, 135
  in risk analysis, 121, 123, 126, 153, 157, 199
  medical, 192–195, 265, 266, 291, 323, 325
  strategies, 192, 193, 196–199
dialectic process, 338–343
disconfirming evidence, 201, 283–285, 308, 314, 315, 389
discovery, scientific, 72
dispersion, 108, 206, 215, 275
Dodson, James, 42
drill-test-analyze strategy, 57, 58, 75, 80, 82, 113, 186, 192, 350, 392
Duncan Dam case study, 149–157
Duncan, James M., 79, 80

Eads, James B., 357, 362
earth-fissure case study, 142–149
earth pressure theory, 348
education, engineering, 78, 187
effective-stress principle, 55, 350–352
Einstein, Albert, 21, 90, 350
Einstein, Herbert, 98, 101, 102, 327
elaboration, creative, 344
Ellsberg paradox, 297
embankment stability symposium, 210–212, 267, 276
epistemic uncertainty, 31–38, 51
equivalence interval, 299, 302
events, defining, 280
event trees, 132–135, 144–146, 245, 270, 272, 277, 278, 413, 414
evidence, 9, 22, 23, 25, 28, 30, 33, 34, 36, 38, 41, 52, 91, 92, 96, 102, 189, 234, 282–289, 295
  as arguments, 286–289
  circumstantial, 95, 217, 237, 251
  disconfirming, 201, 283–285, 308, 314, 315, 389
  in probability assessment, 204, 282–289, 295
  judicial, 33, 34, 187, 215–217, 285, 286
  strength and weight, 190, 215–218, 240, 287–289, 291, 292, 294, 316
exceedance probability, 152, 416
expected cost, 140
expected value, 32, 47, 137, 138, 140, 183
experience, 9, 42, 77, 87, 97–101, 102, 132, 191, 197, 213–215, 261, 309, 324, 326, 386, 392, 404
expert decisionmaking, 376–378
expert elicitation: see formal elicitation
expert systems (see also: artificial intelligence), 48, 377
expertise, 313, 321–393, 395, 405
expertise, creative, 329, 343
experts, characteristics of, 325–330, 338–343
exposure period, 416
external validity, see: coherence

f-N plot, 136, 137
failure modes (see also: Failure Modes and Effects Analysis, risk analysis, reliability analysis, system reliability), 89, 114, 121, 123, 157, 315, 423–429
Failure Modes and Effects Analysis (FMEA), 158–172, 175, 176, 179, 226, 227, 258, 304
failure probability, 5, 6, 107, 113–123, 124, 135, 136, 178, 273
fault trees, 132–135, 226, 244, 270, 280
feedback, 264–269

feel for the problem, 88, 89, 102, 193, 249, 282, 283, 294
Fermat, Pierre de, 20
Feynman, Richard, 90, 225, 246, 247, 251
Fillunger, Paul, 351, 352
first-order second-moment (FOSM) techniques, 112
flowslides, 61, 195, 196
Focht, John (see also: predictive process), 82, 101, 207, 384, 393
Forcheimer, Phillip, 348, 349
formal elicitation, 50, 309–313
forward reasoning, 133, 193, 325, 326, 379, 386
Fracastoro, Girolamo, 24–27, 30, 43
Frölich, Otto, 391
fuzzy-set theory, 411

Galen, 25, 27
Galileo, 23, 32
Gambler's fallacy, 417
Gatun Dam, 369
genius, 331, 337
geology (see also: hypotheses, geologic), v, 28–30, 66, 85–87, 89, 92, 94, 98–100, 187, 188, 207, 254, 255, 345, 385, 387, 388
geomechanical models: see models
Gettysburg, battle of, 356
Goethals, George, 369
graphical presentation, 289, 293, 294
Graunt, John, 39–41
Graves, William S., 371
Great Northern railroad, 364, 365, 368, 330
Grinter Committee, 78
ground improvement case study, 169–175
group communication, 318
group elicitation, 51, 265, 313–319
group interactions, 51, 164, 165, 315–319
guidelines, 381–383

Halley, Edmund, 41, 329
hard-easy effect, 210, 273, 326
Hartley, David, 38, 44, 183
hazard, 13
Hazard Analysis, 158, 226, 227, 237
Hazard and Operability Studies (HAZOP), 158
health risk assessment, 146, 147
heuristics and biases, 112, 182, 183, 191, 200–224, 246, 251, 253, 263–276, 281–295, 301, 302, 306–309, 311, 314, 319, 326
highway engineers, 383, 384
Hill, James J., 364, 366, 370, 372
hindsight bias, 201, 224, 246, 308
hourglass model, 390–392
Hudde, Johannes, 41
human error, 50, 121, 122
Hume, David, 38, 44, 93, 94, 183
Hume's Paradox, 93, 94
Huygens, Christian, 31, 32
hypotheses, 57, 70, 71, 84–86, 89, 102, 198, 316, 318, 326
  formation of, 84–86, 89, 186, 188
  geologic, 85, 86, 89, 186, 188
  in risk analysis, 125, 126, 143–145, 178
  scientific, 57, 70, 71
hypothesis testing (see also: hypotheses), 45, 86, 193, 198
hypothetico-deductive method, 57

ideation, creative, 344
independent events, 412
inductive reasoning, 1, 9, 10, 16, 19, 31, 38, 52, 93–96, 104, 186, 188, 192, 197, 198, 255, 294, 295, 319, 391, 392, 397
  basis for safety, 95, 96
  induction by enumeration, 93, 186
  inductive analogy, 93
  model confirmation, 60
informed consent, 400–402
infrastructure, 403, 404
initiative, 328, 332, 367
initiator, 125, 126, 133, 134, 144–146, 151
insensitivity to predictability, 203, 204, 224
insensitivity to sample size, 205, 206, 224, 288, 291
insight, 71, 329, 335, 350

instinct: see intuition
insufficient adjustment, 201, 216, 224, 287, 295, 303, 308
insurance, 4, 5, 41, 42, 140, 141
internal erosion, 61, 127, 128, 150, 151, 153, 156, 277–279, 304, 306–308, 315, 316, 353, 415, 416
interpretation, 82, 83, 96–99, 102, 103, 187, 192, 238, 249, 255, 391, 392, 397
intersection of events, 409
intuition, 7, 13, 57, 71, 78, 86–88, 91, 102, 197, 234, 242, 285, 294, 309, 356, 379–385
intuitive reasoning, 383, 384
intuitive statistician, 182–184, 222

Jack problem, 204, 205
joint probability, 412
Joralemon, Ira, 342, 378
judgment (see also: expertise, subjective probability, probability), 1, 9, 10, 12, 16, 42, 47, 55, 56, 60, 72, 82–103, 118, 119, 132, 139, 157, 179, 191–193, 257, 260, 261, 294, 295, 322, 327, 383, 395, 398, 402–405
  and experience, 9, 42, 97–101
  and expertise, 324, 383–385, 392, 393
  and geology, 98
  Challenger incident, 239, 383–385, 392, 393
  cognitive processes in, 48, 181–192
  definition, 100–103
  diagnostic, 83–91, 102, 157, 294
  feel for the problem, 88, 89
  hypothesis formation in, 84–86
  induction in, 93–96, 101, 260, 397
  interpretive element, 96–99, 102, 187, 192
  intuition in, 86–88
  relation to subjective probability, 7, 8, 9, 16, 17, 38, 49–52, 179, 257, 294, 295
  role of theory in, 77, 78, 80, 97, 98, 104, 326
  visualization in, 89, 90
justification, scientific, 72

Karensky, Alexander, 370, 372
Kelly, Thomas, 376–378, 385
Kent charts, 300
King, Stephen, 342
knowledge, 8, 9, 15, 21, 22–25, 30, 41, 254, 321, 340, 397, 399
  acquisition of, 330–338
  domain, 323, 324, 329, 330, 338–341, 375, 386, 392
  inductive, 31, 93, 94
  state of, 7, 8, 11, 34, 138, 247, 256, 257, 283, 321, 322, 390, 409
Kolchak, Aleksandr, 371
Kolmogorov, A.N., 20, 44, 46
kriging, 112
Kuhn, Thomas (see also: paradigms), 65–68, 72, 73, 403, 404

landfill case study, 142–149
Laplace, Pierre-Simon, 32, 42, 402
Leibniz, Gottfried von, 20, 33, 34, 43
Leggett, Robert, 87
Lenin, Vladimir Ilyich, 370, 371
Linda problem, 202, 203, 272, 273
liquefaction, 66, 67, 130, 131, 149–157, 169–175, 287, 288, 413, 414
lists, 283–285, 308, 315, 316
Little Round Top, 356
Liverpool & Manchester railway, 333–385
Locke, John, 38, 44, 183
logistic regression, 129–131
Lusser, Robert, 226

Marias Pass, 364, 365
Markov process, 420, 426
McCullough, David, 341, 342, 375
mean, see: statistical properties
memory, 200, 201, 282, 308, 323, 324, 338, 349, 358, 363, 377, 386, 390
mental simulation, 89, 295, 385–389, 390
  in expert performance, 385–389
  in judgment, 89
mercury, 26, 30
Méré, Chevalier de, 19, 20, 31, 33
metacognition, 327

Mill, John Stuart, 44
Mission Dam, 349, 352, 353
Mitchell, James K., 80, 215
models (see also: theory, analysis), 56–70, 107–122, 157, 191
 assumptions of, 58, 69, 74, 120
 completeness of, 61, 62
 confirmation of, 58–60
 consistency, 69, 70
 in reliability analysis, 107, 110, 111, 118, 122
 indeterminacy of, 62, 63
 parameter uncertainty, 107, 114–121
 uniqueness of, 60, 61, 72, 73
model error, 115, 116
model uncertainty, 115–118, 177
Monte Carlo simulation, 105, 113
Morgenstern, N.R., 80
motivational bias, 218–221, 246, 247, 250, 311, 401, 402
Mount Olive cut, 333
Mozart, Wolfgang Amadeus, 330
multiplication rule, 412
mutually exclusive events, 408, 410

naturalistic decisionmaking, 376
New York caisson, 361, 362
Newton, Isaac, 26, 33, 34, 329
Niagara Bridge, 354, 355, 357
nonampliative conclusion, 63, 94
nonlinearity, 62
normalized frequency, 13, 112, 127, 128, 146, 201, 204
normative expertise, 213, 215, 227, 321, 326
null alternative, 140
null hypothesis, 45

observation, 91, 92, 363, 367, 385, 387, 388
observational approach, 85, 238, 388
Olsen, Mark, 349
opinion, 7, 19, 21, 22–24, 30, 51, 52, 398, 401
outcome feedback, 264–169
overconfidence, 206–215, 224, 246, 258, 263–269, 273–276, 284, 301, 308, 312–314, 326
overconfidence index, 274–276, 302

Panama Canal, 363–372, 403
Paracelsus, 25–30, 32, 55, 343
paradigms, 65–68, 75–79, 103, 106, 226, 351, 396
 geotechnical, 66–69, 75–79, 103, 106, 351, 396
 science, 65, 66, 79, 226
parameter state, 413
parameter uncertainty, 107, 114–118, 119–121, 177, 179
Pascal, Blaise, 20, 32, 33
Pascal's triangle, 20
Pascal's wager, 32, 33, 34, 48
pattern templates, 97, 98
patterns, 71, 97–99, 102, 187, 197, 239, 240, 282, 286, 289, 294, 385–389
 in expert performance, 324, 379, 385–390
 in judgment, 71, 97–99, 192
 of anomaly, 97, 239, 294, 379
Peck, Ralph, 76, 77, 81, 85, 88–90, 98, 99, 102, 184, 326, 341, 402
personalistic interpretation, 2, 46–48
Petty, William, 39, 40
pile capacity, 107–109, 118, 267–269
pivot findings, 192, 193, 386, 388, 390
Poincaré, Jules Henri, 70–72, 74, 87, 118, 178, 187, 329, 337, 399
point-estimate method, 112
Poisson, Siméon, 44
posterior probability, 37, 194, 415
practice, 330, 331, 375
precedent, 16, 80, 81, 205–207, 236, 239, 240
prediction error, see: model error
predictive process, 57, 82, 84, 88, 207, 393
predictive validity, 194, 195, 203, 246, 287, 288, 290–293
prior probability, 37, 38, 194, 414
probabilistic indifference, 2, 11, 12

probabilistic methods, 105, 106, 109–113, 125–132, 177, 261, 400
probabilistic seismic hazard analysis (PSHA), 13–15, 51, 151, 152, 304–306
probability (see also: subjective probability, relative-frequency probability, statistics, probabilistic methods, probability distributions, probability of failure), 31–48, 51, 113–123, 124, 135, 136, 178–180, 219–221, 254–259, 399–402
  axioms, 20, 44, 46, 47, 261, 306, 321, 410, 411
  bounds, 407, 420–429
  calculus, 1, 11, 17, 19, 20, 106, 222, 259, 261, 262, 398, 399, 402, 407
  common uses, 11–15, 17
  historic development, 31–48
  medieval meaning, 21–23, 51
  of x successes in n trials, 205
  purposes of, 17, 18, 178–190, 219–221, 254–259, 292, 319, 339–402, 405
  second-order, 304–306
probability aggregation, 312, 313, 317
probability distributions, 107–109, 111, 112, 115, 116, 135, 206, 213–215, 275, 303–305, 310
probability interpretations, 2, 3, 262
probability of failure, 5, 6, 107, 113–123, 124, 135, 136, 178
probability wheel, 296
problem representation, 325, 326, 328, 329, 333–335, 337, 350, 358, 367, 369, 371
progressive failure, 62, 427, 428
proof: see deductive proof
propagation of uncertainty, 51, 112
prototype narrative (see also: causal narrative), 187, 188, 389
pseudostatic methods, 66, 67, 117
psychometrics, 50

qualitative techniques (see also: verbal transformations, Failure Modes and Effects Analysis), 157, 158, 175, 176, 226–228
Quetelet, Adolph, 42, 43

rank-ordering (see also: qualitative techniques), 160, 162
Rappahannock Bridge, 355, 356
Rasmussen report, 49, 112
reading, 338–341
recurrence interval, 416, 417
redundancy, 227, 232, 234, 237, 243, 244–246, 423, 424
reference gambles, 296
reference property, 63
regression to the mean, 417
Reichenbach, Hans, 72
relative-frequency probability, 3–6, 8, 9, 11, 14, 15, 17, 18, 107–123, 138, 397
  definition of, 2, 3
  in reliability analysis, 107–123
  properties of, 9, 10
reliability analysis (see also: system reliability), 106, 107–123, 125, 129, 148, 177–180, 400
reliability block diagram, 423, 424
reliability index, 107
representativeness, 203–206, 217, 246, 272, 287, 290, 293, 314, 326, 389
response, 125–134, 146, 152, 153
return period: see recurrence interval
revealed preferences, 136
revision and weighting, 314
Ripley, Charles, 352
risk, definition of, 123, 242
risk, environmental, 118, 146, 147, 162–167, 169–175
risk analysis (see also: health risk assessment), 106, 122–139, 142–157, 177–180, 201, 400
risk aversion, 34, 136, 137, 141
risk cost, 140, 141
risk perception (see also: acceptable risk), 50–52
risk tradeoffs, 167, 175, 243, 401
Rochester Bridge, 357
Roebling, Emily, 354, 356–358, 365, 370
Roebling, John Augustus, 354, 357, 358, 363

Roebling, Washington Augustus, 354–363, 372–375, 378, 388
Roosevelt, Theodore, 364, 366, 367, 369, 372
Royal Albert Bridge, 357
rule-based reasoning (see also: decision-making, rule-based), 379–381, 383, 384
Russian Railway Service Corps, 370–372

safety, 64, 84, 87, 95, 96, 257
salience, 200
sample population (see also: statistically homogeneous population), 4, 5, 114, 127, 128, 414
Sartre, Jean Paul, 341
satisficing, 377
science, 16, 17, 21, 22, 57, 64–66, 70–74, 78, 79, 147, 254, 255, 396
  career trajectories in, 345–347
  changes in, 64–66, 76
  discovery and justification, 72
  hypotheses in, 57, 70, 71
  intuition in, 57, 72
  objective truth, 16, 70, 71, 73
  statistics in, 43–45
  theories, 57, 64–66, 72, 73
scientific method, 57, 71
second opinion, 265, 401
second-order probability, 304–306
seismic hazard, see: hazard, probabilistic seismic hazard analysis
seismic liquefaction, see: liquefaction
self-monitoring, 327
sensitivity studies, 119, 173, 293
settlement prediction, 98, 99, 213–215, 267
significance testing, 45
signs, 21, 24–31, 38, 41, 43, 52
single-event occurrence, 5, 6, 7, 39, 47, 189, 193, 270, 307
situation awareness, 242, 380, 382, 389–392
societal risk aversion, 136–138
soil behavior, 63, 68, 116, 117
soil dynamics (see also: liquefaction), 66, 67
soil mechanics, 55, 77, 81, 82, 92, 185
spatial variability, 112, 120, 425, 427
stage model, 188, 189
standard deviation, see: statistical properties
Stannard, Robert, 335
state of knowledge, 7, 8, 11, 34, 138, 247, 256, 257, 283, 321, 322, 390, 409
state of nature, 7, 126, 408
stationary processes, 4, 5, 416
statistical properties, 4, 111, 112
statistical reasoning, 184–186, 189–191, 193–195, 202, 205, 235, 239, 241, 246, 262, 270, 278, 282, 283, 286, 288, 290, 291, 308, 311
statistical sample, 4, 40, 41, 114
statistical significance, 36, 45
statistically homogeneous population, 4, 111, 114
statistics (see also: relative-frequency probability), 3, 4, 6, 10, 13, 14, 17, 127
  historic development, 39–45
  hypothesis testing, 45
  in model uncertainty, 117, 118
  in subsurface characterization, 112, 114
Stephenson, George, 331–338, 343, 364, 389
Stephenson's Rocket, 331–333, 335, 336
Stevens, John F., 363–372, 373–375, 386, 388
Stevens Pass, 365
stochastic models, 111, 125, 148
stochastic processes, 3, 4, 111, 306, 416
Stockton & Darlington railway, 333
stop rule, 197, 241, 242
structural reliability, 1, 107
subjective probability (see also: probability, judgment, expertise), 2, 3, 6–9, 11, 14–18, 19, 38, 46–53, 56, 122, 123, 130, 132, 178, 253–319, 321, 395, 399–402, 405
  assessment of, 253–319
  cognitive processes in, 48, 181–192

criteria for validity, 9, 46, 47, 182, 306
definition of, 2, 3
expert judgment in, 49–52, 260, 321, 323, 375, 392, 393
formal elicitation, 50, 309–313
group elicitation, 265, 313–319
heuristics and biases, 200–224
historic development, 32–38, 44
properties of, 9, 10
in nuclear safety, 49–52, 309
variability of, 50–52, 138, 276–280, 299, 300, 302, 304, 306, 313, 399
substantive expertise, 213, 215, 272, 313, 321, 323, 326, 375
Suez Canal, 365
syphilis, 24, 26, 30, 40
system reliability, 113, 157, 423–429
systematic error, 120
systems analysis, 376

Taylor series expansion, 112
Terzaghi, Karl, 30, 55, 64, 71, 74, 75, 77, 82, 86, 87, 92, 97–99, 207, 293, 326, 341, 347–354, 357, 363, 372–375, 377, 386–389, 403
Terzaghi Dam, 352, 353
theory (see also: analysis, models), 56–70, 72, 73, 74–82, 186, 187, 403, 404
as pattern template, 97, 98
geotechnical paradigm of, 75–79, 84, 103, 104
relationship to judgment, 77, 78, 80, 97, 98
role in geotechnical practice, 79–82
thinking, 327, 341, 342, 395–400
total probability theorem, 413
toxicology, see: health risk assessment
Trans-Siberian Railway, 370–372
treatment plant case study, 169–175
Triger (French engineer), 357, 361
Twain, Mark, v, 42, 101

underconfidence, 217, 224, 288
unimodal bounds, 425–428
union of events, 409
unsaturated flow, 67, 68
utility, 33, 34, 141, 142, 296

vagueness, 297, 302–304
Venn diagrams, 408–410
verbal descriptors, 258, 297–302, 316
verbal transformations, 297–302
verification, 58–60
Vermillion Dam, 376, 377
visual thinking, 90
visualization, 89, 90, 187, 271, 295
in expert performance, 324, 358, 362, 363, 369, 385–389
in judgment, 89, 90
Von Braun, Wehrner, 226, 228

Warren, G.K., 356
water heater example, 159–162
weather forecasting, 12, 13, 15, 185, 186, 264–267, 300
WHAT-IF Analysis, 158
Whitman, Robert V., 80
Wilson, Woodrow, 370, 371
Windy Craggy case study, 162–167
wisdom, 97, 98, 254
Wright, John, 357
writing, 337, 338–342, 362